T0338537

HANDBOOK OF HYDRAULIC GEOMETRY

Hydraulic geometry describes the relations between stable channel characteristics and discharge and adjustments made by a stream in response to changes in river discharge and sediment load. This book introduces hydraulic geometry and discusses different theories and their applications in river engineering, thus providing a comprehensive summary for hydraulic engineers, as well as graduate students and researchers in fluvial geomorphology and hydraulic and environmental engineering. Topics covered include the basis of power form of hydraulic geometry relations, validity and stability of power relations, state and assumption of equilibrium, variability of exponents, variation of channel width and velocity, and the effect of stream size and river channel patterns.

VIJAY P. SINGH is a University Distinguished Professor, a Regents Professor, and Caroline and William N. Lehrer Distinguished Chair in Water Engineering at Texas A&M University. He has 47 years of experience in research and teaching in hydrology, groundwater, hydraulics, and water resources. He is a member of National Academy of Engineering, a Distinguished Member of ASCE, an Honorary Member of IWRA, a Distinguished Fellow of AGGS, and an Honorary Member of AWRA. He has received more than 108 national and international awards, including three honorary doctorates.

HANDBOOK OF HYDRAULIC GEOMETRY

Theories and Advances

VIJAY P. SINGH

Texas A&M University

Shaftesbury Road, Cambridge CB2 8EA, United Kingdom

One Liberty Plaza, 20th Floor, New York, NY 10006, USA

477 Williamstown Road, Port Melbourne, VIC 3207, Australia

314–321, 3rd Floor, Plot 3, Splendor Forum, Jasola District Centre, New Delhi – 110025, India

103 Penang Road, #05–06/07, Visioncrest Commercial, Singapore 238467

Cambridge University Press is part of Cambridge University Press & Assessment,
a department of the University of Cambridge.

We share the University's mission to contribute to society through the pursuit of
education, learning and research at the highest international levels of excellence.

www.cambridge.org
Information on this title: www.cambridge.org/9781009222174

DOI: 10.1017/9781009222136

First published 2022

A catalogue record for this publication is available from the British Library.

Library of Congress Cataloging-in-Publication Data
Names: Singh, V. P. (Vijay P.), author.
Title: Handbook of hydraulic geometry : theories and advances / Vijay P. Singh.
Description: First edition | New York, NY : Cambridge University Press, 2022. | Includes bibliographical
references and index.
Identifiers: LCCN 2022027024 (print) | LCCN 2022027025 (ebook) | ISBN 781009222174 (hardback) |
ISBN 9781009222136 (epub)
Subjects: LCSH: Channels (Hydraulic engineering) | Sedimentation and deposition. | Hydrodynamics. |
Stream measurements. | BISAC: SCIENCE / Earth Sciences / Hydrology
Classification: LCC TC175 .S5635 2022 (print) | LCC TC175 (ebook) | DDC 627/.042–dc23/eng/20220808
LC record available at https://lccn.loc.gov/2022027024
LC ebook record available at https://lccn.loc.gov/2022027025

ISBN 978-1-009-22217-4 Hardback

Dedicated to my
wife Anita, who is no more,
son Vinay, daughter Arti,
daughter-in-law Sonali, son-in-law Vamsi
and grandsons Ronin, Kayden, and Davin and, grand daughter Alivia

Contents

Preface

The relations between individual river characteristics – geometric and hydraulic – and discharge over time constitute what is dubbed as hydraulic geometry. These relations have been derived using a variety of concepts – empirical, physical, and statistical. The earliest attempts to derive the relations, which started in the nineteenth century for designing irrigation canal systems in the Indian Subcontinent, were empirical. These attempts culminated in the development of regime theory of hydraulic geometry. The regime or equilibrium concept forms the basis of the theories that developed from the middle of the twentieth century onward. Despite the importance of hydraulic geometry in designing irrigation canals and river training works, a book describing the many theories seems to be lacking. This is what constituted the motivation for this book.

The subject matter of the book comprises 22 chapters. Introducing hydraulic geometry and pointing out its application in river engineering, Chapter 1 goes on to discuss the basis of power form of hydraulic geometry relations, validity and stability of power relations, state and assumption of equilibrium, variability of exponents, variation of channel width and velocity, effect of stream size and river channel patterns, and power relations for drainage basins along with the effect of land use and boundary conditions. The chapter concludes by showing the organization of the book's contents. Physically based approaches to hydraulic geometry relations for width, depth, velocity, and slope require equations of continuity of water, roughness, and sediment transport. Chapter 2 briefly outlines these equations and unit stream power as well as stream power.

The interrelationship between flow, sediment transport, channel resistance, and bank stability determines the regime of an alluvial channel, meaning channel shape and stability. The regime theory and generalized regime theory that predict the size, shape, and slope of a stable alluvial channel through Lacey's and Blench's equations are discussed in Chapter 3.

The average river channel system tends toward an approximate equilibrium between the channel and water and sediment it transports. Under equilibrium, the stream channel depth, width, velocity, and suspended sediment load at a given cross-section can be expressed as power functions of discharge, and these functions constitute the at-a-station hydraulic geometry. In a similar vein, stream channel depth, width, flow velocity, and suspended load along the river vary with discharge as simple power functions under the condition that the frequency of discharge at all cross-sections is equal. These functions constitute the theory of downstream hydraulic geometry. Both at-a-station and downstream hydraulic geometries by the Leopold–Maddock (LM) empirical theory are dealt with in Chapter 4.

A river tends to minimize the variability of factors that govern its hydraulic geometry, leading to the concept of minimum variance. Chapter 5 discusses this concept, including the minimum sum of variances of independent variables, the minimum total variance of components of stream power, the minimum variance of stream power, and the influence of choice of variables.

Chapter 6 discusses the hydraulic geometry of regime channels using dimensional principles and illustrates the application of these principles to channel design, the derivation of width and depth, computation of regime channels, and computational procedure. Chapter 7 describes the hydrodynamic theory for deriving the downstream hydraulic geometry relations of width, depth, velocity, and slope in terms of flow discharge. The theory consists of continuity equation for water, a flow resistance equation such as Manning's equation, a sediment transport equation, and a morphological relation.

The scaling theory for deriving hydraulic geometry with both immobile boundary and mobile bed channels is discussed in Chapter 8. Also discussed are comparisons with stable channel design theories, including the threshold theory, stability index theory, and regime theory, along with the application to channel design. The tractive force theory of hydraulic geometry is treated in Chapter 9, which also contains a discussion of the threshold condition and the validation of assumptions.

Chapter 10 deals with the thermodynamic theory for deriving hydraulic geometry. It discusses the basic hypothesis, variation of flow energy, sediment transport equation, system of regime equations, and computational procedure. The similarity principle is presented in Chapter 11, including a discussion of dominant discharge and comparison with regime relations.

Chapter 12 discusses the theory of channel mobility leading to stable hydraulic geometry. The theory discusses the hypothesis of channel mobility, sediment transport, index of mobility, and hydromorphometric relationships. Chapter 13 presents the maximum discharge and Froude number hypothesis, which states that river morphology is governed by the dominant discharge, saturation of sediment discharge, and maximization of Froude number leading to the minimum amount of energy dissipation. It discusses three cases, including straight and meandering single beds, multiple meandering beds, and meanders.

The stability of river channel adjustment toward equilibrium is controlled by a set of factors governed by the Froude number. Chapter 14 discusses the hypothesis of minimum Froude number, involving river stability and Froude number, modeling and simulation, minimization of Froude number, and testing. Chapter 15 presents the hypothesis of maximum friction factor, which states that the channel geometry evolves to a stable nonplanar shape when the friction factor reaches a local maximum. This hypothesis, however, may not be invariably true.

Chapter 16 presents the maximum flow efficiency hypothesis, which states that the equilibrium state of a channel corresponds to maximum flow. The chapter derives the channel hydraulic geometry for primarily three cross-sections, namely trapezoidal, rectangular, and triangular, and discusses the physical evidence for maximum flow efficiency.

Rivers tend to follow the path of least action for transporting the sediment and water loads imposed on them. Chapter 17 discusses the principle of least action and derives the hydraulic geometry and its validity. Chapter 18 presents the theory of minimum energy dissipation rate and the derivation of river geometry using the theory. Chapter 19 presents the entropy theory, and Chapter 20 presents the minimum energy dissipation and maximum entropy theory, entailing primary morphological equations. The theory leads to four families of downstream hydraulic geometry relations and eleven families of at-a-station hydraulic geometry relations, and the conditions under which the families of hydraulic geometry relations can occur in the field.

The water flowing in a channel has both potential energy due to elevation and kinetic energy due to motion. The potential energy is transformed to kinetic energy, part of which is utilized to entrain and transport sediment, and part is dissipated to overcome boundary and bed friction. The time rate of potential energy expenditure is the stream power. Chapter 21 derives hydraulic geometry using the theory of stream power. Relationships relating channel width, depth, and cross-section as well as discharge to drainage basin area constitute regional hydraulic geometry. Chapter 22 – the concluding chapter – presents the regional hydraulic geometry relationships.

Acknowledgments

The author has drawn from the works of a large number of geomorphologists, hydraulic engineers, and irrigation engineers from the United States and beyond. He has tried to acknowledge these works as specifically as possible; any omission on his part is entirely inadvertent, and he would like to express his apologies in advance. Mr. Jeongwoo Han from the Department of Biological & Agricultural Engineering at Texas A&M University helped prepare figures and solve several problems in the book. His help is deeply appreciated.

1

Introduction

Hydraulic geometry (HG) is a quantitative description of the variation of river characteristics in response to changes in discharge and sediment load. It is impacted by climate, geology, and human interference. Hydraulic geometry relations have been expressed in power form and derived using a multitude of hypotheses. These relations play a fundamental role in the design of alluvial canals, river training works, and watershed management. The objective of this chapter is to introduce preliminary concepts that are deemed important for understanding different aspects of hydraulic geometry.

Notation

a = parameter or proportionality constant
A = cross-sectional area
AHG = at-a-station hydraulic geometry
AMHG = at-many-stations hydraulic geometry
b = exponent
c = parameter or proportionality constant
d = flow depth
D = bed material size
DHG = downstream hydraulic geometry
F = channel shape expressed as a width–depth ratio
f = exponent
HG = hydraulic geometry
J = exponent
k = parameter or proportionality constant
m = exponent

Continued

M = weighted mean percent silt-clay
n = Manning's roughness factor or coefficient
N = proportionality constant or parameter
p = proportionality constant or parameter
Q = flow discharge
Q_2 and $Q_{2.33}$ = discharge corresponding to recurrence interval of 2 and 2.33 years, respectively
Q_b = bankfull discharge
Q_s = sediment load
Q_{susp} = suspended sediment load
R = river bend
RHG = reach-averaged hydraulic geometry
s = proportionality constant or parameter
S = slope
S_b = percentage of silt and clay in the bank alluvium
S_c = percentage of silt and clay in the channel alluvium
γ = specific weight of water
τ_0 = shear stress of water flow on the riverbed
τ_{av} = average shear stress
v = flow velocity
w = channel width
y = exponent
z = exponent

1.1 Definition

River characteristics include geometric characteristics, such as width, depth, hydraulic radius, mean slope, and side (or bank) slope that define the channel form and pattern (meandering, braiding, and sinuosity) and its mean cross-section; and hydraulic variables, which include velocity, discharge, shear stress, tractive force, bed and bank roughness defined by friction, and sediment characteristics defined by grain size, concentration, and load (bed, wash, and suspended). These characteristics together or in some combination result in bank stability status, bed forms (dunes, antidunes, pools, and riffles), braiding, meandering, and sinuosity. The characteristics change in response to a given influx of water and sediment load stemming from the upstream watershed imposed on the river, as well as to erosion from and aggradation of sediment on the channel bed and banks.

The relationships between river characteristics and discharge over time denote the term "hydraulic geometry," which at a site leads to the at-a-station hydraulic geometry (AHG), or which along and between rivers at a comparable discharge frequency leads to the downstream hydraulic geometry (DHG) (Ferguson, 1986). Recently two other forms of hydraulic geometry have been introduced: reach-averaged hydraulic geometry (RHG) (Jowett, 1998) and at-many-stations hydraulic geometry (AMHG) (Gleason and Smith, 2014). In HG, discharge is the key independent control variable, and its variation leads to the variation in river characteristics, subject to the specified boundary conditions, which are quantitatively given by hydraulic geometry. It is implied that the river basin containing the stream network is a hydrologically homogeneous basin, which is seldom the case. The discharge at any point on the river contains potential energy as well as kinetic energy. As water moves down the slope, the potential energy gets converted into kinetic energy in order to perform work for overcoming friction encountered during the movement of water. The rate of conversion depends on the slope and discharge. This process can also be interpreted in terms of stream power (γQS, γ = weight density or specific weight of water, Q = discharge, and S = slope). The river will spend the stream power to convey the discharge and sediment and will try to minimize the rate of expenditure to transport the water and sediment.

Leopold and Maddock (1953) introduced the concept of hydraulic geometry and empirically expressed its relationships for a channel in the form of power functions of discharge as

$$w = aQ^b; \quad d = cQ^f; \quad v = kQ^m, \tag{1.1a}$$

where w is the channel width; d is the flow depth; v is the flow velocity; Q is the flow discharge; b, f, and m are the exponents; and a, c, and k are the parameters or proportionality constants. The hydraulic geometry relationships may also include

$$n = NQ^y; \quad S = sQ^z; \quad Q_s = pQ^J, \tag{1.1b}$$

where n is Manning's roughness factor; S is the slope; Q_s is the sediment load; y, z, and J are the exponents; and N, s, and p are the proportionality constants or parameters. Exponents b, f, m, y, z, and J represent, respectively, the rate of change of hydraulic variables w, d, v, n, S, and Q_s as Q changes; and coefficients a, c, k, N, s, and p are the scale factors that define the values of w, d, v, n, S, and Q_s when $Q = 1$. When any of these relations is plotted on log-log paper, the plot should be a straight line with some scatter, of course, with the exponent defining the slope of the plot and the intercept defining the logarithm of the proportionality constant.

The hydraulic variables, width, depth, and velocity, satisfy the continuity equation:

$$Q = Av, \tag{1.2a}$$

in which A is the flow cross-sectional area. For a rectangular cross-section, $A = wd$, Equation (1.2a) becomes

$$Q = wdv. \tag{1.2b}$$

Inserting Equation (1.1a) in Equation (1.2b) one gets

$$Q^1 = aQ^b kQ^m cQ^f = ackQ^{b+m+f}. \tag{1.2c}$$

Therefore, the coefficients and exponents in Equation (1.1a) satisfy

$$ack = 1, \; b + f + m = 1. \tag{1.3}$$

Since river channels convey water and are shaped by sediment transport, Mackin (1963) reasoned for a deductive approach whereby channel shape and size are deduced from hydraulic laws governing the movement of water and sediment. Different theories, empirical as well physically based, have been proposed for deriving Equations (1.1a) and (1.1b). Ferguson (1986) has provided a review of hydraulics and hydraulic geometry, updating the review given by Richards (1977). Singh (2003) provided a review of several hydraulic geometry theories. Gleason (2015) has classified these theories into empirical, rational, and theoretical. These theories will be discussed in subsequent chapters. Rational and theoretical theories provide an analytical basis for derivation of Equations (1.1a) and (1.1b), which is briefly discussed in Section 1.2.

1.2 Analytical Basis for Hydraulic Geometry Equations

The analytical basis of Equations (1.1a) and (1.1b) is constituted by physical relations that can be grouped into four classes related to fluid dynamics, sediment transport, bank stability, and dynamic equilibrium (Chang, 1988). The relations related to fluid dynamics are continuity equation for steady flow and flow resistance equation; that related to sediment transport is bed load transport equation; those related to bank stability are criteria for bank stability, slope, bank material, and vegetation; and those related to dynamic equilibrium are physical relationships governing the equilibrium state. These relations play interrelated roles in fluvial morphology. It may be noted that the unknowns or degrees of freedom in channel morphology are four, including width, depth, channel slope, and bank slope, which are not independent of each other. These four degrees are equal to the four physical relations mentioned in this paragraph, which should suffice to determine the hydraulic geometry.

Physical relationships for stable width of alluvial rivers are based on the assumption of dynamic equilibrium. For stable width of alluvial channels, Parker (1978) employed a balance in lateral exchange of sediment in flow between bank region and central region, and related the sediment exchange to the mechanics of internal river flow. In a regime channel, steady flows of water and sediment are in balance. For a short reach with constant discharge, the dynamic equilibrium must satisfy equal sediment discharge along the reach, minimum stream power per unit channel length subject to constraints, and streamwise uniformity of power expenditure or energy gradient subject to constraints. These are related to energy-based or power-based hypotheses, which will be discussed in related chapters.

1.3 Types of Hydraulic Geometry

Hydraulic geometry is classified as (1) at-a-site hydraulic geometry and (2) downstream hydraulic geometry. The at-a-site hydraulic geometry entails the mean values of river characteristics at a specific site over a certain period of time, such as a week, a month, a season, or a year, whereas the concept of downstream hydraulic geometry involves the spatial variation in channel form and process at a constant frequency of flow along the channel reach. Two other types of hydraulic geometry have also been proposed: reach-averaged and at-many-stations, which will be briefly discussed in this section.

1.3.1 At-a-Station Hydraulic Geometry

At-a-station hydraulic geometry is concerned with the short-term change (rise or fall) of flow within the channel in response to the change in discharge. Flow characteristics are defined by channel dimensions, shape, slope, and bed roughness, which may be altered by flow, especially floods. The size and shape of a river at a particular site must reflect the balance between erosive and resistive forces. Erosive forces are associated with discharge, velocity, flow depth, and slope, whereas resistive forces are associated with bed material size governed by bed load supply from the upstream and bed forms. The channel geometry tends to stabilize through gradual washing away of collapsed blocks of cohesive bank material, filling-in of scour holes, and incision into riffles. Hydraulic geometry relations change over time because of meandering, development of bars, and general channel adjustment with changing discharge and sediment load. Hydraulic laws and frictional characteristics determine the rate of change of mean velocity with flow depth. The banks and beds of most channels do not change over the short term except when large floods occur.

The analysis of at-a-station hydraulic geometry is typified by the works of Leopold and Maddock (1953) and Wolman (1955), who formalized a set of relations, such as Equations (1a) and (1b) that relate the changes in flow properties (width, mean depth, mean velocity, slope, friction, and sediment load) to mean discharge. This type of analysis describes the regulation of flow adjustments by channel form in response to the variation in discharge and sediment load at a particular cross-section. At-a-station hydraulic geometry relations as applied hold for in-bank flows. However, the reference discharge is normally taken as bankfull discharge.

1.3.2 Downstream Hydraulic Geometry

Knighton (1987) classified the hydraulic geometry–related variables into control, channel, and flow variables, and schematically summarized the downstream trends in these variables. Control variables include discharge, sediment load, bed material load, average sediment size, coarser size, bank material size, and valley slope; channel variables include width, depth, and channel slope; and response variables include stream power, boundary shear, velocity, and flow resistance. Some of these variables are less certain than others. The channel form is mainly controlled by discharge, sediment load, bed and bed material composition, and valley slope. Alluvial channels receive water discharge and sediment load from the upstream watershed, and width, depth, velocity, slope, and friction are adjusted accordingly. Depending on the balance between sediment transport capacity and the sediment load imposed, the channel will either erode or aggrade to a different slope. It has been argued that the slope is adjusted gradually by incision or aggradation, perhaps at the time scale of 10–100 years – the scale often used in canal design or channelization where a preselected slope is maintained. However, slope can change rapidly if there is a change in channel pattern, subject to particular thresholds of slope and discharge. The threshold discharge is often the bankfull discharge, which depends on the channel width and depth, with a trade-off between its magnitude and its frequency. In turn, channel width and depth are related to bankfull discharge, so discharge is the determinant control, but there are other factors such as bank and bed characteristics that affect the width and depth, whereas slope depends on discharge and bed load. Transient changes in channel dimensions at a point are small as compared with downstream trends. Natural channels have heterogeneous material containing fine and coarse grains. Normally, finer grains are mobile even when coarser grains are at or below the threshold. This means that channel intermittently adjusts even with low sediment transport. Rivers steeper than usual tend to be wider, shallower, and swifter than the norm.

The downstream hydraulic geometry involves solving for width, depth, velocity, slope, and friction, given discharge and sediment load along with grainsize. Four relationships are needed for defining the hydraulic geometry, which are the flow continuity, flow resistance, sediment transport equation, and an empirical or metaphysical relation. Richards (1982) has noted that the downstream hydraulic geometry, involving channel process and form, embodies two types of analyses, both of which are expressed as power functions of the form (Rhoads, 1991) given by Equations (1.1a) and (1.1b). Using field data on bankfull hydraulic geometry, bed material grain size, and average channel gradient at closely spaced intervals in 10 alluvial and quasi-alluvial reaches covering a nearly contiguous 260 km segment of the Colorado River in Western Colorado and Eastern Utah, Pitlick and Cress (2002) found that the median surface sediment grain size reduced by a factor of a little more than 2, whereas the average channel slope decreased by a factor of 5. There was a large increase in bankfull depth in comparison with width, indicating almost constant bankfull Shields stress downstream, which for the entire reach was 0.049 – about 50% higher than the threshold for bed load transport.

Analysis of downstream hydraulic geometry involves a modification of the original hydraulic geometry concept and entails variation of channel geometry for a particular reference downstream discharge of a given frequency or mean annual discharge. Implied in this analysis is an assumption of an appropriate discharge that is the dominant flow controlling channel dimensions (Knighton, 1987; Rhoads, 1991). For example, for perennial rivers in humid regions, the mean discharge or a discharge that approximates bankfull flow (Q_b), such as Q_2 or $Q_{2.33}$, is often used in Equations (1.1a) and (1.1b). It should be noted that the coefficients and exponents are not constrained by the continuity equation when the selected discharge substantially differs from the bankfull flow. On the other hand, Stall and Yang (1970) related hydraulic geometry to flow frequency and drainage area. Normally, in downstream analysis $b > f > m$ with typical values of $b = 0.5$, $f = 0.4$, and $m = 0.1$, whereas the regime theory yields $b = 0.5$, $f = 1/3$, and $m = 1/6$.

Sediment movement downstream, unlike discharge, is much more complicated and its pattern is difficult to define at the basin scale, partly because of lack of data. Bhowmik (1984) reported an increase in sediment storage with the increase in floodplain width and surface area. Trimble (1975, 1983) estimated for southern Piedmont and Wisconsin that more than 90% of the sediment eroded from upland slopes was stored in hillslopes, floodplains, and channels, indicating a large difference between supply and transport. Walling (1983) noted that the sediment delivery ratio decreased downstream, dropping below 10%, for drainage areas over 100 km^2. Emmett (1975) expressed a relationship between suspended sediment load (tonnes per day) and discharge (m^3/s) along the upper Salmon River in Idaho as

$$Q_{susp} = 131Q_b^{0.75}. \tag{1.4}$$

Equation (1.4) shows that suspended load increases downstream as bankfull discharge (Q_b) increases but at a lower rate. Similar equations have been reported for other rivers (Bogardi, 1984), but the relationship can vary widely.

The sediment size fraction portrays a complicated picture, because a particular size fraction moving as bed load at one location can be transported entirely as suspended load at another location. Griffiths (1983) found that the ratio of bed load to suspended load for New Zealand rivers declined to approximately 10% in lowland reaches. This implies that the bed load transport is dominant in headwater areas. It is generally agreed that the type, amount, and continuity of sediment transport strongly influence the channel form.

In general, the bed material size, which is derived from slope, tributary, and channel boundary sources, is the largest in the headwater reaches but decreases downstream because of abrasion, weathering, and sorting, as presented by Knighton (1987) in Table 1.1. In an ideal situation, the decrease in size occurs in a gradual transition from boulder to gravel bed to sand bed along the river, alluding to a strong relationship between particle size and slope. However, the transition is often abrupt as sediment size D_{50} approaches 10 mm (Yatsu, 1955).

The bed material size influences the amount and type of sediment transport and resistance. When the channel bed is made of boulders and large grain sizes, the dominant mode of transport is bed load and roughness would be of grain form associated with drag exerted by individual particles (Bathurst, 1978). Also, there will be large energy expenditure for the transport of large particle sizes. As the grain size diminishes to sand along the river, the mode of transport would be more of suspended load; bed forms, such as dunes and antidunes, would develop; and roughness would be skin and form roughness. Another aspect relates to changes in bank material along the river, which impact the river width.

Valley slope is a characteristic of past flow and sediment conditions and is measured along the main valley axis. The slope of the valley floor can vary significantly because of geological processes. Sometimes the slope of the valley can increase because of the marked increase in sediment loads being carried into the valley by tributaries.

1.3.3 Reach-Averaged Hydraulic Geometry

Jowett (1998) employed reach-averaged variables in place of single cross-section based variables for ease of measurement when applying hydraulic geometry to habitat assessment. Wohl and Merritt (2008) found RHG to hold in mountain streams, and Navratil and Albert (2010) also reached the same conclusion but

Table 1.1. *Downstream change in bed material size [for bed material size, D_{50} = median; D_m = mean; D_{90} = size for which 90% of the sample is fine; D_{85} = size for which 85% of the sample is fine; D_{15} = size for which 15% of the sample is finer; β = rate of change of bed material size downstream; α = the proportionality factor; $D = \alpha \exp(\beta L)$; D = bed material size in mm; L = distance in km; B = boulder-bed; G = gravel-bed; S = sand-bed] (after Knighton, 1987)*

Source	Location	River distance (km)	Parameter	Bed material		
				Range (mm)	Type	β
Yatsu (1955)	Kinu, Japan	0–52	D_{50}	20–70	G	−0.0253
		60–100		0.4–0.9	S	−0.0238
	Watarase	0–21		30–80	G	−0.0531
		23–37		0.3–0.9	S	−0.0416
	Tenyu	14		15–50	G	−0.0532
	Kiso	0–15		35–70	G	−0.0348
		15–55		0.4–0.6	S	−0.0104
	Nagara	0–13		25–40	G	−0.0448
		13–49		0.7–1.2	S	−0.0173
	Sho	0–20		20–50	G	−0.0288
	Abe	0–23		15–90	G	−0.0715
	Yahagi	0–35		1–2	S	−0.0247
Hack (1957)	Virginia, Maryland, Calfpasture					
	Tye	0–27	D_{50}	42–75	G	0.0034
	Gillis Falls	0–14		230–680	B/G	−0.0837
		0–13		7–45	G	0.116
Bradley et al. (1972)	Knik River, Alaska	0–26	D_m	44–330	B/G	−0.081

Table 1.1. (*cont.*)

Source	Location	River distance (km)	Parameter	Bed material Range (mm)	Type	β
Rana et al. (1973)	Mississippi, Vicksburg District	440	D_{85}	0.3–0.9	S	−0.0010
			D_{50}	0.2–0.55		−0.00055
			D_{15}	0.18–0.33		−0.00045
Simons and Senturk (1977)	Rhine	200	D_{50}	50–160	G	−0.011
	Mur	140		34–83	G	−0.0195
	Mississippi, from Fort Jackson					
	Rio Grande, from Otowi	1770		0.12–0.72	S	−0.00085
		240		0.14–0.50	S	−0.0057
Church and Kellerhals (1978)	Peace	140	D_{90}	45–280	G	−0.0048
			D_m	25–120		−0.0034
Knighton (1980)	Bollin Dean	0–50	D_m	0.33–67	G/S	−0.118
	Noe	0–20		29–69	G	−0.042
Nordin et al. (1980)	Amazon, from Iquitos	3300	D_{50}	0.15–0.50	S	Slightly negative

found a break in slope at certain discharges. Harman et al. (2008) assessed the model error with RHG for Australian rivers. Conceptually, RHG is similar to at-a-station hydraulic geometry.

1.3.4 At-Many-Stations Hydraulic Geometry

Gleason and Smith (2014) developed the concept of AMHG. Using USGS and other agency field data, they found a strong relationship between the coefficients (log values) and exponents of at-a-station hydraulic geometry over long river reaches (up to about 3,000 km). Thus, they showed that the coefficients and exponents (e.g., log a versus b; log c versus f; and log k versus m) were related and could be predicted from one station to the other. Also, for a given AMHG equation, the AHG parameter can be computed from the other. An important implication is that pairs of width–discharge, depth–discharge, and velocity–discharge are shared by all cross-sections along a river.

1.4 Application of Hydraulic Geometry

The hydraulic geometry relations are of great practical value in the prediction of channel deformation; characterization of entire basins or regions; layout of river training works; design of stable canals and intakes; planning of river flow control works; management of river channels; planning of irrigation schemes; improvement of river works; estimation of flood flows; habitat assessment; in-stream use; runoff routing; hydrologic modeling; and monitoring river systems. Richards (1976) noted that hydraulic geometry relations through their exponents can be employed to discriminate between different types of river sections, which can be used in planning for resource and impact assessment (Allan, 1995).

At-a-station hydraulic geometry is important in a whole range of river management activities, including monitoring streamflow (stage-discharge rating curve), flood routing (Western et al., 1997), when combined with flow duration curves for the construction of water-resources-index duration curves for river-habitat analysis (Jowett, 1998), water quality management, reservoir sedimentation studies, modeling river recovery from point-pollution (Stout, 1979), determining the frequency of sediment movement (Dingman, 2002), estimating minimum flow requirements for fish passage, recreational boating (Mosley, 1982), and assessing available fish habitat (Mosley, 1983).

The downstream hydraulic geometry is important for designing irrigation canals, channelization schemes, predicting response to flow regulation or transfer, and predicting past and future river discharges from channel dimensions. Rosenfield et al. (2007) applied hydraulic geometry for determining optimal flows for stream habitat and characterizing habitat patches within a drainage network.

1.5 Basis of Hydraulic Geometry Relations

The mean values of hydraulic variables of Equations (1.1a) and (1.1b) are known to follow hydraulic laws and the principle of minimum energy dissipation rate (Langbein, 1964; Yang, et al. 1981). Consequently, these mean values are functionally related and correspond to the equilibrium state of the channel. This state is regarded as the one corresponding to the maximum sediment transport capacity, that is, an alluvial channel adjusts its width, depth, slope, velocity, and friction to achieve a stable condition in which it is capable of transporting a certain amount of water and sediment. Leopold and Maddock (1953) have stated that the average river systems tend to develop in such a way that they produce an approximate equilibrium between channel and the transport of water and sediment. However, the approximate equilibrium and hence the average river system is achieved in the long-term, as implied by Lindley (1919) when he used the term "regime" for alluvial channels to state that width, depth, and slope of a channel to carry a given supply of water loaded with a silt charge were fixed by nature. Thus, Thornes (1977) stated that hydraulic geometry is formulated by a set of propositions expressed as power function relationships between discharge and width, depth, velocity, slope, roughness, riffle and pool spacing, and meander wave-length and sinuosity. Application of these relationships in changing channels involves a set of assumptions that are not always valid.

1.6 Equilibrium State

Knighton (1987) observed that at cross-sections undergoing systematic change, the potential for adjustment toward some form of quasi-equilibrium in the short term is related to flow regime and channel boundary conditions. Marked changes in channel form and associated hydraulic geometry can occur over a short period of time in the absence of exceptionally high flows and in a channel with high boundary resistance. This suggests that the approach to quasi-equilibrium or establishment of a new equilibrium position is relatively rapid.

Ponton (1972) found the hydraulic geometry of the Green and Birkenhood River basins in British Columbia, Canada, to significantly depart from the previous state, and attributed the departure to the recent glaciation in the area and the strong control that glacial features exercise on streams. He concluded that the equilibrium throughout the stream was not established and many reaches within each system might have reached a quasi-equilibrium, but these reaches were not yet adjusted to each other because of glacial features. While investigating the influence of a forest on the hydraulic geometry of two mountain streams, Heede (1972) found that dynamic equilibrium was attained in the streams and removal of dead and dying trees would not be permissible where a stream was in dynamic equilibrium and bed material movement would be minimized.

1.7 Equilibrium Assumption

Hydraulic geometry adjusts to the downstream discharge in response to environmental history, bed load, and climate (Rhodes, 1978). The geometric relations of power form have often been applied to small basins, which in many cases have varying geology and climate, or human intervention has disturbed the long-term equilibrium. Thus, the equilibrium assumption can only address "at-a-station" geometry, as each cross-section of the channel adjusts to the discharge of water and sediment in a unique surface and subsurface environment. Furthermore, for extreme discharges (extremely low and high flows) deviating from the mean discharge, even the "at-a-station" interpretation of hydraulic exponents and coefficients will be less than meaningful, because the influence of geology, soil, widening and narrowing floodplains, upstream bog and marsh environments, or network topology on hydraulic geometry is not adequately quantified at the full range of space-time scales.

1.8 Validity of Power Relations

Dury (1976) showed that hydraulic geometry relations of power form were valid for extended sets of data at the 1.58-year mean annual discharge. Chong (1970) found that Equations (1.1a) and (1.1b) were similar over varying environments. Thus, regional generalizations would be acceptable for rivers that have achieved "graded-time" equilibrium (Phillips and Harlan, 1984). However, Park (1977) argued that simple power functions were not the best way to describe the hydraulic geometry. Richards (1973, 1976) reasoned that since the depth and velocity were functions of roughness, when the rate of change in roughness was not uniform, the power function relations for depth and velocity would not reflect the true hydraulic nature. He then described the nonuniform variation of roughness in relation to similar nonuniform changes of depth and velocity with discharge.

Betson (1979) developed procedures for predicting variations in geomorphic relationships in the Cumberland Plateau in Appalachia, United States, for use in streamflow and sediment routing components of a planning-level strip mining hydrology model. He found that at high flows the width and area could be predicted with 25% accuracy about two-thirds of the time. These results were comparable to or better than the results that would normally be obtained from the reconnaissance field surveys used in streamflow routing for planning-level-applications.

Knighton (1974) investigated the downstream and at-a-station variation in the width–discharge relation and its implication for hydraulic geometry. The width of a channel with cohesive banks and no marked downstream variation in bank erodibility increased with discharge in the downstream direction and the rate of increase was principally a function of discharge. The channel width at a cross-section was determined by flow exceeding the threshold of erosion, and its magnitude increased regularly downstream with drainage area. The at-a-station rate of change was controlled by the bank material composition, especially silt-clay content. Thus, at a cross section, the rate of change of width can increase due to the deposition of noncohesive sediment in the form of point bars and central islands, suggesting that the *b* exponent can be used to distinguish meander and braided reaches from straight reaches. The effect of this adjustment is to decrease the mean velocity range. Arguing that the at-a-station variations of hydraulic geometry relations are due to the variations in discharge and the downstream variations are caused by increasing discharge due to increasing drainage area, Stall and Yang (1970) expressed the hydraulic geometry relations in a general form, as was previously done by Stall and Fok (1968), and reasoned that the coefficients in the general relation reflected the influence of physiographic characteristics of the drainage basin.

Allen et al. (1994) employed downstream channel geometry for use in planning level models for resource and impact assessment. For the data on channel dimensions, they found that over a large variety of stream types and physiographic provinces the channel width and depth were predicted with 86% efficiency.

1.9 Stability of Hydraulic Geometry Relations

Hydraulic geometry relations, given by Equations (1.1a) and (1.1b), have been calibrated for a range of environments, using both field observations and laboratory simulations. Without a firm basis, Chong (1970) reasoned that over varying environments rivers behaved in a similar manner and hydraulic geometry relations were stable for rivers that had achieved "graded equilibrium." Parker (1979) found that scale factors, a, c, and k, varied from locality to locality but exponents, b, f, and m, exhibited a remarkable degree of consistency, and seemed independent of location and only weakly dependent on channel type. Examining the factors that produce variations in hydraulic geometry parameters, Rhoads (1991) hypothesized that the parameters were functions of channel sediment characteristics and flood magnitude, and varied continuously rather discretely.

Knighton (1974, 1975) emphasized variations in exponents as opposed to mean values. Rhodes (1978) noted that the exponent values for high flow conditions can be vastly different from those for low flow conditions. Since the physical character and appearance of a river is a product of adjustment to its boundaries to streamflow and sediment regime leading to its hydraulic geometry, it may be desirable to classify rivers into relatively homogeneous stream types. Millar and Quick (1993) discussed the effect of bank stability on the geometry of gravel-bed rivers and showed that the median grain diameter and the modified friction angle of the bank sediment indicated the bank stability and exerted a large influence on the hydraulic geometry. Rosgen (1990) classified rivers into seven major categories that differ in entrenchment, gradient, width/depth ratio, and sinuosity, and each category was divided into six types differentiated by dominant channel materials from bedrock to silt/clay along a continuum of gradient ranges.

For a subalpine stream in a relatively homogeneous environment, Phillips and Harlan (1984) found that hydraulic exponents were not stable over space. The interactions among channel form; discharge; and atmospheric, surface, and subsurface environments in the system produced variables that even in the short run and at a station were neither consistently dependent nor independent. The exponents and coefficients of hydraulic geometry relations of Equations (1.1a) and (1.1b) varied from location to location on the same river and from river to river, as well as from high flow range to low flow range. This is because the influx of water and sediment and the boundary conditions that the river channel is subjected to vary from location to location as well as from river to river. This means that for a fixed influx of water and sediment a channel will exhibit a hierarchy of hydraulic geometry relations in response to the boundary conditions imposed on the channel. It is these boundary conditions that force the channel to adjust its allowable hydraulic variables. For example, if a river is leveed on both sides, then it cannot adjust its width and is, therefore, left to adjust other variables, such as depth, friction, slope, and velocity. Likewise, if a canal is lined, then it cannot adjust its friction.

1.10 Variability of Exponents

Analyzing active channel geometry and discharge relations using data from 318 alluvial channels in the Midwestern United States and 50 Piedmont sites, Kolberg and Howard (1995) examined the variable exponent model of hydraulic geometry for piedmont and Midwestern streams. They showed that the discharge–width exponents were distinguishable, depending on the variations in materials forming the bed and banks of alluvial channels. For example, highly cohesive channels (high silt and clay beds), gravel and cobble streams, and noncohesive sand stream channels had statistically distinguishable exponents for Midwestern streams in the United States. The estimated width–discharge exponents for high silt and clay, gravel, or cobble bed channels deviated significantly from those for alluvial beds with 30% or greater sand content. However, no significant trend was apparent for estimated exponents among other sand-silt-clay channel categories. In the case of Piedmont data, no significant departure of the discharge–width relations was apparent for different groups of stream types based on sediment categories. Both Midwestern and Piedmont data indicated that the width–discharge exponents ranged from 0.35 to 0.46 for groups of streams with width to depth ratios less than 45 range. For groups of streams with width to depth ratios greater than 45, the width–discharge exponents decreased to values below 0.15, suggesting a systematic variation in the exponents and a diminished influence of channel shape. These results were in agreement with the findings of Osterkamp and Hedman (1982). Ferguson (1986) argued that at-a-station hydraulic geometry exponents were functions of cross-sectional channel shape.

Rhoads (1991) reasoned that variations among the channel types were not discrete thresholds but could be viewed as continuous variations. These variations support the assertion of Howard (1980) that thresholds in the hydraulic geometry of alluvial sand and gravel streams exist. However, the implicit control of width–discharge relation by sediment characteristics was not strongly evident for the Virginia and North Carolina Piedmont streams investigated by Kolberg and Howard (1995). Despite the lack of a strong relationship between channel shape and sediment type for sand and gravel Piedmont streams (Schumm, 1960; Miller and Onesti, 1979; Nanson and Huang, 1999), Park (1977) argued that certain hydraulic exponents could be characteristic of different climatic and environmental regimes.

Ellis and Church (2005) investigated the hydraulic geometry of secondary channels of the lower Fraser River (in British Columbia, Canada) divided into upstream, mid, and downstream sub-reaches. They found that at-a-station hydraulic geometry of subtypes stratified along gradients of width, depth, and velocity, but downstream hydraulic geometry conformed to a single power law. However, width and depth exponents deviated from the classical values.

Wilcock (1971) related hydraulic geometry characteristics with the movable fraction of bed load. He found that the channel cross-section did not increase faster downstream than discharge and the hydraulic radius at an upstream station was larger in relation to discharge of a particular station than at a downstream station. At a constant discharge and constant width, an increase in critical shear stress was associated with an increase in velocity and a decrease in depth. With increasing discharge, the critical shear stress will tend to increase only when the rate of increase in velocity (m) equals or exceeds the rate of increase in depth (f). A high rate of increase in velocity is the most important factor for the critical shear stress to increase with discharge, and a high rate of increase in depth is most important for the critical shear stress to decrease.

Using 85 stations from geographically and climatologically homogenous regions in Oklahoma and Kansas, United States, Dodov and Foufoula-Geogiou (2004a) showed that hydraulic exponents of at-a-station hydraulic geometry had a systematic dependence on scale. Likewise, the exponents of the downstream hydraulic geometry depended on the frequency of discharge. Dodov and Foufoula-Georgiou (2004b) provided a physical explanation of the scale dependence and validated the hypothesis that the scale dependence was a result of systematic increase of channel cross-sectional asymmetry over reaches of increasing scale.

1.11 Variation of Channel Width

Klein (1981) analyzed the variation of channel width with downstream discharge and found that the value $b = 0.5$ was a good average. The low b values normally occur for small basins (in lower flows) and for very big basins (in very high flows). Thus, the $b = 0.5$ value, being a good average, tends to smooth out deviations from the average. The value of b ranges from 0.2 to 0.89. Klein (1981) argued that the simple power functions were valid for small basins but did not hold over a wide range of discharges.

1.12 Variation of Channel Velocity

Mackin (1963) noted that in individual segments there were just as many segments with a downstream decrease in velocity as there were with a downstream increase in velocity. Carlston (1969) found in the Susquehanna River that the number of streams with a downstream velocity increase was balanced by an equal number of streams with either a constant velocity or a downstream decrease in velocity. The most common relationship on long segments of rivers was a nearly constant velocity; however, in many smaller streams velocity may increase or decrease downstream because of geological influences present at the mean annual discharge. Large rivers, such as the Mississippi, accommodate a downriver increase in discharge principally through increase in depth, whereas smaller rivers generally accommodate the downriver increase in discharge principally through the increase in width. Leopold (1953) showed that the large-scale floods that moved large quantities of sediment had nearly constant downstream velocity. Saco and Kumar (2002a, b, 2004) discussed the interaction between hydraulic geometry and network geometry that shapes the runoff process at shorter time scales, while the hydraulic geometry is shaped at longer time scales. They showed that the spatial variability of velocities arose due to the spatial variations of hydraulic geometry within a basin, causing a dispersive effect thus impacting the residence time of water and consequently peak response and time to peak. Paik and Kumar (2004) showed that the mean velocity nonlinearly varied with discharge, and as a result the time to peak flow was nonlinearly related to rainfall excess rate and hence the instantaneous response function.

1.13 Effect of Stream Size

Thornes (1970) analyzed the differences in the explained variances of the power function geometry relations for streams of different sizes – smaller as well as bigger (e.g., Susia-Missu and Araguaia Rivers in Brazil) – undergoing significant changes. Smaller streams were unstable and out of phase with the steady state condition in the mainstream. The difference was, therefore, between the minor stream and major tributary of the area, rather than between streams above and below a given discharge. Three possible explanations were advanced: (1) For the long term: Smaller channels undergo greater fluctuations from steady state conditions and are geomorphologically more active with strong slopes. (2) For the medium term: Small channels are substantially impacted by human activities in the form of extensive land clearances. The width to depth ratios are substantially affected by suspended load. (3) For the short term: Seasonal changes affect most significantly the geometric relationships in smaller channels. In smaller streams, there is a marked difference between bankfull discharge and low discharge because of little or no baseflow. This instability may reflect adjustments to long-term erosion of upland areas, changes accommodating the impact of human activities on small basins, or the effects of seasonal differences. The channel form adjusts to the wet season.

1.14 Effect of River Channel Patterns

Carlston (1969) classified rivers into stable and unstable rivers. Normally, straight and meandering rivers are regarded as stable, whereas braided rivers are regarded as unstable. Meandering channels may be stable in the vertical plane but may not aggrade or degrade in the bed and the meanders may move downstream. In addition to changes in the vertical plane, the bed configuration may change for braiding and meandering.

Schumm (1960) classified river patterns into straight, transitional, and meandering, with meandering patterns further classified into tortuous, irregular, and regular meanders. He also classified patterns, based on sinuosity defined as the ratio of thalweg length (length along the line of maximum depth) to valley length.

The sinuosity would be 2.3 for the tortuous pattern, 1.8 for the irregular pattern, 1.7 for the regular pattern, and 1.1 for the straight channel. Leopold and Wolman (1957) classified river patterns into straight, meandering, and braided.

Knighton (1987) classified river channels, based on boundary composition, into two primary types: cohesive and noncohesive. The cohesive type included two secondary types: bedrock channels characterized by no coherent cover of unconsolidated material and generally short reaches; and silt-clay channels characterized by boundaries having a high silt-clay content providing varying degrees of cohesion. The noncohesive type included three secondary types: sandbed channels characterized by live-bed channels composed largely of sandy material (0.063–2 mm), which is transported at a wide range of discharges; gravel-bed channels characterized by gravel (2–64 mm) or cobbles (64–256 mm), which are transported only at higher discharges; and boulder-bed channels composed of very large particles (>256 mm), which are moved infrequently and gradually grade into bedrock channels.

Investigating into river channel patterns, Chitale (1970, 1973) found that the coefficients and exponents in Equations (1.1a) and (1.1b) depended on the type of pattern. He classified channels into three groups: (1) single, (2) multi-thread, and (3) transitional channels. Single channels were further subdivided into (a) meandering, (b) straight, and (c) transitional between meandering and straight. Meandering channels were distinguished as regular or irregular and simple or compound. Multithread channels were divided into (a) braided and (b) branching-out channels. For small streams, including straight, shoaled, and meandered, the slope varied inversely with 0.12 power of discharge, whereas the power of discharge changed to -0.21 for critical straight-line water-surface slope. For braided channels, the power changed to -0.44. The meander length changed with discharge raised to the power of 0.5, but it changed with the range of discharge used. Similarly, the surface width changed with discharge raised to the power of 0.42.

In a study on the Lower Yellow River, China, Li and Li (2004) noted that in addition to riverbed material the relative movement of the river bank also affected the channel pattern. For a given flow, the river width reflects the change due to bank erosion. Rivers that have very high or small resistance have small curvatures. Rivers that have moderate resistance to erosion may have very large or very small curvatures. Rivers with moderate bank movement have larger curvatures. Nanson and Hickin (1983) described the relationship between the ratio of river bend, R, to river width, w, and the fluctuating speed of the riverbed (annual regressive bank width), indicating the compactness of the bend. The compactness of a compact bend is about 2.0 and that of a loose bend is about 6.0, and the dividing line between the two is about 3.0. They found geomorphic thresholds of different channel patterns of the Lower Yellow River:

$$50 > \frac{\sqrt{b}}{d} > 24 \qquad \text{braided}$$

$$24 > \frac{\sqrt{b}}{d} > 18 \qquad \text{wandering}$$

$$18 > \frac{\sqrt{b}}{d} > 7 \qquad \text{straight-slightly curved}$$

$$7 > \frac{\sqrt{b}}{d} > 3.5 \qquad \text{meandering.}$$

Discussing the relationship between river morphology and bend curvature, Qi et al. (2002) showed that rivers with large bend curvatures had narrow and deep channels and those with shallow and wide channels were always straight. When w/d decreases with the increase in discharge, the river channel tends to be narrow and deep, and when w/d increases the channel tends to be wide and shallow. A meandering reach is narrow and deep, whereas a wandering reach is wide and shallow. There is a strong relation between discharge and channel form, which shows that river morphology is a key factor that controls the transition from one channel pattern to another and affects the sediment transport capacity of the river. Narrow and deep channels tend to have higher sediment transport capacity and lead to greater elevation difference between main channel and floodplain, creating favorable conditions for the development of river bends. It may be inferred that channel type controls the formation of different channel patterns. An alluvial river tends to be straight at high flows because of relatively weak restrictions on the channel boundaries, and tends to be bending at low flows because of the formation of sand waves, formation of new shores, and bending of the dynamic axis of flow. This is followed by the deposition of sediment on one bank and scouring on the opposite bank, generating concave and convex banks: the initiation of meandering. The channel characteristics result from the sediment-water conditions. It has been found that channel slope is inversely proportional to the content of clay and silt (grain size smaller than 0.074 mm), meaning that the coarser the sediment composition the wider and shallower the channel. Also, there is a strong relationship between the ratio of sediment concentration at high flow and low flow to the exponent of sediment concentration-discharge power relation.

The discharge-slope relationship has also been used to identify different river patterns. Using data from 42 rivers in China that included slightly bending, braided, forked, and meandering patterns, Li and Li (2004) found

$$S = 0.0015 Q^{-0.364}. \tag{1.5}$$

For rivers similar in size, rivers steeper in slope more easily develop a wandering pattern. For rivers with the same slopes, rivers with smaller discharge tend more toward a meandering shape. Schumm (1973) showed that the slope change controls the stream pattern in a geomorphic threshold. Schumm (1960) expressed channel shape and sediment as

$$F = 255M^{-1.08}, \tag{1.6}$$

in which F is the channel shape expressed as a width–depth ratio (also called aspect), and M is the weighted mean percent silt-clay defined as

$$M = \frac{S_c \times w + S_b \times 2d}{w + 2d}, \tag{1.7}$$

where S_c is the percentage of silt and clay in the channel alluvium, and S_b is the percentage of silt and clay in the bank alluvium. The channel shape depends on the value of M.

For distinguishing different channel patterns, Begin (1981) employed the ratio of shear stress of water flow on the riverbed to average shear stress. Li and Li (2004) found for the same 42 rivers:

$$\frac{\tau_0}{\tau_{av}} = 666.67Q^{0.364}S, \tag{1.8}$$

in which τ_0 and τ_{av} represent shear stress on the riverbed and the average shear stress of the 42 rivers, respectively. The units are in the MKS system. Then, they concluded that

$$\frac{\tau_0}{\tau_{av}} < 0.75 \quad \text{meandering}$$

$$0.75 < \frac{\tau_0}{\tau_{av}} < 1.5 \quad \text{straight-slightly curved}$$

$$\frac{\tau_0}{\tau_{av}} > 1.5 \quad \text{braided and wandering}$$

$$2 < \frac{\tau_0}{\tau_{av}} < 1.5 \quad \text{wandering}$$

$$\frac{\tau_0}{\tau_{av}} > 2.0 \quad \text{mostly braided}$$

$$\frac{\tau_0}{\tau_{av}} = 2.0 \quad \text{wandering concentrated.}$$

Investigating changes in reach morphology and hydraulic geometry, Knighton (1972) discussed the interaction of the processes of erosion and deposition with channel hydraulics. A braided channel reach may gradually morph into a meandering reach by means of a slight modification to the flow pattern and without any change in the independent variables. In braided reaches, large differences in the flow behavior occur between juxtaposed channels, and streamflow may be concentrated in the channel that offers the least resistance. Using data from 137 meandering and braided rivers, Millar (2000) showed bank vegetation exercised a significant influence on alluvial channel patterns.

1.15 Effect of Hyper-concentrated Floods on Channel Geometry Adjustment

There are rivers in the world where hyper-concentrated floods with sediment concentration of more than 200–300 kg/m^3 often occur during the flood season, such as the Lower Yellow River in China and Kosi River in India. Such floods alter the geometry of braided reaches. Li et al. (2018) investigated the adjustment of both at-a-station channel geometry and downstream (reach scale) channel geometry due to such floods in a braided reach of the Lower Yellow River. They noted that of the several factors influencing the reach-scale adjustment index, the mean sediment transport was a key factor. The adjustment index was defined as the ratio of the pre- and post-flood characteristic parameter of channel geometry, which was defined by the ratio of reach-scale bankfull wetted perimeter to the wetted cross-sectional area where the wetted perimeter was approximated as the sum of the mean width plus twice the flow depth.

1.16 Effect of Dam Removal

The impact of dam construction on river ecology, hydrology, and geomorphology has been studied over the years. However, these days, dam removal in the United States is being considered as a viable alternative in river management, and its impacts are receiving much attention from the scientific community, public, and policy makers. A major concern is the fate of sediment stored in reservoirs. Dam removal exposes the sediment to erosion by incoming flow and subsequent transport downstream. Downstream sediment transport can affect local structures, downstream biota, sediment characteristics, and channel geometry adjustment. Further, if the stored sediment is contaminated then the downstream communities may be impacted, which may have legal implications (Doyle et al. 2002). In a study on channel adjustments following two dam removals on two low-gradient, fine-to-coarse grained rivers in Wisconsin, Doyle et al. (2003) found that large quantities of fine sediment were eroded from channels and were deposited 3–5 km downstream. One river exhibited rapid upstream changes, bed degradation, minimal bank erosion, and sediment deposition on channel margins and new floodplain. The other river experienced head-cut migration that governed channel adjustment, exhibited by a deep, narrow channel formed downstream of the head cut, with negligible changes upstream of the head-cut. Chin et al. (2002) surveyed a 19-km channel reach downstream of Somerville Dam on the Yegua River (a tributary of the Brazos River), Texas, and found that the dam closure led to a decrease of channel capacity by 65% in a 34-year period, which corresponded to an approximately 85% reduction in annual flood peaks. Channel depth changed the most, with an average reduction of 61% and channel width reduced by about 9%, indicating cohesive bank material along with growth of riparian vegetation during low-flow periods.

1.17 Power Relations for Drainage Basins

For subbasins of the Sangamon and Vermilion Rivers, McConkey and Singh (1992) found that power function relations had poor performance for low flows, and observed that the variation in discharge was consistently dependent on drainage area and annual flow duration. Because of the existence of riffles and pools in natural streams, the hydraulic geometry relations were less reliable, especially at low flows, which typically create more critical conditions for fish species and various aquatic life forms. The power function relations were reasonable for watersheds greater than 100 square miles (256 km^2) but may not be reliable for smaller drainage area streams.

1.18 Effect of Land Use

Lane and Foster (1980) analyzed adjustments of stream channels due to changes in discharge and channel characteristics resulting from changing land use. Their results showed that channel widths increased as discharge and hydraulic resistance increased and that narrower channels resulted from larger critical stresses. Therefore, the land use that caused these changes caused a readjustment in streams. For example, changes in the amount of sediment eroded from the channel boundary as the boundary adjusted to changing discharge reflected changes in land use. Investigating the influence of forest on hydraulic geometry, Heede (1972) reported that streams needed additional logs to adjust to slope, and average step length between logs and gravel bars was related to channel gradient and median bed material size. The average flow velocity ranged from 0.5–2.5 fps. Millar (2000) investigated the influence of bank vegetation, quantified in terms of friction angle, on alluvial channel patterns of alluvial gravel-bed rivers, including meandering, wandering, and braided. He concluded that bank vegetation exercised a significant control on alluvial channel patterns.

1.19 Boundary Conditions

The boundary conditions that a channel has to satisfy vary with the type of channel. Based on their boundary conditions and hydrologic input, open channels can be classified (Yu and Wolman, 1987) into three groups: (a) channels with rigid boundaries and uniform flow, (b) channels with erodible beds and banks, and (c) channels in equilibrium and regime. In the first case, there is only one flow depth above critical depth associated with uniform flow. If the discharge is known, the geometry is determined. In the second case, geometry, resistance, flow, and sediment transport are interrelated. The channel geometry is determined by the laws of fluid flow, sediment transport, and bank stability. In the third case, the hydraulic geometry is determined by the stability of channel bank, availability of sediment load (bed and bank material, and cohesive material) for transport, and vegetative cover, in addition to the magnitude and variability of flow. The channel hydraulic geometry is assumed to be determined by discharge. Yu and Wolman (1987) investigated the effect of variable flow conditions characteristic of natural rivers on the channel hydraulic geometry.

1.20 Organization of Contents

The hydraulic geometry relations of Equations (1.1a) and (1.1b) are derived using a variety of hypotheses. Each hypothesis leads to unique relations between channel form parameters and discharge, and the relations corresponding to one hypothesis are not necessarily identical to those corresponding to other hypotheses but are of the same form. The hypotheses include (1) regime theory, (2) Leopold–Maddock empirical theory, (3) minimum variance theory, (4) dimensionless principles, (5) hydrodynamic theory, (6) scaling theory, (7) tractive force theory, (8) thermo-dynamic theory, (9) similarity principle, (10) mobility theory, (11) maximum sediment discharge and Froude number hypothesis, (12) principle of minimum Froude number, (13) principle of maximum friction, (14) maximum flow efficiency theory, (15) principle of least action, (16) theory of minimum energy dissipation rate, (17) entropy theory, (18) energy dissipation rate and entropy theory, (19) theory of stream power, and (20) regional geometry. Gleason (2015) classified these theories and the resulting approaches into rational, extremal, and empirical, whereas Chang (1988) classified them into empirical and rational. These hypotheses for deriving hydraulic geometry are discussed in separate chapters.

References

Allan, J. D. (1995). Channels and flow, the transport of materials. In: *Stream Ecology, Structure and Function of Running Waters*, pp. 8–12, Chapman & Hall Publishing, London.

Allen, P. M., Arnold, J. G., and Byars, B. W. (1994). Downstream channel geometry for use of in planning-level models. *Water Resources Bulletin*, Vol. 30, No. 4, pp. 663–671.

Bathurst, J. C. (1978). Flow resistance of large-scale roughness. Journal of Hydraulics Division, ASCE, pp. 1587–1604.

Begin, Z. B. (1981). The relationship between flow-shear and stream pattern. *Journal of Hydrology*, Vol. 52, Nos. 3 and 4, pp. 307–319.

Benson, M. A. (1962). *Factors influencing the occurrence of floods in a humid region of diverse terrain*. U.S. Geological Survey Water Supply Paper 1580-B, Washington, DC.

Betson, R. P. (1979). A geomorphic model for use in streamflow routing. *Water Resources Research*, Vol. 15, No. 1, pp. 95–101.

Bhowmik, N. G. (1984). Hydraulic geometry of floodplains. *Journal of Hydrology*, Vol. 68, pp. 369–401.

Bogardi, J. (1984). *Mobile-Bed Fluviology*. University of Alberta Press, Edmonton.

Bradley, W. C., Fahnestock, R. K., and Rowekamp, E. T. (1972). Coarse sediment transport by flood flows on Knik River, Alaska. *Geological Society of America Bulletin*, Vol. 83, pp. 1261–1284.

Carlston, C. W. (1969). Downstream variations in the hydraulic geometry of streams: Special emphasis on mean velocity. *American Journal of Science*, Vol. 267, pp. 499–509.

Chang, H. H. (1988). *Fluvial Processes in River Engineering*. John Wiley & Sons, New York.

Chin, A., Horrie, D. L., Trice, T. H., and Given, J. L. (2002). Adjustment of stream channel capacity following dam closure, Yegua Creek, Texas. *Journal of the American Water Resources Association*, Vol. 38, No. 6, pp. 1521–1531.

Chitale, S. V. (1970). River channel patterns. *Proceedings of Hydraulics Division, ASCE*, Vol. 96, No. HY1, pp. 201–222.

Chitale, S. V. (1973). Theories and relationships of river channel patterns. *Journal of Hydrology*, Vol. 19, pp. 285–308.

Chong, S. E. (1970). The width, depth and velocity of Sungei Kimla, Perak. *Geographica*, Vol. 6, pp. 72–63.

Church, M. and Kellerhals, R. (1978). On the statistics of grain size variation along a gravel river. *Canadian Journal of Earth Sciences*, Vol. 15, pp. 1151–1160.

Dingman, S. L. (2002). *Physical Hydrology*. Prentice Hall, Upper Saddle River, NJ.

Dodov, B. and Foufoula-Geogiou, E. (2004a). Generalized hydraulic geometry: Derivation based on multiscaling formalism. *Water Resources Research*, Vol. 40, W06302, https://doi.org/10.1029/2003WR002082.

Dodov, B. and Foufoula-Geogiou, E. (2004b). Generalized hydraulic geometry: Insights based on fluvial instability analysis and a physical model. *Water Resources Research*, Vol. 40, W12201, https://doi.org/10.1029/2004WR003196.

Doyle, M. W., Stanley, E. H., and Harbor, J. M. (2002). Geomorphic analogies for assessing probable channel response to dam removal. *Journal of the American Water Resources Association*, Vol. 38, No. 6, pp. 1567–1579.

Doyle, M. W., Stanley, E. H., and Harbor, J. M. (2003). Channel adjustments following two dam removals in Wisconsin. *Water Resources Research*, Vol. 39, No. 1, https://doi.org/10.1029/2002WR001714.

Dury, G. H. (1976). Discharge prediction, present and former, from channel dimensions. *Journal of Hydrology*, Vol. 30, pp. 219–245.

Ellis, E. R. and Church, M. (2005). Hydraulic geometry of secondary channels of lower Fraser River, British Columbia, from acoustic Doppler profiling. *Water Resources Research*, Vol. 41, W08421, https://doi.org/10.1029/2004WR0003777, pp. 1–15

Emmett, W. W. (1975). *The channels and waters of the Upper Salmon River area, Idaho*. U.S. Geological Survey Professional Paper 870A, Washington, DC.

Ferguson, R. I. (1986). Hydraulics and hydraulic geometry. *Progress in Physical Geography*, Vol. 10, p. 1031,

Gleason, C. J. (2015). Hydraulic geometry of natural rivers: A review and future directions. *Progress in Physical Geography*, Vol. 39, No. 3, pp. 337–360.

Gleason, C. J. and Smith, L. C. (2014). Towards global mapping of river discharge using satellite images and at-many station hydraulic geometry. *Proceedings of the National Academy of Sciences*, Vol. 111, pp. 4788–4791.

Griffiths, G. A. (1983). Stable channel design in alluvial rivers. *Journal of Hydrology*, Vol. 65, pp. 259–270.

Hack, J. T. (1957). Studies of longitudinal stream profiles in Virginia and Maryland. U.S. Geological Survey Professional Paper 294-B, pp. 45–67.

Harman, C., Stewardon, M., and DeRose, R. (2008). Variability and uncertainty in reach bankfull hydraulic geometry. *Journal of Hydraulic Geometry*, Vol. 351, pp. 13–25.

Heede, B. D. (1972). Influences of a forest on the hydraulic geometry of two mountain streams. *Water Resources Bulletin*, Vol. 8, No. 3, pp. 523–530.

Howard, A. D. (1980). Thresholds in river regimes. In: *Thresholds in Geomorphology*, edited by D. R. Coates and J. D. Vitek, pp. 227–258, Allen and Unwin, Winchester, MA.

Jowett, I. G. (1998). Hydraulic geometry of New Zealand rivers and its use as a preliminary method of habitat assessment. *Regulated Rivers: Research and Management*, Vol. 14, pp. 451–466.

Klein, M. (1981). Drainage area and the variation of channel geometry downstream. *Earth Surface Processes and Landforms*, Vol. 6, pp. 589–593.

Knighton, A. D. (1972). Changes in braided reach. *Geological Society of America Bulletin*, Vol. 83, pp. 3813–3922.

Knighton, A. D. (1974). Variation in width-discharge relation and some implications for hydraulic geometry. *Geological Society of America Bulletin*, Vol. 85, pp. 1069–1076.

Knighton, A. D. (1975). Variations in at-a-station hydraulic geometry. *American Journal of Science*, Vol. 275, pp. 186–218.

Knighton, A. D. (1980). Longitudinal changes in size and sorting of stream-bed material in four English rivers. *Geological Society of America Bulletin*, Vol. 91, pp. 55–62.

Knighton, A. D. (1987). River channel adjustment: The downstream dimension. In: *River Channels: Environment and Process*, edited by K. S. Richards, pp. 95–128, Basil Blackwell, Oxford.

Kolberg, F. J. and Howard, A. D. (1995). Active channel geometry and discharge relations of U.S. piedmont and midwestern streams: The variable exponent model revisited. *Water Resources Research*, Vol. 31, No. 9, pp. 2353–2365.

Lane, L. J. and Foster, G. R. (1980). Modeling channel processes with changing land use. *Proceedings, ASCE Symposium on Watershed Management*, Vol. 1, pp. 200–214.

Langbein, W. B. (1964). Geometry of river channels. *Journal of the Hydraulics Division, ASCE*, Vol. 90, No. HY2, pp. 301–311.

Leopold, L. H. (1953). Downstream change of velocity in rivers. *American Journal of Science*, Vol. 25, pp. 606–624.

Leopold, L. B. and Maddock, T. J. (1953). Hydraulic geometry of stream channels and some physiographic implications. *U.S. Geological Survey Professional Paper*, Vol. 252, p. 55.

Leopold, L. B. and Wolman, M. G. (1957). River channel patterns: braided, meandering and straight. *U.S. Geological Survey Professional Paper*, Vol. 282-B, pp. 39–59.

Li, J., Xia, J., Zhou, M., Deng, S., and Wang, Z. (2018). Channel geometry adjustment in response to hyperconcentrated floods in a braided reach of the Lower Yellow River. *Progress in Physical Geography*, Vol. 42, No.3, pp. 352–368.

Li, Z. and Li, Z. (2004). Geomorphic thresholds for channel evolution in the Lower Yellow River. *International Journal of Sediment Research*, Vol. 19, No. 3, pp. 191–201.

Lindley, E. S. (1919). Regime channels. *Proceedings, Punjab Engineering Congress*, Vol. 7, p. 68.

McConkey, S. A. and Singh, K. P. (1992). Alternative approach to the formulation of basin hydraulic geometry equations. *Water Resources Bulletin*, Vol. 28, No. 2, pp. 305–312.

Mackin, J. H. (1963). Rational and empirical methods of investigation in geology. In: *The Fabric of Geology*, edited by Albritton, C. C., Jr., Addison-Wesley Publishing Co., Reading, MA, pp. 135–163.

Millar, R. G. (2000). Influence of bank vegetation on alluvial channel patterns. *Water Resources Research*, Vol. 36, No. 4, pp. 1100–1118.

Millar, R. G. and Quick, M. C. (1993). Effect of bank stability on geometry of gravel rivers. *Journal of Hydraulic Engineering*, Vol. 119, No. 2, pp. 1343–1363.

Miller, T. K. and Onesti, L. J. (1979). The relationship between channel shape and sediment characteristics in the channel perimeter. *Geological Society of America Bulletin*, Vol. 90, pp. 310–304.

Mosley, M. P. (1982). Analysis of the effect of changing discharge on channel morphology and instream uses in a braided river, Ohau River, New Zealand. *Water Resources Research*, Vol. 8, No. 4, pp. 800–812.

Mosley, M. P. (1983). Flow measurements for recreation and wildlife in New Zealand rivers: A review. *Journal of Hydrology (N.Z.)*, Vol. 22, pp. 152–174.

Nanson, G. C. and Hickin, E. J. (1983). Channel migration and incision on Beatton River. *Journal of Hydraulic Engineering*, Vol. 109, pp. 327–337.

Nanson, G. C. and Huang, H. Q. (1999). Anabranching rivers: Divided efficiency leading to fluvial diversity. In: *Varieties of Fluvial Form*, edited by A. J. Miller and A. Gupta, Wiley, Chichester.

Navratil, O. and Albert, M. B. (2010). Non-linearity of reach hydraulic geometry relations. *Journal of Hydrology*, Vol. 388, pp. 280–290.

Nordin, C. F., Meade, R. H., Curtis, W. F., Bosio, N. J., and Landin, P. M. B. (1980). Size distribution of Amazon River bed sediment. *Nature*, Vol. 286, pp. 52–53.

Osterkamp, W. R. and Hedman, E. R. (1982). *Perennial-streamflow characteristics related to channel geometry and sediment in Missouri River basins*. U.S. Geological Survey Professional Paper 1242, pp. 37, Washington, DC.

Paik, K. and Kumar, P. (2004). Hydraulic geometry and nonlinearity of the network instantaneous response. *Water Resources Research*, Vol. 40, W03602, pp. 1–7.

Park, C. C. (1977). World-wide variations in hydraulic geometry exponents of stream channels: An analysis and some observations. *Journal of Hydrology*, Vol. 33, pp. 133–146.

Parker, G. (1978). Self-formed rivers with equilibrium banks and mobile bed: Part II. The gravel river. *Journal of Fluid Mechanics*, Vol. 76, No. 3, pp. 457–480.

Parker, G. (1979). Hydraulic geometry of active gravel rivers. *Journal of Hydraulic Division, Proc. ASCE*, Vol. 105, No. HY9, pp. 1185–1201.

Phillips, P. J. and Harlan, J. M. (1984). Spatial dependency of hydraulic geometry exponents in a subalpine stream. *Journal of Hydrology*, Vol. 71, pp. 277–283.

Pitlick, J. and Cress, R. (2002). Downstream changes in the channel geometry of a large gravel bed river. *Water Resources Resseacrh*, Vol. 38, No. 10, 1216, https://doi:10.1029/2001WR000898.

Ponton, J. R. (1972). Hydraulic geometry in the Green and Birkenhead river basins, British Columbia. In: *Mountain Geomorphology: Geomorphological Processes in the Canadian Zcordillera*, edited by H. O. Slaymaker and H. J. McPherson, pp. 151–160, Tantalus Research Limited, Vancouver.

Qi, P., Liang, G., Sun, Z., and Qi, H. (2002). The forming conditions of alluvial river channel patterns. *International Journal of Sediment Research*, Vol. 17, No. 1, pp. 83–88.

Rana, S. A., Simons, D. B., and Mahmood, K. (1973). Analysis of sediment sorting in alluvial channels. *Journal of Hydraulics Division, ASCE*, Vol. 99, pp. 1967–1980.

Rhoads, B. L. (1991). A continuously varying parameter model of downstream hydraulic geometry. *Water Resources Research*, Vol. 27, No. 8, pp. 1865–1872.

Rhodes, D. D. (1978). Worldwide variations in hydraulic geometry exponents of stream channels: an analysis and some observations: Comments. *Journal of Hydrology*, Vol. 33, pp. 133–146.

Richards, K. S. (1973). Hydraulic geometry and channel roughness: A nonlinear system. *American Journal of Science*, Vol. 273, pp. 877–896.

Richards, K. S. (1976). Complex width-discharge relations in natural river sections. *Geological Society of America Bulletin*, Vol. 87, pp. 199–206.

Richards, K. S. (1977). Channel and flow geometry: a geomorphologic perspective. *Progress in Physical Geography*, Vol. 1, pp. 65–102.

Richards, K. S. (1982). *Rivers: Form and Process in Alluvial Channels*. Metheun, London.

Rosgen, D. L. (1990). A classification of natural rivers. *Catena*, Vol. 22, pp. 169–190.

Rosenfield, J. S., Post, J., Robins, G., and Hatfield, T. (2007). Hydraulic geometry as a physical template for the river continuum: Application to optimal flows and longitudinal trends in salmonid habitat. *Canadian Journal of Fish Aquaculture Science*, Vol. 64, pp. 755–767.

Saco, P. M. and Kumar, P. (2002a). Kinematic dispersion in stream networks: 1. Coupling hydraulic and network geometry. *Water Resources Research*, Vol. 38, No. 11, pp. 26-1.

Saco, P. M. and Kumar, P. (2002b). Kinematic dispersion in stream networks: 2. Scale sizes and self-similar network organization. *Water Resources Research*, Vol. 38, No. 11, pp. 27-1.

Saco, P. M. and Kumar, P. (2004). Kinematic dispersion effects of hillslope velocities. *Water Resources Research*, Vol. 40, WO1301.

Schumm, S. A. (1960). The shape of alluvial channels in relation to sediment type. U.S. Geological Survey Professional Paper 352B, pp. 17–30.

Schumm, S. A. (1973). Geomorphic thresholds and complex response of drainage systems. *Journal of Sediment Research*, Vol. 3, pp. 39–43.

Simons, D. B. and Senturk, F. (1977). *Sediment Transport Technology*. Water Resources Publications, Highlands Ranch, CO.

Singh, V. P. (2003). On the theories of hydraulic geometry. *International Journal of Sediment Research*, Vol. 18, No. 3, pp. 196–218.

Stall, J. B. and Fok, Y.-S. (1968). *Hydraulic Geometry of Illinois Streams*. University of Illinois Water Resources Center, Research Report No. 15.

Stall, J. B. and Yang, C. T. (1970). *Hydraulic Geometry of 12 Selected Stream Systems of the United States*. University of Illinois Water Resources Research Center, Research Report No. 32.

Stout, H. P. (1979). Prediction of oxygen deficits associated with effluent inputs to the rivers of the Forth catchment. *Proceedings of the Institution of Civil Engineers*, Part 3, Vol. 67, pp. 51–64.

Thornes, J. B. (1970). The hydraulic geometry of stream channels in the Xingu-Araguaia headwaters. *The Geographical Journals*, Vol. 136, pp. 376–382.

Thornes, J. B. (1977). Hydraulic geometry and channel change. In: *River Channel Changes*, edited by K. J. Gregory, pp. 91–100, John Wiley & Sons, New York.

Trimble, S. W. (1975). Denudation studies: Can we assume stream steady state? *Science*, Vol. 188, pp. 1207–1208.

Trimble, S. W. (1983). A sediment budget for Coon Creek basin in the Driftless Area, Wisconsin, 1853–1977. *American Journal of Science*, Vol. 283, pp. 454–474.

Walling, D. (1983). The sediment delivery problem. *Journal of Hydrology*, Vol. 65, pp. 209–237.

Western, A. W., Finlayson, B. L., McMahon, T. A., and O'Neill, I. C. (1997). A method for characterizing longitudinal irregularity in river channels. *Geomorphology*, Vol. 21, pp. 39–51.

Wilcock, D. N. (1971). *Investigation into the relation between bedload transport and channel shape. Geological Society of America Bulletin*, Vol. 82, pp. 2159–2176.

Wohl, E. and Merritt, D. M. (2008). Reach-scale channel geometry of mountain streams. *Geomorphology*, Vol. 93, pp. 168–185.

Wolman, M. G. (1955). *The natural channel of Brandywine Creek, Pennsylvania*. U. S. Geological Survey Professional Paper 271, Washington, DC.

Yang, C. T., Song, C. C., and Woldenberg, M. T. (1981). Hydraulic geometry and minimum rate of energy dissipation. *Water Resources Research*, Vol. 17, pp. 877–896.

Yatsu, E. (1955). On the longitudinal profile of the graded river. *Transactions of the American Geophysical Union*, Vol. 36, pp. 655–663.

Yu, B. and Wolman, M. G. (1987). Some dynamic aspects of river geometry. *Water Resources Research*, Vol. 23, No. 3, pp. 501–509.

2

Governing Equations

Physically based approaches to hydraulic geometry relations for width, depth, velocity, and slope require equations of continuity of water, roughness, and sediment transport. Different methods have been employed for different expressions of roughness and sediment transport. Without delving into their underlying theories, this chapter briefly outlines these expressions, as they will be invoked in subsequent chapters. Also, unit stream power, stream power as well as entropy have been employed, which are also briefly discussed.

Notation

a = bedload layer thickness

A = cross-sectional area

a_0 = constant

a_1 = constant

a_s = hypothetical height of the suspended sediment

b_f = bed form factor

e = coefficient

C = Chezy coefficient: a measure of resistance to flow

C_{cr} = constant

C_{ec} = coefficient

C_I = constant of proportionality

C_s = average concentration of sediment in the channel

C_* = constant of proportionality

C_3 = exponent

d = mean flow depth (cross-sectionally averaged)

D = particle size

D_{gr} = dimensionless grain size

D_s = Nikuradse's sand roughness or mean diameter of sand grains
E_s = specific energy head
E_{sc} = critical value of Shields entrainment function
f = Darcy–Weisbach friction factor
F_d = drag force, proportional to the shear stress
F_{gr} = grain resistance component
F_r = Froude number
F_* = overall resistance of grains (form plus grain)
f_z = Chezy's friction factor
$f(x)$ = probability density function
g = acceleration due to gravity
H = total head of water at any point
$H(X)$ = entropy of the system
h_f = loss of energy
K = coefficient
K_0 = proportionality factor
K_1 = a function of sediment size
L = length of the channel reach
M_n = mobility number (Shield's)
m_c = correction factor associated with the degree of meandering
M_i = initial motion parameter, the value of F_{gr} when sediment transport begins
n = Manning's roughness coefficient
n_0 = basic n for a straight clean regular channel
n_s = transitional sediment size factor
n_2 = additional component of roughness due to the variation in shape and size of the channel from one cross-section to another
p = static pressure
P = wetted perimeter
$p_i, i = 1, 2, \ldots, N$ = probabilities
q = water discharge per unit width
Q = water discharge
Q_b = bedload discharge
Q_s = sediment transport rate
q_B = volume of grain material transported per second per unit width
q_{B*} = dimensionless bed load transport
q_c = critical transport or initiation of the suspended
q_{cr} = critical discharge per unit width
q_s = sediment transport per unit time per unit width
q_T = total sediment transport
q_{top} = sediment transported up to a height
R = hydraulic radius ($= A/P$)

Continued

R_n = Reynolds number

r_v = ratio of flow velocity at height $y = 0.2$ mean depth d to velocity at $y = 0.8\ d$

s = density ratio

S = slope

S_f = friction slope or energy line slope or energy slope

S_r = grain resistance responsible for bed load movement

S_w = water surface or hydraulic grade line slope or hydraulic gradient

S_0 = channel bottom or bed slope

SP = stream power per unit channel length

SP_b = stream power for bed load transport

SP_r = stream power per unit width to overcome the resistance generated by sand grain roughness and bed forms

SP_s = stream power per unit width to transport suspended sediment

SP_1 = power dissipated by minor factors, such as bends

t = time

USP = unit stream power

USP_* = dimensionless form of unit stream power

v = flow velocity

v_* = shear velocity

v_b = velocity of particles

v_{cr} = critical mean flow velocity at the incipient motion

v_{*cr} = critical shear velocity

V_{ol} = net volume of particles

w = width of flow area

w_T = top width ($= dA/dd$)

x = downstream distance

Y = potential energy per unit weight of water in a stream

Y_s = stream fall

z = elevation from the datum to the channel bed

Z = ratio of flow depth to particle diameter

α_e = kinetic energy distribution factor

α = coefficient that accounts for the relationship of grain roughness to particle size

γ = weight density or specific weight of water

γ_s = specific weight of sediment

μ = dynamic viscosity

υ = kinematic viscosity

ϕ = angle of repose of sediments

ρ = fluid density

ρ_m = density of mixture

ρ_s = density of sediment

τ = tractive force, equal to the component of the gravity force parallel to the channel bed that acts on the body of water

τ_b = effective bed shear stress (total bed shear stress minus the critical shear stress) due to the movement and transport of bedload transport

τ_{avg} = average shear stress

τ_c = critical value for the initiation of sediment motion

τ_{cr} = critical shear stress or critical Shields parameter for initiation of suspension

τ_{form} = form drag due to bed features

τ_{grain} = grain resistance due to the granular texture

τ_0 = bed shear

τ_{0*} = dimensionless bed shear stress

τ_s = shear stress of suspended sediments

τ_{ws} = shear stress due to the total fluid-sediment mixture

θ_b = bed slope angle

θ_{cr} = critical value of θ for the initiation of sediment motion

ω = terminal fall velocity of sediment particles

ω_d = particle fall velocity of diameter D

2.1 Introduction

A river transports water and sediment. In its regime state, a dynamic equilibrium is achieved, meaning the river self-adjusts its channel pattern, geometry, roughness, and slope (Chang, 1979). During the adjustment, the aggradation and degradation of sediment particles are balanced with no net transport. The river can develop three channel patterns: straight, meandering, and braided. A channel is regarded as straight if it does not display a sinuous course but its thalweg wanders back and forth from bank to bank (Leopold et al., 1964) and its slope is about the same as the valley slope. Straight channels are found either on flat slopes or on steep slopes. River banks are straight for reaches but not the entire river length. Meandering is the most common pattern. The slope of meandering channels is less than the valley slope but greater than the flat slope. A channel is regarded as braided if its flow is subdivided through the development of bars or islands, is usually not sinuous, and has a large width to depth ratio. The channel pattern and geometry are a response to water discharge and sediment load imposed on the river and are important for river training and design.

The flow of water and sediment in an alluvial channel is governed by the law of conservation of mass and Newton's laws of motion. These are nature's laws for alluvial channels. For design of canals the fundamental laws can be divided into two groups. The first group consists of (1) conservation of mass of water, (2) conservation of mass of sediment, and (3) Newton's law of motion for water. The second group consists of (1) Newton's law for the motion of sediment in suspension and that moving on the bed, and (2) Newton's law for the motionless sediment in the banks. For deriving channel geometry, it is assumed that the flow in channels or canals can be adequately described by one-dimensional equations for steady uniform flow of incompressible water and sediment.

2.2 Continuity Equation

Consider a prismatic river reach in which the flow of water is denoted by water discharge Q and the sediment transport rate Q_s. Let the cross-sectional area be denoted by A, width by w, flow depth by d, and mean (cross-sectionally averaged) flow velocity by v. Then, one can write the continuity of flow as

$$Q = vA. \tag{2.1}$$

If the cross-sectional area is rectangular then

$$Q = vwd, \qquad A = wd. \tag{2.2}$$

Sometimes, the continuity equation is also written as

$$Q = vRP, \tag{2.3}$$

where P is the wetted perimeter and R is the hydraulic radius ($= A/P$). For wide rectangular channels, $P \sim w$ and $R \sim d$. Then, Equation (2.3) reduces to Equation (2.2).

2.3 Flow Characterization

Flow in rivers and canals is subcritical and its critical and supercritical states are only transient. Likewise, the flow is turbulent and is seldom laminar. Thus, hydraulic geometry is derived for subcritical and turbulent flow. To characterize flow, two dimensionless numbers, Froude number and Reynolds number, are employed and are now defined.

Froude number (F_r) is defined as the ratio of inertial force to gravitational force:

$$F_r = \frac{v}{\sqrt{gd}}, \tag{2.4}$$

and is used to determine the nature of flow. Here g is the acceleration due to gravity. If F_r is less than one, then flow is characterized as subcritical, which is most prevalent in nature. If it equals one then flow is critical and if it is greater than one then flow is supercritical. Sometimes, it is also defined as

$$F_r = \frac{v}{\sqrt{gR}}. \tag{2.5}$$

If the channel is rectangular or wide then Equation (2.5) reduces to Equation (2.4), because $R \sim d$.

Reynolds number (R_n) is defined as the ratio of inertial force to viscous force:

$$R_n = \frac{\rho v d}{\mu} = \frac{v d}{\upsilon}, \tag{2.6}$$

where μ is the dynamic viscosity and is equal to the product of fluid density (ρ) and kinematic viscosity (υ). R_n is used to determine if flow is laminar ($R_n < 1{,}000$), transient (R_n is between 1,000 and 3,000), or turbulent ($R_n > 3{,}000$). Turbulent flow is most prevalent in nature, and R_n is usually greater than 5,000. It is noted that these limits of R_n are only approximate values.

2.4 Energy Equation

In deriving hydraulic geometry, the law of conservation of energy is often not directly employed. However, it is implicit through energy slope. It may therefore be relevant to express the energy equation here. The law of conservation of energy can be expressed in terms of the head of water at any point where the total head (H) equals the sum of the elevation above a datum, pressure head, and velocity head, as shown in Figure 2.1.

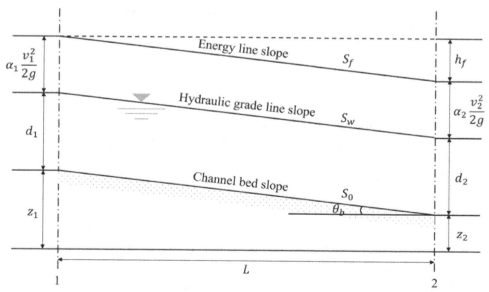

Figure 2.1 Energy head of flow at a point in a channel

$$H = z + d + \alpha_e \frac{v^2}{2g}, \tag{2.7}$$

where z is the elevation from the datum to the channel bed and α_e is the kinetic energy distribution factor. It may be noted that $d = p/\gamma$ is the pressure head, where p is static pressure and γ is the weight density or specific weight of water. Often, $\left(\frac{p}{\gamma}\right) + z$ is called the piezometric head or hydraulic head. Thus, the total energy head is the sum of piezometric head or hydraulic head and kinetic energy head. The specific energy head (E_s) measured with respect to the channel bottom is then defined as the sum of pressure energy head and kinetic energy head:

$$E_s = d + \alpha_e \frac{v^2}{2g}. \tag{2.8}$$

When water moves from section 1 to section 2, it has to do work to overcome friction during its movement, which requires the expenditure of energy denoted as the loss of energy, h_f. Referring to Figure 2.1, the law of conservation of energy can be expressed as

$$z_1 + d_1 + \alpha_1 \frac{v_1^2}{2g} = z_2 + d_2 + \alpha_2 \frac{v_2^2}{2g} + h_f, \tag{2.9}$$

where subscripts 1 and 2 refer to sections 1 and 2, respectively. Equation (2.9) reduces to Bernoulli equation for $\alpha_1 = \alpha_2 = 1$, and $h_f = 0$. The left side of Equation (2.9) expresses the total head at section 1, and the right side expresses the total head plus the head loss h_f and can be written as

$$H_1 = H_2 + h_f, \tag{2.10}$$

where H_1 is the total energy head at section 1, and H_2 is the total energy head at section 2.

Equation (2.9) allows to define three slopes that are often referred to in hydraulic geometry discussions. The channel bottom or bed slope, denoted as S_0, can be defined as

$$S_0 = \frac{z_1 - z_2}{L}, \tag{2.11}$$

where L is the length of the channel reach or the distance between sections 1 and 2, z_1 is the elevation of the channel bed at section 1, and z_2 is the elevation of the channel bed at section 2.

The water surface or hydraulic grade line slope or simply hydraulic gradient, S_w, can be defined as

$$S_w = \frac{(z_1 + d_1) - (z_2 + d_2)}{L}. \tag{2.12}$$

The energy line slope or energy slope (S_f) can be defined as

$$S_f = \frac{H_1 - H_2}{L} = \frac{h_f}{L}. \tag{2.13}$$

The energy slope is often referred to as the friction slope, assuming the loss of energy is primarily due to friction.

2.5 Gradually Varied Flow Equation

The equation for gradually varied flow in a channel can be derived by differentiating Equation (2.7) with respect to the downstream distance x, and using Equations (2.11) and (2.13) and can be expressed as

$$\frac{dd}{dx} = \frac{S_0 - S_f}{1 - \frac{Q^2 w_T}{gA^3}}, \tag{2.14}$$

where w_T is the top width (dA/dd), S_0 is the bed slope, S_f is the friction or energy slope, and g is the acceleration due to gravity. If the channel is rectangular, then Equation (2.14) reduces to

$$\frac{dd}{dx} = \frac{S_0 - S_f}{1 - \frac{v^2}{gd}}. \tag{2.15}$$

Equation (2.15) can also be expressed in terms of Froude number as

$$\frac{dd}{dx} = \frac{S_0 - S_f}{1 - F_r^2}. \tag{2.16}$$

Equation (2.14) or (2.15) can also be expressed in terms of the conveyance factor defined based on the resistance equation: Manning's or Chezy's.

2.6 Flow Resistance Equations

Flow resistance plays a fundamental role in river hydraulics and sedimentation engineering, and is one of the basic parameters in physically based approaches to hydraulic geometry. Leopold et al. (1964) divided the hydraulic roughness of natural channels into three elements: (1) skin resistance: a function of the roughness of surficial material; (2) internal distortion: caused by boundary features, such as bars, boulders, bends, and bed undulations that generate eddies, and secondary circulation; and (3) spill resistance: caused by flow accelerations and decelerations. There are three basic equations that are used in hydraulic geometry discussions and are briefly discussed here. Spill resistance is not considered here because it is negligible in straight, clean, and symmetrical river reaches in which flow is fully turbulent, uniform, and subcritical. In many cases, an overall measure of resistance is considered as described in this section. The formulas describing such resistance are referred to as uniform flow formulas, as they are used to determine the normal depth or flow depth of uniform flow in a channel. A flow resistance formula is a stage-discharge predictor.

2.6.1 Chezy's Equation

The Chezy equation can be expressed as

$$v = C\sqrt{RS}; \quad Q = CA\sqrt{RS}, Z = CA\sqrt{R}, \tag{2.17a}$$

in which C is the Chezy coefficient, a measure of resistance to flow, Z is the conveyance, and S is the energy slope, often taken as bed slope. Equation (2.17a) states that the ratio of the velocity to the square root of the product of hydraulic radius and slope would be constant for any channel and the constant will be Chezy's coefficient, which has dimensions of $L^{0.5}/T$. If C is divided or scaled by the square root of the acceleration due to gravity g then $C_* = C/\sqrt{g}$ becomes dimensionless. One can also look at Equation (2.17a) as follows:

$$v^2 = C^2 RS; v^2 = \frac{C^2}{g} gRS = \frac{1}{f_z} gRS; \frac{1}{f} = \frac{C^2}{g}, \tag{2.17b}$$

where f_z is regarded as Chezy's dimensionless friction factor and is typically 1/8 of the Darcy–Weisbach friction factor f. Chow (1959) gave typical values of Darcy–Weisbach friction between 0.016 and 0.04, which will lead to values of f_z as 0.002 and 0.005. Equation (2.17b) allows us to express shear stress in terms of cross-sectional average velocity, as will be shown later.

2.6.2 Manning's Equation

The Manning equation can be expressed as

$$v = \frac{c}{n} R^{\frac{2}{3}} S^{\frac{1}{2}}; Q = \frac{cA}{n} R^{\frac{2}{3}} S^{\frac{1}{2}}, \quad Z = \frac{c}{n} A R^{\frac{2}{3}}, \tag{2.18}$$

where n is Manning's roughness coefficient having the dimensions of $TL^{-1.3}$, Z is the conveyance, and $c = 1.486$ in the FPS system of units and $c = 1$ in the MKS system of units. Comparison of Equations (2.17) and (2.18) shows that

$$C = c\frac{R^{\frac{1}{6}}}{n}. \qquad (2.19)$$

It may be remarked that since Manning's n is regarded as a function of sediment grain size, Chezy's C can also be expressed likewise in conjunction with Equation (2.19).

The Manning equation is also known as the Manning–Strickler equation. The Manning coefficient (n) is related to the equivalent particle size denoted as D (in feet) by Strickler's empirical formula as (Henderson, 1966):

$$D = \left(\frac{n}{0.34}\right)^6 \qquad (2.20a)$$

or

$$n = 0.34D^{\frac{1}{6}}. \qquad (2.20b)$$

However, different values in place of 0.34 in Equation (2.20a or 2.20b) have been employed. If D is taken as D_{50}, then the values that have been used are 0.041 and 0.07, and if D is taken as D_{90} then 0.066, 0.038, and 0.024 have been used. If D is measured in meters, then

$$n = \frac{D^{\frac{1}{6}}}{21.1}. \qquad (2.21a)$$

Myer-Peter and Muller (1948) expressed

$$n = \frac{D_{90}^{\frac{1}{6}}}{26}, \qquad (2.21b)$$

in which D_{90} is in meters, where 90% of the material is finer. Lane and Carlson (1953) reported

$$n = \frac{D_{75}^{\frac{1}{6}}}{39}. \qquad (2.21c)$$

Manning's n has also been expressed as

$$n = \frac{r_v - 1}{r_v + 0.95}D^{\frac{1}{6}}, \qquad (2.22)$$

in which r_v is the ratio of flow velocity at 0.2 mean depth d to velocity at 0.8 d.

Another way to estimate Manning's n is by the Cowan method (Cowan, 1954), which expresses:

$$n = (n_0 + n_2)m_c, \qquad (2.23)$$

Table 2.1. *Values of n_0*

Class of bed material	Median bed material size D_{50} in mm	Value of n_0	Remark
Medium gravel	8–16	0.024	The values of n_0 for bed material size
Coarse gravel	16–32	0.028	less than 32 mm are based om
Very coarse gravel	32–64	0.032	estimates given by Chow (1959)
Cobbles	64–128	0.036	

where n_0 is the basic n for a straight clean regular channel; n_2 is the additional component of roughness due to the variation in shape and size of the channel from one cross-section to another; m_c is the correction factor associated with the degree of meandering (m_c is equal to 1.00 if the sinuosity is less than 1.2; equal to 1.15 if the sinuosity is between 1.2 and 1.5; and equal to 1.30 if the sinuosity is greater than 1.5). Bray (1979) reported the values of n_0 as given in Table 2.1.

For gravel-bed rivers, the bulk of resistance is caused by the texture of the bed due to the coarser fractions of materials. Therefore, various modifications of rough turbulent flow equations are used. One variant is the Strickler form of the Manning equation where roughness is evaluated from bed grading (Strickler, 1924). Using data from rivers in California, Limerinos (1970) proposed the Manning roughness coefficient as

$$n = \frac{0.113 R^{\frac{1}{6}}}{1.16 + 2 \log \left(\frac{R}{D_{84}}\right)}, \tag{2.24}$$

in which D_{84} is the grain size so that 84% of the bed material is finer, and R can be replaced by d:

$$n = \frac{0.113 d^{\frac{1}{6}}}{1.16 + 2 \log \left(\frac{d}{D_{84}}\right)}. \tag{2.25}$$

Using a large data base (sample size = 520), Dingman and Sharma (1997) developed an expression for Manning's n and compared this with the equations developed by Jarrett (1984) and Riggs (1976):

$$\text{Dingman and Sharma (1997)}: n = 0.32 \, A^{0.00} R^{-0.16} S_w^{0.38} \tag{2.26}$$

$$\text{Jarrett (1984)}: n = 0.217 \, A^{-0.173} R^{0.267} S_w^{0.156} \tag{2.27}$$

$$\text{Riggs (1976)}: n = 0.210 \, A^{-0.33} R^{0.667} S_w^{0.095}, \tag{2.28}$$

where A is the cross-sectional area (m^2), R is the hydraulic radius (m), and S_w is the water surface slope.

2.6.3 Einstein–Chien Equation

Einstein and Chien (1952) proposed an expression for the average flow velocity as

$$v = C_* \sqrt{8g} \left(\frac{d}{D}\right)^{C_3} d^{\frac{1}{2}} S_0^{\frac{1}{2}}, \tag{2.29}$$

where C_* and C_3 are, respectively, the coefficient and exponent, and S_0 is the bed slope. Equation (2.29) follows the power-form approximation of the Keulegan equation to be discussed later.

2.6.4 Darcy–Weisbach Equation

The Darcy–Weisbach equation can be expressed in terms of head loss H as

$$H = f \frac{L}{4R} \frac{v^2}{2g}, \tag{2.30}$$

where f is the friction factor, L is the length, and R is the hydraulic radius. The energy slope can then be expressed as

$$\frac{H}{L} = S_f = \tan \theta_b = f \frac{1}{4R} \frac{v^2}{2g}. \tag{2.31}$$

Equation (2.31) is also expressed as

$$f = \frac{8gRS_f A^2}{Q^2}. \tag{2.32}$$

Equating the friction slope of Equation (2.31) with the one from Manning's equation, one can write

$$\frac{n^2 v^2}{c^2 R^{\frac{4}{3}}} = f \frac{1}{4R} \frac{v^2}{2g}. \tag{2.33}$$

Equation (2.33) leads to

$$n = \frac{c f^{\frac{1}{2}} R^{\frac{1}{6}}}{\sqrt{8g}}. \tag{2.34}$$

The bed shear τ_0 can be expressed as

$$\tau_0 = \gamma R S_0 \cos \theta_b = \gamma R S_0 \simeq \gamma d S_0, \tag{2.35a}$$

where the hydraulic radius is approximately equal to flow depth d for wide channels and bed slope is small, so θ_b is approximately 0. Combining Equation (2.17b) with Equation (2.35), the bed shear can be expressed as

$$\tau_0 = \rho f_z v^2 \quad \text{or} \quad \tau_0 = \frac{f}{8} \rho v^2. \tag{2.35b}$$

Interestingly, $\tau_0 = 2v^2$ if $f_z = 0.002$ and $\tau_0 = 5v^2$, where $\rho = 1{,}000 \ \mathrm{Ns^2/m^4}$, v is m/s, and τ_0 is N/m^2. Here N denotes Newton.

The depth of flow, d, in a stream is related to its mean annual discharge Q as (Leopold and Maddock, 1953):

$$d = a_0 Q^{a_1} , \tag{2.36}$$

where a_0 and a_1 are constants. Combining Equations (2.35) and (2.36),

$$S_0 = \frac{\tau_0}{\gamma a_0} Q^{-a_1}. \tag{2.37}$$

Using data from 359 streams ranging in patterns from highly sinuous (greater than or equal 1.5) to low sinuous (less than 1.5) and braided, Begin (1981) reported

$$S_0 = 0.00506 Q^{-0.327}. \tag{2.38}$$

Equation (2.38) shows that $\tau_0/(\gamma a_0) = 0.00506$, and $a_1 = 0.327$. On average,

$$\frac{\tau_{avg}}{\gamma a_0} = \frac{S_0 Q^{0.327}}{0.00506}. \tag{2.39}$$

From Equation (2.13) and Equation (2.31), one can write

$$\frac{v}{v_*} = \sqrt{\frac{8}{f}}, \tag{2.40}$$

where v_* is the shear velocity. Then, shear velocity can also be written as

$$v_* = \sqrt{g R S_0 \cos \theta_b} \simeq \sqrt{g R S_0}. \tag{2.41}$$

Inserting Equation (2.41) in Equation (2.40), the result is

$$\frac{v}{\sqrt{g R S_0}} = \sqrt{\frac{8}{f}}. \tag{2.42}$$

Equation (2.42) can be expressed as

$$\frac{v^2 f}{8gdS_0} = \frac{F_r^2 f}{8S_0} = 1.$$ (2.43)

The term $\frac{fF_r^2}{8S_0}$ is known as the kinematic wave number for open-channel flow. The kinematic wave number is important in regime considerations, as it illustrates that for a constant Froude number the ratio f/S_0 is also constant.

For relatively high in-bank flow in natural gravel bed rivers, Bray (1979) found a reliable equation for velocity as

$$v = 9.6\, d^{0.67} S_0^{0.32}.$$ (2.44)

When cast as the Manning equation, Equation (2.44) would yield the Manning roughness coefficient as

$$\frac{1}{n} = 9.6 S_0^{0.18}.$$ (2.45)

Manning's n has also been expressed as

$$n = 0.39 S_0^{0.38} d^{-0.16}.$$ (2.46)

2.6.5 Friction Factor Equations

A large array of values of f have been reported. The friction factor f depends on the size, gradation, and density of the sediment material (sand); the fluid density; and the configuration of sand on the bed. Typical values of f for one regime canal in Pakistan ranged from 0.031 to 0.049 as the discharge varied from 1,840 to 3,060 ft^3/s. Friction factor f has been computed using a number of methods, some of which are utilized more frequently in hydraulic geometry relations. These methods are given here.

Manning–Strickler Equation: This equation can be expressed as

$$\sqrt{\frac{8}{f}} = 8.3 \left(\frac{d}{D_{90}}\right)^{\frac{1}{6}},$$ (2.47)

where D_{90} is the sediment particle size such that 90% of the particles are finer.

Henderson Equation: Henderson (1966) defined the friction factor for fully rough flow as

$$f = 0.13 \left(\frac{D}{d}\right)^{\frac{1}{3}},$$ (2.48)

where D is a characteristic bed material size, typically D_{50}. The grain shear will dominate flow resistance for immobile stream beds.

Bathurst Equation: Bathurst (1985) reported

$$\sqrt{\frac{8}{f}} = 4 + 5.62 \log\left(\frac{R}{D_{84}}\right). \tag{2.49}$$

Bray Equation: Using field data, Bray (1979) obtained

$$f = 0.54 \left(\frac{D_{50}}{R}\right)^{0.56}. \tag{2.50}$$

Keulegan Equation: Keulegan (1938) derived

$$\sqrt{\frac{8}{f}} = 6.25 + 5.75 \log\left(\frac{d}{D_{50}}\right), \tag{2.51}$$

where D_{50} is the median particle size.

Hey Equation: Hey (1979) obtained the Keulegan-type equation:

$$\sqrt{\frac{8}{f}} = 6.25 + 5.75 \log\left(\frac{d}{3.5D_{84}}\right). \tag{2.52}$$

Griffiths Equation: Griffiths (1979) obtained the Keulegan-type equation:

$$\sqrt{\frac{1}{f}} = 0.76 + 1.98 \log\left(\frac{R}{D_{50}}\right). \tag{2.53}$$

Equation (2.53) was based on 186 field data sets from gravel-bed rivers. For $4 < R/D_{50} < 200$, Equation (2.53) was approximated as

$$f = 0.50 \left(\frac{D_{50}}{R}\right)^{0.52}. \tag{2.54}$$

Griffiths (1981) expressed f for a rigid bed channel as

$$\frac{1}{\sqrt{f}} = 1.54 \left(\frac{R}{\left(\frac{\gamma_s}{\gamma} - 1\right)D_{50}}\right)^{0.287}, \tag{2.55}$$

and for a mobile bed channel, where bed load transport occurs, as

$$\frac{1}{\sqrt{f}} = 2.41 \left(\frac{v}{\sqrt{g\left(\frac{\gamma_s}{\gamma} - 1\right)D_{50}}}\right)^{0.340}, \tag{2.56}$$

where γ is the weight density of water, and γ_s is the weight density of sediment.

Griffiths (1981) obtained a set of equations for coarse gravel bed rivers. For rigid boundary gravel bed rivers, he obtained (MKS units):

$$\frac{1}{\sqrt{f}} = 1.33 \left(\frac{R}{D_{50}}\right)^{0.287}. \tag{2.57}$$

A better calibrated equation was reported as

$$\frac{1}{\sqrt{f}} = 0.760 + 1.98 \log_{10}\left(\frac{R}{D_{50}}\right), \tag{2.58}$$

for which velocity was computed from $f = \left(\frac{8gRS}{v^2}\right)$. For channels with active bed load transport, he obtained

$$\frac{1}{\sqrt{f}} = 2.21 \left(\frac{v}{\sqrt{gD_{50}}}\right)^{0.340}. \tag{2.59}$$

Using regression analysis of data, Sonkar and Ram (2014) reported logarithmic as well as power type resistance equations for different sizes of bed materials, such as gravel, cobbles, and boulders. These equations are:

Logarithmic Type:

$$\sqrt{\frac{8}{f}} = 3.28 + 5.76 \log\left(\frac{R}{D_{50}}\right) \tag{2.60}$$

$$\sqrt{\frac{8}{f}} = 3.40 + 5.96 \log\left(\frac{R}{D_{65}}\right) \tag{2.61}$$

$$\sqrt{\frac{8}{f}} = 3.50 + 6.10 \log\left(\frac{R}{D_{84}}\right) \tag{2.62}$$

$$\sqrt{\frac{8}{f}} = 3.60 + 6.16 \log\left(\frac{R}{D_{90}}\right), \tag{2.63}$$

with regression coefficients of 0.807, 0.643, 0.68, and 0.73, respectively.

The Keulegan type expression for f has been given as

$$\frac{1}{\sqrt{f}} = C_0 \log\left(\frac{C_1}{C_2}\frac{d}{D}\right), \tag{2.64}$$

in which

$$C_0 = 2.03; \ C_1 = 11.75; \ C_2 = 3.16; \ D = D_{84}$$
$$C_0 = 2.03; \ C_1 = 11.75; \ C_2 = 3.78; \ D = D_{84}$$
$$C_0 = 2.03; \ C_1 = 11.10; \ C_2 = 3.50; \ D = D_{84}$$
$$C_0 = 2.03; \ C_1 = 13.46; \ C_2 = 3.50; \ D = D_{84}$$
$$C_0 = 2.16; \ C_1/C_2 = 3.83; \ D = D_{90}$$
$$C_0 = 2.36; \ C_1/C_2 = 1.27; \ D = D_{50}.$$

Equation (2.64) has also been expressed as

$$\frac{1}{\sqrt{f}} = C_0 \log\left(12.2\frac{d}{D}\right). \tag{2.65}$$

In Equation (2.65), d/D denotes the relative submergence.

Power Type:

$$\sqrt{\frac{8}{f}} = 3.76 \left(\frac{d}{D_{50}}\right)^{0.288} \tag{2.66a}$$

$$\sqrt{\frac{8}{f}} = 4.15 \left(\frac{d}{D_{65}}\right)^{0.278} \tag{2.66b}$$

$$\sqrt{\frac{8}{f}} = 4.98 \left(\frac{d}{D_{84}}\right)^{0.288} \tag{2.66c}$$

$$\sqrt{\frac{8}{f}} = 5.12 \left(\frac{d}{D_{90}}\right)^{0.260}, \tag{2.66d}$$

with regression coefficients of 0.81, 0.68, 0.75, and 0.76, respectively. Equations (2.66a)–(2.66d) can be expressed as

$$\frac{1}{\sqrt{f}} = C_* \log\left(\frac{d}{D}\right)^{C_3}, \tag{2.67}$$

which is a power approximation of the Keulegan (1938) equation, which is helpful in obtaining analytical solutions of hydraulic geometry. When plotted on the log-log paper, exponent C_3 is the slope of the relative submergence line. If Equation (2.65) is plotted on the log-log paper then the relative submergence line is also defined, and equating the two lines yields exponent C_3 as

$$C_3 = \frac{1}{\ln\left(\frac{12.2\,d}{D}\right)}.$$ (2.68)

Equation (2.68) shows that as submergence (d/D) increases, the value of C_3 decreases, tending to zero when d/D tends to infinity. Then, both the Darcy–Weisbach friction factor and Chezy's friction coefficient (C) remain constant as $f = 8g/C^2$. For the intermediate values of relative submergence, such as $d/D = 200$, Equation (2.68) will approximate $C_3 = 1/6$, which is comparable to the Manning–Strickler relation. For small values of relative submergence ($d/D < 10$), the value of C_3 increases rapidly for coarse bed material channels, as discussed by Julien and Wargadalam (1995), who reported the following: $C_3 = 0.5$ obtained by Leopold and Wolman (1957) for $0.7 < d/D < 10$, $C_3 = 0.25$ obtained by Ackers (1964) for $3 < d/D < 13$, $C_3 = 0.25$ obtained by Kellerhals (1987), $C_3 = 0.44$ obtained by Charlton et al. (1978) for $2 < d/D < 10$, $C_3 = 0.281$ obtained for gravel-bed streams by Bray (1979), and $C_3 = 0.46$ obtained for very steep cobble and boulder-bed streams by Mussetter (1989) for $0.2 < d/D < 4$.

2.6.6 Shear Stress in Alluvial Channels

In an alluvial channel with smooth bed, the flow in the vertical direction can be characterized into three regions, including laminar sublayer, transient, and fully turbulent. Let the thickness of the laminar sublayer be z_0 and thickness of the transient zone be z_1. Thus, the turbulent zone starts at $z_0 + z_1$ and ends at the free water surface. The laminar sublayer is a thin layer starting from the bed up, wherein the velocity varies linearly and the local shear stress (τ_0) is provided by Newton's law of viscosity:

$$\tau_0 = \mu \frac{dv}{dz} \quad \text{or} \quad v = \frac{\tau_0 z}{\mu},$$ (2.69)

where v is the velocity of flow so that $\frac{dv}{dz}$ is the velocity gradient.

The Reynolds number as the laminar flow transitions into turbulent flow, that is, from $z = 0$ to $z = z_0$, has been experimentally found to be 11.6, that is, $\frac{v z_0}{v} = 11.6$. Then, the thickness of the sublayer can be expressed as

$$z_0 = 11.6 \frac{v v}{v_*}.$$ (2.70)

On rough beds where roughness elements protrude into the turbulent zone, the laminar sublayer may be distributed on the rough surface. The size of roughness elements permits the boundary surface into the hydraulically smooth regime, transition regime, and hydraulically rough regime. For the hydraulically smooth regime, the shear Reynolds number has been experimentally found to be less than 5:

$$\frac{D_s v_*}{\upsilon} < 5, \tag{2.71}$$

where D_s is Nikuradse's sand roughness or mean diameter of sand grains.

For the transition zone, in which protrusions increase the resistance, the shear Reynolds number is between 5 and 70. In the hydraulically smooth regime, the shear Reynolds number is greater than 70. The velocity profile is a function of boundary roughness. In the hydraulically rough regime, channels usually have bed ripples or dunes and hence the overall resistance of the bed consists of the form drag due to these bed features and grain resistance due to the granular texture (Yang et al. 2005).

For a wide channel, the bed shear stress can be expressed as

$$\tau_0 = \tau_{form} + \tau_{grain}, \tag{2.72}$$

where τ_0 is given by Equation (2.35) and shear velocity is given by Equation (2.40). In both Equations (2.35) and (2.41), the slope is the hydraulic or energy gradient. The grain resistance for coarse material can be determined from the rough turbulent equation expressed as

$$\tau_{grain} = \frac{\rho v^2}{\sqrt{32 \log \left(\frac{\alpha d}{D} \right)}}, \tag{2.73}$$

in which α is the coefficient that accounts for the relationship of grain roughness to particle size.

Now the question arises about the type of bed features, which include flat bed, ripples, bars, dunes, antidunes and surface waves, dunes and ripples, and chutes and pools, as shown in Figure 2.2 (Simons and Richardson, 1966). Plane or flat bed is a bed surface that does not have elevations or depressions larger than the largest bed sediment sizes. Ripples are small bed forms having wave lengths less than 30 cm and heights less than 5 cm. The profiles of these ripples are approximately triangular in shape with long gentle upstream slopes and short steep downstream slopes. Bars represent bed forms that have lengths of the same order as the channel width or larger and heights comparable to average depth of the generating flow. The bars are classified as point bars, alternate bars, middle bars, and tributary bars, as shown in Figure 2.3. Dunes are bed forms smaller than bars but larger than ripples, and their profiles are out of phase with the water surface profile. Transition represents the transitional bed configurations generated by flow conditions that are intermediate between the conditions that generate dunes and

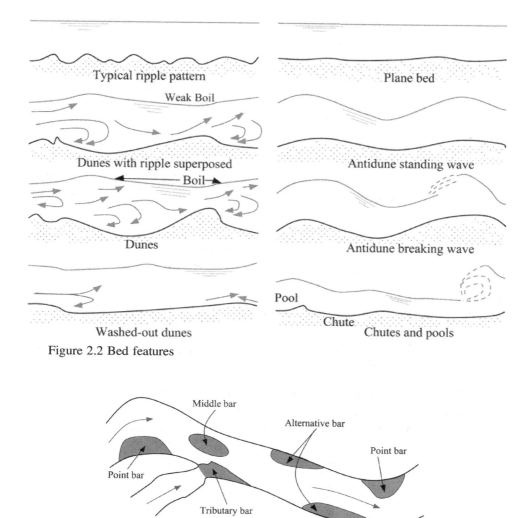

Figure 2.2 Bed features

Figure 2.3 Different types of bars

those that generate flat beds. In the real world, part of the bed may be covered by dunes and the remainder by a plane bed. Antidunes, also called standing waves, exist where bed and water surface profiles are in phase. While the flow is traversing in the downstream direction, the sand waves and water surface waves are traversing in the upstream direction. Chutes and pools are bed forms that occur on relatively large slopes with high velocities and sediment concentration and comprise large elongated mounds of sediment.

Ripples and dunes and combinations thereof are referred to as "lower regime" and antidunes along with plane bed with high sediment transport rate are referred to as "upper regime." Different parts of the channel bed may be in one regime or the other, and the distinction between the two regimes is not clear cut, for velocity increases or decreases.

The dimensionless bed shear stress τ_{0*}, indicating sediment mobility, can be expressed as

$$\tau_{0*} = \frac{v_*^2}{gD(\gamma_* - 1)} = F_*, \tag{2.74}$$

in which $\gamma_* = \gamma_s/\gamma$, γ_s = specific weight of sediment, and γ specific weight of water, and F_* is the overall resistance of grains (form plus grain). The grain resistance component F_{gr} can be expressed as

$$F_{gr} = F_*^{n_s} \left\{ \frac{v}{\sqrt{32 \log \left(\frac{10d}{D}\right)} \sqrt{gD(\gamma_* - 1)}} \right\}^{1-n_s}, \tag{2.75}$$

in which n_s represents the transitional sediment size factor. The White–Bettess–Paris (White et al., 1987) equation for lower regime can be written as

$$\frac{F_{gr} - M_i}{F_* - M_i} = 1 - 0.76 \left\{ 1 - \exp\left[-\left(\log D_{gr}\right)^{1.7} \right] \right\} \tag{2.76}$$

where M_i is the initial motion parameter, the value of F_{gr} when sediment transport begins, and D_{gr} is the dimensionless grain size.

Engelund (1966) reported a simple lower regime equation as:

$$\tau_{0*} = \frac{v}{\sqrt{\left[32gD(\gamma_* - 1)\right] \log \left(\frac{\alpha d}{D}\right)}}, \tag{2.77}$$

where α is approximately equal to 5.

Ackers (1983) and White et al. (1973) derived an expression for upper regime bed form resistance as

$$\frac{\left(F_{gr} - M_i\right) + 0.5\left(F_{gr} - M_i\right)^4}{F_* - M_i} = 1.07 - 0.18 \log D_{gr}. \tag{2.78}$$

The next point is one of transition from lower regime to upper regime. White et al. (1987) used a dimensionless form of stream power (USP_*) to mark the transition:

$$USP_* = \frac{vS}{D_{gr}(gv)^{\frac{1}{3}}}, \tag{2.79}$$

where υ is the kinematic viscosity. If USP_* is less than 0.011 then there is lower regime flow and if USP_* is greater than 0.020 then there is upper regime flow, and the values in between correspond to the transitional zone.

Hogg et al. (1988) showed that during major floods in the Indus River near Sukkur, the Manning n reduced to about 0.011 with the washing out of dunes and the bed becoming plane and then increasing to about 0.03 for the duned bed.

2.7 Sediment Transport

Sediment transport in two-dimensional free surface flow can be characterized by mass density of water ρ (kg/l), kinematic viscosity of fluid υ (m^2/s), mass density of sediment ρ_s (kg/l), sediment grain size or diameter D (m), mean flow depth d (m), shear velocity of flow v_* (m/s), and acceleration due to gravity g (m/s^2). Four dimensionless numbers associated with these quantities are often defined:

$$s = \frac{\rho_s}{\rho}. \tag{2.80}$$

$$Z = \frac{d}{D}. \tag{2.81}$$

$$M_n = \frac{v_s^2}{(s-1)gD}. \tag{2.82}$$

$$D_{gr} = \left[g \frac{(s-1)}{v^2} \right]^{\frac{1}{3}} D. \tag{2.83}$$

Here D_{gr} is the dimensionless grain size; M_n is the mobility number (Shield's); s is the density ratio; and Z is the ratio of flow depth to particle diameter. The friction factor depends on the roughness of the surface indicated by grain size, the mobility number, and density ratio. The value of Z will be significant if the friction factor depends on the distribution of sediment in the bulk of flow.

By definition, $C_s = Q_s/Q$, where C_s is the average concentration of sediment in the channel, and Q_s is the sediment discharge. Sediment transport equations normally involve parameters at the incipient motion, which is the condition under which the sediment particles begin to move. The criteria for defining the condition for incipient motion are based on shear stress, shear velocity, or energy. These are briefly outlined here.

2.7.1 Criteria for Incipient Motion

Shear Stress: The shear stress, also called tractive force, that acts on the channel bed is the pull of water on the area that is in contact with water. In sediment transport equations for hydraulic geometry, the assumption of uniform flow is often made. Under this assumption, the tractive force, τ, is equal to the component of the gravity force parallel to the channel bed that acts on the body of water and can be expressed as

$$\tau = \gamma d S_f. \tag{2.84}$$

White (1940) reasoned that the drag force F_d was proportional to the shear stress and the square of particle diameter:

$$F_d = C_I \tau D^2, \tag{2.85}$$

where C_I is the constant of proportionality. Utilizing Equation (2.85), the particle overturning moment and resisting moment can be computed and then equating the two moments, the critical shear stress, τ_{cr}, can be expressed as

$$\tau_{cr} = C_{cr}(\gamma_s - \gamma)d, \tag{2.86a}$$

where C_{cr} is a constant, and γ_s is the specific weight of sediment. Chang (1979) expressed critical shear stress as

$$\tau_{cr} = 0.0125 + 0.019 \, D, \tag{2.86b}$$

where D is particle size in mm and shear stress is in pounds per square foot.

Critical Velocity: Yang (1973, 1977) developed a velocity approach to determine the critical velocity at the incipient motion, based on four assumptions: (1) The drag and lift forces acting on a sediment particle are proportional to the relative velocity between fluid and particle, (2) the flow velocity can be described by the logarithmic law, (3) the resistance force acting on a particle is equal to the product of friction coefficient and the difference between the submerged weight of the particle and the lift force acting on the particle, and (4) the incipient motion is initiated when the drag force is equal to the resistance force. Yang (1973) specified the following criteria for the incipient motion:

$$\frac{v_{cr}}{\omega} = \frac{2.5}{\log\left(\frac{v_* D}{v}\right) - 0.66} + 0.66, \quad 1.2 < \frac{v_* D}{v} < 70 \tag{2.87a}$$

$$\frac{v_{cr}}{\omega} = 2.05, \quad 70 \leq \frac{v_* D}{v}, \tag{2.87b}$$

where v_{cr} is the critical mean flow velocity at the incipient motion, v is the kinematic viscosity, and ω is the terminal fall velocity of sediment particles.

Shear Velocity: Shear velocity, also called friction velocity, can in general be expressed as

$$v_* = \sqrt{\frac{\tau}{\rho}}. \tag{2.88}$$

One can also determine shear velocity from Manning's equation as

$$v_* = v \left[\frac{n}{c} \frac{\sqrt{g}}{R^{\frac{1}{6}}} \right], \tag{2.89}$$

in which $c = 1$ for MKS units and $c = 1.49$ for FPS units.

Energy: The energy needed for the vertical movement of sediment particles is proportional to the square of shear velocity (Yalin, 1963; Yang, 1973) and is due to turbulent fluxes. At the incipient motion, the following can be expressed:

$$\frac{\rho K^3 v_{*cr}^2}{(\rho_s - \rho)gK^4} = \frac{\tau_{cr}}{(\gamma_s - \gamma)K} = \text{constant}, \tag{2.90}$$

where K is a coefficient, v_{*cr} is the critical shear velocity, τ_{cr} is the shear stress at the incipient motion, and ρ_s is the density of sediment.

2.7.2 Transport Equations

Sediment transport rate depends on a number of factors, including flow depth, velocity, and discharge; energy slope; shear stress; stream power; turbulence intensity; bed configuration; sediment particle size and gradation; channel patterns; vegetation; and water temperature. Sediment transport equations have been derived by considering dominant factors as independent variables. Yang (1977) classified these equations. based on regime theory, shear stress, velocity, discharge, slope, and stream power; and a brief outline of those sediment transport equations that have been used in deriving hydraulic geometry relations is given here. Low (1980) indicated volumetric sediment transport per unit time per unit width (m^3/s/m) as

$$q_s = 40 \left[\frac{v_*^2}{gD(s-1)} \right]^3 \omega D, \tag{2.91}$$

in which D is in m, and ω is the fall velocity of particles.

DuBoys Equation: The DuBoys equation is for bed load transport that is part of sediment load transported by rolling or sliding along the channel bed. The bed load is dominated by excess shear stress. Thus, the DuBoys equation is based on the shear stress approach and can be expressed as

$$q_s = K_1 \tau (\tau - \tau_0), \tag{2.92}$$

where K_1 is a function of sediment size. Straub (1954) experimentally estimated K_1 as

$$K_1 = \frac{0.391}{D^{\frac{3}{4}}}. \tag{2.93}$$

Shields Equation: The Shields equation for bed load is also based on the shear stress approach and can be expressed as

$$q_s = 10 \frac{\gamma q}{\gamma_s - \gamma} S_f \gamma \frac{\tau - \tau_{cr}}{(\gamma_s - \gamma)D}, \tag{2.94}$$

in which γ is the specific weight of water, γ_s is the specific weight of sediment, q is water discharge per unit width, and τ_{cr} is the critical shear stress.

Schoklitsch Equation: The Schoklitsch equation is based on the discharge approach and can be written as

$$q_s = 7,000 \frac{S_f^{\frac{3}{2}}}{D^{\frac{1}{2}}} (q - q_{cr}), \tag{2.95}$$

where q and q_{cr} are water discharge and critical discharge per unit width.

Meyer-Peter–Muller (MPM) Equation: The MPM equation (Meyer-Peter and Muller, 1948) is based on the discharge approach and can be expressed as

$$q_s = \left[\frac{D}{2.5} \left(\frac{q^{\frac{2}{3}} S_f}{D} - 17 \right) \right]^{\frac{3}{2}}, \tag{2.96}$$

where D for a mixture of nonuniform material should be replaced by D_{35}. Equation (2.96) should be applied to coarse material with particle size greater than 5 mm. Meyer-Peter and Muller (1948) also expressed the transport equation as an energy or slope type equation:

$$\gamma R S_r = 0.047 \, (\gamma_s - \gamma)D + 0.25 \, \rho^{\frac{1}{3}} q_s^{\frac{2}{3}}, \tag{2.97}$$

where S_r is the grain resistance responsible for bed load movement. Wong and Parker (2006) modified the Meyer-Peter and Muller formula as

$$q_s = 3.97 \left[g \left(\frac{\gamma_s}{\gamma} - 1 \right) D_{50}^3 \right]^{0.5} \left[\frac{dS}{\left(\frac{\gamma_s}{\gamma} - 1 \right) D_{50}} - E_{sc} \right]^{1.5}, \tag{2.98}$$

where E_{sc} is the critical value of the Shields Entrainment Function (SEF).

Engelund and Hansen Equation: Engelund and Hansen (1967) proposed

$$f\frac{q_T}{\sqrt{\left(\frac{\gamma_s}{\gamma}-1\right)dD^3}} \sim \left[\frac{dS_f}{\left(\frac{\gamma_s}{\gamma}-1\right)D}\right]^{\frac{5}{2}}, \tag{2.99}$$

in which q_T is the total sediment transport. They, then, expressed

$$f\frac{q_T}{\sqrt{\left(\frac{\gamma_s}{\gamma}-1\right)dD^3}} = 0.077 \left[\frac{dS_f}{\left(\frac{\gamma_s}{\gamma}-1\right)D}\right]^2 \sqrt{\left[\frac{dS_f}{\left(\frac{\gamma_s}{\gamma}-1\right)D}\right]^2} + 0.15. \tag{2.100}$$

For small values of $\frac{dS_f}{\left(\frac{\gamma_s}{\gamma}-1\right)D}$, Equation (2.100) asymptotically approaches

$$f\frac{q_T}{\sqrt{\left(\frac{\gamma_s}{\gamma}-1\right)dD^3}} \sim \left[\frac{dS_f}{\left(\frac{\gamma_s}{\gamma}-1\right)D}\right]^2, \tag{2.101}$$

which is for bed load transport only. For large values of $\frac{dS_f}{\left(\frac{\gamma_s}{\gamma}-1\right)D}$, Equation (2.101) asymptotically approaches

$$f\frac{q_T}{\sqrt{\left(\frac{\gamma_s}{\gamma}-1\right)dD^3}} \sim \left[\frac{dS_f}{\left(\frac{\gamma_s}{\gamma}-1\right)D}\right]^3, \tag{2.102}$$

which is for transport in which suspended sediment is dominant.

Engelund (1966) expressed the Meyer-Peter formula for dimensionless bed load transport as

$$q_{B*} = 8(\theta - 0.047)^{\frac{3}{2}}, \tag{2.103}$$

where

$$\theta = \frac{\tau_0}{(s-1)\gamma D}, \tag{2.104}$$

and

$$q_{B*} = \frac{q_B}{\sqrt{(s-a)gD^3}}, \tag{2.105}$$

in which q_B is the volume of grain material transported per second per unit width; γ is the unit weight of water; s is the relative density of grain material; and a is the bedload layer thickness.

Einstein–Brown Equation: The Einstein–Brown equation (Einstein, 1942; Brown, 1950) can be expressed as

$$0.456\phi = \exp(-0.391\psi), \quad \psi > 5.5 \tag{2.106}$$

$$\phi = C_{eb}\left(\frac{1}{\psi}\right)^3, \quad \psi \leq 5.5, \tag{2.107}$$

where C_{eb} is a coefficient,

$$\phi = \frac{q_s}{\gamma_s F\left[g\left(\frac{\gamma_s}{\gamma} - \right)D^3\right]^{\frac{1}{2}}} \tag{2.108}$$

$$\psi = \frac{\gamma_s - \gamma}{\tau_0}D \tag{2.109}$$

$$F = \left[\frac{2}{3} + \frac{36v^2}{gD^3\left(\frac{\gamma_s}{\gamma} - 1\right)}\right]^{\frac{1}{2}} - \left[\frac{36v^2}{gd^3\left(\frac{\gamma_s}{\gamma} - 1\right)}\right]^{\frac{1}{2}}. \tag{2.110}$$

Smart Formula: Smart (1984) proposed

$$q_s = 4.5[\theta - \theta_{cr}]DvS^{0.6}, \tag{2.111}$$

where

$$\theta = \frac{v_*^2}{gD\left(\frac{\gamma_s}{\gamma} - 1\right)}, \tag{2.112}$$

and θ_{cr} is the critical value of θ for the initiation of sediment motion.

2.8 Stream Power

Yang (1976) defined unit stream power (USP) as

$$\frac{dY}{dt} = \frac{dx}{dt}\frac{dY}{dx} = vS_f, \tag{2.113}$$

in which Y is the potential energy per unit weight of water in a stream, v is the average flow velocity, S_f is the energy slope, t is time, and x is the distance along the channel. Thus, unit stream power is defined as the product of average velocity and energy slope or the time rate of potential energy expenditure per unit mass of water. Chitale (1973) expressed potential energy per unit mass as the product of stream fall (Y_s) times unit weight of volume of water. Then, the rate of potential energy expenditure per unit mass of water with respect to time equals $\frac{Y_s\gamma}{t}$. Since $Y_s/L = S$ and $t = L/v$, one can write $\frac{Y\gamma}{t} = Sv\gamma$, where L is the length of the channel course, and S is the slope.

Stream power (SP) per unit channel length is defined as

$$SP = \gamma Q S_0. \tag{2.114}$$

The stream power has the dimensions of energy per unit time per unit length and is equal to unit stream power integrated over the channel cross-sectional area. The total power per unit width to transport the mixture of water and sediment given by Equation (2.114) can also be expressed as

$$SP = \rho g d v S_0 = \gamma R S_0 v = \tau_0 v, \tag{2.115}$$

in which R is the hydraulic radius. Equation (2.115) is obtained by dividing Equation (2.114) by the channel width. It may be noted that the mass of sediment is relatively small by comparison with the mass of water. A large part of SP is consumed to overcome the resistance generated by sand grain roughness and bed forms. Denoting the stream power per unit width to overcome this resistance by SP_r, Equation (2.115) can be simplified as

$$SP_r = K_0 SP, \tag{2.116}$$

where K_0 is the proportionality factor. Inserting Equation (2.115) in equation (2.116),

$$SP_r = K_0 \rho g S_0 Q / w, \tag{2.117}$$

in which water discharge $Q = vdw$.

One can also express the power to transport suspended sediment as well as to transport bed load. The power to transport suspended sediment, which is related to flow turbulence in terms of shear stress of suspended sediments (τ_s), can be expressed as

$$SP_s = \tau_s v. \tag{2.118}$$

Following Einstein and Chien (1952), the stream power per unit width to transport suspended sediment can be expressed as

$$SP_s = \frac{(\rho_s - \rho)}{\rho_s} C_{ec} d\omega = \tau_s v \quad \text{for} \quad S_0 \le \frac{\omega}{v}, \tag{2.119}$$

where ω is the average settling velocity of particles, and C_{ec} is a coefficient. Holtorff (1983) expressed

$$C_{ec} dv = \frac{\rho_s g Q_s}{w}, \tag{2.120}$$

in which w is the mean width of the cross-section, and Q_s is the discharge of the suspended sediments. Combining Equations (2.119) and (2.120),

$$SP_s = (\rho_s - \rho)g\frac{Q_s\omega}{wv}. \tag{2.121}$$

The stream power for bed load transport can be expressed as

$$SP_b = \tau_b v_b, \tag{2.122}$$

in which τ_b is the effective bed shear stress (total bed shear stress minus the critical shear stress) due to the movement and transport of bedload transport, and v_b is the velocity of particles. Equation (2.122) can be written as

$$SP_b = gQ_b\left[\frac{\rho_s - \rho}{w}\right]\tan\phi, \tag{2.123}$$

where ϕ is the angle of repose of sediments, and Q_b is the bedload discharge. Equation (2.123) assumes that fluid drag on the particles ($\tau_b A$, $A =$ net area of the mass of particles) in equilibrium with the mean longitudinal frictional force exerted by the bed on the sediment particles $[(\rho_s - \rho)gV_{ol}\tan\phi, V_{ol} =$ net volume of particles], and that the lift force is negligible for the initiation of longitudinal movement. Assuming $d_b = V_{ol}/A$,

$$\tau_b = (\rho_s - \rho)gd_b\tan\phi. \tag{2.124}$$

Equation (2.124) can be inserted in Equation (2.122) to obtain $Q_b/(v_b w)$.

From the energy conservation principle, the total stream power SP is consumed to mainly overcome the sand grain and bed form resistance to erode the bottom material, place the eroded material in the load for transport, and transport the suspended sediment load, and is dissipated by friction due to bends, expansion, and other intrusions. Thus, the total stream power can then be expressed as

$$SP = SP_1 + SP_s + SP_b + SP_r, \tag{2.125}$$

where SP_1 is the power dissipated by minor factors, such as bends, etc.

Inserting Equations (2.114), (2.118), (2.122), and (2.117) into Equation (2.125),

$$\rho QS_0 v(1 - K_0) = (\rho_s - \rho)(Q_b v\tan\phi + Q_s w), \tag{2.126}$$

with

$$s = \frac{\rho_s - \rho}{\rho}; \quad K = \frac{(1 - K_0)}{s}. \tag{2.127}$$

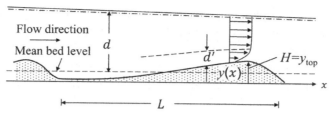

Figure 2.4 Dune on a bed

Then,

$$K = \frac{vQ_b \tan\phi + \omega Q_s}{QvS_0}.$$ (2.128)

Equation (2.128) is important for computing bed load and suspended sediment load. It is, however, important to determine the values of parameters K_0 and K. Pacheco-Ceballos (1990) discussed a procedure to evaluate these parameters which is described later.

Consider a dune, as shown in Figure 2.4, where sediments are deposited by rolling on the front downstream from the crest of the dune. Fredsoe (1982) expressed

$$\frac{y}{H} = \frac{q_{top}}{q_b},$$ (2.129)

where H is the total height of the dune, q_{top} is the sediment transported up to a height y, and q_b is the bed load sediment transport. If the bedload layer thickness is equal to a below which sediment is transported as bedload, then Equation (2.129) can be cast as

$$\frac{2D}{a} = \frac{q_c}{q_a},$$ (2.130)

where q_c is the critical transport on the initiation of suspended sediment, since the thin layer of two particles can be regarded as the minimum for suspension (Einstein, 1950), and D is the mean diameter of bed particles.

The transport of sediment is considered mainly by bedload, for which the Peter-Meyer–Muller formula can be used. Equation (2.130) yields

$$\frac{a}{2D} = \left[\frac{\tau - \tau_c}{\tau_{crs} - \tau_c}\right]^{1.5} \quad \text{for} \quad a \geq 2D,$$ (2.131)

in which τ_{cr} is the critical Shields parameter for the initiation of suspension. When the bed shear stress value τ exceeds the critical value for the initiation of sediment motion τ_c, the sediment particles will move by rolling and sliding in contact with the bed; when $\tau = \tau_c$, the particles will enter into the initiation stage, $a = 2D$; and a critical condition for the initiation of suspension will occur when $y = 2D$. For the increasing value of shear stress τ, the particles will move in suspension. When $\tau < \tau_{cr}$ only a small fraction of sediment will be set in motion. Following van Rijn (1984), the value of τ_{cr} can be evaluated as follows:

$$\frac{v_{*cr}}{\omega_d} = \frac{4}{D_*} \quad \text{for} \quad 1 \le D_* \le 10, \tag{2.132}$$

and

$$\frac{v_{*cr}}{\omega_d} = 0.4 \quad \text{for} \quad D_* > 10, \tag{2.133}$$

where D_* is the particles parameter evaluated as

$$D_* = D\left[\frac{sg}{v}\right]^{\frac{1}{3}}, \tag{2.134}$$

where ω_d is the fall velocity of particles of diameter D, and

$$\tau_{cr} = \frac{v_{*cr}^2}{sgD}. \tag{2.135}$$

Now, Equation (2.116) can be cast as

$$K_0 SP = SP - SP_{sediment}. \tag{2.136}$$

Equation (2.136) can be written as

$$K_0 \tau_0 v = \tau_0 v - \tau_{ws} v_b - \tau_{ss} v, \tag{2.137}$$

where τ_{ws} is the shear stress due to the total fluid–sediment mixture. For a river of sediments with uniform velocity,

$$\tau_{ws} = \frac{\rho_m g a S_0}{2} \tag{2.138}$$

$$\tau_{ss} = \frac{e\rho_m g a_s S_0}{2}, \tag{2.139}$$

in which a_s is the hypothetical height of the suspended sediment and e is a coefficient. Equations (2.138) and (2.139) hold for dunes or ripples having a triangular shape.

For plane bed,

$$\tau_{ws} = \rho_m g a S_0 \tag{2.140}$$

$$\tau_{ss} = e\rho_m g a_s S_0. \tag{2.141}$$

Then, in general form one can express

$$\tau_{ws} = \frac{\rho_m g a S_0}{b_f},\tag{2.142}$$

$$\tau_{ss} = \frac{e\rho_m g a_s S_0}{b_f},\tag{2.143}$$

in which ρ_m is the density of mixture, and b_f is the bed form factor. Inserting Equations (2.124), (2.143), and (2.114) in Equation (2.137),

$$K_0 = 1 - \frac{\rho_m}{\rho d b_f}\left(a\frac{v_b}{v} + ea_s\right),\tag{2.144}$$

and

$$K = \frac{\rho_m}{s\rho d b_f}\left(a\frac{v_b}{v} + ea_s\right).\tag{2.145}$$

2.9 Entropy

Entropy is regarded as a measure of uncertainty, and information is gained only if there is uncertainty about an event. Consider a phenomenon that produces a number of events. More information is needed to characterize an event if it occurs rarely but less information is needed if it occurs frequently. In a similar manner, consider a system that is characterized by a set of states. If the system occupies a particular state more frequently then less information will be needed to characterize that system state. Suppose a system is characterized by a random variable X, which takes on N values denoted as x_i $I = 1, 2, \ldots, N$, with probabilities p_i, $I = 1, 2, \ldots, N$. Then, the entropy of the system, $H(X)$, can be defined (Shannon, 1948) as

$$H(X) = -\sum_{i=1}^{N} p_i \log p_i,\tag{2.146}$$

where log can be to the base 2, e, or 10 corresponding to the units of entropy as bit, Napier, or dogit (or decibel), respectively. It can be shown from Equation (2.146) that the maximum value of entropy will be when $p_i = 1/N$, that is, all states are equally likely to occur. If X is a continuous random variable with probability density function as $f(x)$, then the entropy will be

$$H(X) = -\int_{-\infty}^{x} f(x) \log f(x) dx.\tag{2.147}$$

Equation (2.147) also leads to the maximum value of entropy if $f(x)$ is a uniform distribution.

References

Ackers, P. (1964). Experiments on small streams in alluvium. *Journal of Hydraulics Division*, Vol. 90, No. 4, pp. 1–37.

Ackers, P. and White, W. R. (1983). Sediment transport: new approach and analysis. *Proceedings of Hydraulics Division, ASCE*, Vol. 99, No. HY1, pp. 2041–2060.

Bathurst J. C. (1985). Flow Resistance estimation in mountain rivers. *Journal of Hydraulic Engineering. ASCE*, Vol. 111, No. 4, pp. 625–643.

Begin, Z. B. (1981). The relationship between flow-shear stress and stream patters. *Journal of Hydrology*, Vol. 52, pp. 307–319.

Bray, D. I. (1979). Estimating average velocity in gravel bed rivers. *Journal of Hydraulics Division, ASCE*, Vol. 105, pp. 1103–1122.

Brown, C. B. (1950). Sediment Transport. Chapter 12 in *Engineering Hydraulics*, edited by H. Rouse, Wiley, New York.

Chang, H. H. (1979). Minimum stream power and river channel patterns. *Journal of Hydrology*, Vol. 41, pp. 303–327.

Charlton, F. G., Brown, P. M., and Benson, R. W. (1978). *The hydraulic geometry of some gravel rivers in Britain*. Report IT 180, Hydraulics Research Station, Wallingford.

Chitale, S. V. (1973). Theories and relationships of river channel patterns. *Journal of Hydrology*, Vol. 19, pp. 285–308.

Chow, V. T. (1959). *Open Channel Hydraulics*. McGraw-Hill Book Company, New York.

Cowan, W. L. (1954). Estimating hydraulic roughness coefficients. *Agricultural Engineering*, Vol. 37, No. 7, pp. 473–475.

Dingman, S. L. and Sharma, K. P. (1997). Statistical development and validation of discharge equations for natural channels. *Journal of Hydrology*, Vol. 199, pp. 13–35.

Einstein, H. A. (1942). Formulae for transportation of bed-load. *Transactions, ASCE*, Vol. 107, pp. 561–577.

Einstein, H. A. (1950). *The bed-load function for sediment transportation in open channel flows*. Technical Bulletin No. 1026, U.S. Department of Agriculture, Soil Conservation Service, Washington, DC.

Einstein, H. A. and Chien, N. (1952). *Second approximation to the solution of suspended load theory*. Institute of Engineering Research, University of California, Berkeley, Issue 2, Series, 47.

Engelund, F. (1966). Hydraulic resistance of alluvial streams. *Proceedings of Hydraulics Division, ASCE*, Vol. 92, No. HY2, pp. 315–326.

Engelund, F. and Hansen, E. (1967). *A nomograph on sediment transport in alluvial streams*. Technical University of Denmark, Copenhagen. p. 63.

Fredsoe, J. (1982). Shape and dimensions of stationary dunes in rivers. *Journal of Hydraulic Research*, Vol. 108, No. 8, pp. 932–946.

Griffiths, G. A. (1979). *Rigid boundary flow resistance of gravel rivers*. Ministry of Works Development, Water and Soil Division, Report WS127, p. 20.

Griffiths, G. A. (1981). Flow resistance in coarse gravel bed rivers. *Journal of Hydraulics Division, ASCE*, Vol. 107, No. HY7, pp. 899–918.

Henderson, F. M. (1966). *Open Channel Flow*. Macmillan, New York, p. 522.

Hey, R. D. (1979). Flow resistance in gravel bed rivers, *Journal of Hydraulics Division, ASCE*, Vol. 105, pp. 365–379.

Hogg, I. G. G., Guganesharajah, K., Gunn, P. D. S., and Ackers, P. (1988). *The influence of river regime on the flood management of Sukkur barrage, Pakistan*. International Conference on River Regime, Hydraulics Research Station, Wallingford.

Holtorff, G. (1983). The evolution of meandering channels. *Proceedings of the Second International Symposium on River Sedimentation*, pp. 692–705, Nanjing, China.

Jarrett, R. D. (1984). Hydraulics of high gradient streams. *Journal of Hydraulic Engineering*, Vol. 110, pp. 1519–1529.

Julien, P. Y. and Wargadalam, J. (1995). Alluvial channel geometry: Theory and applications. *Journal of Hydraulic Engineering*, Vol. 121, No. 4, pp. 312325.

Kellerhals, R. (1987). Stable channels with gravel-paved beds. *Journal of Waterways Division, ASCE*, Vol. 93, No. 1, pp. 63–84.

Keulegan, G. B. (1938). Laws of turbulent flow in open channels, *Journal of Research, National Bureau of Standards, U.S.*, Vol. 21, pp. 707–741.

Lane, E. W. and Carlson, E. J. (1953). *Some Factors Affecting the Stability of Canals Constructed in Coarse Granular Materials*. International Association for Hydraulic Research, Minneapolis, MN.

Leopold, L. B. and Maddock, J. T. (1953). The hydraulic geometry of stream channels and some physiographic implications. U.S. Geological Survey Professional Paper 252, pp. 1–57.

Leopold, L. B. and Wolman, M. G. (1957). *River-channel patterns: braided, meandering and straight*. USGS Professional Paper 282-B, U.S. Geological Survey, Washington, DC, pp. 38–85.

Leopold, L. B., Wolman, M. G., and Miller, J. P. (1964). *Fluvial Processes in Geomorphology*, W. H. Freeman, San Francisco.

Limerinos, J. T. (1970). Determination of the Manning coefficient from measured bed roughness in natural channels.

Low, H. S. (1980). Effect of sediment density on bed-load transport. *Journal of Hydraulic Engineering*, Vol. 115, No. 1, pp. 124–138.

Meyer-Peter, E. and Muller, R. (1948). *Formula for bed load transport*. International Association of Hydraulic Research, Second meeting, Stockholm, p. 39.

Mussetter, R. A. (1989). Dynamics of mountain streams. Ph.D. Dissertation, Department of Civil Engineering, Colorado State University, Fort Collins.

Pacheco-Ceballos, R. (1990). Transport of sediments: Analytical solution. *Journal of Hydraulic Research*, Vol. 27, No. 4, pp. 501–518.

Riggs, H. C. (1976). A simplified slope-area method for estimating flood discharges in natural channels. *Journal of Research, U.S. Geological Survey*, Vol. 4, pp. 285–291.

Shannon, C. E. (1948). A mathematical theory of communications, I and II. *Bell System Technical Journal*, Vol. 27, pp. 379–443.

Simons, D. B. and Richardson, E. V. (1966). *Resistance to flow in alluvial channels*. U.S. Geological Survey Professional Paper 422-J, Washington, DC.

Smart, G. M. (1984). Sediment transport formula for steep channels. *Journal of Hydraulic Engineering*, Vol. 110, No. 3, pp. 267–276.

Sonkar, R. K. and Ram, S. (2014). Flow resistance in gravel bed rivers. *IJSRD - International Journal for Scientific Research & Development*| Vol. 2, No. 07, 2014 | ISSN (online): 2321-0613.

Straub, I. G. (1954). *Terminal Report on Transportation Characteristics: Missouri River Sediment*. University of Minnesota, St Anthony Falls Hydraulics Laboratory, Sediment Series No. 4. Minneapolis.

Strickler, A., (1924), *Beiträge zur Frage der Geschwindigheitsformel und der Rauhigkeitszahlen für Strome, Kanale und Geschlossene Leitungen. Mitteilungen des Eidgenössischer Amtes für Wasserwirtschaft*, Bern, Switzerland.

van Rijn, L. C. (1984). Sediment transport, Part II. *Journal of Hydraulic Engineering*, Vol. 110, No. 11, pp. 1613–1641.

White, C. M. (1940). The equilibrium of grains on the bed of an alluvial channel. *Proceedings of Royal Society, London, Series A*, Vol. 174, p. 322.

White W. R., Bettess, R., and Wang, S. (1987). Frictional characteristics of alluvial streams in lower and upper regimes. *Proceedings, Institution of Civil Engineers*, Vol. 83, No. 2, pp. 685–700.

White, W. R., Milli, H., and Crabe, A. D. (1973). *Sediment Transport: An Appraisal of Available Methods*. Vol. 2: Performance of theoretical method when applied to flume and field data. Report INT 119, Hydraulic Research Station, Wallingford.

Wong, M., and Parker, G. (2006). Reanalysis and correction of bed-load relation of Meyer-Peter and Müller using their own database. *Journal of Hydraulic Engineering*, Vol. 132, No. 11, pp. 1159–1168.

Yalin, M. S. (1963). An expression for bed-load transportation. *Journal of the Hydraulics Division*, Vol. 89, No. 11, pp. 221–250.

Yang, C. T. (1973). Incipient motion and sediment transport. *Journal of Hydraulics Division, ASCE*, Vol. 99, pp. 1679–1704.

Yang, C. T. (1976). Minimum unit stream power and fluvial hydraulics. *Journal of Hydraulics Division, ASCE*, Vol. 102, No. HY7, pp. 919–934.

Yang, C. T. (1977). The movement of sediment in rivers. *Geophysical Surveys*, Vol. 3, pp. 39–68.

Yang, S. Q., Tan, S. K., and Lin, S. Y. (2005). Flow resistance and bed form geometry in a wide alluvial channel. *Water Resources Research*, Vol. 41, W09419, doi: 10.1029/2005WR004211.

3

Regime Theory

The interrelationship between flow, sediment transport, channel resistance, and bank stability determines the regime of a channel in alluvium, meaning channel shape and stability. Thus, the regime theory predicts the size, shape, and slope of a stable alluvial channel under given conditions. This chapter presents regime relations for channel width, depth, and gradient.

Notation

A = flow cross-sectional area

B_I = braiding index

B_s = shear strength of bank material

C = Chezy's coefficient

C_d = dimensionless constant

$C_i, i = 1, 2, \ldots, 5$ = coefficients

C_l = function

C_s = bed sediment concentration

C_w = function

d = flow depth over the flat channel bed

D = sediment diameter

D_m = mean bed material diameter

d_B = mean flow depth over the bed

d_m = maximum depth

d_r = mean flow depth at riffles

d_{rm} = maximum flow depth at riffles

F = Froude number

F_B = bed factor

Fs = side factors

Continued

---

f_s = Lacey's silt factor (sediment size parameter)

f_{vs} = silt factor

f_{RS} = silt factor

g = acceleration due to gravity

K = coefficient

K_1 = coefficient

k_s = Nikuradse equivalent grain size roughness

L_m = wave length of meanders

m = critical velocity ratio

M = weight mean silt-clay

M_{bw} = meander belt width

M_d = distance from meander to meander along the river axis

M_l = meander length

M_w = meander width

n = exponent

N_L = resistance coefficient in Lacey resistance equation

n = exponent

n_m = Manning's roughness coefficient

ρ = mass density of water

P = wetted perimeter of the flow section

q = discharge rate per unit width

Q = volumetric rate of discharge

Q_b = bankfull discharge

Q_s = sediment load transport

q_T = total sediment transport rate per unit width

r = radius of curvature

R = hydraulic radius (A/P)

$r_{w\text{-}d}$ = width-depth ratio

S = slope

S_c = critical straight-line water surface slope above which meandering occurs

S_u = sinuosity

S_v = valley slope

TW = top width

u = exponent

USP = unit stream power

v_0 = velocity for the same depth given by the Kennedy formula

v_s = fall velocity of sediment particles

w = width of bed

w_r = riffle values of mean water line riffles

w_M = mean width

X_M = sediment loading

y_1 = hydraulic depth (A/w)

Z = riffle spacing
γ = weight density of water
γ_s = weight density or specific weight of sediment
γ_{s*} = difference between weight density of sediment and that of water
σ_D = standard deviation of log-normal size/frequency distribution
σ_{rD} = standard deviation at riffle depth
v = cross-sectional mean flow velocity
υ = kinematic viscosity

3.1 Introduction

In hydraulics, a regime channel is neither scoured nor filled. In practice, it is not possible for a channel not to scour or fill, because a channel constructed in alluvium tends to gradually scour or fill in bed and banks. Thus, a regime channel can be understood to be a channel in which the amount of scour is equal to the amount of fill and is in dynamic equilibrium or stable condition. Even this condition is difficult to satisfy in the short term but can be achieved in the long term, and that is what is meant by a stable channel, that is, the channel has reached the state of equilibrium. Thus, the width, depth, and gradient of a channel carrying sediment-laden flow are considered fixed, but these channel dimensions correspond to a long-term average. Blench (1952) provided an excellent discussion of what is meant by regime and regime channel, whereas Stevens and Nordin (1987) made a critique of the regime theory for alluvial channels. Wolman (1955) summarized several key points common in regime, regimen, equilibrium, and poised. First, they all imply a certain amount of stability in time. Second, they imply flexibility in the adjustments that the channel itself makes. Third, there are no fixed criteria to judge whether a given stream is or is not in regime or is poised for or exhibits an average equilibrium. Fourth, they do not imply that adjustments of slope are more important than the adjustments of cross-section or vice versa. It may be noted that perfect equilibrium is seldom achieved owing to a multitude of irregularities occurring in natural materials.

Since the design and development of irrigation canal systems in the Indian Subcontinent in the twentieth century, a large number of studies have been reported on regime channels, and as a result the regime theory has been developed. Ackers (1992) has given an excellent discussion of the theory and this chapter follows his work. Erosion and deposition of sediment are inherent to rivers, and through these processes they alter landscape over long periods of time. During the process of alteration, several space and time scales can be discerned from the perspective of sediment transport: micro-, meso-, macro-, and mega-scale. At the micro-scale, the mechanics of entrainment and transport of individual particles within a bed of identical material is emphasized, which is essential for understanding drag and lift forces. At this scale, flume experiments are often employed for investigating transport mechanisms.

At the meso-scale, physically based equations are developed for flow resistance and sediment transport. These equations apply to average channel conditions and are inherently general. Using these equations in concert with the continuity principle, channel processes can be modeled and spatial and temporal changes in channel width, slope, and velocity can be simulated. Because input of water and sediment load stemming from the watershed can also change in response to erosion and deposition, practical applicability of these equations is thus limited.

Macro-scale approaches are employed for large-scale development of river systems over long periods of time. Simpler transport functions are used to simulate channel morphological change. Based on systematic changes occurring in the dominant channel-forming flow and its associated sediment load, the average bankfull dimensions of the river are simulated over time and space.

At mega-scale, considering the sequence of runoff episodes, sediment supply and sediment transport, dynamic and discontinuous aspects of channel development at the drainage-basin scale are modeled. The modeling approaches are black-box approaches and provide little physical insight into the processes controlling sediment transport and channel adjustment.

Ackers (1992) emphasized that the description of rivers should encompass bed forms as well as planform, and described the boundary features of rivers at three different scales. The first is the ripple scale, where ripples relate to the particle size. It is like a microscale. The second is at the scale of dunes and bars, which relate to the channel depth. This can be considered as a macro-scale. The third is at the scale of shoals and meanders, which are related to channel width. This can be considered as a mega-scale.

For effective management of rivers, channel stability must be duly considered. Channel response to any natural or manmade changes depends on the degree of change and the prevailing stability of the channel. Consequently, controls on erosion and deposition and channel response to the changes in the watershed must be identified. This is accomplished by considering the interaction between channel form, flow regime, and sediment transport. These factors are inherently interrelated. A macro-scale model can be employed to outline this interaction, which is important for the long-term (over 10 years or longer) channel development. The interaction between sediment transport processes and flow regime influences total sediment load and the imbalance between sediment input and output at a section that controls erosion, deposition, and change in channel geometry. When considering long-term changes in the average dimensions of the channel, it is necessary to consider only the dominant discharge.

Natural rivers carry widely varying flows, but the stream size and geometry are determined by a certain range of flows or a formative flow. This formative flow is called the dominant discharge. The reason it is called formative is because at this discharge, sediment movement and bank erosion are most active and the channel adjustment is caused by such a flow. In other words, such a flow controls the channel geometry.

Bankfull discharge is generally considered to be the dominant steady flow that would generate the same regime channel shape and dimensions as would the natural sequences of flows. This is because investigations on the magnitude and frequency of sediment transport have determined that for stable rivers the flow that in the longer term transports most material has the same frequency of occurrence as bankfull flow. For stable gravel-bed rivers, this is considered to be a 1.5-year flood.

For computing the dominant discharge, one can consider the product of sediment load and the probability of occurrence of the corresponding discharge, providing the flow rate with the maximum sediment movement, which is approximately equal to mean annual flood. Nixon (1959) showed that under temperate climate, in many watersheds the bankfull flow occurred for several days a year and was approximately equal to mean annual flow. Two further points can be made. First, the bankfull flow controls the generation of meander length, which is approximately proportional to the square root of this discharge and was emphasized by Inglis (1947). Second, sediment transport may diminish at discharges greater than bankfull discharge, because a portion of the discharge goes overbank, overall resistance increases, and overbank spill reduces the erosive tendency of flow.

3.2 Regime Theory

Regime theory was developed in the late nineteenth century to design and operate extensive irrigation canal systems in India. The canals were excavated in fine sand-bed material, and sediment entered the canals through canal head works. Channel width, depth, and slope are the dependent variables, and channel-forming discharge, bed gradation, and sediment-inflow concentration are the independent variables. The dependent variables are determined using regression analysis, indicating that regime methods should be applied to channels that are similar to those used in regression analysis. Also, they are applied to low-energy systems and where discharge is relatively uniform with time, sediment transport is low, and Froude number is less than 0.3.

The regime of a channel in alluvium depends on the interrelationship between flow, sediment transport, channel resistance, and bank stability. It is this interrelationship that determines the channel shape and its stability, and the equations, which are mostly empirical, expressing the channel dimensions, including width, depth, and gradient, constitute the regime theory. Thus, the objective of regime theory is to predict the size, shape, and slope of a stable alluvial channel under given conditions. A channel is characterized by its width, depth, and slope. The regime theory empirically relates these characteristics to the water and sediment discharge transported by the channel. For deriving regime channel dimensions, empirical measurements are made on channels, and empirical equations are fitted to the observed data. While fitting, channel characteristics are related primarily to discharge, but allowance is also made for variations in other variables, such as sediment size. For practical purposes, rivers tend to be in equilibrium (in regime) or in quasi-equilibrium (quasi-regime), for its characteristics have not changed over a long period of time. Canals usually maintain constant discharge, and regime relations may, therefore, be established using field data. The regime theory is composed of three parts: (1) concepts of alluvial channels in regime, (2) data base, and (3) formulations (Stevens and Nordin, 1987).

3.2.1 Lacey's Equations

The regime theory can be traced to Kennedy (1895), who, using observations from Bari Doab, the Punjab, India, developed an empirical relation between cross-sectional mean flow velocity (v) and mean flow depth over the bed (d_B) as:

$$v = 0.55md_B^{0.64}, \tag{3.1}$$

where m is the critical velocity ratio, often assumed unity. The canal had a range of discharges from 26 to 1,700 ft^3/s and depths from 2.2 to 7.0 ft, but all had more or less the same Froude number = 0.21. The nonsilting velocities in the upper Bari Doab canal system had been achieved by aggradation and widening. Equations similar to Equation (3.1) were reported by Kennedy, where the exponent and the multiplication factor were, respectively, 0.64 and 0.84 for Upper Bari Doab Canal, Punjab; 0.57 and 0.91 for Shwebo Canal, Burma; 0.55 and 0.67 for Godawari Western Delta, Madras; and 0.52 and 0.93 for Kistna Western Delta, Madras; these, reported by Lindley, were, respectively, 0.57 and 0.96 for Lower Chenab Canal, Punjab; and these, reported by Ghaleb, were, respectively, 0.39 and 0.39 for canals in Egypt.

About a quarter century later, Lindley (1919), using 800 observations from the Lower Chenab canal system, developed

$$v = 0.57 \, d_B^{0.57} \tag{3.2a}$$

$$v = 0.59 w^{0.353} \tag{3.2b}$$

$$w = 7.84 \, d_B^{1.61}, \tag{3.3}$$

where w is the width of bed. Equation (3.3) was not based on observations but was derived using Kutter's resistance formula, where $n = 0.0225$, which was not different from Manning's roughness coefficient n_m. For Lower Chenab canal, the multiplier in Equation (3.3) was found to be 3.8.

Lacey (1929) developed three functions of velocity (m/s), wetted perimeter (m), and gradient or slope (energy gradient of open channel), with allowance for bed sediment size as:

$$v = 0.625 \, (f_s R)^{\frac{1}{2}} \tag{3.4}$$

$$P = 4.84 \, Q^{\frac{1}{2}} \tag{3.5}$$

$$S = 0.000304 \, f_s^{\frac{5}{3}} Q^{-\frac{1}{6}}, \tag{3.6}$$

where R (m) is the hydraulic radius (A/P), P is the wetted perimeter of the flow section (m), A is the flow cross-sectional area (m^2), Q is the volumetric rate of discharge (m^3/s), depth (m), width (m), and f_s is the silt factor (sediment size parameter). Using $Q = vA$ and $A = PR$, the slope was expressed as

$$S = \frac{f_s^{\frac{5}{3}}}{1{,}790 Q^{\frac{1}{6}}}. \tag{3.7}$$

Since there is no explicit expression for sediment transport, it is assumed that a regime canal designed with Lacey's equations would carry the sediment supplied to it.

The silt parameter was expressed as a function of sediment size as

$$f_s = (2{,}500D)^{\frac{1}{2}}, \tag{3.8}$$

where D is the sediment diameter (m). Now recall the continuity equation

$$v = Q/A \quad \text{or} \quad v = Q/(RP), \, A = RP. \tag{3.9}$$

Using Equations (3.5) and (3.9), Equation (3.4) can be expressed as

$$R = \left(\frac{Q}{9.15 f_s} \right)^{\frac{1}{3}}. \tag{3.10}$$

The hydraulic radius R was derived as

$$R = 0.48 \left(\frac{Q}{f_s}\right)^{\frac{1}{3}}.$$ (3.11)

Lacey (1929) amended the velocity equation by including the resistance function as

$$v = \left(\frac{1}{N_L}\right) R^{\frac{3}{4}} S^{\frac{1}{2}},$$ (3.12)

where

$$N_L = 0.0225 f_s^{\frac{1}{4}},$$ (3.13)

where N_L is the resistance coefficient in the Lacey resistance equation. Equation (3.12) is similar to the Manning equation with roughness coefficient $n_m = 0.0225$ for sand bed channels, but the exponent for R is 3/4 rather than 2/3, which is appropriate for beds with ripples and dunes. Equation (3.12) can also be expressed in terms of unit stream power (*USP*), which equals the product of velocity and slope ($= vS$) as

$$vS = USP = \left(\frac{1}{N_L}\right) R^{\frac{3}{4}} S^{\frac{3}{2}}.$$ (3.14)

Equation (3.14) shows that the unit stream power can be expressed in terms of hydraulic radius, slope, and roughness factor.

Lacey (1930) used Lindley's data from the Lower Chenab Canal system. The data consisted of measured cross-sections and longitudinal slopes and were combined into 33 data sets. The radii ranged from 0.86 to 9.10 ft and the slopes from 0.000127 to 0.00042. The average velocity was computed with Chezy's equation:

$$v = C(RS)^{1/2},$$ (3.15)

with C as Chezy's coefficient, which was determined by the Ganguillet–Kutter (G.–K.) formula and Kutter's $n = 0.0225$ for all sets. The computed discharges were from 7.20 to 9.500 ft^3/s. A plot of v versus R yielded

$$v = 1.13 R^{1/2}.$$ (3.16)

Lacey (1930) assumed that this v was close to the nonsilting velocity for the Lower Chenab canal and reasoned that it was different from that for the Upper Bari Doab canal, because the silts were different and were classified by the critical velocity ratio, that is, the ratio of regime velocity for a given depth to the velocity for the same depth that the Kennedy formula would produce. Lacey (1930) assigned a value of 1.0 for the standard silt and defined his silt factor as the square of the velocity ratio $\left[\frac{v_0}{v} = f_s^{0.5}\right]$, where v_0 is the velocity for the same depth given by the Kennedy formula, and he assigned a value of unity to the Bari Doab canal System. Thus, Lacey's regime equation became

$$v = 1.17(f_s R)^{\frac{1}{2}},\qquad(3.17)$$

where v is the nonsilting or regime velocity and f_s is Lacey's silt factor. The ratio of velocities was $v_0/v = 1.13/1.17 = 0.966$. The silt factor was $(1.13/1.17)^2 = 0.933$ for the Chenab canal System. Equation (3.17) is an expression for sediment transport in a regime channel if the sediment load (Q_s) equals

$$Q_s = KPv^3,\qquad(3.17a)$$

where K is a coefficient that would vary with sediment sizes or slopes, and P is the wetted perimeter. Then, the average sediment concentration C_s (Q_s/Q) can be expressed as

$$C_s = \frac{Q_s}{Q} = \frac{KPv^3}{PRv} = \frac{Kv^2}{R}.\qquad(3.17b)$$

From Equations (3.17) and (3.17b), Lacey's silt factor f_s can be written as

$$f_s = 0.73\frac{C_s}{K}.\qquad(3.17c)$$

From 15 sets of measured discharges and cross-sections from the Madras and Godavari Western Delta canal systems, $f_s^{0.5} = 0.675$, and the silt factor was $f_s = 0.456$. The discharges and radii varied from 33 to 6,100 ft.^{3}/s and 1.76 to 10.23 ft., respectively. The Froude number was approximately 0.14. Thus, it could be said that the critical velocities in the Punjab and Madras canals were 0.966 and 0.675, respectively. If the silt factor in a given regime canal is 0.5, the hydraulic mean depth of the canal would be $1/f_s$, two times the hydraulic depth of a standard silt channel having the same mean velocity.

Bose (1936) derived an alternative slope equation with the exponent of Q as 0.21 in place of 1/6 for canals, where the Madras canals were wider and the Bari Doab canals were narrower.

Lane (1937) observed that the wetted stream perimeter (P or approximately width) was a function of bed slope and discharge Q as

$$P = 2.67Q^{0.5}.\qquad(3.18)$$

Natural rivers cover a wide range of discharge and slope, while the range of values for canals is relatively small. Therefore, Lacey's relation cannot be applied to natural streams, because it contradicts the finding of Lane (1937), which states that steep slope streams tend to be wider and shallower than streams of the same discharge on a flat slope. This implies that the width–depth ratio is at least partly a function of the slope and not of discharge alone. A very large width–depth ratio is an indication of the river's tendency to braid.

Lacey (1946) suggested a more general equation

$$Sv = K\left[(RS)^{0.5}\right]^n,$$ (3.19)

in which K is a coefficient, and n is an exponent. Inglis (1947) derived three most fundamental expressions for velocity, width/depth ratio (w_M/d_B, w_M = mean width, and d_B = mean streamflow depth), and gradient as functions of discharge Q, sediment loading X_M, and sediment size D, thus showing the sensitivity to sediment. The velocity was shown to depend on Q to the power of 1/12, w_M/d_B on Q to the power of 7/12, and S on Q to the power of 5/12.

Lacey's equation may satisfy one or more of the following characteristics: (1) It is simple. For example, for $D_{50} = 0.2$ mm and $Q = 1,000$ ft.^{3}/s, one gets immediately $f_s = 0.71$, $P = 84.4$ ft., $R = 5.3$ ft., and $S = 0.00010$. (2) No information on incoming sediment load is needed. (3) It requires no information on the type or strength of bank materials. The width depends only on discharge.

From extensive investigations of stable channels for discharges between 0.4 and 5.4 cfs at Hydraulic Research Station, Wallingford, Ackers (1964) reported the best-fit geometry relations as

$$A = 1.60Q^{0.85}$$ (3.20)

$$v = 1.00Q^{0.15}$$ (3.21)

$$w = 3.6Q^{0.42}$$ (3.22)

$$d = 0.28Q^{0.42},$$ (3.23)

where A is the cross-sectional area in ft.^{2} and w and d are in ft., and v is in ft./s. It may be noted that the exponent 0.85 of Q for A in Equation (3.20) is close to 0.833 in Lacey's equation obtained from the product of Equations (3.11) and (3.18). In regime equations, the exponent of Q for w is 0.5 and the exponent of Q for d is 1/3, while these exponents are, respectively, 0.42 from Equation (3.22) and 0.42 from Equation (3.23). Ackers (1964) reported that the ratios w/d and P/hydraulic mean depth were almost independent of discharge.

The design equations can be summarized with f_s designated f_{vs} as

$$f_{vs} = \frac{0.73v^2}{R}$$ (3.24a)

$$R = 4.47 \left(\frac{Q}{f_s}\right)^{\frac{1}{3}}, \tag{3.24b}$$

and Equation (3.8) for slope and Equation (3.18). A rough estimate of f_s was given by D (inches) $= f_s^2/64$. Lacey (1946) used another equation for velocity by relating silt factor to roughness factor as

$$v = 16.1 \left(R^2 S\right)^{\frac{1}{3}}, \tag{3.25a}$$

from which f_s designated as f_{RS} as

$$f_{RS} = 190 \left(R S^2\right)^{\frac{1}{3}}. \tag{3.25b}$$

The silt factors can be related by eliminating R from Equations (3.24a) and (3.25b) as

$$f_{RS} f_{vs}^3 = 5 \times 10^6 (vS)^2. \tag{3.25c}$$

3.2.2 Blench's Equations

In 1939 Blench introduced bed and side factors, F_B and F_s, to isolate the influence of bed material from the composition of the alluvium forming the banks. He normalized the slope function as friction factor as a function of a width-based Reynolds number:

$$\frac{v^2}{d} = F_B \tag{3.26}$$

$$\frac{v^3}{w} = F_s \tag{3.27}$$

$$\frac{w}{d} = v\frac{F_B}{F_S} \tag{3.28}$$

$$\frac{v^2}{gdS} = C_d \left(\frac{vw}{v}\right)^{\frac{1}{4}}, \tag{3.29}$$

where d is the flow depth over the flat channel bed, g is the acceleration due to gravity, v is the kinematic viscosity, w is the mean width (A/d), and C_d is a dimensionless constant, often taken as 3.63. In the case of canals, the bed is usually rough, whereas, in analogy with Blasius equation, Equation (3.27) would be for smooth turbulent flow. Inglis (1947) argued that regime parameters were sensitive to sediment loading. Equation (3.29) was expressed as

$$v = 3.63 \left(\frac{d}{X}\right)^{\frac{1}{4}} \sqrt{gdS}; \quad X = \frac{\sqrt{vF_S}}{F_B}. \tag{3.30}$$

Blench (1951) provided a physical rationale for his regime equations. He stated that Equation (3.18) was equivalent to

$$\frac{P}{Rv} = 7.11 \quad \text{or} \quad \frac{\frac{v^2}{R}}{\frac{v^3}{P}} = 7.11 \tag{3.31}$$

and had a physical meaning. Equation (3.31) states that the ratio of bed to side factor is the same for all channels. By substituting w for P and d for R one can express

$$\frac{v^2/d}{v^3/w} = 7.11, \tag{3.32}$$

where the numerator has a physical significance, since this denotes the force per unit mass. Likewise, the denominator that is associated with sides can be considered as a measure of tractive force acting on the sides, provided the sides are smooth. This therefore leads to Equation (3.31), in which vw/v is the Reynolds number in terms of width. Blench (1951) provided design equations as

$$w = \left(\frac{F_B Q}{F_S}\right)^{0.5} \tag{3.33}$$

$$d = \left(\frac{F_s Q}{F_B^2}\right)^{\frac{1}{3}} \tag{3.34}$$

$$S = \frac{F_B^{\frac{5}{6}} F_S^{\frac{1}{12}} Q^{-\frac{1}{6}}}{K\left(1 + \frac{C_d}{233}\right)} \tag{3.35}$$

$$S = \frac{F_B^{\frac{7}{8}}}{K w^{\frac{1}{4}} d^{\frac{1}{8}}\left(1 + \frac{C_d}{233}\right)} \tag{3.36}$$

$$S = \frac{F_B^{\frac{11}{18}} Q^{-\frac{1}{12}}}{K w^{\frac{1}{6}}\left(1 + \frac{C_d}{233}\right)}, \tag{3.37}$$

where

$$K = \frac{3.63g}{v^{\frac{1}{4}}}. \tag{3.38}$$

For all practical purposes, $C_d/233$ can be taken as 0, and C_d is parts per hundred thousand by weight. Therefore, Equation (3.35) will reduce to

$$S = \frac{F_B^{\frac{5}{6}} F_s^{\frac{1}{12}} Q^{-\frac{1}{6}}}{\frac{C_d g}{v^{\frac{1}{4}}}}.$$ (3.39)

The value of $3.63g/v^{1/4}$ was reported as 2,080 for $v = 10^{-5}$. He suggested

$$F_B = 2\sqrt{D_m},$$ (3.40)

where D_m is the mean bed material diameter in mm. The value of F_B was found to be from 0.6 to 1.25 with a common value of approximately 1.0. For F_S, he suggested a value of 0.3 for glacial till material (tough clay banks), 0.2 for silty sand-loam material (or silty, clay, and loam banks), 0.1 for material with little cohesion or friable banks, and between 0.05 and 1.0 for sand sides of rivers below tide level.

Vanoni (1975) discussed in the American Society of Civil Engineers (ASCE) Manual 54 (ASCE 1975) the regime equations modified by Blench (1951) using data from Indian canals with sand beds and slightly cohesive to cohesive banks. These equations can be used for the design of canals with sand beds with sediment inflow less than 30 milligrams per liter, involving the computation of width, depth, and slope as functions of bed-material grain size, channel-forming discharge, bed-material sediment concentration, and bank composition. The regression equation for slope is

$$S = \frac{F_B^{0.875}}{\frac{3.63g}{v^{0.25}} w^{0.25} d^{0.125} \left(1 + \frac{C_s}{2,330}\right)},$$ (3.41)

where

$$F_B = 1.9\sqrt{D_{50}},$$ (3.42)

and C_s is the bed sediment concentration (ppm).

3.3 Generalized Regime Theory

Introducing *non-dimensional* quantities, Lacey and Pemberton (1972) showed that $v/R^{0.5}$ and $R^{0.5}/S$ were both constant, independent of Q, throughout a system of given sedimentology but would vary from system to system. From field observations, they found that a log-log plot of $v/R^{0.5}$ versus $R^{0.5}S$ was similar to a plot of $w^{0.5}$ versus sediment particle diameter, as shown in Figure 3.1 as done by Ackers (1980). This analogy led to a straight-line approximation within different size ranges:

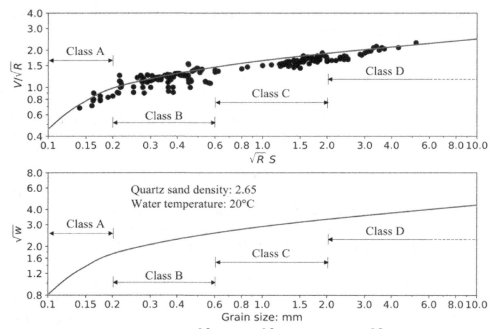

Figure 3.1 Log-log plot of $v/R^{0.5}$ versus $R^{0.5}S$ and a plot of $w^{0.5}$ versus particle sediment particle diameter

$$\frac{v}{R^{0.5}} = \text{constant} \times \left(\sqrt{RS}\right)^{u},$$ (3.43)

where u is exponent.

For $0.1 < D < 0.2$ mm, $u = 1$,

$$v \propto RS.$$ (3.44)

For $0.2 < D < 0.6$ mm, $u = 0.5$,

$$v \propto R^{\frac{3}{4}}S.$$ (3.45)

Equation (3.43) is like Manning's equation. For $0.6 < D < 2$ mm, $u = 0.33$,

$$v \propto \left(R^2 S\right)^{\frac{1}{3}}.$$ (3.46)

Equation (3.46) is similar to the equation derived by Bray (1982a, b) for gravel bed rivers and by Malhotra (1939) for irrigation canals. For $D > 2$ mm, $u = 0.25$,

$$v \propto R^{\frac{5}{8}}S^{\frac{1}{4}}.$$ (3.47)

Further, the regime function/fall velocity analogy yielded the following proportionalities:

$$R^{\frac{1}{2}}S \propto D \qquad (3.48)$$

$$\frac{v^2}{R} \propto w. \qquad (3.49)$$

The generalized regime functions for different sediment sizes are still empirical. However, for broader applicability, it would be desirable to provide a physical basis for the three controlling processes and insights into the conditions imposed due to sediment size and density, sediment supply, water temperature, and discharge. The processes were: (i) the relationship of bed sediment transport to hydraulic parameters of channel flow; (ii) the relationship of hydraulic resistance to bed configuration, which, in turn, depends on sediment type and transport; and (iii) the erosive resistance of banks as a function of flow characteristics.

Depending on the sediment load and the stability of channel boundary under normal flow, channels can be distinguished as threshold channels, alluvial channels, and transition channels. The design of a channel depends on the flow of water and sediment transport and their interaction with channel boundary, and this interaction determines the type of technique that should be employed for design.

A threshold channel can be understood to be a channel whose bed is composed of coarse material and its boundary material does not move during design flow because the forces applied by flow are less than the threshold for the movement of boundary material. Such a channel is a fixed boundary channel. Fine sediment is not regarded as part of a channel bed and may pass through the channel as wash load. Because the boundary material is not easily erodible and there is little interaction between bed and suspended sediment transport, such channels are amenable to quick adjustment of their geometry. Examples of such channels are the channels formed during dam breaks, high runoff due to recession of glaciers, and channels armored by the reduction of sediment supply and degradation.

An alluvial channel is understood to be a channel whose bed and banks are formed of material transported by the stream under present flow conditions, and there is an exchange of material between the inflow of sediment load and the bed and banks. Alluvial channels carry large amounts of sediment, which tends to be coarse, and adjust their width, depth, slope, and planform in response to changes in water or sediment discharge. Such a channel is the mobile boundary channel.

As the name suggests, a transition channel is between threshold and alluvial channels; however, the transition is not always obvious. There are instances where the transition occurs such that one reach of the channel may be alluvial, while another threshold. Likewise, a threshold reach can morph into an alluvial reach by the flattening of slope, or a channel may be alluvial at low discharges when there is an adequate sediment supply and threshold at high discharges. Conversely, a channel may be a threshold channel at low flows but alluvial during very high discharges.

Using data from canals in India and the United States, Simons and Albertson (1963) modified the regime method by expanding the range of conditions applicable to the original equations. Eliminating the need for computing bed, bank, or sediment concentration factors and not considering the inflowing sediment concentration as an independent variable, they developed three sets of equations for three classes of channels, based on the composition of streambed and streambanks:

$$P = C_1 Q^{0.512} \tag{3.50}$$

$$R = C_2 Q^{0.361} \tag{3.51}$$

$$A = C_3 Q^{0.873} \tag{3.52}$$

$$v = C_4 \left(R^2 S \right)^{\frac{1}{3}} \tag{3.53}$$

$$\frac{w}{d} = C_5 Q^{0.151}, \tag{3.54}$$

where Q is the channel-forming discharge (ft.^{3}/s), P is the perimeter (ft.), R is the hydraulic radius (ft.), A is the channel cross-sectional area (ft.^{2}), v is the mean channel velocity (ft./s), w is the average channel width (ft.), d is the average flow depth (ft.), S is the slope, and C_i, $i = 1, 2, \ldots, 5$, are coefficients. These coefficients were suggested for three types of bed and bank sediment as shown in Table 3.1.

The relationships between channel geometry and slope for the three channel types were:

$$d = 1.73 R \qquad (1 < R < 7) \tag{3.55}$$

$$d = 2.11 + 0.934 R \quad (7 < R < 12). \tag{3.56}$$

Table 3.1. *Coefficients for modified regime equations*

Coefficient	Sand bed and sand banks	Sand bed and cohesive banks	Cohesive bed and cohesive banks
C_1	3.30	2.51	2.12
C_2	0.37	0.43	0.51
C_3	1.22	1.08	1.08
C_4	13.9	16.1	16.0
C_5	6.5	4.3	3.0

Note: * A soil is classed as cohesive if the plasticity index is >7

$$w = 0.9P \tag{3.57}$$

$$w = 0.92TW - 2.0, \tag{3.58}$$

where *TW* is the top width. The modified regime equations were derived for the data whose limits are given in Table 3.2.

Application of the modified regime method entails the following steps:

(1) Determine the primary independent variable, which is the channel-forming discharge.
(2) Classify the bed and bank materials as either sand bed and sand banks, sand bed and cohesive banks, or cohesive bed and cohesive banks for determining the coefficients for the appropriate class.
(3) Compute inflow to the design reach.
(4) Calculate sediment transport rate using an appropriate sediment transport equation.
(5) Evaluate if the modified regime method is applicable.
(6) Determine the channel geometry.
(7) Compute the acceptable safe slope.
(8) Check the slope calculated with the modified regime equations. Use Manning's equation with a realistic roughness coefficient and consistent cross-sectional geometry.
(9) Calculate the sediment transport capacity of the design reach and compare it with the upstream supply. If the capacity is adequate, the design is complete, or else redo the calculations.

Example 3.1: Design a stable rectangular channel with cohesive bed and cohesive banks using the modified regime approach. The channel-forming discharge is 500 ft.^{3}/s and Manning's n_m is assumed as 0.025.

Solution: The channel is designed using the following steps.

(1) Compute the channel perimeter, P, using Equation (3.50) and coefficient C_1 from Table 3.1:

$$P = 2.12Q^{0.512} = 2.12\,(500)^{0.512} = 51.1 \text{ ft.}$$

(2) Compute the hydraulic radius, R, using Equation (3.51) and C_2 from Table 3.1:

$$R = 0.51Q^{0.361} = 0.51\,(500)^{0.361} = 4.81 \text{ ft.}$$

(3) Compute flow cross-sectional area, A, using Equation (3.52) and C_3 from Table 3.1:

$$A = 1.08\,Q^{0.873} = 1.08\,(500)^{0.873} = 245.26 \text{ ft.}^2$$

Table 3.2. *Limits of data sets used in the development of modified regime equations*

Data source	Median bed material size (mm)	Banks	Discharge (ft.^{3}/s)	Sediment concentration (ppm)	Slope (L/L)	Bedforms
United States and Indian canals	0.318 to 0.465	Sand	100 to 400	<500	0.000135 to 0.000388	Ripple to dunes
	Cohesive 0.029 to 0.36	Cohesive	5 to 88,300	<500	0.000059 to 0.00034	Ripple to dunes
	Cohesive 0.06 to 0.46	Cohesive	137 to 510	<500	0.000063 to 0.000114	Ripple to dunes

(4) Compute the mean velocity, v:

$$v = \frac{Q}{A} = \frac{500 \text{ ft.}^3/\text{s}}{245.26 \text{ ft.}^2} = 2.04 \text{ ft./s}$$

(5) Compute the depth using Equation (3.55)

$$d = 1.73R = 1.73 \times 4.8 = 8.30 \text{ ft.}$$

(6) Compute the Froude number, F:

$$F = \frac{v}{\sqrt{gd}} = \frac{2.55}{\sqrt{32.2 \times 8.3}} = 0.16$$

Check if $F < 0.3$.

(7) Compute the bottom width, w, using Equation (3.57):

$$w = 0.9P = 0.9 \times 51.1 = 46.00 \text{ ft.}$$

(8) Because the channel is rectangular, the top width TW is the same as the bottom width w.

(9) Compute the width-to-depth ratio, w/d, using Equation (3.54) and coefficient C_5 from Table 3.2 as

$$\frac{w}{d} = C_5 Q^{0.151} = 3.0 \, (500)^{0.151} = 7.67$$

(10) Compute the regime slope using Equation (3.53) and coefficient C_4 from Table 3.1: $2.04 = C_4 \left(R^2 S\right)^{\frac{1}{3}} = 16.0 \left(4.81^2 S\right)^{\frac{1}{3}}$
This yields $S = 0.00$

(11) Compute Manning's n_m from Manning's equation:

$$n_m = \frac{1.486}{Q} A R^{\frac{2}{3}} S^{\frac{1}{2}} = \frac{1.486}{500} \times 196.1 \times 4.81^{\frac{2}{3}} \times 0.04^{\frac{1}{2}} = 0.33$$

(12) Compute the channel slope assuming uniform flow:

$$S = \left(\frac{n_m Q}{1.486 A R^{\frac{2}{3}}}\right)^2 = \left(\frac{0.33 \times 500}{1.486 \times 196.1 \times 4.81^{\frac{2}{3}}}\right)^2 = 0.039$$

(13) Select the design slope: There are two values of roughness coefficients, one $(n_m = 0.33)$ obtained as shown and the other specified $(n_m = 0.025)$. The way to select the slope is as follows. Obtain the data from channels physiographically similar to the design channel. If the regime slope fits the data then the calculated regime slope can be selected. If the data is not available, then one can choose the slope from the uniform flow equation.

(14) Compute the sediment transport capacity of the design channel and compare it with the upstream supply. If the design sediment transport capacity is smaller than the supply sediment transport capacity, then either the design must be modified or sediment removal must be provided for.

3.4 Process-Based Regime Equations

Regime equations represent three fundamental equations that govern the channel geometry. There are several physically based sediment transport and alluvial resistance functions (Ackers, 1983, 1988). The sediment transport equations are based on scientific principles considering dimensional consistency, similarity principles, and mechanics of bed material transport and turbulent suspension. They are mainly based on laboratory investigations, and their application to field conditions like sediment regime of irrigation canals would necessarily involve extrapolation, often beyond the range of original calibration.

Design of a channel can be based on generally accepted sediment transport and resistance functions. However, the width function is still a tricky part. For practical considerations, it may not be entirely unreasonable to employ a regime type equation with its coefficient determined from bank characteristics as given by Simons and Albertson (1960):

$$w = K_1 Q^{\frac{1}{2}}, \tag{3.59}$$

where coefficient K_1 is given as 6.34 for sand bed and banks, 4.71 for sand bed and cohesive banks, 3.08 for sand bed and cohesive banks with heavy sediment load, 2,000–8,000 mg/l, 3.98 for cohesive bed and banks, and 3.17 for coarse noncohesive material.

Bakker et al. (1986) found a value of $K_1 = 4.7$ for the Punjab canals and $K_1 = 4.0$ for the Sind canals. One can also determine the width function by optimization, noting that a canal tends toward the maximum efficiency in transporting the fluxes of water and sediment imposed on the canal. The geometry will adjust such that the slope is minimum consistent with the sediment transport rate and channel resistance.

Inglis (1947) introduced sediment concentration into geometric equations and Lacey (1957–1958) suggested dimensionless equations as

$$v \propto \frac{\sqrt{g}}{\sqrt{D}} dS \tag{3.60}$$

$$\frac{v^3}{w} \propto g^{\frac{3}{2}} D^{\frac{1}{2}} \tag{3.61}$$

$$\frac{v^2}{gd} \propto (C_s v_s)^{\frac{1}{2}}(vg)^{-\frac{1}{6}}, \tag{3.62}$$

in which C_s is the sediment concentration (by weight), and v_s is the fall velocity of sediment particles. Equations (3.60)–(3.62) are called Inglis–Lacey equations, which would define a stable channel for any sediment diameter D and concentration C_s. In Equation (3.62), the left side is Froude number and the right side is a sediment concentration parameter. It shows that Froude number is proportional to the one fourth power of sediment concentration, meaning Froude number increases with increasing sediment concentration.

Chien (1957) analyzed the regime theory on the basis of bed-load functions, discussed the limitations of the theory, and indicated that the dimensions and slope of an alluvial channel for a given discharge depended on sediment load and bed material size. He showed that one silt factor depended on sediment concentration and the other silt factor depended on the sediment size. Recasting the Lacey regime theory as

$$\frac{v^2}{R} = 1.325 f_{vs} \tag{3.63}$$

$$R^{\frac{1}{2}}S^{\frac{1}{2}} = 0.0052 f_{RS} \tag{3.64}$$

and Equation (3.18), where f_{vs}, depending on the bed material size, and f_{RS} are the silt factors. Neglecting the bank friction, and considering the bank material size as 0.25 mm, hydraulic radius varying from 2 ft. to 25 ft., and bed slope as 0.001 to 0.004, he found

$$f_{vs} = 0.061 \left(\frac{q_T}{q}\right)^{0.715} \tag{3.65}$$

$$f_{RS} = 1.18 \left(\frac{q_T}{q}\right)^{0.052}, \tag{3.66}$$

in which q_T is the total sediment transport rate per unit width, and q is the discharge rate per unit width.

For the hydraulic radius varying from 2 ft. to 30 ft.; and slope varying from 0.001 to 0.005 for the 0.25 mm bed material size, from 0.003 to 0.0025 for the 2.5 mm bed material size, and from 0.001 to 0.01 for the 25 mm bed material size, with D in mm and $q_T/q < 200$ ppm, he expressed

$$f_{RS} = 2.2 D^{0.45} \left(\frac{q_T}{q}\right)^{0.052}. \tag{3.67}$$

He emphasized that for conditions different from those the regime theory derived for f_{vs} also depended on channel hydraulic characteristics, which would make application of the theory difficult. If the bank friction was included, it would make a slight difference in the values of silt factors.

Lacey (1957) showed that for certain combinations of variables the sediment load did not factor in, as shown next. Lacey's basic equation for silt factor f_s can be expressed as

$$v = 1.155 \sqrt{f_s R}. \tag{3.68}$$

Equation (3.63) can be written for f_s as

$$f_s = 0.75 \frac{v^2}{R}. \tag{3.69}$$

If R is replaced by d in Equation (3.69) then f_s will be a function of Froude number. Lacey proposed

$$f_s = 1.75 \sqrt{D}, \tag{3.70}$$

where D is in mm. He expressed flow equation with no silt factor or roughness factor as

$$v = 16.0 \, R^{\frac{2}{3}} S^{\frac{1}{2}}. \tag{3.71}$$

Combining Equations (3.68), (3.69), and (3.71), the silt factors can be obtained as

$$f_{vs} = 0.75 \frac{v^2}{R} \tag{3.72}$$

$$f_{RS} = 192 \, R^{\frac{1}{3}} S^{\frac{2}{3}} \tag{3.73}$$

$$f_{\frac{RS}{v}} = 3.072 \frac{RS}{v}. \tag{3.74}$$

The silt factor given by Equation (3.74) can be expressed as

$$f_{\frac{RS}{v}} = \frac{f_{RS}^{\frac{3}{2}}}{f_{vr}^{\frac{1}{2}}}. \tag{3.75}$$

For a sand bed with a mean diameter of 0.25 mm, the silt factors can be written as

$$f_{vs} = \text{const.} \left(\frac{q_T}{q} \right)^{\frac{1}{2}} \tag{3.76}$$

$$f_{RS} = \text{const.} \left(\frac{q_T}{q} \right)^{\frac{1}{6}}. \tag{3.77}$$

Equation (3.76) is valid for the entire range but Equation (3.77) holds for sediment concentration greater than 1,000 ppm. It can be inferred that RS/v is independent of sediment concentration q_T/q.

The flow equation can be written as

$$v = 3.072 \frac{RS}{f_s}. \tag{3.78}$$

Using Equation (3.70) and inserting Equation (3.78), the flow equation can be expressed as

$$v = \text{const. } g^{\frac{1}{2}} \frac{RS}{D^{\frac{1}{2}}}. \tag{3.79}$$

Equation (3.79) is the Lacey–Malhotra equation.

Introducing $X = q_T/q$ and terminal velocity of particle in water v_s, the Inglis–Lacey equations can be expressed as

$$v = \text{const. } D^{\frac{1}{12}}(Xv_s)^{\frac{1}{12}}Q^{\frac{1}{4}} \tag{3.80}$$

$$d = \text{const. } D^{\frac{1}{6}}(Xv_s)^{-\frac{1}{3}}Q^{\frac{1}{3}} \tag{3.81}$$

$$S = \text{const. } D^{\frac{5}{12}}(Xv_s)^{\frac{5}{12}}Q^{-\frac{1}{6}} \tag{3.82}$$

$$w = \text{const. } D^{-\frac{1}{4}}(Xv_s)^{\frac{1}{4}}Q^{\frac{1}{2}}. \tag{3.83}$$

From Equations (3.80)–(3.83), the terminal velocity and bed load can be eliminated, yielding

$$v^4 = \text{const. } wdS. \tag{3.84a}$$

Eliminating X and v_s from Equations (3.80)–(3.82), one can express

$$\frac{v}{dS} = \frac{\text{const.}}{\sqrt{D}}. \tag{3.84b}$$

Also,

$$\frac{v^3}{w} = \frac{\text{const.}}{\sqrt{D}}. \tag{3.84c}$$

Equations (3.84b) and (3.84c) do not contain sediment load but do include sediment size.

Using measurements on velocity, depth, width, slope, and sediment load made in a reach of canal in Pakistan with flow discharge at least 90% of design discharge, Tarar and Choudri (1979) modified the Lacey equations and found

$$f_{vs} = 0.75 \frac{v^2}{d}, d = \frac{A}{w} \tag{3.85a}$$

$$f_{RS} = 192 \left(dS^2 \right)^{\frac{1}{3}} \tag{3.85b}$$

$$f_{D50} = 1.76 D_{50}^{\frac{1}{2}}. \tag{3.85c}$$

If there is a difference between f_{vs} and f_{RS}, the mean value is recommended:

$$f_m = \left(f_{RS} f_{vs} \right)^{\frac{1}{2}}. \tag{3.86a}$$

The other equations used for design are Equation (3.67) for width and

$$S = \frac{f_m^{\frac{5}{3}}}{1.830 \, Q^{\frac{1}{6}}} \tag{3.86b}$$

for slope. Stevens and Nordin (1987) argued for not using Equation (3.86a).

For stable canal design, a unique relation between w and d is imposed in order to obtain channel geometry and slope. Based on extensive data on canals in India, Ranga Raju and Garde (1988) gave values of the width to depth ratio (w/d) as 4.5, 5.0, 6.5, 9.0, 12.0, 15.0, and 18.0 for respective discharge (m³/s) values of of 5, 10, 15, 50, 100, 200, and 300. Raga Raju and Misri (1979) derived an expression for flow depth by combining Manning's equation with the continuity equation:

$$d = \left[\frac{1.818Q}{\left(\frac{w}{d} + 0.5 \right) m} \right]^{0.378}, \tag{3.87}$$

in which m is the critical velocity ratio as defined in the Kennedy Equation (3.1) for velocity, which depends on the sediment size, and the aspect ratio is a function of $SQ^{0.02}/(n_m m)^2$, where n_m is the Manning coefficient.

Using dimensional analysis, Kondap and Garde (1979) found a dimensionally homogeneous equation for stable canal width and cross-section for a vast set of canal data covering discharge Q from 0.104 to 427 m³/s, sediment size from 0.02 to 7.6 mm, slope from 0.051×10^{-3} to 0.593×10^{-3}, and total sediment load from 9.0 to 2090 ppm:

$$\frac{w}{D} = 0.212 \left(\frac{g^{\frac{1}{2}} D^{\frac{5}{2}}}{v} \right)^{0.231} \left(\frac{Q}{D^2 \left(\frac{\gamma_s * D}{\rho} \right)^{\frac{1}{2}}} \right)^{0.548} \tag{3.88}$$

$$\frac{d}{D} = 130 \left(\frac{g^{\frac{1}{2}} D^{\frac{3}{2}}}{v} \right)^{-0.235} \left(\frac{Q}{D^2 \left(\frac{\gamma_s * D}{\rho} \right)^{\frac{1}{2}}} \right)^{0.304} \tag{3.89}$$

$$\frac{A}{D^2} = 2.21 \left(\frac{Q}{D^2 \left(\frac{\gamma_s^* D}{\rho} \right)^{\frac{1}{2}}} \right)^{0.855} \tag{3.90}$$

$$\frac{S}{\left(\frac{\gamma_s^*}{\rho} \right)} = 0.212 \left(\frac{y_1}{D} \right)^{-1.095} \left(\frac{v}{\left(\frac{\gamma_s^* D}{\rho} \right)^{\frac{1}{2}}} \right)^{1.50}, \tag{3.91}$$

in which $\gamma_{s*} = \gamma_s - \gamma$ is the weight density of sediment minus the weight density of water, ρ is the mass density of water, v is the kinematic viscosity, and y_1 is the hydraulic depth $= A/w$. For the data used, the sediment load did not have an appreciable effect of w, d, A, or S. These equations were further simplified as

$$w \propto Q^{0.548} D^{-0.0235} \tag{3.92}$$

$$d \propto Q^{0.384} D^{-0.1276} \tag{3.93}$$

$$A \propto Q^{0.855} D^{-0.1375}. \tag{3.94}$$

3.5 Natural Channels

Description of natural channels involves bed forms, such as ripples and dunes, and plan form. The boundary features can be considered as having three different scales, including ripples related to particle size, dunes and bars related to channel depth, and shoals and meanders related to channel width. The first two scales are especially relevant to straight channels, such as canals, but also apply to natural streams, which are seldom straight. This implies that the third scale distinguishes them from manmade channels.

Manmade canals have three degrees of freedom of adjustment (width, depth, and slope) for achieving stability, which should be reconsidered for natural channels because plan form is another determinant of fluvial geometry. For example, if a stream is passing through a valley of known slope, then it cannot adjust its slope. Likewise, if a stream is passing through a plane of known gradient then it will lengthen its path through meandering, meaning the plan form becomes integral to the slope adjustment. Hey (1978) suggested nine degrees of freedom that natural channels seek to achieve a stable condition, including average bankfull velocity, hydraulic mean depth, maximum bankfull depth, slope, wave length of bed forms, mean of bed forms, bankfull wetted perimeter, channel sinuosity, and arc length of meanders. The implication here is that nine equations will be needed to define the hydraulic geometry. However, fewer degrees would suffice for characterizing the overall hydraulics.

There are several criteria that have been employed in fluvial hydraulics for categorizing river patterns. Depending on the freedom to adjust their channel, all rivers can be classified into two groups: alluvial channels and bedrock channels. Bedrock channels are formed between rock outcrops and the material constituting their bed and banks determines their morphology, which is not easily adjustable. Alluvial rivers, that are of primary interest here, adjust their morphology in response to water and sediment load they receive from their watershed. They flow through a channel whose bed and banks are formed by this sediment load. Schumm (1963a, 1971) employed discharge and the sediment load to classify alluvial streams, as shown in Table 3.3 (after Schumm, 1963b). These two variables determine the channel morphology. In order of increasing sediment transport and overall slope, streams have also been classified as (Kellerhals et al. 1976): straight; straight channels with shoals on alternate sides; fixed meanders, that is, with little bank erosion; free meanders, with active bank movement; braided, multithread channels; and mountain torrents.

Depending on the ratio of the length along the deep water channel to the length measured along the valley (straight line slope), natural rivers can be categorized as straight, sinuous, meandering, and braided, as shown in Figure 3.2 (Leopold and Wolman, 1957, 1960; Simons et al. 1975). Desloges and Church (1989) noted that many rivers depart from these categories (Mollard, 1973), for plan form geometries vary through a continuum from meandered to braided. They provided morphological data for two Canadian rivers: Fraser and Bella Coola, as shown in Table 3.4. For Fraser River the data was over the lower 150 km with a wandering channel morphology on a cobble-gravel fan in the first 50 km before transitioning into its single-thread sand bed reach. The wandering reach was wider, shallower than confined cobble-bed reach upstream or sand-bed single-thread reach downstream. Typical River reaches over long distances are usually unstable and can be created by manmade cutoffs or by contraction works such as dikes or revetments. The rivers were classified as sinuous if the ratio of the length along the deep water channel to the length along the valley was below 1.5; meandering refers to more tortuous channels, whereas a braided river has several branching and rejoining channels, and not a single channel and is relatively wide with ill-defined unstable banks. Brice (1960) suggested a braiding index (B_I) as

$$B_I = \frac{2 \times \text{total length of bars in reach}}{\text{reach length at midchannel}}, \tag{3.95}$$

which accounts for the extra bank length caused by braiding.

The probable channel configurations in order of increasing sediment transport and overall slope include straight channels, straight channels with shoals on alternate sides, fixed meanders with little bank erosion, free meanders with active

Table 3.3. *Alluvial channel classification (after Schumm, 1963b)*

Mode of sediment transport and type of channel	Channel sediment (M) percent	Bedload (%) of total load	Channel stability		
			Stable (Graded stream)	Depositing (Excess load)	Eroding (Deficiency of load)
Suspended load	>20	<5	Stable suspended load. Width–depth ratio<10; sinuosity usually >2.0; gradient relatively gentle	Depositing suspended load channel. Major deposition on banks causes narrowing of channel; initial sediment deposition minor.	Eroding suspended -load channel. Streambed erosion predominant; initial channel widening minor.
Mixed load	5–20	3–11	Stable mixed-load channel. Width–depth ratio>10; <40; sinuosity usually <1.3; gradient moderate.	Depositing mixed-load channel. Initial major deposition on banks followed by streambed deposition.	Eroding mixed-load channel. Initial streambed erosion followed by channel widening.
Bed load	<5	>11	Stable bed load channel. Width–depth ratio>40; sinuosity usually <1.5; gradient relatively steep.	Deposition bed load channel. Streambed deposition and island formation.	Eroding bed load channel. Little streambed erosion; channel widening predominant.

Table 3.4. *Morphological data for selected reaches of Bella Coola and Fraser Rivers (after Desloges and Church, 1989)*

Reach	Width (m)	Mean depth (m)	Width–depth ratio	Thalweg gradient	Valley gradient
Bella Coola River[*]					
Unstable	171	1.4	122	0.0033	0.0034
Stable	104	2.1	50	0.0019	0.0026
Fraser River[*]					
Confined	268	10.1	27	0.0006	0.00075
Unstable	517	6.6	78	0.00048	0.00068
Sand-bed (stable)	540	12.6	43	0.00005	0.00010

[*] average of several reaches

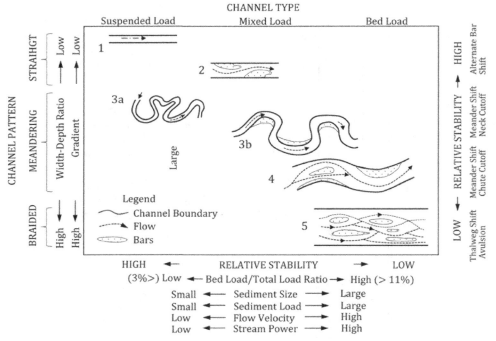

Figure 3.2 Categories of natural rivers as straight, sinuous, meandering, and braided (Leopold and Wolman, 1957, 1960; Simons et al. 1975)

bank erosion, and braided and multithreaded channels, and mountain torrents. Leopold and Wolman (1960) indicated that most rivers had the same ratio of curvature radius to channel width in the range of two to three. In fully developed turbulent flow, Hey (1978) expressed the wave length of meanders (L_m) as a function of channel width as

$$L_m = 2\pi w. \tag{3.96}$$

For width/median bed material grain size equal to or greater than a thousand,

$$L_m = 20w. \tag{3.97}$$

This equation is different from the one by Leopold and Wolman (1960):

$$L_m = 10.9w^{1.01} \text{ ft.} \tag{3.98}$$

Hey (1975) proposed a linear relation between wave length with radius of curvature *r* to channel width:

$$r = 2.4w, \tag{3.99}$$

whereas Leopold and Wolman (1960) reported

$$L_m = 4.7r^{0.98}. \tag{3.100}$$

Sinuous streams tend to exhibit certain characteristics, such as a low width–depth ratio (r_{w-d}), a high percentage of silt-clay in the perimeter, a high percentage of silt-clay in the banks (*B*), and a lower gradient than straight channels for the same discharge. Schumm (1963a) expressed the relation between width–depth ratio and sinuosity (S_u):

$$S_u = 3.5 \, r_{w-d}^{-0.27}. \tag{3.101}$$

Relatively wide and shallow channels tend to be straight, whereas relatively narrow and deep channels tend to deviate from a straight course. The relationship between sinuosity and the weighted mean silt-clay (*M*) (percentage) in the perimeter was expressed as

$$S_u = 0.94 \, M^{0.25}. \tag{3.102}$$

Bettess and White (1983) developed a framework for quantitative predictions of meandering and braiding of alluvial streams, based on the slope of the stream and the slope of the valley. Braiding can lead to quasi-equilibrium among discharge, sediment load, and channel capacity. It is primarily caused by the supply of sediment more than the channel carrying capacity leading to the deposition of part of the sediment load or steep slopes producing a wide shallow channel conducive to the formation of bars and islands. Braiding is also caused when banks, such as sand and gravel, are easily eroded. The stream widens during high flows and bars form during low flows, which may stabilize by armoring and vegetation forming islands. A meandering channel comprises alternating bends with an *S*-shape in the plan view housing deep pools and shallow crossings in the short straight reach connecting the bends. The thalweg goes from a pool through a crossing to the next pool.

There is a close connection between the general gradient and dominant sediment size in the river bed. Steep rivers in mountainous terrain contain boulders and coarse gravel, further down in the submountain region they have gravel beds, and still further down in the plains their beds will be of sand and silt. Coarse bed streams will usually have a wide range of sediment sizes. There will be considerable sorting of sediment between different parts of the bed and in flow depth. Further, there is frequent armouring, with a one- or two-grain thick layer of coarse sediment overlying the bulk of the bed composed of a wide range of sediment sizes.

The bed roughness of gravel rivers is different from that of sand bed rivers, since ripples do not occur in gravel bed rivers, and dunes are replaced by bars that may also be diagonal. These features cause a succession of chutes and pools, so flow may be critical at the bar crests (chutes) and return to subcritical in the pools beyond. Diagonal bars cause angled flows akin to alternate shoals of sand bed rivers creating meanders. Furthermore, coarse bed rivers have steeper slopes and greater velocities than sand bed rivers, and are prone to having upper regime bed features, such as antidunes. The resistance of gravel bed rivers is therefore treated differently.

Using laboratory model tests, Ackers and Charlton (1970a, b) reported a threshold value of sediment discharge below which channels carrying small discharge would remain straight and above which they may meander, if the valley slope does not restrict the development of hydraulic gradient necessary to transport discharge and sediment. They categorized small streams as straight, straight but exhibiting well developed shoals, or meandering based on the following criteria:

Straight $\quad S < 0.0015 Q^{-0.12}$

Shoaled $\quad 0.0015 Q^{-0.12} < S < 0.0021 Q^{-0.12}$

Meandered $\quad S > 0.0021 Q^{-0.12}$.

For fairly uniform sands with medium particle sizes between 0.15 mm and 0.70 mm and discharge less than 4 cfs, Ackers and Charlton (1970b) gave the critical straight-line water surface slope above which meandering occurs as $S_c = 0.0020 Q^{-0.12}$. For discharges greater than 0.2 cfs as bankfull flow and dominant discharge and medium sand sizes less than 1.00 mm, Ackers and Charlton (1970b) defined the critical water surface slope as $S_c = 0.0018 Q^{-0.21}$. However, natural channels seldom are straight for more than 10 times the width. Meandering is associated with bank erosion. Yang (1971) argued that the river channels tend to minimize the time rate of energy expenditure. Combining it with the Manning equation and continuity equation, the slope for a given discharge would decrease because of meandering development, but width would increase and generate bends.

Kellerhals (1967) analyzed the roughness, cross-section, slope, and reduction in load for 12 reaches of five gravel rivers and five laboratory gravel channels consisting of coarsest material ranging in size from 0.5 in. to 18 in. for discharge varying from 1 cfs to 100,000 cfs. He found regime type equations (in feet-second units):

$$w = 1.8Q^{0.5} \tag{3.103}$$

$$d = 0.166Q^{0.4}k_s^{-0.12} \tag{3.104}$$

$$v = 3.34Q^{0.1}k_s^{0.12} \tag{3.105}$$

$$S = 0.12Q^{-0.4}k_s^{0.92}, \tag{3.106}$$

where d was defined as the cross-sectional area divided by water surface width w, v is mean velocity, and k_s is the Nikuradse equivalent grain size roughness.

For cases with fixed w, as in laboratory channels, Kellerhals (1967) found

$$d = 0.266Q^{0.8}k_s^{-0.12}w^{-0.8} \tag{3.107}$$

$$v = 3.76Q^{0.2}k_s^{0.12}w^{-0.2} \tag{3.108}$$

$$S = 0.0752Q^{-0.6}k_s^{0.92}w^{0.8}. \tag{3.109}$$

Equations (3.107)–(3.109) are significantly different from the regime equations of Lacey (1933–1934, 1946, 1947) and Blench (1957) because of the differences between the mechanisms of sand and gravel transport.

Using a statistical analysis, Griffiths (1980) derived hydraulic geometry relationships for width, depth, velocity, bed slope, suspended sediment concentration, and hydraulic roughness (in terms of the Darcy–Weisbach friction factor) of New Zealand gravel rivers at mean annual water discharge (MKS system):

$$w = 7.09Q^{0.48} \tag{3.110}$$

$$d = 0.21Q^{0.43} \tag{3.111}$$

$$v = 0.61Q^{0.11} \tag{3.112}$$

$$C_s = 8.18Q^{0.31} \tag{3.113}$$

$$S = 0.02Q^{-0.49} \tag{3.114}$$

$$f = 1.21Q^{-0.38}, \tag{3.115}$$

in C_s is the suspended sediment concentration (ppm).

Using data from 62 stable gravel-bed river reaches in the United Kingdom, Hey and Thorne (1986) derived equations relating reach average values of width (w), mean depth (d), maximum depth (d_m), average slope (S), average bankfull reach velocity (v), wetted perimeter (P), hydraulic radius (R) in terms of bankfull discharge (Q_b), bed sediment load transport (Q_s), median sediment diameter (D_{50}), shear strength of bank material (B_s), standard deviation of log-normal size / frequency distribution (σ_D) [$\sigma = \frac{1}{2} \log \left(\frac{D_{84}}{D_{16}} \right)$ where D_{16} and D_{84} denote grain sizes for which 16% and 84% of the bed material are finer; σ_D = average bed material size variability for the reach], and an index of vegetation density. Likewise, riffle values of mean water line riffles with (w_r), mean flow depth at riffles (d_r), and maximum flow depth at riffles (d_{rm}) were related to Q_b, Q_s, B_s standard deviation at riffle depth (σ_{rD}), and D_{r50} and vegetation type. All quantities are in meters except B_s, which is in kPa, defined as half the compression strength obtained as the average of the upper, middle, and lower sections of the cohesive layer. Based on the bank vegetation and density, vegetation was categorized into four types: (1) grassy banks with no trees and bushes, (2) 1–5% tree/shrub cover, (3) 5–50% tree/shrub cover, and (4) greater than 50% tree/shrub cover or incised into flood plain. The sinuosity (S_u) was defined as the ratio of channel length to valley length, and this permitted computation of valley slope (S_v) as the product of sinuosity and channel slope. They derived the following equations:

Width: The simple width function was given as

$$w = 3.67 \, Q_b^{0.45}, \tag{3.116}$$

which was similar to that derived by Charlton et al. (1978). Inclusion of vegetation types culminated in the following:

$$w = 4.33 \, Q_b^{0.50} \qquad \text{vegetation type I} \tag{3.117}$$

$$w = 3.33 \, Q_b^{0.50} \qquad \text{vegetation type II} \tag{3.118}$$

$$w = 2.37 \, Q_b^{0.50} \qquad \text{vegetation type III} \tag{3.119}$$

$$w = 2.34 \, Q_b^{0.50} \qquad \text{vegetation type IV.} \tag{3.120}$$

These equations closely agree with those given by Nixon (1959), Kellerhals (1967), and Simons and Alberton (1963).

Mean depth: The simple depth relation was given as

$$d = 0.33 \, Q_b^{0.35}. \tag{3.121}$$

This equation compared well with those of Nixon (1959) and Simons and Albertson (1963). With the inclusion of bed material size, the equations were

$$d = 0.22 \, Q_b^{0.37} D_{50}^{-0.11} \qquad \text{vegetation types I-IV} \qquad (3.122)$$

$$d = 0.28 \, Q_b^{0.36} D_{85}^{-0.07} \qquad \text{vegetation types I-IV.} \qquad (3.123)$$

These equations are different from those obtained by Bray (1982b) and Kellerhals (1967). The inclusion of both material sizes led to

$$d = 0.20 \, Q_b^{0.36} D_{50}^{-0.56} D_{85}^{0.35} \qquad \text{vegetation types I-IV.} \qquad (3.124)$$

SLOPE: The simple equation was given as

$$S = 0.008 \, Q_b - 0.20. \qquad (3.125)$$

This agrees with that given by Simons and Albertson (1963) but is not reliable. The addition of bed material size significantly improved the result,

$$S = 0.098 \, Q_b^{-0.31} D_{50}^{0.71} \qquad (3.126)$$

$$S = 0.038 \, Q_b^{-0.28} D_{84}^{0.62}. \qquad (3.127)$$

The inclusion of bed load led to

$$S = 0.087 \, Q_b^{-0.45} D_{50}^{-0.09} D_{84}^{0.84} Q_s^{0.10} \qquad \text{vegetation types I-IV.} \qquad (3.128)$$

Velocity: The velocity equation was reported as

$$v = 1.70 \, Q_b^{0.10} Q_s^{0.03} D_{50}^{0.18}. \qquad (3.129)$$

Hydraulic radius: The hydraulic radius was given as

$$R = 0.15 \, Q_b^{0.41} Q_s^{-0.02} D_{50}^{-0.14}. \qquad (3.130)$$

Wetted perimeter: The wetted perimeter was reported as

$$P = 4.53 \, Q_b^{0.49} Q_s^{-0.0}. \qquad (3.131)$$

Riffle geometry: The meander arc length or riffle spacing (Z) was given as

$$Z = 6.31 \, w \qquad \text{vegetation types I-IV.} \qquad (3.132)$$

Sinuosity was found to be

$$S_u = \frac{S_y}{S} \qquad \text{vegetation types I-IV.} \qquad (3.133)$$

Riffle width was given as

$$r_w = 1.034 \, w. \qquad (3.134)$$

Riffle mean depth was reported as

$$r_d = 0.951\, d. \tag{3.135}$$

Riffle maximum depth was given as

$$r_{dm} = 0.912\, d_m. \tag{3.136}$$

Riffle velocity was reported as

$$r_v = 1.033\, v. \tag{3.137}$$

Likewise for pools, the following relations were obtained:

$$\text{width} \qquad p_w = 0.966\, w \tag{3.138}$$

$$\text{mean depth} \qquad p_d = 1.049\, d \tag{3.139}$$

$$\text{velocity} \qquad p_v = 0.967\, v \tag{3.140}$$

$$\text{maximum depth} \qquad p_{dm} = 1.088\, d_m \tag{3.141}$$

$$\text{median bed material size} \qquad p_{D_{50}b} = 0.81\, D_{50}. \tag{3.142}$$

These equations are valid for gravel-bed material size (D_{50}) ranging from 0.014 to 0.176 m; composite banks of fine cohesive alluvium overlying gravel; bankfull discharge (Q_b) ranging from 3.9 m^3/s to 424 m^3/s; bed material transport rate (Q_s) ranging from 0.001 kg/s to 114.14 kg/s; and relative roughness (d/D_{50}) from 5 to 90.

3.6 Applications

Extensive data used in the development of the regime theory point to the laws pertaining to the self-formation of regime-type channels. Blench (1951) indicated seven other applications, including (1) determination of width between incised river banks; (2) determination of scour between bridge piers; (3) determination of scour downstream of piers, along groins, and at spur heads; (4) aggradation upstream from reservoirs; (5) degradation downstream from reservoirs; (6) estimation of dredging; (7) determination of model scales; and (8) gravel bed response to environmental change. Following Blench (1951), these applications are briefly discussed here.

3.6.1 Width between Incised River Banks

Under the condition that a river is flowing full with maximum discharge like a canal and is fairly straight, the width between incised banks can be estimated using

(3.18), which is a special case of Equation (3.33). Equation (3.18) was employed for determining spans of several Indian railway bridges over large rivers and is now used to choose the spans of bridges and barrages in Indian plains rivers.

3.6.2 Scour between Bridge Piers

Under the condition that the river approach to a bridge is fairly straight, the depth of scour between piers can be estimated using Equation (3.34), which when combined with Equation (3.33) can be expressed as

$$d = \left(\frac{q^2}{F_B}\right)^{\frac{1}{3}}, \qquad (3.143)$$

where $q = Q/w$, which should be determined for the particular river. For example, it may be assumed that two thirds of the discharge passes through half the span of the bridge. For oblique flow, the depths are doubled for design.

3.6.3 Scour Downstream from Piers

For maximum flood discharge Q, the maximum scour depth can be estimated using Equation (3.34). Field data show that the exponent of Q as 1/3 provides a good fit. Of course, the values of empirical coefficients will vary from one location to another and with the type of structure.

3.6.4 Aggradation Upstream from Reservoirs

When there is an obstruction of flow through a dam reservoir, there will be a backwater buildup upstream like an M1 flow profile, where the flow depth will be greater than the normal depth and velocity will be smaller than the normal velocity for the same discharge. This will lead to the settlement of sediment, aggradation will therefore occur, and the reservoir will start filling up. This will disturb the river regime and raise bed slope, and the river will try to adjust to a new regime. When a barrage is built on a river, it is observed that riverbed rises parallel to the old slope.

3.6.5 Degradation Downstream from Reservoirs

When a dam reservoir is built on a river, the sediment is trapped and deposits in the reservoir, and the water flowing downstream of the reservoir is sediment starved and degrades the riverbed. The river then adjusts its bed to a lower level.

Degradation of the riverbed is observed at the location where there is a hydraulic jump on the spillway floor.

3.6.6 Estimation of Dredging

The Lacey theory does not apply to off-regime conditions. However, it does seem to yield good results if some kind of mean discharge is employed in place of bankfull or maximum discharge, as was reported by USBR (1948).

3.6.7 Model Scales

Equations (3.33)–(3.41) show that if the bed factor and side irregularities of the prototype are reproduced in the model then the scale ratios of coefficients will be unity. Then, one can write the scale ratios of width, depth, and slope, with subscript r denoting the ratio, as

$$w_r = Q^{\frac{1}{2}} \tag{3.144}$$

$$d_r = Q^{\frac{1}{3}} \tag{3.145}$$

$$S_r = Q^{-\frac{1}{6}}. \tag{3.146}$$

Equations (3.144) and (3.145) show that

$$(\text{depth scale}) = (\text{width scale})^{\frac{2}{3}}. \tag{3.147}$$

Equation (3.147) is empirical. If the bed material is different, then the coefficients may not be unity and an allowance will have to be made for their impact.

3.6.8 Meandering

Meander geometry is related to stream discharge. It is implied here that meander is free that develops in alluvium and is free to erode laterally without any valley restrictions and where waveforms are not distorted by heterogeneities in the alluvium. Meandering patterns are channel characteristics and meander geometry is characterized by wave length (M_l), meander belt width (M_{bw}), meander width (M_w), and distance from meander to meander (M_d) along the river axis. Meander width is synonymous with meander belt width. Then, two ratios are often defined: ratio of meander belt width (M_{bw}) to M_d and the ratio of meander belt width to stream width w. Jefferson (1902) stated that the meander wave length equaled twice the value of M_d. Carlston (1965) noted that this ratio was distinctly biased. Jefferson (1902) found the average value of ratio $(M_{bw}/2M_d)$ as 1.43 for 23 streams, which by taking inverse would lead to the average wave length value of 0.7 times the average meander width (M_w). Carlston (1965) found the meander width to be about 0.75 times the meander wave length.

The meander geometry has been related to mean annual discharge as $Q^{0.5}$. To define the influence of meandering on rivers, the meander width (M_w) and meander length (M_l) can be expressed as

$$M_w = C_w Q^{\frac{1}{2}} \tag{3.148}$$

$$M_l = C_l Q^{\frac{1}{2}}, \tag{3.149}$$

where Q is a representative discharge, usually maximum discharge; coefficients C_w and C_l are a function of F_B and F_S, and river discharge hydrograph.

Jefferson (1902) found the meander belt width for three rivers on flood plains as

$$M_{bw} = 75 Q^{0.5}, \tag{3.150}$$

where Q is the mean annual discharge. For the three rivers, the wave length was expressed as

$$M_l = (2M_d) = 46 Q^{0.5}. \tag{3.151}$$

For these rivers, stream width was expressed as

$$w = 4.5 Q^{0.5}. \tag{3.152}$$

For 23 rivers on flood plains, the mean ratio for M_{bw}/w was 17.6, and the mean ratio of $M_{bw}/(2M_d)$ or M_{bw}/M_l for 21 rivers was 1.43. Then

$$\frac{M_l}{w} = \frac{17.6}{1.43} = 12.6. \tag{3.153}$$

Inserting Equation (3.152) in Equation (3.153),

$$M_l = 57 Q^{0.5}. \tag{3.154}$$

Using Equation (3.150) and $M_{bw}/(2M_d)$ or $M_{bw}/M_l = 1.4$, the meander length becomes

$$M_l = \frac{75}{1.4} Q^{0.5} = 54 Q^{0.5}. \tag{3.155}$$

For 31 free meanders, Carlston (1965) found $M_{bw}/M_l = 0.75$. Using this value in Equation (3.150),

$$M_l = \frac{75}{0.75} Q^{0.5} = 100 Q^{0.5}. \tag{3.156}$$

However, using data from meanders and 31 gaging stations with mean annual discharge ranging from 31 cfs to 562,800 cfs, Carlston (1965) found

$$M_l = 106.1Q^{0.46} \tag{3.157}$$

and for bankfull discharge ($Q_b = Q_{1.5}$),

$$M_l = 8.2Q_b^{0.62}. \tag{3.158}$$

If the discharge was mean monthly maximum discharge (Q_{mmm}), then

$$M_l = 80Q_{mmm}^{0.46}. \tag{3.159}$$

The meander width for 31 samples was found to be

$$M_w = 65.8Q^{0.47}. \tag{3.160}$$

For the bankfull stage, which is wider than for mean annual discharge, the meander wave length was related to stream width as

$$M_w = 10.9w^{1.01}. \tag{3.161}$$

It was observed that a decrease in slope led to a decrease in meander wave length and width. The dominant discharge controls the free meander wave length and should therefore be collected accurately.

3.6.9 Channel Design

Channel design is a broad term that encompasses not only the design of a new channel but also the stabilization or realignment of an existing channel, and stable channel is implied in the design. Engineering measures, such as diversion of flow, construction of a dam, contraction of a channel or confining the channel between jetties or training walls to increase the navigable depth, closure of river forks, and cutoff, influence the characteristics of streams. The impact of these engineering measures determines the way the channel is to be stabilized or realigned.

A stable channel is defined as a channel whose planform, cross-section, and longitudinal profile are sustainable over time. It is also defined as a channel where aggradation and degradation of sediment are in balance or are negligibly small. In other words, there is no net scour or deposition of sediment and the channel may be self-sustaining. However, natural streams aggrade or degrade over time or in response to major storm events, causing floods and generating heavy sediment load. On the other hand, there can also be reasons for bed degradation. Channel design should take account of these changes occurring in the channel. Likewise, depending on the hydrologic changes occurring in the watershed, the channel can also migrate without necessarily losing its stable character. The magnitude and frequency of migration should be taken into consideration when designing a channel for reducing the migration.

There are a number of techniques available, depending on the objective of channel design, environmental considerations, type of channel, availability of data, and economic considerations. Not all techniques are applicable under all conditions, and each technique has its strengths and weaknesses. Some techniques are for fixed-bed channels and some for mobile-bed channels. The purpose of this chapter is to illustrate the methodology for channel design.

The choice of a design technique is also influenced by the duration of flow that occurs in the stream over the year. In hydrology, streams are classified as perennial, intermittent, or ephemeral. A perennial stream flows continuously; an intermittent stream, also called seasonal, flows only at certain times of the year when it receives water from some surface source; and an ephemeral stream flows only in direct response to precipitation. Osterkamp and Hedman (1982) define a perennial stream as a stream that has a measurable surface discharge more than 80% of the time. An intermittent stream flows continuously for periods of at least 30 days.

Design variables vary with the type of channel. For fixed-boundary channels, the independent hydraulic design variables are design discharge and channel roughness, but roughness can be a dependent variable if there is a choice of boundary materials. The objective here is efficient flow.

The dependent design variables are width, depth, and slope, which are adjusted to achieve the desired flow conveyance. The channel boundary is considered fixed at the design discharge, and bed-material sediment inflow is regarded as negligible.

For alluvial channels with mobile bed, the bed-material sediment inflow is significant. Since these channels adjust their geometry to efficiently transport sediment without significant degradation or aggradation, the objective is to achieve a state of dynamic equilibrium by selecting appropriate channel geometry. Independent design variables include inflow discharge hydrograph, bed-material gradation, streambank characteristics, and sediment inflow. The dependent design variables are width, depth, slope, and planform. Since hydraulic roughness is a function of bed material, bank roughness may be considered a dependent variable.

A threshold channel is designed for design discharge and a specified channel boundary material, so that the channel boundary is stable and channel dimensions of width, depth, and slope are obtained. Commonly used design methods are the allowable velocity method and the allowable shear stress method. The velocity method is usually applied for channels lined with different materials, while the shear stress method is applied for gravel-bed channels.

Table 3.5. *Methods of channel design (after Ackers, 1988)*

Authors	Year	Functions used	Comments/Remarks
Raymond	1951	1. Simplified Meyer-Peter sediment transport equation 2. Manning resistance equation 3. -	Graphical solution Width unspecified
Lane	1955	1. Continuity equation 2. Flow resistance equation 3. Bed threshold equation 4. Bank threshold equation	Analytical solution
Chien	1956	1. Einstein bed load transport function 2. Einstein–Barbarossa Equation 3. -	Graphical solution Suggested regime width
Chitale	1966	1. Laursen bed load transport function 2. Lacey/Strickler resistance equation 3. Limiting tractive force expression	Wide channel is assumed but was unchecked Design chart
Englund and Hansen	1967	1. Engelund–Hansen total load function 2. Engelund–Hansen resistance equation 3.	Design chart Lower regime bed form Width unspecified
Smith	1970 1974	1. Colby sand transport function 2. Modified Einstein–Brown resistance equation 3. Regime width	Computer program Confirmed by Lower Chenab canal data
Chang	1979	1. Various sediment transport equations 2. Various resistance equations 3. Minimum stream power concept	Lower and upper bed forms could yield dual solution Design charts
Ramette	1979 1980	1. Meyer-Peter bed load transport function 2. Strickler or Engelund–Hansen resistance equation 3. Maximum bed load and scour energy concept	Algebraic solution Extended to meandering and braided natural channels
Ackers	1980	1. Ackers–White equation sediment transport equation 2. White et al. resistance equation 3. Regime width	Numerical solution with graphical representation Confirmed by Lower Chenab data

Table 3.5. (*cont.*)

Authors	Year	Functions used	Comments/Remarks
White et al.	1981	1. Ackers–White sediment transport equation 2. White et al. resistance equation 3. Optimal principle	Tables developed from computer solution
Bakker et al.	1986	1. Ackers–White equation 2. van Rijn bed roughness equation 3. Lacey width with revised coefficients	Computer program, CADIS Confirmed by data from Pakistan canals (ACOP)

Alluvial channels are designed for design discharge so that it can transport the incoming water and sediment load and lead to design variables of width, depth, slope, and planform. The design method involves the use of resistance and sediment transport equations along with fluvial geomorphic principles.

In alluvial channels, bed and banks are composed of unconsolidated sediment particles that allow for a continual exchange of these particles with flow. The processes involved during the interaction between flow and channel boundary are erosion, sediment transport, sedimentation, and resuspension, and these are impacted by the inflow of water and sediment loads. It is these processes that determine the shape and size of the channel. There are a variety of methods for designing alluvial channels, which can be classified as regime, analogy, hydraulic geometry, extremal, and analytical methods.

The preceding discussion on physical processes shows that regime equations for alluvial channels result from three physical process functions, which lead to improved methods for irrigation canal design, as shown in Table 3.5 (Ackers, 1988).

3.6.10 Gravel-Bed River Response to Environmental Change

Eaton and Millar (2017) discussed the strengths and limitations of a regime approach to predicting gravel-bed response to environmental change, such as land use changes, construction of hydropower dams, and climate change. The principal idea to characterize channel response is the concept of grade change, which entails a change in channel configuration owing to a change in sediment supply and governing conditions of hydrological regime. They used the UBC (University of British Columbia) regime model, which is based on bank stability criterion and pattern stability criterion. Subject to these constraints, the model predicted both stable channel cross-sectional dimensions and the number of anabranches that the stream must have. The model can be employed to quantify the response of gravel bed streams to changes in flow regime, riparian vegetation, and sediment supply.

References

Ackers, P. (1964). Experiments on small streams in alluvium. *Journal of the Hydraulics Division*, Vol. 90, No. 4, pp. 1–37.

Ackers, P. (1980). *Use of sediment transport concepts in stable channel design*. 1st Workshop on Alluvial River Problems, Roorkee, India.

Ackers, P. (1983). *Sediment Transport Problems in Irrigation System Design. Developments in Hydraulic Engineering-I*, edited by P. Novak, Applied Science Publishers, London, pp. 151–195.

Ackers, P. (1988). Alluvial channel hydraulics. *Journal of Hydrology*, Vol. 100, pp. 177–204.

Ackers, P. (1992). 1992 Gerald Lacey Memorial Lecture: Canal and river regime in theory and practice: 1929–92. *Proceedings, Institution of Civil Engineers, Water, Maritime and Energy*, Vol. 96, pp. 167–176.

Ackers, P. and Charlston, F. G. (1970a). The geometry of small meandering streams. *Proceedings, Institution of Civil Engineers, Supplement Xii*, 289.

Ackers, P. and Charlston, F. G. (1970b). Meander geometry arising from varying flows. *Journal of Hydrology*, Vol. 11, No. 3, pp. 230–252.

Bakker, B., Vermas, H., and Choudri, A. M. (1986). *Regime theories updated or outdated*. International Commission on Irrigation and Drainage, Darves Bornoz Special Session, Lahore, Report 9.

Bettess, R. and White, W. R. (1983). Meandering and braiding of alluvial channels. Proceedings, Institution of Civil Engineers, Part 2, pp. 525–538.

Blench, T. (1951). Regime theory for self-formed sediment-bearing channels. *Proceedings, ASCE*, Vol. 77, Separate 70, pp. 1–18.

Blench, T. (1952). Regime theory for self-formed sediment bearing channels. *Transactions, American society of Civil Engineers*, Vol. 117, pp. 383–408.

Blench, T. (1957). *Regime Behavior of Canals and Rivers*. 138 pp., Butterworths, London.

Bose, N. K. (1936). *Silt movement and design of channels*. Punjab Engineering Congress, Paper 192.

Bray, D. L. (1982a). Flow resistance in gravel bed rivers. In: *Fluvial Processes, Engineering and Management*, edited by Hey et al., John Wiley, Chichester, pp. 109–133.

Bray, D. L. (1982b). Regime equations for gravel-bed rivers. In: *Gravel-Bed Rivers*, edited by R. D. Hey, J. C. Bathurst, and C. R. Thorne, John Wiley, Chichester, pp. 517–542.

Brice, J. C. (1960). Index for description of channel braiding. *Geological Society of America Bulletin*. Vol. 71, p. 1833.

Carlston, C. W. (1965). The relation between free meander geometry to stream discharge and its geometric implications. *American Journal of Science*, Vol. 263, pp. 864–885.

Chang, H. H. (1979). Maximum stream power and channel patterns. *Journal of Hydrology*, Vol. 41, pp. 303–327.

Charlton, F. G., Brown, P. M., and Benson, R. W. (1978). *The hydraulic geometry of some gravel bed rivers in Britain*. Report IT 180, Hydraulics Reseacrh Station, Wallingford.

Chien, N. (1956). Graphic design of alluvial channels. *Transactions of ASCE*, Vol. 121, pp. 1267–1287.

Chien, N. (1957). A concept of the regime theory. *Transactions, American Society of Civil Engineers*, Vol. 122, pp. 785–793.

Chitale, S. W. (1966). *Design of Alluvial Channels*. ICID, Delhi, Q20, E17.

Desloges, J. R. and Church, M. A. (1989). Canadian landform examples-13. *The Canadian Geographer*, Vol. 33, No. 4, pp. 360–264.

Eaton, B. and Millar, R. (2017). Predicting gravel bed river response to environmental change: The strengths and limitations of regime-based approach. *Earth Surface Processes and Landforms*, Vol. 42, pp. 994–1008.

Engelund, F. and Hansen, E. (1967). *A monogram on sediment transport in alluvial streams*. Teknisk Verlag, Copenhagen.

Griffiths, G. A. (1980). Hydraulic geometry relationships of some New Zealand gravel bed rivers. *Journal of Hydrology (New Zealand)*, Vol. 19, No. 2, pp. 106–118.

Hey, R.D. (1975). Geometry of river meanders. *Nature*, Vol. 262, pp. 482–484.

Hey, R.D. (1978). Determinate geometry of river channels. *Journal of Hydraulics Division, ASCE*, Vol. 104, No. HY6, pp. 868–885.

Hey, R. D. and Thorne, C. R. (1986). Stable channels with mobile gravel beds. *Journal of Hydraulic Engineering, ASCE*, Vol. 112, No. 3, pp. 671–689.

Inglis, C. C. (1947). *Meanders and their bearing in river training*. Maritime Paper No. 7, Institution of Civil Engineers, London.

Jefferson, M. S. W. (1902). Limiting width of meander belts. *National Geography Magazine*, Vol. 13, pp. 373–384.

Kellerhals, R. (1967). Stable channels with gravel-paved beds. *Journal of the Waterways and Harbors Division, ASCE*, Vol. 93, No. WW1, pp. 63–84.

Kellerhals, R. (1976). Stable channels with gravel paved beds. *Journal of the Waterways and harbors Division, ASCE*, Vol. 83, No. WW1, pp. 63–83.

Kellerhals, R., Church, M., and Bray, D. I. (1976). Classification and analysis of river processes. *Journal of Hydraulics Division, ASCE*, Vol. 107, No. HY7, pp. 813–829.

Kennedy, R. G. (1895). On the prevention of silting in irrigation canals. *Proceedings, Institution of Civil Engineers*, Vol. 119, pp. 281–290.

Kondap, D. M. and Garde, R. J. (1979). Design of stable channels. ICOLD Special Issue, *Journal of Irrigation and Power*, Vol. 36, No. 4, October.

Lacey, G. (1929–1930). Stable channels in alluvium. *Proceedings, Institution of Civil Engineers*, Vol. 229, pp. 259–384.

Lacey, G. (1933–1934). Uniform flow in alluvial rivers and canals. *Minutes of the Proceedings, Institution of engineers, London*, Vol. 237, pp. 421–453.

Lacey, G. (1946). A general theory of flow in alluvium. *Journal of Institution of Civil Engineers, London*, Vol. 27, pp. 16–47.

Lacey, G. (1947). A general theory of flow in alluvium. *Journal of Institution of Civil Engineers, London*, Vol. 28, pp. 425–451.

Lacey, G. (1957). Discussion of "A concept of the regime theory, by N. Chien, (1957)." *Transactions, American Society of Civil Engineers*, Vol. 122, pp. 794–797.

Lacey, G. (1957–1958). *Flow in alluvial channels with sand be mobile beds*. Proceedings, Institution of Civil Engineers, London, Vol. 9, pp. 146–164.

Lacey, G. and Pemberton, W. (1972). A general formula for uniform flow in self-formed alluvial channels. *Proceedings, Institution of Civil Engineers, Part 2*, Vol. 53, pp. 373–381.

Lane, E. W. (1937). Stable channels in erodible materials. *Transactions, ASCE*, Vol. 102, pp. 123–142.

Lane, E. W. (1955). Design of stable channels. *Transactions, ASCE*, Vol. 120, No. 2776, pp. 1234–1279.

Leopold, L. B. and Wolman, M. G. (1957). *River channel patterns, braided, meandering and straight*. U.S. Geological Survey, Washington, DC.

Leopold, L. B. and Wolman, M. G. (1960). River meanders. *Bulletin of the Geological Society of America*, Vol. 71, pp. 760–794.

Lindley, E. S. (1919). Regime channels. *Proceedings, Punjab Engineering Congress*, Vol. 7, p. 63.

Malhotra, (1939). *Annual Report (Technical)*. Central Board of irrigation, India, 1939–40 (reported in Inglis, 1948).

Mollard, I. D. (1973). *Airphoto interpretation of fluvial features*. Proceedings of the 9th Canadian Hydrology Symposium, Edmonton, Alberta; National Research Council of Canada, Associate Committee of Hydrology, Ottawa, pp. 341–380.

Nixon, M. (1959). A study of bankfull discharge of rivers in England and Wales. *Proceedings, Institution of Civil Engineers*, Vol. 12, pp. 157–174.

Osterkamp, W. R. and E. R., Hedman (1982). *Perennial-streamflow characteristics related to channel geometry and sediment in Missouri River basins, U.S. Geol.* Survey Prof. Paper 1242, 37 pp., Washington, DC.

Ramette, M. (1979). Une approche rationale de la morphologie fluviale. *Houivve Blanche*, Vol. 34, No. 8, pp. 491–498.

Ramette, M. (1980). A theoretical approach to fluvial processes. *Proceedings, International Symposium on River Sedimentation, Chinese Society of Hydraulic Engineering*, Beijing, pp. 601–622.

Ranga Raju, K. G. and Garde, R. J. (1988). Design of stable canals in alluvial material. *International Journal of Sediment Research*, Vol. 3, No. 1, pp. 10–37.

Vol. 38, No. 2, pp. 191–194.

Raymond, J. P. (1951). *Study of saturated rivers by means of geographical representation*. IAHR, Report 4th Meeting, Bombay.

Schumm, S. A. (1963a). Sinuosity of alluvial rivers on the Great Plains. *Geological Society of America Bulletin*, Vol. 74, pp. 1089–1106.

Schumm, S. A. (1963b). *A tentative classification of alluvial river channels*. U.S. Geological Survey Circular 47, Washington, DC.

Schumm, S. A. (1971). Fluvial geomorphology: The historical perspective. Chapter 4 in *River Mechanics*, Vol. 1, edited by H. W. Shen, Fort Collins, CO.

Simons, D. B. and Alberton, M. L. (1960). Uniform water conveyance channels in alluvial material. *Journal of Hydraulics Division, ASCE*, Vol.85, No. HY5, pp. 33–71.

Simons, D. B. and Alberton, M. L. (1963). Uniform water conveyance channels in alluvial material. *Transactions, ASCE*, Vol. 128, Part 1, No. 3399, pp. 65–167.

Simons, D. B., Lagasse, P. F., Chen, Y. H., and Schumm, S. A. (1975). *The river environment: A reference document, prepared for U.S. Department of the Interior, Fish and Wildlife Service*, Twin Cities, MN.

Smith, K. V. H. (1970). Similarity in unlined irrigation canals systems. *Proceedings, of Hydraulics Division, ASCE*, Vol. 96, No. HY1, p. 123.

Smith, K. V. H. (1974). *Comparison of prediction techniques with records of observations on the Lower Chenab canal system*. University of Southampton.

Stevens, M. A. and Nordin, C. F. (1987). Critique of the regime theory for alluvial channels. *Journal of Hydraulic Engineering*, Vol. 113, No. 11, pp. 1350–1380.

Tarar, R. N. and Choudri, A. M. (1979). *Behavioral evaluation of some Pakistan canals: Part1. Presented at the June 26–29, International Symposium on Mechanics of Alluvial Channels, held in Lahore*, Pakistan.

USBR (1948). *Proceedings, Federal Interagency Sedimentation Conference.* U.S. Bureau of Reclamation, Washington, DC, p. 114.

Vanoni, V. A. (1975). *Sedimentation Engineering. American Society of Civil Engineers (ASCE)* Manual 54, 746 pp., ASCE Press, Reston, VA.

White, W. R., Paris, E., and Bettess, R. (1981). River regime based on sediment transport concepts. Hydraulics Research Station, Wallingford, Report IT 201.

Wolman, M. G. (1955). *The natural channel of Brandywine Creek, Pennsylvania.* Geological Survey Professional Paper 271, pp. 1–56, U.S. Department of the Interior, Washington, DC.

Yang, C. T. (1971). On river meanders. *Journal of Hydrology*, Vol. 13, pp. 231–253.

4

Leopold–Maddock (LM) Theory

The average river-channel system tends toward an approximate equilibrium between the channel and water and sediment it transports. Both discharge and sediment load principally depend on the drainage basin. Under equilibrium, the stream channel depth, width, velocity, and suspended sediment load at a given cross-section can be expressed as power functions of discharge, and these functions constitute the at-a-station hydraulic geometry (AHG). In a similar vein, stream channel depth, width, flow velocity, and suspended load along the river vary with discharge as simple power functions under the condition that the frequency of discharge at all cross-sections is equal. These functions are similar even for rivers having very different physiography and constitute the theory of downstream hydraulic geometry (DHG). The power functions for both types of hydraulic geometry – at-a-station and downstream – form the Leopold and Maddock (LM) (1953) theory, which is discussed in this chapter. The discussion is divided according to the type of geometry.

For the same discharge frequency along the river, depth, width, and velocity of flow increase with discharge downstream. The increase in velocity downstream results, despite the decreasing particle size downstream, from the increase in depth, which overcompensates for the decrease in slope. Further, depth, width, and velocity are functions of sediment load transported by the channel. The suspended load per unit volume of water tends to decrease slightly downstream. These functions determine the changes in the channel shape downstream, and the average channel system thus tends towards an approximate equilibrium between the channel shape and the transport of water and sediment.

Notation

a = constant of proportionality
A = cross-sectional area
A_c = stereotype cross-sectional area
AHG = at-a-station hydraulic geometry
a_0 = coefficient
b = exponent
B = bed width
c = constant of proportionality
d = mean depth
D = sediment particle size
DHG = downstream hydraulic geometry
d_e = depth of expected channel
d_F = depth of former channel
d_m = maximum depth of flow
d_m^b = bankfull maximum water depth in cross-section
f = exponent
F_s = sediment factor
h = exponent
h_r = roughness height
J = slope of suspended sediment curve
k = constant of proportionality
K = generalized conductance coefficient
l = wave length
l_F = former wave length
m = exponent
n = roughness parameter
p = numerical constant
P = wetted perimeter
p_0 = exponent
q_0 = exponent
r = exponent
R = hydraulic radius
r_B = silt-clay content of banks
r_M = silt-clay content of perimeter
Q = discharge
Q_0 = mean observed discharge
Q_e = mean stereotype discharge
Q_s = sediment transport rate
$Q_{2.33}$ = mean annual flow

Continued

Q_s = sediment load
S = slope of the energy grade line $\cong$ slope of the river profile.
S_e = stereotype slope
S_o = observed slope
SD = standard deviation
v = mean velocity
w = water-surface width
w^b = bankfull width
x = cross-section distance from the channel center
z = vertical distance to the bed
φ = exponent

4.1 Introduction

Hydraulic geometry involves relations of width, depth, velocity, slope, and roughness, which are primarily expressed in terms of discharge, although sediment size and sediment transport do affect slope as well roughness. These relations are of fundamental value for channel design, including canals for irrigation; river training works; stream restoration; and so on. In general, one can regard discharge, sediment size, and sediment load as independent variables. Then, width, depth, velocity, slope, and roughness can each be expressed as a function of these independent variables. Depending on the data representing different types of materials, ranges of discharge, and ranges of sediment load, different functions have been derived, but the underlying concept remains the same. These functions have been derived empirically as well as conceptually. Leopold and Maddock (1953) considered hydraulic geometry of natural rivers as two types, (1) at-a-station hydraulic geometry and (2) downstream hydraulic geometry, and developed an empirical theory that constitutes the focus of this chapter.

4.2 At-a-Station Hydraulic Geometry

In their classic work, Leopold and Maddock (1953) showed that at a given cross-section, stream channel characteristics defined by depth, width, velocity, and suspended sediment load are related to discharge as simple power functions under approximate equilibrium, which seems to exist even in headward upgraded tributaries and at a given cross-section for all discharges up to the bankfull discharge. The power functions are such that an increase in discharge results in an increase of each of the variables, because river cross-sections tend to be approximately rectangular, trapezoidal, triangular, or semi-elliptical. Using field measurements of average width, depth, velocity, discharge, and suspended-sediment load, it was shown that width, depth, and velocity were functions of suspended load, which was regarded as an index of the total sediment load. For a given discharge, an increase in width at a constant velocity corresponded to a decrease in suspended sediment load. This is because the flow depth would decrease if the slope and roughness remained the same and would carry less suspended sediment. On the other hand, for constant discharge and velocity, an increase in width corresponded to an increase in bed load. This is because there is greater bed surface in contact with water, and reduced depth would be prone to greater settlement of sediment. At a given stream cross-section, suspended-sediment load per unit of discharge increases rapidly with increased discharge.

It was further postulated that both discharge and sediment load were independent of the stream channel and depended on the nature of the drainage basin. These characteristics are important determinants of the shape of a channel cross section, and the relation between hydraulic characteristics of the channel and suspended load provides a logical explanation of the observed channel shape. Field data showed that these relations were similar for several rivers and applied to both rivers and stable (regime, meaning neither scouring nor aggrading) canals. The implication is that the average river-channel system tends toward an approximate equilibrium between the channel and water and sediment it transports. This chapter discusses the empirical LM theory and draws heavily from Leopold and Maddock (1953).

4.2.1 Geomorphologic Data and Their Measurement

For deriving their hydraulic geometry relations, Leopold and Maddock (1953) assembled concurrent measurements of mean velocity, width, shape and area of the cross section, and suspended sediment as well as discharge, where stage or gage height defined by the elevation of water surface above an arbitrary datum, velocity, and width were measured. It may be pertinent to learn about the measurement of data they used. The depth and velocity were measured at various distances from the bank, and the cross-sectional area of flow was determined from these depth measurements and their spacing. The depth was the mean depth associated with the particular rate of discharge and was determined by dividing the measured cross-sectional area of flow by the corresponding width of water surface. This provided a mean depth equal to the depth of a rectangular channel having the same surface width as the actual channel at the given stage and the same cross-sectional area. It was noted that the suspended-load data were primarily derived from streams in the

Western United States. Velocity was the quotient of discharge divided by the cross-section area of flow and was the mean velocity of the cross-section. A large volume of data was available for this mean velocity, although this may be less than meaningful for sediment transport. Sediment was sampled by a depth integrating suspended load sampler.

A gaging station at a cross-section was installed at a location where the relation of discharge to a stage was as stable as possible and discharge was plotted against gage height as a smooth curve. The gages were, thus, installed either upstream from a "section control" where critical velocity was achieved over some outcrop or gravel bar, or in a reach of channel where the channel properties determined the stage-discharge relation, that is, a relatively straight and uniform reach generally devoid of unusual riffles on bars.

In periods of low flow, current-meter measurements were normally made by wading. Otherwise, current-meter measurements were often made from bridges. The cross-section of a stream of a gaging station above a control section was not representative of an average reach of channel, especially high discharges. More representative sections were found at gaging stations having channel control. These sections were often equipped with a cable car from which the current meter may be operated. The data were collected up to bankfull discharge. Data exhibiting usual effects introduced by construction, such as bridge abutments, were excluded.

4.2.2 Bankfull Discharge

Bankfull discharge is often regarded as the channel-forming discharge and corresponds to the bankfull depth. This discharge occurs with a frequency of about 1.5 (Leopold, 1994; Johnson and Heil, 1996). It is used in deriving downstream hydraulic geometry, design of stream restoration, design of highway crossing, stream planform, and classification of streams. Intuitive as it may appear, determination of bankfull disharge in the field is not easy and is subjective, because it depends on the determination of bankfull depth, which is sensitive to the selection of bankfull elevation, which in turn is defined as the elevation of the active floodplain surface. However, Leopold and Skibitzke (1967) noted that they were able to satisfactorily estimate bankfull discharge from a single boat trip down the river. There is subjectivity in defining bankfull depth and bankfull discharge, which results in a range of values, depending on the observer.

Gordon et al. (1992) discussed examples where it was difficult to accurately define bankfull elevation: (1) where the bank tops are not at the same elevation, (2) where streams are unstable, (3) where a break between stream banks and floodplain is not clear, and (4) where complexities, such as benches and terraces,

Table 4.1. *Return periods of bankfull discharge*

Location	Return period of Q_{bf}	Source
England and Wales	0.46	Nixon (1959)
Belgium and surrounding sites	0.7–5.3	Petit and Pauquet (1997)
Yampa River basin, Northwest Colorado and Southeast Wyoming	1.18–1.40 (mode)	Andrews (1980)
Queensland, Australia	1.1–1.85	Dury et al. (1963)
Channels with active floodplain, Western United States	1.32 (mode = 1.5)	Williams (1978)
Pacific northwest (Oregon, Washington, and Idaho)	1.4 (average)	Castro and Jackson (2001)
Eastern half of United States	1.5 (average)	Leopold et al. (1995)
North America	1.58	Dury (1981)
Perennial streams in Green River basin, Wyoming	1.7 (median)	Lowham (1982)
Central Pennsylvania	2.3	Brush (1961)
Cumberland Basin in Eastern New South Wales	4–10	Pickup and Warner (1976)

occur. In such instances, interpretation of bankfull depth can lead to different estimates of bankfull discharge. Different methods have been used to determine bankfull depth and bankfull discharge. Williams (1978) identified eight methods for determining bankfull discharge at cross-sections and eight methods for a channel reach, and discussed sources of error for each method. Using the flood frequency method, Leopold (1994) determined that bankfull discharge had a recurrence interval of 1.5 years, whereas Dury (1976) determined the recurrence interval to be 1.58 years. For urbanized areas, the bankfull discharge can occur more frequently with a return period of 1.5 years. Some have found the bankfull discharge to have a recurrence interval of 2 to 5 years. Wilkerson (2008) compiled published bankfull discharge return periods that ranged from 0.46 years to 10 years, depending on the location, as shown in Table 4.1, and discussed a method for bankfull discharge prediction. Johnson and Heil (1996) discussed uncertainties in estimating bankfull conditions.

4.2.3 Effective Discharge

The effective discharge is defined as the discharge that transports the maximum fraction of annual sediment load over a period of years. The effectiveness of discharge over a period of years is determined by its magnitude and frequency (Wolman and Miller, 1960). The relative significance of a given discharge is measured by its sediment transport rate (Q_s), which varies with discharge (Q) raised to a power as

$$Q_s \propto Q^\varphi, \tag{4.1}$$

where exponent φ is greater than one. This implies that if discharge is increased at a specific rate, the sediment transport rate will increase at a greater rate. The frequency of effective discharge varies from one stream to another, varies longitudinally along the stream, and decreases as the frequency of large floods increases. Andrews (1980) computed effective and bankfull discharges in the Yampa River basin of Colorado and Wyoming and found that the effective discharge had a recurrence interval of 1.18 to 3.26 years for annual flood series. He also found that the effective discharge and the bankfull discharge were nearly equal.

Goodman (2004) presented different methods for determining effective discharge, including field observations; sediment transport rating curve; direct method; sediment rating curve and fitted frequency curve of flow; sediment predictor equation, constant width, and fitted frequency distribution of flow; sediment predictor equation, variable width, and fitted frequency distribution of flow (Nash, 1994).

4.2.4 Frequency of Discharge

The mean daily discharge was the average of all rates of water flow for a day, and was expressed in cubic meters per second or cubic feet per second. The frequency of a daily mean discharge of a given river gaging station was determined by counting the number of occurrences of the rate throughout the period of record. The number of occurrences was usually expressed in terms of cumulative frequency or flow-duration curve, as shown in Figure 4.1 [USGS 08116650 Brazos River near Rosharon, Texas; drainage area: 45,339 square miles; contributing drainage area: 35,773 square miles; datum of gage: 0 feet above NGVD29]. The data for constructing Figure 4.1 are given in Table 4.2. To construct such a curve, the lowest and the highest values of discharge were first noted. Then, the range of discharge values was divided into a convenient number of classes, say 5 to 10. The number of days in which the measured daily mean discharge belonged to each class was counted. The numbers in each class were then accumulated from the largest discharge class to the smallest, and each cumulative figure was divided by the total number of days of record. The resulting quotients represented the percent of time given discharges were equaled or exceeded at the station. The discharge was on the vertical axis (ordinate) on the log scale and percent of time flow equaled or exceeded discharge indicated was on the horizontal axis (abscissa).

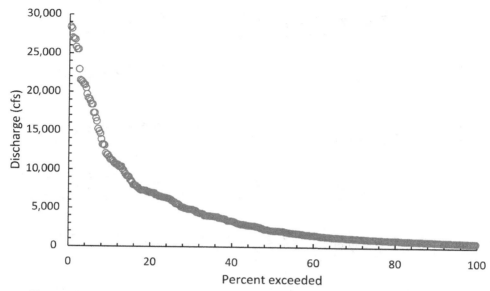

Figure 4.1 Cumulative frequency curve of equaled or exceeded discharge (or flow duration curve)

The figures representing the percent of time were then plotted against discharge to provide the flow-duration curve. Suppose a discharge of 1,000 cubic meters per second (CMS) or more occurred 10% of the time. That meant that through a long period, 10% of the days, or on average 10 days in every 100 days, experienced a discharge of 1,000 CMS or more.

The cumulative frequency, or duration, of a given daily discharge was expressed as the percent of time that this time was equaled or exceeded. Discharge rates of the same cumulative frequency might differ from one stream to another. For example, a point on a large river and a point on a small tributary may have the same frequency but the discharge in the tributary may only be a small fraction of the discharge in the large river.

The median discharge was discharge that was equaled or exceeded 50% of the time, that is, half the days had flow greater than the median, and half the days had less flow. The mean annual discharge was generally greater than the median discharge, and the mean rate of discharge was equaled or exceeded on somewhat less than half the number of days at most points on a river.

Now consider the variation of discharge from a low value to a high value at a given river cross-section, and the variation of discharge at different cross-sections downstream. To that end, consider two cross-sections, say one and two, along the river. Leopold and Maddock (1953) observed that the width and depth of the channel increased at a given river cross-section with increase in discharge, and

Table 4.2. *Daily discharge (cfs) from January 1, 2020 to December 31, 2020*

Day	Jan.	Feb.	Mar.	Apr.	May.	Jun.	Jul.	Aug.	Sep.	Oct.	Nov.	Dec.
1	763	1,410	3,760	27,000	7,100	17,300	1,840	2,760	749	3,920	804	2,800
2	806	1,320	4,060	25,500	6,550	13,200	1,540	3,310	799	3,380	912	2,080
3	782	1,230	4,010	22,900	6,440	9,890	1,290	1,960	768	2,970	924	2,270
4	749	1,160	3,960	21,000	7,770	8,370	1,110	2,160	705	2,730	852	2,120
5	765	1,150	3,900	21,400	7,340	7,400	1,060	2,300	642	2,740	832	1,650
6	811	1,160	3,600	21,500	6,100	6,500	996	1,820	1,090	3,200	868	1,360
7	876	1,120	3,130	18,400	4,830	5,560	953	1,420	11,900	3,030	854	1,280
8	878	1,050	2,830	15,200	4,040	4,420	838	1,190	20,900	2,640	831	1,200
9	876	1,000	5,150	13,200	3,660	3,910	784	1,030	20,500	2,260	780	1,120
10	868	989	7,050	11,600	3,960	3,740	736	993	14,600	1,700	720	1,050
11	1,140	945	6,980	10,400	4,010	3,350	714	855	9,590	1,590	669	924
12	1,250	915	6,160	11,300	3,620	2,660	697	770	6,560	1,740	691	996
13	1,180	911	4,690	10,600	3,380	2,080	803	600	4,890	1,520	682	2,060
14	1,200	909	3,700	9,400	2,920	1,740	2,100	597	4,690	1,780	702	7,100
15	1,290	926	3,670	10,700	2,910	1,470	2,110	580	6,290	2,070	667	3,250
16	1,360	1,110	4,990	18,800	6,350	1,290	1,740	599	7,510	1,970	633	1,720
17	1,300	7,030	5,580	21,200	4,890	1,170	1,930	682	7,310	2,110	594	1,340
18	1,190	8,690	5,480	19,100	3,990	1,090	2,590	705	6,750	1,890	637	1,110
19	1,520	6,650	4,760	16,600	4,030	1,030	2,900	675	6,450	1,610	652	1,260
20	1,430	4,860	4,300	14,900	4,360	1,040	2,990	653	6,410	1,620	620	4,970
21	1,250	3,980	4,990	13,100	5,610	1,060	2,660	628	7,310	1,700	604	2,930
22	1,190	3,500	5,780	11,200	6,970	1,620	2,450	661	16,200	1,480	587	1,790
23	2,780	3,360	6,320	10,700	6,730	2,720	2,430	799	17,300	1,300	592	1,390
24	2,550	4,460	10,400	11,000	5,940	2,280	2,050	776	13,900	1,030	681	1,330
25	1,520	9,060	19,200	11,900	6,240	4,320	2,050	724	11,300	955	703	1,380
26	1,420	9,160	25,500	12,100	7,330	10,100	2,310	839	7,940	980	738	1,210
27	1,520	7,010	26,900	10,800	8,000	7,350	2,120	1,280	5,880	908	843	1,050
28	1,510	5,130	25,800	9,190	7,960	4,380	1,780	813	5,350	845	3,380	994
29	1,630	4,070	26,800	8,530	10,400	2,760	1,830	663	4,840	796	7,180	967
30	1,500		28,400	7,730	18,400	2,210	2,900	610	4,320	722	5,130	969
31	1,450		28,200		19,700		1,910	655		759		2,080

variations in discharge followed a pattern in time that was characteristic of the position of the cross-section and the river. Different discharges at a given cross-section varied in frequency.

Under the condition of low discharge, it was postulated that every point along the river was experiencing a discharge that was small for that point; or, if all points along a river system were experiencing a relatively low discharge, the frequency of discharge at any one point was about the same as the frequency at any other point. Of course, the discharge, not its frequency, would generally be much greater near the mouth of a river draining a large area than at some point upstream. The same postulate was made for the condition of high discharge.

The term "at a station" referred to a given station where different discharges had different frequencies. The term "in a downstream direction" referred to cross-sections situated along a stream. "Change of discharge downstream" meant different discharges of the same frequency at cross-sections situated along the stream, but the discharges at the cross-sections were different. The differences in discharge both at a station and in a downstream direction were associated with differences in width, depth, and velocity.

4.2.5 Variations in Hydraulic Characteristics

The characteristics of channel shape at certain river cross-sections were examined, and then at a given cross-section, the width and depth of the channel and, therefore, the change in velocity with the amount of water flowing in the cross-section were analyzed. The changes in width, depth, and velocity in response to changes in discharge were observed to have certain characteristics that applied to many natural river cross sections. The river-channel characteristics, including slope and roughness, in relation to observed data on suspended-sediment load shed light on the interaction of hydraulic characteristics related to channel shape and sediment load and are therefore useful for the study of fluvial processes. The relations between observed channel characteristics and suspended-sediment load involved the process of adjustment between channel shape and sediment load, because the mechanism of adjustment between the capacity of a stream to carry a sediment load and available load was important for understanding graded rivers.

The width was plotted against discharge on a log-log paper, and depth and velocity were plotted against discharge on a log-log paper, as shown in Figure 4.2. Data for Figure 4.2 are given in Table 4.3. The plots of width, depth, and velocity against discharge were nearly straight lines and this was typical of a large number of gaging stations, but there was a considerable scatter of points on both sides of the straight lines. Part of the scatter could be attributed to the temporary scour and fill of the stream bed. At lower discharges measured by wading rather than from a

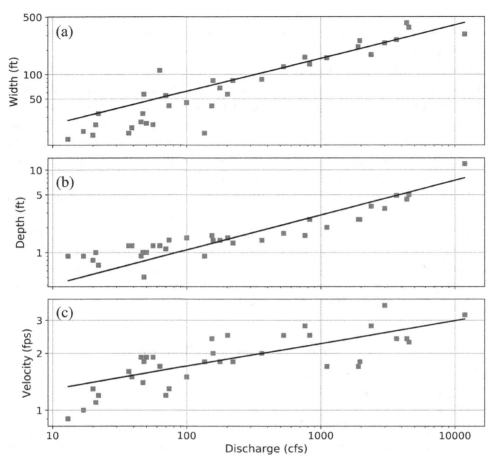

Figure 4.2 (a) Plot of width against discharge; (b) plot of depth against discharge; and (c) plot of velocity against discharge (after Leopold and Maddock, 1953)

bridge or cable, part of the scatter of points may be because successive measurements were not made at exactly the same location in the cross-section.

The relation of hydraulic geometry of natural river cross-section to discharge was then described as

$$w = aQ^b \tag{4.2}$$

$$d = cQ^f \tag{4.3}$$

$$v = kQ^m, \tag{4.4}$$

where Q is the discharge; w is the water-surface width; d is the mean depth; v is the mean velocity, and b, f, and m are the exponents, and a, c, and k are the constants of

Table 4.3. *Hydraulic data for Figure 4.2*

Station	Mean annual discharge (cfs)	Width (ft)	Mean velocity (fps)	Mean depth (ft)
Yellowstone River at Corwin Springs, Montana	2,999	242	3. 6	3.4
Yellowstone River at Billings, Montana	6,331	2.53	4.8	6.2
Wind River near Dubois, Wyoming	177	68	1.8	1.4
Wind River near Burris, Wyoming	828	134	2.5	2.5
Wind River near Crowheart, Wyoming	1,315	155	3. 5	2.4
Wind River at Riverton, Wyoming	1108	159	1.7	2
Pop Agle River near Riverton, Wyoming	683	123	3.5	3.3
Bighorn River at Thermopolis, Wyoming	1908	218	1.7	2.5
North Fork Owl Creek near Anchor, Wyoming	16.2	17	1.9	0.6
Owl Creek near Thermopolis, Wyoming	45.5	26	1.9	0.9
Gooseberry Creek near Grass Creek, Wyoming	20.8	19.5	3	0.6
Bighorn River at Manderson, Wyoming	1960	258	1.8	2.5
Medicine Lodge Creek near Hyattville, Wyoming	38	24	3	0.9
Graybull River at Meeteetse, Wyoming	364	87	2	1.4
Graybull River near Basin, Wyoming	200	77	3.8	1.3
Bighorn River at Kane, Wyoming	2,383	175	2.8	3.6
Bighorn River near St. Xavier, Montana	3,676	264	2.4	4.9
Bighorn River near Hardin, Montana	4,535	371	2.3	5
Little Bighorn River at Stateline, Wyoming and Montana	154	41	2.4	1.6
Little Bighorn River below Pass Creek near Wyola, Montana	202	57	2.5	1.5
Little Bighorn River near Crow Agency, Montana	294	96	3.5	1.2
Bighorn Rive near Custer, Montana	4,390	420	2.4	4.4
Yellowstone River near Sidney, Montana	11,860	308	3. 2	11.9
Red Fork near Barnum, Wyoming	50	25	1.9	1

Table 4.3. (cont.)

Station	Mean annual discharge (cfs)	Width (ft)	Mean velocity (fps)	Mean depth (ft)
Middle Fork Powder River above Kaycee, Wyoming	74	41	1.3	1.4
Middle Fork Powder River near Kaycee, Wyoming	157	84	2	1.4
North Fork Powder River near Haselton, Wyoming	13	16	0.9	0.9
North Fork Powder River near Mayoworth, Wyoming	39	22	1.5	1.2
South Fork Powder River near Kaycee, Wyoming	48	57	1.8	0.5
Powder River at Sussex, Wyoming	136	19	1.8	0.9
Middle Fork Crazy Woman Creek near Greub, Wyoming	20	18	1.3	0.8
North Fork Clear Creek near Buffalo, Wyoming	21	24	1.1	1
North Fork Clear Creek near Greub, Wyoming	22	33	1.2	0.7
Crazy Woman Creek near Arvada, Wyoming	63	112	1.7	1.2
Powder River at Arvada, Wyoming	434	23	3.2	0.8
North Fork Crazy Woman Creek near Buffalo, Wyoming	16	42	1.8	0.9
North Fork Crazy Woman Creek near Greub, Wyoming	17	20	1	0.9
Rock Creek near Buffalo, Wyoming	37	19	1.6	1.2
South Piney Creek at Willow Park, Wyoming	47	33	1.4	1
Piney Creek at Kearney, Wyoming	100	45	1.5	1.5
Piney Creek at Ucross, Wyoming	70	55	1.2	1.1
Clear Creek near Arvada, Wyoming	222	84	1.8	1.3
Little Power River near Broadus, Montana	56	24	1.9	1.2
Powder River at Moorhead, Montana	528	124	2.5	1.7
Power River near Locate, Montana	763	163	2.8	1.6

proportionality or coefficients. Using data collected in a laboratory sand (0.67 mm and 2 mm) channel, Wolman and Brush (1961) related cross-sectional area (A) (ft^2) to discharge (cfs) as a power function:

$$A = a_0 Q^{0.67}, \tag{4.5}$$

where a_0 is a coefficient. They plotted data from Platte River at Grand Inland, Nebraska, and found a similar power relationship.

Recall that

$$Q = \text{cross-sectional area} \times \text{velocity}, \quad \text{or} \quad Q = wdv. \tag{4.6a}$$

With the substitution of Equations (4.2)–(4.4) into Equation (4.6a), it can be observed that

$$Q = aQ^b cQ^f kQ^m = ackQ^{b+f+m}. \tag{4.6b}$$

Therefore, it follows from Equation (4.6b) that

$$b + f + m = 1.0 \tag{4.7a}$$

and

$$a \times c \times k = 1.0. \tag{4.7b}$$

Exponents b, f, and m represent the slopes of the three lines and the sum of these slopes must equal unity. Constants a, c, and k represent the intercepts of the lines and equal, respectively, the values of w, d, and v at discharge of unity. For actual data, the product of $a \times c \times k$ is not computed at unit discharge, because the values of discharge are generally much larger than unity. Since the product of the values of width, depth, and velocity at any given discharge must equal that discharge, it may be better to consider the product as follows. For example, in Figure 4.2, at 1,000 cfs: $w = 157$ ft, $d = 2.83$ ft, and $v = 2.25$ fps; their product $w \times d \times v$ equals 1,000. Equations (4.2)–(4.4) represent hydraulic relations for representative natural channel cross-sections. Since width and depth vary widely for a given discharge from one cross-section to another, the intercept values a, c, and k will also vary. Based on 20 river cross-sections representing a large variety of rivers in the Great Plains and the Southwest, Leopold and Maddock (1953) found the average values of exponents b, f, and m to be: $b = 0.26$, $f = 0.40$, and $m = 0.34$. However, for seven gaging stations on Brandywine Creek in Pennsylvania, Wolman (1955) found the mean values of $b = 0.4$, $f = 0.41$, and $m = 0.55$. These values illustrate that the values of b, f, and m can have large variability. Although these exponent values are biased toward semiarid conditions in the western United States, they nonetheless provide some indication of the order of magnitude of typical values. From 73 New Zealand river reaches, Jowett (1998) found these at-a-station hydraulic geometry exponents $\pm SD$ (SD = standard deviation) as $b = 0.176 \pm 0.066$, $f = 0.306 \pm 0.088$, and $m = 0.427 \pm 0.090$. Typically, $f > m > b$ with mean values of 0.4, 0.34, and 0.26 (Leopold and Maddock, 1953).

At most river cross-sections, the depth increases with discharge somewhat faster than does width ($f > b$). Rhodes (1977) stated $b < f$ at 90% of sites, indicating a drop in width/depth ratio as discharge increased and a concave-upwards cross-section. Further, a great majority of sites exhibited a more rapid increase in velocity compared to depth (greater m/f ratio) than expected for channels with constant roughness. For cohesive silty banks, where exponent b is small, exponents f and m depend on the roughness and resistance to scour of the bed material. When roughness decreases with increasing discharge, exponent m becomes bigger. The ratio of width to depth decreases with increasing flow through the channel. The relative rates of increase of width and depth are functions of channel shape. A channel of triangular shape is the only channel for which width to depth remains constant with changing discharge. Trapezoidal and elliptical channels are characterized by the decrease of the width to depth ratio, and in rectangular channels this ratio decreases with increasing discharge faster than in trapezoidal channels. The relative rates of increase of width, depth, and velocity depend on the shape of the channel, the slope of the water surface, and the roughness of the wetted perimeter. Park (1977) and Rhodes (1977, 1978) showed the scatter of at-a-station hydraulic geometry exponents obtained from field observations. Richards (1973) raised the question on the validity of power relations for hydraulic geometry. Knighton (1974, 1975) noted that at-a-station hydraulic geometry may not be constant in time, and Phillips and Harlin (1984) noted that it may not be constant in space. Knighton (1974, 1975) suggests that it is the local conditions that control the values of b, f, and m, not the regional climatic or hydrological factors.

4.2.6 Relation of Channel Shape to Frequency of Discharge

Leopold and Maddock (1953) selected discharges at each gaging station that were equaled or exceeded 1, 4, 10, 30, and 50% of the time, as well as the mean annual discharge. For each discharge frequency, curves were plotted that represented the change of hydraulic geometry with discharge. They observed considerable scatter of points on the depth-discharge and velocity-discharge curves but the width-discharge curves had much less scatter. Within each family, the curves generally tended to be parallel, and this tendency was more consistent in the width-to-discharge relations than in the velocity-to-discharge relations. Further, there was a pattern of interrelationship of stream characteristics that included relative frequency of discharge. This is important to define important aspects of the hydraulic geometry of stream channels and how the channel adjusts its shape to carry the sediment load supplied to it.

4.2.7 Suspended Sediment and Discharge at a Particular Cross-Section

Sediment transported by a stream may be divided into suspended load and bed load. Depending on the flow conditions, sediment may be part of the suspended load at one time and part of the bed load at another time. Moreover, the capacity of a channel to carry sediment of one size is at least partially independent of its capacity to carry that of another size. That is why wash load consisting of particles finer than those making up the stream bed is differentiated from the bed-material load.

The relation of suspended load (in units of weight per unit of time, as tons per day) to discharge (of the water-sediment mixture in cfs or m^3/s), given in Table 4.4, sometimes called the sediment rating curve, when plotted on logarithmic paper, at a particular gaging station exhibits considerable scatter around a mean line, as shown in Figure 4.3. At a given discharge, increased amounts of suspended load supplied to a reach tend to change channel characteristics that lead to increasing capacity to transport, especially through the adjustment of the ratio of velocity to depth. At a station, suspended load increases more rapidly than does water discharge, as the steeply sloping mean line shows. If water and suspended sediment increase at the same rate, the sediment concentration would be constant.

The rapid increase of suspended load with discharge can be attributed to the following reasons: (1) Sheet, rill, and gully erosion, caused collectively by rainfall and runoff, occurs in the watershed and represents the sediment supplied to the major stream channels that must transport it. For each increment of water, a large increment of debris is generated. (2) Large discharges are caused by large rainfall events that cause a larger percentage of surface runoff. Large runoff rates imply greater depths and velocities of water flowing in eroding rills. These factors cause greater erosion with the increase in the duration and intensity of rain. High discharges and consequent high velocity tend to scour the channel bed, and result in increased suspended-sediment concentration. On the other hand, changes in velocity and depth during a flood passage may be caused by changes in the concentration of suspended sediment. This can be explained as follows. At certain stations, deposition on the bed occurs under the condition of high velocities that accompany the rapidly increasing suspended loads during flood rises. These velocities are higher than those accompanying the scour of the bed when suspended loads decrease during flood recession. Thus, velocity and depth may be

Table 4.4. *Arroyo discharge measurements during flash-flood flow at particular cross-sections in New Mexico and Wyoming (from Leopold and Maddock, 1953)*

Station	Discharge (cfs)	Suspended sediment load (tons/day)
Meet John Wash, near Casper, Wyoming 2	3.4	86
	2.3	38
	1.14	–
	0.8	10
	0.31	5.3
	0.14	0.9
Cafiada Ancha. near El Rancho Montoso, New Mexico	40	5,800
	173	62,000
	2.8	260
Ancha Chiquita, near Rancbo Montoso, New Mexico	21.5	–
	11.2	–
	3.1	–
Rio Santa Fe at Old Albuquerque Road, New Mexico 3	126	17,500
	79	8,500
	40	3,600
Rio Oalisteo at Cerrillos, New Mexico	66	5,200
	800	110,000
Tributary to Hermanas Arroyo, near Las Dos, New Mexico	2.1	155
	0.14	3.9
Arroyo de los Chamisos, near Mt. Carmel Chapel, New Mexico	120	–
	57	10,500
	9.5	1,600

considered dependent variables and sediment concentration- an independent variable. The relation between suspended sediment load and discharge can then be expressed as

$$Q_s = pQ^J, \qquad (4.8)$$

where Q_s is the sediment load or discharge (tons per day), and p and J are numerical constants. Leopold and Maddock (1953) found the slope, J, of the suspended sediment curve to be greater than unity, and typical values of J were in the range of 2.0 to 3.0. Emmett (1975) reported for Upper Salmon River, Idaho, the value of J as 0.75 and $p = 13.1$ for Q in m^3/s and Q_s in tonnes/day.

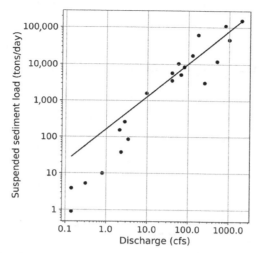

Figure 4.3 Relation between suspended sediment load and discharge

4.2.8 Width, Depth, Velocity, and Suspended Sediment at a Given Discharge

The influence of suspended sediment on channel shape can be determined, considering the condition of constant discharge. At constant width, an increase in velocity leads to an increase in suspended sediment and a decrease in depth by virtue of the continuity relation ($Q = wdv$ = constant). Conversely, at constant velocity, an increase in width would lead to a decrease in suspended load and a decrease in depth, as shown in Figure 4.4. A wider river with a particular velocity normally carries a smaller suspended load than a narrow river with the same velocity and discharge. For the same width, the cross-section with higher velocity would carry larger suspended load than the one with lower velocity. However, these relations would depend on roughness, slope, and particle size.

4.2.9 Role of Channel Roughness and Slope in the Adjustment of Channel Shape to Sediment Load

Recall Manning's equation

$$v = \frac{1.5}{n} d^{\frac{2}{3}} S^{\frac{1}{2}},$$ (4.9)

in which v is the velocity, d is the depth equal to the hydraulic radius, S is the slope of the energy line, and n is the roughness parameter. Equation (4.9) shows that an increase in velocity and a decrease in depth would require an increase in ($S^{\frac{1}{2}}/n$),

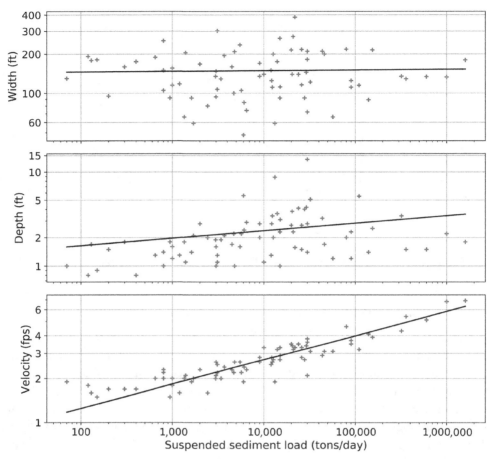

Figure 4.4 Relation between suspended sediment load and channel shape factors at constant discharge

that is, an increase in channel slope or a decrease in roughness or both. It was noted earlier that at constant width and discharge, increased sediment load would be associated with increased velocity. This is also true under the condition of varying discharge. To that end, substituting Equations (4.3) and (4.4) into Equation (4.9), one obtains

$$kQ^m = \frac{1.5}{n}\left(cQ^f\right)^{\frac{2}{3}}S^{\frac{1}{2}}. \tag{4.10}$$

Because coefficients c and k are numerical constants for a given set of conditions, one can express from Equation (4.10)

$$Q^m \propto Q^{\frac{2}{3}f}\frac{S^{\frac{1}{2}}}{n}. \tag{4.11}$$

If f and n do not change with discharge then

$$m = \frac{2}{3}f \qquad (4.12)$$

or

$$\frac{m}{f} = \frac{2}{3}, \qquad (4.13)$$

where $\frac{S^{\frac{1}{2}}}{n}$ increases with discharge, $\frac{m}{f} > \frac{2}{3}$ and when $\frac{S^{\frac{1}{2}}}{n}$ decreases with discharge, $\frac{m}{f} < \frac{2}{3}$. Leopold and Maddock (1953) found the average values of m and f at a station as $m = 0.34, f = 0.40$, and $m/f = 0.85$. They found that $\frac{S^{\frac{1}{2}}}{n}$ increased with discharge at a station. The slope of the water surface tends to remain approximately constant, whereas the increase in the concentration of suspended load with discharge is accompanied with decreasing roughness. Hence, $\frac{S^{\frac{1}{2}}}{n}$ tends to increase with discharge, which explains that the mean at-a-station value of m/f exceeds 2/3.

Vanoni (1941) showed that an increase in suspended load tended to decrease turbulence and decrease channel resistance, thus increasing velocity. Thomas (1946) reported that increased sediment concentration resulted in decreased Manning's n. Large changes in suspended sediment concentration at a cross-section must be reflected in roughness, because changes in water surface slope are sufficient to account for changes in the velocity–depth relation.

When a flood occurs, the water surface slope changes and the slope tends to be steeper during flood rise than during recession. The shape of a flood hydrograph leads to the most pronounced change of slope coinciding closely with the passage of flood crest. Changes in velocity–depth relations may coincide with changes in suspended load–discharge relations at a discharge well below the peak. At-a-station adjustments in velocity–depth relations are caused primarily by changes in roughness with changes in sediment load rather than by changes in water surface slope. Lower values of roughness are associated with larger values of suspended-sediment concentration.

The bed roughness is affected by particle size, bed configuration, and sediment load. The effect of sediment load on roughness is itself affected by particle size and bed configuration. At a station with a specific bed configuration, an increase in discharge may be associated with a slight increase or a slight decrease in roughness. However, the bed may change with increasing discharge to forms, such as smooth, riffles, bars, dunes, and antidunes. Langbein (1942) noted that antidunes formed when the Froude number approached one (critical flow). Also, the Froude number would be lower if the particles were finer and the transition would occur from smooth bed to dunes and from dunes to antidunes. For a given discharge, the increase in the value of n caused by the change in bed configuration may be greater than that by a considerable increase in discharge on a constant bed configuration. It can be concluded that for the same slope and discharge decreasing particle size tends to increase roughness when the bed configuration remains the same, and with no change in particle size the change in bed configuration considerably changes bed roughness. The increase of suspended load with discharge at a station necessitates an increase in velocity for the transport of sediment load in comparison with depth or a relatively large value of m/f. Leopold and Maddock (1953) found this value to be 0.85, greater than 2/3. This requires $\frac{S^{\frac{1}{2}}}{n}$ to increase with increasing discharge, which is achieved by the decrease in Manning's n associated with the increase in sediment concentration with discharge.

4.3 Downstream Hydraulic Geometry

Frequency of discharge is important for understanding how the width, depth, and velocity of a natural river change in the downstream direction. The mean annual discharge is equaled or exceeded about the same percent of time in a large number of rivers. It approximately represents the discharge that equaled or exceeded one day in every four over a long period.

For most rivers, the depth, width, and velocity tend to increase progressively as power functions of discharge. When plotted against mean discharge for all gaging stations on all tributaries, the width, depth, and velocity corresponding to the mean annual discharge can be expressed as Equations (4.2)–(4.4). Proceeding downstream along a given river, the discharge tends to increase because of the progressively increasing drainage area, meaning width, depth, and velocity increase as well, though there are some streams in which discharge decreases downstream, especially in arid areas. Leopold and Maddock (1953) found the average values of exponents as $b = 0.5$, $f = 0.4$, and $m = 0.1$.

It may be noted that in the plots of Equations (4.2)–(4.4) each gaging station provides one point. There is, however, considerable scatter of points in the graphs. In a given river basin where all cross-sections are experiencing the same frequency of discharge, the corresponding values of depth, width, and velocity at different cross-sections having the same discharge tend to be similar, regardless of where in the watershed or on what tributary the cross-sections may be. In the downstream direction the rates of increase in width, depth, and velocity relative to discharge are of the same order of magnitude for rivers of different sized drainage basins and of widely different physiographic regions. Jowett (1998) found the downstream hydraulic geometry exponents as $b = 0.5$, $f = 0.25$, and $m = 0.25$ for 73 river reaches and $b = 0.48$, $f = 0.30$, and $m = 0.22$ for 98 reaches combining 25 reaches used by Griffiths (1980), who found $b = 0.48$, $f = 0.43$, and $m = 0.11$. He also compared with $b = 0.45$–0.55, $f = 0.20$–0.35, and $m = 0.20$–0.35 found by

Henderson and Ibbitt (1996), and $b = 0.4$–0.5, $f = 0.3$–0.4, and $m = 0.1$–0.2 found by Park (1977). Normally, $b > f > m$ with typical values of $b = 0.5$, $f = 0.4$, and $m = 0.1$, which compare with 1/2, 1/3, and 1/6 of the regime theory.

4.3.1 Variation of Hydraulic Characteristics

It may be reemphasized that along the channel in the downstream direction, the plots of width, depth, and velocity using data from different cross-sections must correspond to the same frequency of discharge or mean annual discharge. In reality, this discharge may not occur exactly at all stations but should be approximately close for a large number of days each year. In general, moving downstream, width, depth, and velocity increase as discharge increases. Of course, there can be reaches, especially in arid regions, in which width, depth, and velocity decrease going downstream.

In the downstream direction the rates of increase of width, depth, and velocity with respect to discharge exhibit the same order of magnitude for rivers having drainage basins of different areas and widely differing physiography. Further, the rates of increase of width, depth, and velocity with discharge downstream may be similar for different rivers having markedly different width to depth ratios at any particular discharge. The velocity tends to increase with mean annual discharge downstream for most rivers. Sometimes it is held that the velocity of a stream is greater in the headwaters than in lower reaches. The impression of greater velocity upstream is due to the notion that slopes are steeper in upper reaches than in lower reaches. However, the Manning relation shows that velocity also depends on depth besides slope as shown by Equation (4.9), in which d is the depth $\cong$ hydraulic radius for natural river cross-sections, and S is the slope of the energy grade line $\cong$ slope of the river profile.

Since velocity increases downstream with mean annual discharge, the increase in depth overcompensates for the decreasing river slope. The magnitude of this rate of change becomes clear by comparison of the exponents of depth and slope. In Equation (4.9) the velocity depends on depth to the power of 2/3 and on slope to the power of 1/2. Thus, the rate of increase of depth downstream tends to overcompensate for the decreasing slope and tends to provide a net increase of velocity at mean annual discharge in the downstream direction.

4.3.2 Relation of Channel Shape to Frequency of Discharge

For more infrequent discharges, the depth–discharge curve and the velocity–discharge curve exhibit considerable scatter of points around their mean lines, but the width–discharge curve shows much less scatter. When curves are plotted for different frequencies of discharges, the width-to-discharge relations tend to be approximately parallel to each other in comparison with the velocity-to-discharge relations. With less frequent discharges (flood flows), the velocity increases somewhat faster with discharge downstream than with low flows. However, Leopold and Maddock (1953) observed that in the same stream in the Yellow River basin in Wyoming, velocity remained nearly constant downstream despite a rapid decrease in slope.

The interrelationship of stream characteristics that include relative frequency of discharge shows a pattern. To explain the pattern, consider main streams in two basins A and B and two stations, one at the headwaters and the other downstream. For basin A, at a gaging station near the headwaters of a stream the width increases with the increase of discharge, and the same is observed for a downstream station. If the discharge at the headwater gage is of the same cumulative frequency (say 50% of the time) as the discharge at the downstream gage, then the width increases with the increase in discharge downstream, corresponding to this cumulative frequency of 50%. If discharges at the two gages have the same cumulative frequency (1% of the time), then the width increases downstream at this discharge frequency of 1%. In a similar vein, the depth increases with the increase in discharge at the upstream gage, as well as at the downstream gage.

If the depth–discharge plot at the downstream gage is parallel to the width–discharge plot for a given frequency (occurring 50% of the time) and the slope is 0.5 (exponent $b = 0.5$), then the slope of the velocity–discharge curve would be zero ($m = 0$). At that frequency the velocity does not change downstream. The velocity increases with the increase in discharge at the upstream gage, as well as at the downstream gage. For 1% frequency the velocity is higher at both stations than at the more frequent discharge but does not change downstream. In basin B the width-to-depth ratio increases downstream for any frequency. This increase is reflected by the steeper slope of the width–discharge relation than that of the depth–discharge relation. If the slope of the width–discharge line is 0.5, then the steepening of slope of the velocity–discharge line at the downstream gage more or less coincides with that at the upstream gaging station.

Wolman (1955) summarized changes in width, depth, velocity, slope, and roughness in the downstream direction for discharge of different durations (percent of time flow is equaled or exceeded) on Brandywine Creek in Pennsylvania, as in Table 4.5 (Wolman, 1955).

4.3.3 Relation of Suspended Sediment to Discharge in a Downstream Direction

The average sediment concentration tends to decrease in the downstream direction. This may be partly because the percent of land does not contribute to sediment

Table 4.5. *Exponents of discharge of different frequencies for width, depth, velocity, slope, and roughness (after Wolman, 1955)*

Frequency (duration of time flow is equaled or exceeded)	Rate of change with discharge of				
	Width b	Depth f	Velocity m	Slope	Roughness
Principal stations					
50	0.34	0.45	0.32	−0.80	−0.40
15	0.38	0.42	0.32	−0.92	−0.51
2	0.45	0.43	0.17	−0.97	−0.32
Bankfull	0.42	0.45	0.05	−1.07	−0.28
Principal stations and headwater tributaries					
50	0.57	0.40	0.03	–	–
2	0.56	0.40	0.02	–	–

load, and therefore the percent of sediment free water tends to increase somewhat with drainage area. It should, however, be noted that individual rivers may differ in this respect. Field observations show that the average annual production of sediment per square mile of drainage area decreases with increasing drainage area. The sediment production per unit area is greater for smaller drainage areas than for larger drainage areas. This means that on average there should be a tendency toward smaller concentration of suspended sediment downstream. The concentration of suspended sediment decreases somewhat with discharge of the same frequency. The relation of suspended sediment load to discharge can be expressed by Equation (4.8), in which Q is the discharge of the same frequency.

4.3.4 Width, Depth, Velocity, and Suspended Sediment at a Given Discharge

For a constant discharge, at constant width of channel, an increase in velocity is accompanied by an increase in suspended sediment. This requires a decrease in depth. On the other hand, if velocity is constant, an increase in width is accompanied by a decrease in suspended load. This is also associated with a decrease in depth. Thus, decreasing width at constant velocity or increasing velocity at constant width each results in increased capacity for suspended load at constant discharge. Field data show that a wide river having a particular velocity is observed to carry a smaller suspended load than a narrower river having the same velocity and discharge. If two rivers have equal width at the same discharge then the one having the higher velocity carries a larger suspended load.

4.3.5 Width, Depth, and Suspended Load at a Variable Discharge

For a given rate of increase of discharge to width (i.e., value of exponent b), the slope of the curve relating suspended load to discharge (i.e., value of J) will depend on the rate of increase of velocity with discharge defined by m. Because $b + f + m = 1$, f is known if b and m are known. A plot of J versus (m/f) shows an increasing trend for a given value of b. This shows that the rate of increase of suspended-sediment load with discharge is a function of the ratio (m/f). For a given value of b, the steeper the slope of the suspended sediment−discharge line, the greater the slope of the velocity−discharge line.

In the downstream direction the suspended sediment concentration decreases slightly. A value of $J = 1.0$ means that suspended load is increasing at the same rate as discharge and represents equal sediment concentration. Therefore, the observed value for rivers should be somewhat less than 1.0. For the values of m, f, and b in the downstream direction, where $b = 0.5$, and $m/f = 0.25$, it is observed that $J = 0.8$, indicating a slight decrease in a downstream direction for a given frequency. For cross-sections of different rivers having different channel characteristics, the observed suspended sediment load varies directly as a function of velocity and depth, and inversely as a function of the water surface width. The suspended load varies more with velocity than with depth and width. Observed suspended sediment curves show that the value of J increases with m/f. For a given width, at a given discharge an increase in suspended sediment requires an increase in velocity and a reduction in depth.

A given discharge at a downstream station may represent a frequent flow or a low flow condition, whereas to attain the same discharge the upstream station must experience a flood flow. A given discharge is equaled or exceeded more often at a downstream station than at the upstream station. Generally, low flow is likely to carry a smaller concentration of suspended sediment and a flood flow a larger concentration. At a given discharge the downstream station has a larger width and a lower velocity. Thus, a smaller suspended-sediment load requires for equilibrium larger width or lower velocity or both.

4.3.6 Relations between Width, Depth, Velocity, and Bed Load at a Given Discharge

At constant discharge, an increase in velocity at constant width is associated with an increase of both suspended load and bed load transport. At constant velocity and discharge, an increase in width is associated with a decrease in suspended load and increase in bed load transport. Lane (1937) observed that if a channel was supplied a heavy bed load then in order to be stable it must move this load along, which would require a higher velocity along the bed but the same velocity along the banks, and this could only occur with a wider and shallower channel. That is, an alluvial channel that is relatively wide and shallow probably carries a relatively large bed load. Based on the Indian literature, the broad, shallow channel is the type of cross-section best adapted for the transport of heavy bed load. It can be argued that despite a large suspended load in a river, only the bed load is of real significance in fluvial geomorphology, assuming it places the greatest tax on the energy of the stream.

The bed material size influences the flow depth. A decreasing bed material size tends to reinforce the effect of discharge on depth, as Simons and Albertson (1963) showed that sand-bed canals with non-cohesive banks had width and depth coefficients more than twice those of gravel channels. Bank material characteristics control the strength and stability of channel banks and hence the adjustment of channel width. Bank strength depends on the degree of cohesion, which can be expressed by the silt-clay content (r_B) of the banks or perimeter (r_M). For sand-bed rivers in Great Plains, Ferguson (1973) obtained

$$w = 33.1 \, Q_{2.33}^{0.58} r_B^{-0.66} \tag{4.14}$$

and Schumm (1971) developed

$$w = 5.54 \, Q_{2.33}^{0.58} r_M^{-0.37} \tag{4.15}$$

$$d = 5.54 \, Q_{2.33}^{0.42} r_M^{0.35}, \tag{4.16}$$

where $Q_{2.33}$ is the discharge occurring with a frequency of 2.33 years, which is approximately equal to mean annual flow. For more cohesive banks, r_B or r_M will increase, which will increase the influence of suspended load over bed load transport. This will lead to a slower increase in width but more rapid increase in depth than expected of the effect of discharge alone.

4.3.7 Channel Shape Adjustment during Individual Floods

During the passage of a flood, the channel of an alluvial stream is scoured and filled with considerable rapidity. A lower suspended load is accompanied by a lower velocity and a correspondingly higher depth. At a constant discharge and width, a decrease in suspended-sediment load is accompanied by a decrease in velocity and an increase in depth. For the same value of b, a decrease in J is associated with a decrease in the m to f ratio, while the value of J increases. Analysis of scour and fill of a river bed during flood shows that changes in the bed occur simultaneously with changes in the rate of change of suspended-sediment concentration.

Observed changes in the stream bed result from the changes in sediment load brought from upstream. The underlying hydrodynamic factors tend to promote a mutual adjustment between channel shape and sediment load carried into the reach. The change in sediment load that results in a change of channel shape involves both bed load and suspended load. With no change of channel width, a decrease of suspended load at a given discharge is accompanied by an increase in depth by bed scour that results in a decrease in velocity. The decrease in velocity provides the adjustment of capacity for carrying the load of particular sediment size-distribution supplied by the watershed. In response to a decrease in load, the channel shape becomes adjusted through scour to the lower capacity required for quasi-equilibrium.

If high suspended-sediment concentration results from the scouring action of high velocities, it is implied that high velocity in a reach scours the channel in that reach. The increase in sediment transport results from the local bed scour and should then account for the increase in sediment concentration. Then, increasing velocity will be associated with bed scour and decreasing velocity with bed fill. The scour and fill of the bed of the main stem of an alluvial river during flood appears to adjust channel shape in response to varying sediment load and the adjustment occurs rapidly. A time lag measured in days and perhaps weeks is required for the adjustments to take place.

4.3.8 Significance of Channel Roughness and Slope in Adjustment of Channel Shape to Sediment Load

At constant discharge, suspended-sediment load is related to the shape factors: width, velocity, and depth. At constant width and discharge, increased suspended-sediment load would be associated with increased velocity. When Q and w are constant, the product $v \times d$ must be constant. An increase in velocity, therefore, requires a decrease of depth. Recalling the Manning equation, an increase in velocity and a decrease in depth require an increase in the factor $S^{1/2}/n$; this means that increased velocity and decreased depth would require an increase in channel slope or a decrease of roughness or both.

Under varying discharge, similar results are obtained. From the depth–discharge and velocity–discharge functions, one can express

$$Q^m \propto Q^{\frac{2f}{3}} \frac{S^{\frac{1}{2}}}{n}.$$ (4.17)

If S and n do not change, then

$$m = \frac{2}{3}f \quad \text{or} \quad \frac{m}{f} = \frac{2}{3},$$ (4.18)

where $S^{\frac{1}{2}}/n$ increases with discharge, $\frac{m}{f} > \frac{2}{3}$ and when $S^{\frac{1}{2}}/n$ decreases with discharge, $\frac{m}{f} < \frac{2}{3}$.

The average values of m and f for rivers studied by Leopold and Maddock (1953) were $m = 0.1, f = 0.4$, and $\frac{m}{f} = 0.25$. For the rivers under study, the factor $S^{1/2}/n$ increases with discharge at a station and decreases downstream. In the downstream direction the value of roughness coefficient, n, tends to remain about constant, while slope decreases. This explains that $m/f < 2/3$.

An increase in suspended load tends to decrease channel resistance and thus causes an increase in velocity. This may be a result of decreased turbulence. Increased concentration is associated with decreased values of Manning's roughness factor. When a river cross-section is considered, the large changes in suspended-sediment concentration that occur with changes in discharge are not accounted for by variations in the velocity–depth relations alone; changes in roughness must occur.

During the passage of a flood through a river cross-section, the slope of the water surface changes. In a given reach, slope tends to be steeper during flood rise than during recession. The steeper slope on the flood rise is because the rise of a cross-section precedes the rise of any cross-section downstream. However, the normal shape of a flood hydrograph tends to make the most marked change of slope coincide closely with the passage of the flood crest in a given reach.

The reduction in width implies a larger capacity for suspended load at a given velocity and discharge, and a smaller capacity for bed load. With decreased velocity, the new channel will have a lowered capacity for both types of load at a given discharge. Even during a period of years, adjustment of a channel reach in response to a change in suspended load may be made primarily by changes in channel shape through changes in roughness rather than in channel slope.

The three factors that affect bed roughness include particle size, bed configuration, and sediment load. Due to abrasion, in most rivers a decrease in particle size of the bed material in the downstream direction occurs. A change in the caliber of load affects bed roughness and through bed roughness affects other stream characteristics.

A distinction is made between roughness resulting from ripples or dunes on the channel bed and roughness due to the individual particles of bed material. The roughness due to channel irregularities includes not merely semi-permanent obstructions, such as rocks and bars, but transitory conditions of bed ripples and waves, which change with discharge.

The Manning roughness factor in pipes represents approximately a permanent characteristic of the boundary surface. This is true for certain open channels, drainage canals, and even irrigation canals where the banks and bed are fixed. However, in most sediment-laden channels the roughness is not a permanent characteristic but must change with the configuration of the bed. It is then logical that the smooth bed of an alluvial river will have a lower Manning's n than where the bed has dunes and antidunes. An increase in bed roughness will result in a decrease of velocity with respect to depth. Changes in the velocity–depth relation constitute part of the adjustment of channel shape to load, and variations in roughness affect capacity for load.

Decreasing sediment size results in a decrease of roughness due to the individual particle. However, in hydrodynamically rough channels, the roughness due to bed configuration may be more important than that due to particle size, and decreased particle size may result in larger or more effective bed ripples or waves and thus in increased roughness. At a given discharge, the change in bed configuration may increase the value of n more than may a considerable increase in discharge during which the bed configuration remains the same.

For the same slope and discharge, decreasing particle size tends to increase the roughness even when the condition of the bed is the same. Second, with no change in particle size, a change in bed configuration may precipitously change bed roughness, and its effect may be of the same or greater order than the effect of differences in particle size.

Headwater tributaries flowing on gravel or boulders move large bed debris only in exceptional flows. The channel roughness in such streams is determined by the sizes of individual rocks or cobbles. The downstream decrease in the size of gravel or pebble bed generally results in a downstream decrease in the value of channel roughness. The decrease in the value of roughness is interpreted as a result of decreasing particle size but may also be influenced somewhat by a downstream decrease in suspended-sediment concentration.

The value of channel roughness is controlled by the size of individual particles when this size is large, that is, gravel, cobbles, or even coarse sand. When the sediment size is large, the channel roughness tends to decrease downstream as the particle size decreases. When stream beds are composed of fine material, the bed configuration (ripples, dunes, or waves) tends to control the channel roughness. In such streams, roughness changes little in the downstream direction, for discharge of equal frequency occurs along the reach.

In summary, the velocity–depth relations in the downstream direction are required for the transportation of the supplied sediment load. These relations are provided by the concomitant change of roughness and slope downstream. Roughness is governed by the sediment size, the suspended-sediment concentration, and bed configuration, all of which are nearly independent of the channel. That part of the required relation of velocity to depth not provided by the downstream change in roughness is provided by the change in slope, because slope is the dependent factor, which can be adjusted by stream processes.

The characteristic rapid increase of suspended load with discharge at a station requires for transportation of that load a relatively rapid increase of velocity in comparison with depth, or a relatively large value of *m/f*. In the downstream direction, the sediment load does not increase as fast as discharge. This means that the concentration of sediment decreases slightly downstream. The transportation of this load requires that the depth increases with discharge much faster than does velocity, or the *m/f* ratio must be small.

4.3.9 Stable Irrigation Canal: An Analogy to a Graded River

An irrigation canal constructed in alluvial material tends to gradually scour or fill its bed and banks until a cross-section is achieved that remains in equilibrium with water and sediment flowing through it. A regime channel is a channel that is neither scoured nor filled. Lacey (1930, 1939) utilized measurements of channel characteristics on reaches of Indian canals that had achieved stable cross-sections. He proposed a number of basic equations:

$$P = 2.67Q^{\frac{1}{2}} \tag{4.19}$$

$$v = 1.15\, F_s^{\frac{1}{2}} R^{\frac{1}{2}}, \tag{4.20}$$

where P is the wetted perimeter, Q is the discharge, v is the mean velocity, R is the hydraulic radius $\approx$ mean depth, and F_s is a sediment factor. In the present study, the exponent of width to discharge is 0.5. If $P \approx w$, the two relations are identical. Because of limited data, Lacey postulated that this width–discharge relation was fundamental and independent of sediment size. Lacey's Equation (4.20) states

$$v \propto R^{\frac{1}{2}}. \tag{4.21}$$

Because $R \approx d =$ mean depth of wide rivers, this relation is comparable to the *m/f* relation for alluvial rivers. Using Equations (4.3) and (4.4),

$$\left(\frac{d}{c}\right)^{\frac{1}{f}} = Q = \left(\frac{v}{k}\right)^{\frac{1}{m}} \tag{4.22}$$

or

$$v \propto d^{\frac{m}{f}}. \tag{4.23}$$

In the downstream direction for river data, $m/f = 0.25$, whereas the Lacey exponent is 0.5. The difference in the exponents may be due to the differences in downstream sediment concentration. Comparison of regime canals and natural rivers is valid only in the downstream direction. Field data show that a larger value of *m/f* corresponds to a larger value of J for any value of b (the rate of increase of width with discharge). The larger value of m/f implies that regime canals will show a rate of increase of suspended sediment load with discharge along the canal length.

4.3.10 Sediment and Longitudinal Profile

The longitudinal profile of a stream may depend on the change in particle size, which in turn is associated with sediment concentration, slope, flow velocity, and roughness. In a mountain stream, the bed is covered with large sediment sizes, including boulders and gravel, in the headwaters where the slope is steep, becomes sandy some distance downstream, and is dominated by clay and silt near the mouth. Along the length the bed slope decreases, although the cause and effect relation between slope and sediment size is not established. As sediment size and slope decrease, the velocity also decreases. This would imply that bed roughness would decrease. It can be surmised that the diminishing sediment size downstream increases the percentage of sediment load, the suspended load, and the total load remaining constant.

From Manning's equation, it can be stated that if constant depth and velocity carry the constant suspended load then a decrease in slope would correspond to a decrease in bed roughness. However, in fine-grained materials, a decreasing sediment particle size leads to an increase in bed roughness if slope and discharge are constant. Roughness is controlled by suspended sediment concentration, not by particle size, and larger concentration is associated with smaller roughness. However, when sediment particle size becomes small enough, further decrease in particle size corresponds to an increase in roughness.

Different longitudinal profiles correspond to different rates of change of velocity and depth.

The longitudinal profile of a stream is governed by a system of eight equations containing eight unknowns, which can be summarized as follows.

(1) Continuity Equation (4.6).
(2) Sediment load–discharge relation governed by watershed characteristics: $L = F(Q)$, where F is a function.
(3) Particle size changes occurring downstream: $D = F(Q)$.
(4) Width–depth relation involving the relative mobility of bed and bank: $w = F(d)$.
(5) Relation between exponents J, m/f, and b: $J = F\left(\frac{m}{f}, b\right)$.
(6) Relation between Manning's roughness coefficient n and flow depth d, sediment particle size D, discharge Q, and slope $S : n = F(d, D, Q, S)$.
(7) Relation between change in roughness with respect to change in discharge and change in sediment load with respect to change in discharge: $\frac{dn}{dQ} = \frac{dL}{dQ}$.

(8) Manning's equation: $v = \frac{1.5}{n} d^{\frac{2}{3}} S^{\frac{1}{2}}$. This equation represents the equivalence between the loss of energy due to friction and the decrease in energy potential. Here slope is regarded as a dependent variable that is adjusted by the stream in response to velocity, depth, and roughness.

The last six equations involve factors that depend on stream processes. The stream profile is limited by its head and mouth, and its slope is determined by physiography and interactions among processes. When a smooth longitudinal profile is developed, an approximate equilibrium is achieved. Consider, for example, a shallow and wide channel. For a constant discharge, the shear stress will be greater in the horizontal plane than in the vertical plane. Conversely, in a narrow and deep channel, the shear will be greater on the banks than on the bed. These two shapes will have characteristic distributions of velocity. The relative erodibility of bed and bank leads to the channel shape, permitting a velocity distribution that would be in approximate equilibrium with erodibility characteristics.

Different longitudinal profiles exhibit different concavities in response to different relations of velocity and depth to discharge. However, for a profile to have an approximate equilibrium, a particular rate of increase of velocity and depth is essential, as long as the drainage basin is able to supply water and sediment discharge in a characteristic manner. Channel roughness changes with sediment load and particle size but its changes downstream are small because of the interactions among opposing factors. The particular rate of increase of velocity and depth is needed to transport the sediment load and its change. If roughness is constant, the slope must generally decrease in order to enable the needed velocity–depth relations. This explains why longitudinal profiles of natural rivers transporting sediment are concave. Tanner (1971) represented the long profile of a river as a composite of segments each represented as nearly exponential. On a semi-log paper, each segment was a straight line, with a slight upward concavity near the upper end of the segment and a slight upward concavity near the lower end. He then utilized five models to represent the river profile, including classical profile equation, Lagrange linear equation, Leopold and Langbein (1962) random walk model, mixed random walk model, and Devdariani's (1967) heat analogy model. None of these models was able to satisfactorily represent the river profile.

4.4 Variation of Exponents

The exponents of hydraulic geometry relations given by Equations (4.2)–(4.4) vary from location to location on the same river and from river to river, from high flow range to low flow range, with variations in materials forming the bed and banks of alluvial channels, and channel aspect and shape. Rhoads (1991) hypothesized that

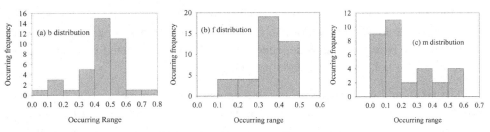

Figure 4.5 Histograms of occurring frequency of exponents

the exponent parameters were functions of channel and sediment characteristics and flood magnitude, and that the parameters varied continuously rather than discretely. Park (1977) showed that exponents, b, f, and m seemed independent of location and only weakly dependent on channel type. Using 54 surveys of 17 river reaches of rivers in Australia, Stewardson (2005) stated that much of the variation in hydraulic exponents could be attributed to the location of cross-sections and measurement error. Using values of b, f, and m reported in the literature, Singh et al. (2003) plotted frequency histograms, as shown in Figure 4.5. Phillips and Harlin (1984) found the hydraulic exponents to be not stable over space. Knighton (1974) observed variations in exponents, and Rhodes (1978) concluded that the exponent values varied from high flow conditions to low flow conditions.

For midwestern and piedmont data, Kolberg and Howard (1995) indicated that the b exponent ranged from 0.35 to 0.46 for groups of streams with the width to depth ratio less than 45 and decreased to values below 0.15 for groups of streams with the width to depth ratios greater than 45. Klein (1981) found the value of b ranging from 0.2 to 0.89, with low b value normally occurring for small basins (in low flows) and for very big basins (in very high flows), which was less than the one for intermediate basins, and the value $b = 0.5$ was a good average.

Dingman (1984) showed that in natural channels where width increased with flow depth the value of b would be higher and the values of f and m lower than derived for rectangular channels. Ferguson (1986) and Dingman (2007) found considerable variation in the values of exponents and coefficients.

4.5 *b-f-m* Diagram

The *b-f-m* diagram, proposed by Rhodes (1977), as shown in Figure 4.6, helps interpret at-a-station hydraulic geometry relations and classify channel patterns. The discussion here closely follows Rhodes (1977). The construction of the diagram assumes that the sum of exponents b, f, and m is one, as shown by Leopold and Maddock (1953). Under this assumption, a set of hydraulic geometry relations can be represented by a single point. However, the sum may not always

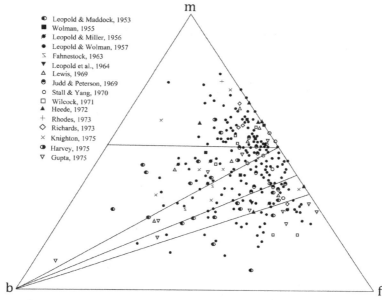

Figure 4.6 *b-f-m* diagram

be unity because of several factors, such as errors in data, fitting of regression line, validity of power relation, and lack of adjustment of the channel to certain discharges. Velocity and cross-sections are subject to measurement errors. Regression lines can be fitted using the least squares method, but the sum of exponents to be unity cannot be guaranteed. The validity of power relations has been questioned by Richards (1973). One of the reasons is that the channel roughness may change with discharge causing nonlinearity, due to which the power relation may not be adequate. It has been shown by Lewis (1966) that the hydraulic geometry relations developed for intermediate to high discharges do not hold for small discharges. Even for moderate to large discharges the channel width responds differently due to channel irregularities (Richards, 1976).

If the sum of b, f, and m values is unity, the exponents can be plotted on a triangular coordinate system (Rhodes, 1977), as shown in Figure 4.7. The *b-m-f* diagram can be used to interpret channel form and process, and fill and scour dynamics (Merigliano, 1997). The m/f ratio can be used to interpret channel shape and sediment transport. Rhodes (1977) employed five criteria to divide the diagram into various fields that provide information on similarities and relationships between different data sets. The lines of division correspond to constant values of width–depth ratio $(b = f)$, competence $(m = f)$, and Froude number $(m = f/2)$, velocity-cross-sectional area ratio (related to the Darcy–Weisbach friction factor) $(m = b + f)$, and slope–roughness ratio, which is related to Manning's equation

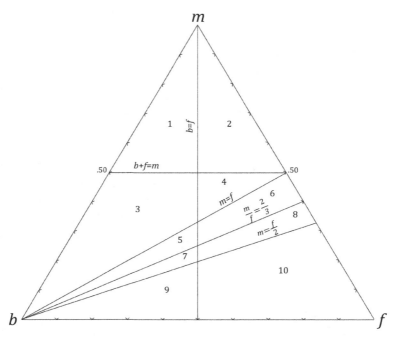

Figure 4.7 Divided *b-m-f* diagram

$(m/f = 2/3)$, as shown in Figure 4.7. The various subdivisions are briefly described in what follows.

>   **Subdivision 1:** This is bounded by $b = f$ and $m = b + f$. The line $b = f$ relates to changes in the width–depth ratio (w/d) with increasing discharge, because $\frac{w}{d} \propto Q^{b-f}$. Thus, if $b = f$, then the ratio w/d becomes constant. Likewise, if $b > f$, the left side of the line, the ratio increases with increasing discharge and if $b < f$, the right side of the line, the ratio decreases with increasing discharge. This provides information on channel shape, relative stability of bed and bank material, and channel adjustment to bed load transport. Schumm (1960) showed that channel shape and the relative stability of bed and bank materials were interrelated. Changes in channel width and depth with increasing discharge may be caused either by erosion of banks and bed or by filling of a stable channel. When $b > f$, the implication is that the stream banks are more easily eroded than is the bed, and when $b < f$ the reverse situation will happen.

For a stable channel under normal discharge conditions, the case $b < f$ points to a tendency for the channel cross-section to become proportionally deeper and narrower, that is, smaller w/d ratio, with increasing discharge. The case $b > f$ implies that the width–depth ratio increases with increasing discharge, meaning a wider but shallower channel. For stable channels, rectangular and parabolic channels will exhibit a decreased width–depth ratio, with the rate of decrease being greater for rectangular shape than for other shapes. The width–depth ratio reflects a change in channel cross-section shape with increasing discharge and changes in the width–depth relationship but not the absolute values. For example, a cross-section may exhibit a relatively large width–depth ratio for all stages even if $b > f$.

The width–depth ratio with discharge has important implications for sediment transport. Wide and shallower channels are more conducive to the transport of a large bed load (Leopold and Maddock, 1953). Using Newton's law of viscosity, Rhodes (1977) explained the underlying reasoning. Since the shear stress is directly proportional to the change in velocity with height, greater shear will result if the velocity changed more rapidly with depth above the bed, that is, the isovels (lines of equal velocity) would be more closely spaced. With increasing discharge, closer spacing of isovels will lead to increased bed load transport. Channels plotted on the left half of the *b-m-f* diagram possess a higher potential for bed load transport than those on the right side. Leopold and Maddock (1953) stated that deep narrower channels were more conducive to greater suspended sediment load than wide shallower channels.

Subdivision 2: Here $m = f$. The *b-m-f* diagram is bounded by $m = f$ or $f = 1$. Wilcox (1971) stated that if the rate of increase in velocity (m) equaled or exceeded the rate of increase in depth (f) the competence would increase. The channels corresponding to the points above the line $m = f$ would exhibit an increase in competence with increasing discharge. The line $m = f$ is related to fluvial sediment transport. The ratio m/f must be greater than 1 if suspended load is to increase for a given width. Leopold and Maddock (1953) related the ratio of the rate of change of velocity to the rate of change of depth with the rate of increase of suspended sediment, and the rate of increase of suspended sediment with discharge was inversely proportional to the rate of increase of width.

Subdivision 3: This is bounded by $m = f/2$. Using the expression of Froude number and noting the critical condition, it can be shown that $Q^{f/2} \propto Q^m$. For $m = f/2$, the Froude number does not vary with discharge. For $m > f/2$, the points above the line, the Froude number increases with increasing discharge, and if $m < f/2$, the points below the line, the Froude number decreases with increasing discharge. The points below the line will not correspond to supercritical flow on a rising limb, which will not be able to accommodate some kinds of sediment transport.

Subdivision 4: This is bounded by $m = b + f$ or $m = 0.5$. Points above the line correspond to channels with rapid increase in velocity with discharge. If $m > b + f$, then velocity increases faster than the cross-sectional area. Points on the line indicate that velocity increases at the same rate as the cross-sectional area, and channels are quite stable and experience decreased resistance. In this case, $m = b + f$. If the velocity increases faster than the cross-sectional area then flow resistance must decrease rapidly with increasing discharge. In this case the channel must be stable, with no significant erosion of bed and banks. The rate of increase of the width–depth ratio (*b*/*f*) can be utilized as an index of channel stability, with decreasing value corresponding to stable sections. Channels plotting on the line $b + f = m$ are stable and have decreased resistance with increased discharge.

Subdivision 5: This is bounded by $m/f = 2/3$. Using Manning's equation and assuming $S \propto Q^h$ and Manning's coefficient $n \propto Q^y$, one can express $Q^m \propto Q^{(\frac{2}{3})f+(\frac{1}{2})h-y}$ or $m = (\frac{2}{3})f + \frac{h}{2} - y$. If $m > 2f/3$, then $h/2-y > 0$ or $S^{\frac{1}{2}}/n$ must increase with discharge. If $m < 2f/3$, $h/2-y < 0$ or $S^{\frac{1}{2}}/n$ decreases with increasing discharge. If $h = 0$, then resistance decreases. If $m = 2f/3$, then $h/2 = y$ and $S^{\frac{1}{2}}/n$ do not vary with discharge. Since slope changes little with discharge, the roughness can be assumed to change the most. At a given station, roughness may decrease with discharge during the rising stage due to reduced flow distortion, the increased suspended sediment concentration increasing effective viscosity, and the presence of large roughness elements in small channels. However, roughness may increase with discharge due to the increase of dunes or greater turbulence because of higher vegetation along the banks. Of the factors that may contribute to changes in roughness, it is difficult to determine which factor contributes in a particular case and even more difficult to determine the relative contribution of each factor.

The five subdivisions of *b-m-f* diagram shown in Figure 4.6 delineate ten areas as shown in Figure 4.7. All cross-sections corresponding to a point within a particular field should exhibit similar hydrologic and morphologic responses to changes in discharge. Likewise, all channels plotting within the same field will differ only in their rates of responses to changes in discharge. Since hydraulic geometry exponents reflect the rate of change in hydraulic geometry variables with respect to change in discharge, the values of these exponents relative to each other with increasing or decreasing discharge provide information on channel bed and bank stability, erosion, widening or narrowing of channels (or width-depth ratio), changes in roughness, changes in slope, and processes of fluvial sediment transport. For example, the *b*-to-*f* ratio reflects the change in the shape of flow cross-section with changing discharge. The *m*/*f* ratio relates to roughness. The ratio $m/(b + f)$ relates to channel stability and flow resistance.

The divided *b-f-m* diagram can be used to group and compare the hydraulic geometry of stream channels. The diagram serves as an aid in interpreting the exponents of at-a-station hydraulic geometry. The expected relationships between increasing discharge and fluvial parameters for different types of channels can be summarized (Rhodes, 1977), as shown in Table 4.6.

Rhodes (1977) warned about using mean values, since at-a-station hydraulic geometry is a sensitive indicator of bank and bed material conditions at a particular location, and as these conditions change, the channel changes its morphology and in turn its hydraulic geometry.

Table 4.6. *Likely direction of morphologic and hydrodynamic parameter changes with increasing discharge for different channel types*

Channel type	Width–depth ratio (w/d)	Competence	Froude number	Velocity– area (v/A)	Slope– roughness ($S^{0.5}/n$)
1	Increases	Increases	Increases	Increases	Increases
2	Decreases	Increases	Increases	Increases	Increases
3	Increases	Increases	Increases	Decreases	Increases
4	Decreases	Increases	Increases	Decreases	Increases
5	Increases	Increases	Increases	Decreases	Increases
6	Decreases		Increases	Decreases	Increases
7	Increases		Increases	Decreases	Decreases
8	Decreases		Increases	Decreases	Decreases
9	Increases		Decreases	Decreases	Decreases
10	Decreases		Decrease	Decreases	Decreases

It seems plausible to explore a relationship between the exponents of hydraulic geometry and channel pattern, straight, braided, or meandering, for channel morphology is a major determinant for the hydraulic geometry of a cross-section. For a straight channel, the cross-section should be symmetrical, and with increasing stage the width should increase slowly in comparison with depth $(b < f)$. If the channel is meandering, its cross-section at bends should be asymmetric, and the asymmetry should lead to a higher rate of increase of width but f may still be greater than b. Braided channels are usually wide and shallow and have unstable banks. With discharge the width–depth ratio increases $(b > f)$. Rhodes (1977) observed that all channels with b-to-f ratio greater than one were braided. Cross-sections of straight reaches generally had low values of b (<0.23). Meandering channels also have low b values. Both straight and meandering patterns respond similarly to increasing discharge, but braided channels respond distinctly differently. The b-m-f diagram may help interpret changes in at-a-station hydraulic geometry (Knighton, 1975).

4.6 Analytical Determination of Exponents and Coefficients

It is hypothesized that at-a-station hydraulic geometry (AHG) can be determined by cross-section geometry and hydraulic relations (Dingman, 1984; Ferguson, 1986). To analytically derive the AHG, Dingman (2007) employed a general form of the cross-sectional geometry represented by a parabolic section that can specialize into triangular, Lane-type B stable section, convex, and rectangular sections. Although much of the literature assumes river cross-sections to be rectangular, it is worthwhile to derive here the AHG for the general cross-section

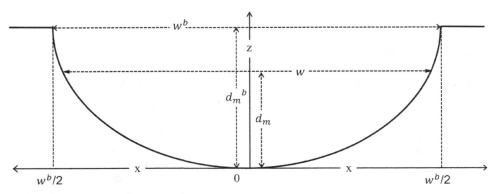

Figure 4.8 Parabolic channel shape

following Dingman (2007). In the derivation, there are three main steps involved: (1) symmetrical general channel cross-section, (2) relation of average depth and water-surface width to maximum depth and bankfull width, and (3) generalized power-law hydraulic relation.

Let the channel cross-section be parabolic in shape, as represented in Figure 4.8, where the maximum depth of flow is denoted as d_m, the water surface width as w, bankfull width as w^b, and bankfull maximum water depth in the cross-section as d_m^b. The cross-section distance from the channel center is represented by x and the vertical distance from the x-axis is denoted by z such that z represents the vertical distance to the bed and hence the value of $z = 0$ at the center. Thus, the channel cross-section can be presented as

$$z = d_m^b \left(\frac{2}{w^b} \right)^r, \quad 0 \leq x \leq \frac{w^b}{2}, \tag{4.24}$$

where r is an exponent. As shown in Figure 4.9, Equation (4.24) leads to a triangular section if $r = 1$, to the Lane-type B stable section if $r \approx 1.75$ (Henderson, 1966), a parabolic section if $r = 2$, a convex cross-section if $r < 1$, and a rectangular section if $r \to \infty$. Jowett (1998) found $r \leq 2$ for 73 cross-sections of New Zealand rivers and $r < 1$ for 25 of these cross-sections.

From Equation (4.24), Dingman (2007) expressed the mean flow depth as

$$d = \left(\frac{r}{1+r} \right) d_m \quad \text{or} \quad d_m = \left(\frac{1+r}{r} \right) d \tag{4.25}$$

and water surface width as

$$w = w^b \left(\frac{d_m}{d_m^b} \right)^{\frac{1}{r}} = w^b \left(\frac{1}{d_m^b} \right)^{\frac{1}{r}} \left(\frac{1+r}{r} \right)^{\frac{1}{r}} d^{\frac{1}{r}}. \tag{4.26}$$

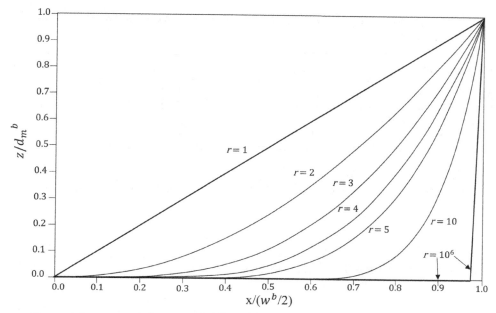

Figure 4.9 Plot of z/d_m^b versus $\left(\frac{x}{w^b}\right)^r$ or various values of r

Dingman (2007) expressed a generalized power-law hydraulic relation for velocity as

$$v = Kd^{p_0}S^{q_0},$$ (4.27)

where p_0 and q_0 are the exponents, and K is the generalized conductance coefficient. Equation (4.27) leads to commonly used hydraulic relations as shown in Table 4.7.

Now the AHG equations are derived. First, the flow depth–discharge relation is derived. Inserting Equations (4.25)–(4.27) in the continuity Equation (4.6a), the result is

$$Q = w^b \left(\frac{1}{d_m^b}\right)^{\frac{1}{r}} \left(\frac{1+r}{r}\right)^{\frac{1}{r}} KS^{q_0} d^{1+\frac{1}{r}+p_0}.$$ (4.28)

Equation (4.28) can be recast in terms of d as

$$d = \left[\left(\frac{1}{w^b}\right)(d_m^b)^{\frac{1}{r}}\left(\frac{r}{1+r}\right)^{\frac{1}{r}}\left(\frac{1}{K}\right)\left(\frac{1}{S^{q_0}}\right)\right]^{\left(\frac{r}{1+r+rp_0}\right)} Q^{\frac{r}{1+r+rp_0}}.$$ (4.29)

Table 4.7. *Special cases of Equation (4.27) [K values are in SI units; h_r is the roughness height.]*

Hydraulic relation	Value of K	Value of p	Value of q
Manning's equation	$1/n$	0.667	0.500
Chezy's equation	$0.552\ C$	0.500	0.500
Bjerklie et al. (2005)	7.33	0.667	0.333
Power law	$22.874/h_r^{0.5}$	1.00	0.500

From Equation (4.29), the exponent of Q for flow depth can be expressed as

$$f = \frac{r}{1+r+rp_0} \tag{4.30}$$

and coefficient c of the depth-discharge as

$$c = \left(\frac{1}{w^b}\right)^{\frac{r}{1+r+rp_0}} \left(d_m^b\right)^{\frac{1}{1+r+rp_0}} \left(\frac{r}{1+r}\right)^{\frac{1}{1+r+rp_0}} \left(\frac{1}{K}\right)^{\frac{r}{1+r+rp_0}} \left(\frac{1}{S^{q_0}}\right)^{\frac{r}{1+r+rp_0}}. \tag{4.31}$$

The width–discharge relation can now be derived. With d replaced by the depth–discharge relation $\left(d = cQ^f\right)$ in Equation (4.6a) with velocity replaced by Equation (4.27), one can write

$$Q = w\left(cQ^f\right)\left[K\left(cQ^f\right)^{p_0} S^{q_0}\right]. \tag{4.32}$$

Equation (4.32) can be cast as

$$w = \left[\left(\frac{1}{c}\right)^{1+p_0} \left(\frac{1}{K}\right)\left(\frac{1}{S^{q_0}}\right)\right] Q^{1-f-fp_0}. \tag{4.33}$$

With the use of Equations (4.30) and (4.31), Equation (4.33) can be expressed as

$$w = \left[\left(\frac{1}{c}\right)^{1+p_0} \left(\frac{1}{K}\right)\left(\frac{1}{S^{q_0}}\right)\right] Q^{\frac{1}{1+r+rp_0}}. \tag{4.34}$$

From Equation (4.34), exponent b of the width–discharge relation can be expressed as

$$b = \frac{1}{1+r+rp_0} \tag{4.35}$$

and coefficient a of the relation as

$$a = \left[\left(\frac{1}{c} \right)^{1+p_0} \left(\frac{1}{K} \right) \left(\frac{1}{S^{q_0}} \right) \right]. \tag{4.36}$$

Coefficient c in Equation (4.36) can be replaced by Equation (4.31), leading to

$$a = \left(w^b \right)^{\frac{r+rp_0}{1+r+rp_0}} \left(\frac{1}{d_m^b} \right)^{\frac{1+p_0}{1+r+rp_0}} \left(\frac{1+r}{r} \right)^{\frac{1+p_0}{1+r+rp_0}} \left(\frac{1}{K} \right)^{\frac{1}{1+r+rp_0}} \left(\frac{1}{S^{q_0}} \right)^{\frac{1}{1+r+rp_0}}. \tag{4.37}$$

The velocity–discharge relation can now be expressed by expressing Equation (4.6b) as

$$Q = aQ^b c Q^f v. \tag{4.38}$$

Equation (4.38) can cast as

$$Q^{1-b-f} = acv. \tag{4.39}$$

Equations (4.6b) and (4.7a) lead to

$$v = \frac{1}{ac} Q^{1-b-f}. \tag{4.40}$$

It is noted that coefficients a and c are given, respectively, by Equations (4.37) and (4.31).

Exponent m of the velocity–discharge ratio can be obtained by inserting Equations (4.30) and (4.35) in Equation (4.40) as

$$m = \frac{rp_0}{1+r+p_0}. \tag{4.41}$$

Coefficient k of the velocity–discharge relation can be obtained by inserting Equations (4.37), and (4.31) in Equation (4.40) as

$$k = \left(\frac{1}{w^b} \right)^{\frac{rp_0}{1+r+rp_0}} \left(d_b^* \right)^{\frac{p_0}{1+r+rp_0}} \left(\frac{r}{1+r} \right)^{\frac{p_0}{1+r+rp_0}} K^{\frac{1+r}{1+r+rp_0}} S^{\frac{q_0(1+r)}{1+r+rp_0}}. \tag{4.42}$$

It is now interesting to note that

$$\frac{b}{f} = \frac{\frac{1}{1+r+rp_0}}{\frac{r}{1+r+rp_0}} = \frac{1}{r} \quad \text{or} \quad \frac{f}{b} = r. \tag{4.43}$$

Exponents b, f, and m given, respectively, Equations (4.35), (4.30), and (4.41) satisfy the continuity Equation (4.6b) as shown here:

$$b + f + m = \frac{1}{1 + r + rp_0} + \frac{r}{1 + r + rp_0} + \frac{rp_0}{1 + r + rp_0} = 1. \qquad (4.44)$$

Dingman (2007) discussed the variability of exponents and coefficients associated with AHG. Exponents vary with cross-section shape r and the depth exponent in the hydraulic relation. For all hydraulic relations, b decreases but f and m increase with the cross-section tending to be more rectangular. Further, b is a stronger function of r than are f and m. Comparison with empirical values reported in the literature showed that empirical values of b $(0.04 < b < 0.26)$ and f $(0.28 < f < 0.42)$ fell within the theoretical ranges, and in all cases $f > b$ (for $r > 1$). However, $f > m > b$ was not always true. The empirical range of m was from 0.33 to 0.55.

Equations (4.37), (4.31), and (4.42), respectively, show that coefficients a, c, and k depend on geometric and hydraulic relations in a complicated manner. Dingman (2007) investigated the variation of these coefficients with r for specifying baseline values of bankfull width, bankfull depth, K, and slope. Coefficient a increases with r. Exponent p only slightly affects the coefficients. However, the influence of p on k depends on the value of q.

Dingman and Afshari (2018) verified the analytical AHG relations using 109 survey cross-sections from 34 reaches of rivers in New Zealand whose bankfull width varied from 8 to 284 meters each with 4–13 measured flows. These relations were found to be in good agreement with empirical relations derived using regression analysis.

4.7 Comparison with Stable Irrigation Channels

The discharge and sediment load as well as the volume, density, and size of bed material in any reach are connected to the reaches above and below it, because the river is a continuum. If the controlling variables remain more or less constant over a long period of time then it is expected that the river system would attain stability. Wolman (1955) compared the Brandywine Creek data with several formulas derived from canals and rivers in India, as shown in Table 4.8, and found a great deal of similarity.

4.8 Reconstruction of Discharges

Dury (1976) developed equations for predicting discharge from channel dimensions including slope of existing channels and modified these equations for predicting discharge of former channels. These equations were derived as follows. First, the wave length (l, m) was related to bed width (w, m) as

Table 4.8. *Comparison between formulas for Brandywine Creek, Pennsylvania, and formulas for stable channels in alluvial material*

Author	Location or other data	Formula	Remarks
Lindley (1919)	Canals, India	$P = 1.984Q^{0.506}$	
Lacey (1939)	Punjab, India	$v = 1.138R^{0.4995}$	
	Madras, India	$v = 0.79R^{0.508}$	Finer silt
	Rivers, India	$v = 16.0R_s^{\frac{2}{3}}S^{0.50}$	
	Punjab, India	$P = 2.80Q^{0.50}$	
	Jamrao, India	$P = 2.5Q^{0.50}$	
	Rivers and torrents, India	$P = 2.63Q^{0.506}$	
	Canals, India	$v = 1.55f_s^{0.50}R^{0.50}$	
	Canals, India	$S = 0.000542f_s^{\frac{5}{3}}Q^{-1/6}$	
Punjab Research Institute (Lacey, 1939)	India	$R = 0.47Q^{0.50}$	
	India	$S = 0.00209D^{0.86}Q^{-0.39}$	
Woods, F. W. (Lane, 1937)	Canals	$B = d^{2.305} - 0.5d$	
Moleworth and Yenudunia (Lane, 1937)	Canals	$d = (9060S + 0.0725)w^{0.50}$	
Schoklitsch (1937)	Rivers	$w = a_0Q^{0.60}$	
Wolman (1955)	Brandywine Creek	$w = aQ^{0.57}$	Tributaries included
	Brandywine Creek	$w = aQ^{0.40}$	Average between principal stations
	Brandywine Creek	$d = 0.089w^{0.70}$	Tributaries included
	Brandywine Creek	$v = kd^{-0.75}$	Tributaries included

Symbols: a = a coefficient; a_0 = a coefficient; B = bed width (ft.); d = mean depth for wide channels (ft.); D = diameter of silt particles (mm); f_s = silt factor; P = wetted perimeter (ft.); Q = discharge (cfs); R = hydraulic radius (ft.); S = slope in ft./ft.; v = men velocity (fps); w = width in feet, approximately equal to P and B.

$$l = 11w. \tag{4.45}$$

Second, the width–discharge relation was expressed as

$$w = 2.99Q^{0.55}. \tag{4.46}$$

Third, using Equations (4.45) and (4.46), the wave length was expressed as a function of discharge as

$$l = 11w = 32.89Q^{0.55}. \tag{4.47}$$

Fourth, slope was related to discharge as

$$S = 0.0396Q^{-0.74}. \tag{4.48}$$

Fifth, for discharge prediction,

$$Q = \frac{1}{2}\left[\left(\frac{w}{2.99}\right)^{1.81} + 0.83A_c^{1.09}\left(\frac{S_0}{S_e}\right)^{0.25}\right], \tag{4.49}$$

where A_c is the stereotype cross-sectional area, S_o is the observed slope, and S_e is the stereotype slope. The expected slope as the stereotype slope was expressed as

$$S_e = 0.0198\left[\left(\frac{w}{2.99}\right)^{1.81} + 0.83A_c^{1.09}\right]^{-0.74}. \tag{4.50}$$

With the inclusion of wave length, Equation (4.50) was further refined as

$$S_e = 0.0132\left[\left(\frac{w}{2.99}\right)^{1.81} + \left(\frac{l}{32.857}\right)^{1.81} + 0.83A_c^{1.09}\right]^{-0.74}. \tag{4.51}$$

Sixth, the discharge prediction equation was formulated as

$$Q = \frac{1}{3}\left[\left(\frac{w}{2.99}\right)^{1.81} + \left(\frac{l}{32.857}\right)^{1.81} + 0.83A_c^{1.09}\left(\frac{S_0}{S_e}\right)^{0.25}\right]. \tag{4.52}$$

Seventh, the corresponding values for Equation (4.52) were expressed as

$$Q_e = 0.915\, Q_o, \tag{4.53}$$

in which Q_0 is the mean observed discharge and Q_e is the mean stereotype discharge ($Q_{1.58}$).

These equations can be applied to ungaged basins. Leopold et al. (1964) indicated that the ratio of mean discharge and $Q_{1.58}$ is about 0.115, which will be subject to regional variation. Dury (1976) expressed Manning's roughness for the stereotype (existing) stream as

$$n = 0.14 A_c^{-0.22}.$$ (4.54)

The depth-wavelength was written as

$$d = 0.0452 l^{0.654}.$$ (4.55)

When width was included, Equation (4.55) was revised as

$$d = \frac{1}{2} \left[0.217 w^{0.654} + 0.452 l^{0.654} \right].$$ (4.56)

The ratio of former wave length (l_F) to former width (w_F) was found to be

$$l_F = 10.31 w_F.$$ (4.57)

The ratio of former depth (d_F) to expected depth (d_e) was expressed as

$$\frac{d_F}{d_e} = S_{in}^{-0.2863} \left(\frac{w_F}{w} \right)^{-0.36},$$ (4.58)

where S_{in} is the sinuosity. Incorporation of wave length led to

$$\frac{d_F}{d_e} = S_{in}^{-0.2863} \left(\frac{l_F}{l} \right)^{-0.36}.$$ (4.59)

When both width ratio and wavelength ratio were included, Equation (4.59) was revised as

$$\frac{d_F}{d_e} = S_{in}^{-0.2863} \left\{ \frac{1}{2} \left[\left(\frac{w_F}{w} \right) + \left(\frac{l_F}{l} \right) \right] \right\}^{-0.36}.$$ (4.60)

These equations can be employed to compute former depth.

References

Andrews, E. D. (1980). Effective and bankfull discharges of streams in the Yampa River basin, Colorado and Wyoming. *Journal of Hydrology*, Vol. 46, pp. 311–330.

Bjerklie, D. B., Moller, D., Smith, L. C., and Dingman, S. L. (2005). Estimating discharge in rivers using remotely sensed hydraulic information. *Journal of Hydrology*, Vol. 309, pp. 191–209.

Brush, L. M (1961). *Drainage basins, channels, and flow characteristics of selected streams in Central Pennsylvania*. U.S. Geological Survey Professional Paper 282-F, pp. 145–181, Washington, DC.

Castro, J. M. and Jackson, P. I. (2001). Bankfull discharge recurrence intervals and regional hydraulic geometry relationship patterns in Pacific Northwest, USA. *Journal of American Water Resources Association*, Vol. 37, No. 5, pp. 1249–1262.

Chang, H. H. (1988). *Fluvial Processes in River Engineering*. Krieger Publication Company, Melbourne, FL.

Devdariani, A. (1967). The profile of equilibrium and a regular regime. Soviet Geology Review and Translation, March 1967, pp. 168–183 (from Voprosy Geograffi, Quantitative Methods in Geomorphology, No. 63, 1963, pp. 33–48).

Dingman, S. L. (1984). *Fluvial Hydrology*. W.H. Freeman, New York.

Dingman, S. L. (2007). Analytical derivation of at-a-station hydraulic geometry relations. *Journal of Hydrology*, Vol. 334, pp. 17–27.

Dingman, S. L. and Afshari, S. (2018). Field verification of analytical hydraulic geometry relations. *Journal of Hydrology*, Vol. 564, pp. 850–872.

Dury, G. H. (1976). Discharge prediction, present and future, from channel dimensions. *Journal of Hydrology*, Vol. 30, pp. 219–245.

Dury, G. H. (1981). Magnitude-frequency analysis and channel morphometry. In: *Fluvial Geomorphology*, edited by M. Morisawa, Allen and Unwin, London, pp. 91–121.

Dury, G. H., Hails, J. H., and Robbie, M. B. (1963). Bankfull discharge and the magnitude-frequency series. *Australian Journal of Science*, Vol. 26, pp. 123–124.

Emmett, W. W. (1975). *The channels and waters of the Upper Salmon River area, Idaho. (hydrologic evaluation of the Upper Salmon River area)*. Geological Survey Professional Paper 870-A, U.S. Department of the Interior, Geological Survey, Washington, DC, pp. 116.

Ferguson, R. I. (1973). Channel pattern and sediment type. *Area*, Vol. 5, pp. 38–41.

Ferguson, R. I. (1986). Hydraulics and hydraulic geometry. *Progress in Physical Geography*, Vol. 10, pp. 1–31.

Goodman, P. (2004). Analytical solutions for estimating effective discharge. *Journal of Hydraulic Engineering*, Vol. 130, No. 8, pp. 729–738.

Gordon, N. D., McMahon, T. A., and Finlayson, B. L. (1992). *Stream Hydrology: An Introduction for Ecologists*. pp. 526, Wiley & Sons, New York.

Griffiths, G. A. (1980). Hydraulic geometry relationships of some New Zealand gravel-bed rivers. N.Z. *Journal of Hydrology*, Vol. 19, pp. 106–118.

Henderson, F. M. (1966). *Open Channel Flow*. Macmillan, New York.

Henderson, R. and Ibbitt, R. (1996). When the going gets tough, the tough get going: field work in a sub-alpine basin on the West Coast, South Island. *Water Atmosphere*, Vol. 4, No. 4, pp. 8–10.

Johnson, P. A. and Heil, T. M. (1996). Uncertainty in estimating bankfull conditions. *Water Resources Bulletin*, Vol. 39, No. 6, pp. 1283–1291.

Jowett, I. G. (1998). Hydraulic geometry of New Zealand rivers and its use in as a preliminary methods of habitat assessment. *Regulated Rivers: Research and Management*, Vol. 14, pp. 451–466.

Klein, M. (1981). Drainage area and the variation of channel geometry downstream. *Earth Surface Processes & Landforms*, Vol. 6, pp. 589–593.

Knighton, A. D. (1974). Variation in width-discharge relation and some implications for hydraulic geometry, *Geological Society of America Bulletin*, Vol. 85, pp. 1069–1076.

Knighton, A. D. (1975). Variation of at-a-station hydraulic geometry. *American Journal of Science*, Vol. 275, pp. 186–218.

Kolberg, F. J. and Howard, A. D. (1995). Active channel geometry and discharge relations of U. S. piedmont and midwestern streams: The variable exponent model revisited, *Water Resources Research*, Vol. 31, No. 9, pp. 2353–2365.

Lacey, G. (1930). Stable channels in alluvium. *Proceedings of Institution of Civil Engineers*, Vol. 119, pp. 281–290.

Lacey, G. (1939). *Regime Flow in Incoherent Alluvium*. Central Board of Irrigation (India), Publication 20, Simla, India.

Lane, E. W. (1937). Stable channels in erodible channels. *Transactions of the ASCE*, Vol. 102, pp. 1236–1260.

Langbein, W. B. (1942). Hydraulic criteria for sand waves. Transactions of American Geophysical Union, Part 2, pp. 615–618.

Langbein, W. B. and Leopold, L. B. (1964). Quasi-equilibrium states in channel morphology. *American Journal of Science*, Vol. 262, pp. 782–794.

Leopold, L. B. (1994). *A View of the River*. Harvard University Press, Cambridge, MA.

Leopold, L. B. and Langbein, W. B. (1962). *The concept of entropy in landscape evolution.* U.S. Geological Survey Professional Paper 500-A, pp. 20, Washington, DC.

Leopold, L. B. and Maddock, T. (1953). *The hydraulic geometry of stream channels and some physiographic implications*. Geological Survey Professional Paper 252, U.S. Geological Survey, Washington, DC.

Leopold, L. B. and Skibitzke, H. E. (1967). Observations on unmeasured rivers. *Geografiska Annales*, Vol. 49A, pp. 247–255.

Leopold, L. B., Wolman, M. G., and Miller, J. F. (1964). *Fluvial Processes in Geomorphology*. Freeman, San Francisco.

Leopold, L. B., Wolman, M. G., and Miller, J. P. (1995). *Fluvial Processes in Geomorphology*. Dover, New York.

Lewis, L. A. (1966). The adjustment of some hydraulic variables at discharges less than one cfs. *Professional Geographer*, Vol. 18, pp. 230–234.

Lindley, E. S. (1919). Regime channels. *Proceedings of the Punjab Engineering Congress*, Vol. 7, pp. 63–74.

Lowham, H. W. (1982). *Streamflow and channels of the Green River basin, Wyoming*. U.S. Geological Survey Water Resources Investigations Report 81-71, pp. 73, Washington, DC.

Merigliano, M. F. (1997). Hydraulic geometry and stream channel behavior: an uncertain link. *Journal of American Water Resourcs Association*, Vol. 33, No. 6, pp. 1327–1336.

Nash, D. B. (1994). Effective sediment transporting discharge from magnitude frequency analysis. *Journal of Geology*, Vol. 101, pp. 79–95.

Nixon, M. (1959). A study of bankfull discharges of rivers in England and Wales. *Proceedings of Institution of Engineers*, Vol. 12, No. 2, pp. 157–174.

Park, C. C. (1977). World-wide variations in hydraulic geometry exponents of stream channels: An analysis and some observations. *Journal of Hydrology*, Vol. 33, pp. 133–146.

Petit, F. and Pauquet, A. (1997). Bankfull discharge recurrence interval in gravel-bed rivers. *Earth Surface Processes and Landforms*, Vol. 22, 685–693.

Phillips, P. J. and Harlin, J. M. (1984). Spatial dependency of hydraulic geometry exponents in a subalpine stream. *Journal of Hydrology*, Vol. 71, pp. 277–283.

Pickup, G. and Warner, R. F. (1976). Effects of hydrologic regime on magnitude and frequency of dominant discharge. *Journal of Hydrology*, Vol. 29, pp. 51–76.

Rhoads, B. L. (1991). A continuously varying parameter model of downstream hydraulic geometry. *Water Resources Research*, Vol. 27, No. 8, pp. 1865–1872.

Rhodes, D. D. (1977). The *b-f-m* diagram: Graphical representation and interpretation of at-a-station hydraulic geometry. *American Journal of Science*, Vol. 277, pp. 73–96.

Rhodes, D. D. (1978). Worldwide variations in hydraulic geometry exponents of stream channels: An analysis and some observations: Comments. *Journal of Hydrology*, Vol. 33, pp. 133–146.

Richards, K. S. (1973). Hydraulic geometry and channel roughness: A nonlinear system. *American Journal of Science*, Vol. 273, pp. 877–896.

Richards, K. S. (1976). Complex width–discharge relations in natural river sections. *Geological Society of America Bulletin*, Vol. 87, pp. 199–206.

Schoklitsch, A. (1937). *Hydraulic Structures: A Text and Handbook*. Translated by Samuel Shulits, The American Society of Mechanical Engineers, New York.

Schumm, S. A. (1960). *The shape of alluvial channels in relation to sediment type*. Professional Paper 352B, pp. 17-30, U.S. Geological Survey, Washington, DC.

Schumm, S. A. (1971). Fluvial geomorphology: The historical perspective. In: *River Mechanics*, Vol. 1, edited by H. W. Shen, pp. 4.1–4.30, Fort Collins, Colorado.

Simons, D. B. and Albertson, M. L. (1963). Uniform water conveyance channels in alluvial material. *Transactions, ASCE*, Vol. 128, pp. 65–107.

Singh, V. P. Yang, C. T., and Deng, Z. Q. (2003). Downstream hydraulic geometry relations: II. Calibration and Testing. *Water Resources Research*, Vol. 3, No. 12, 1337, doi:10.1029/2003WR002484.

Stewardson, M. (2005). Hydraulic geometry of stream reaches. *Journal of Hydrology*, Vol. 306, pp. 97–111.

Tanner, W. F. (1971). The river profile. *Journal of Geology*, Vol. 79, pp. 482–492.

Thomas, A. R. (1946). Slope formulae for rivers and canals. *Journal of Central Board of Irrigation (India)*, Vol. 3. No. 1, pp. 40–49.

Vanoni, V. A. (1941). Some experiments on the transportation of suspended load. Transactions of American Geophysical Union, Part 3, pp. 608–620.

Wilcox, D. N. (1971). Investigation into the relation between bed load transport and channel shape. *Geological Society of America Bulletin*, Vol. 82, pp. 2159–2176.

Wilkerson, G. V. (2008). Improved bankfull discharge prediction using 2-year recurrence-period discharge. *Journal of American Water Resources Association*, Vol. 44, No.1, pp. 243–258.

Williams, G. P. (1978). Bankfull discharge of rivers. *Water Resources Research*, Vol. 14, No. 6, pp. 1141–1154.

Wolman, M. G. (1955). *The natural channel of Brandywine Creek, Pennsylvania*. Geological Survey Professional Paper 271, pp. 1–56, U.S. Department of the Interior, Washington, DC.

Wolman, M. G. and Brush, L. M. (1961). *Factors controlling the size and shape of stream channels in coarse noncohesive sands*. Geological Survey Professional Paper 282-G, pp. 183–210, U.S. Department of the Interior, Washington, DC.

Wolman, M. G. and Miller, J. P. (1960). Magnitude and frequency of forces in geomorphic processes. *Journal of Geology*, Vol. 68, pp. 54–74.

5

Theory of Minimum Variance

The concept of minimum variance is a statistical concept, and its premise is that a river tends to minimize the variability of factors that govern its hydraulic geometry. This concept has been applied in different ways, and this chapter discusses the theory of minimum variance from different viewpoints.

Notation

a = constant of proportionality

b = exponent

c = constant of proportionality

d = depth

$d(x, y)$ = distance between two vectors

f = exponent

f_c = friction

H = head loss

J = exponent for Darcy–Weisbach friction factor

k = constant of proportionality

m_i = exponent of discharge Q_i corresponding to variable x_i

P = joint probability represented as $\{p_i\, p_2 \ldots p_n\}$

p_i = probability of variable x taking on the value x_i

Q = discharge

r and ρ = coefficient of correlation

r^2 = coefficent of determination

s = amount of adjustment

S = slope of the energy line approximated by bed slope

SP = stream power

Continued

USP = unit stream power
v = velocity
w = width
x = vector
X = set of n variables represented as $\{x_i, i = 1, 2, \ldots, n\}$
$\bar{x}_i$ = mean
$\|x\|$ = length of x
y = vector
z = exponent for slope
$\alpha_i = (b_i, f_i, m_i)$
β = regression coefficient
$\lambda = (m - b)/(f - b)$
σ_i^2 = variance
τ = shear stress
γ = weight density of water

5.1 Introduction

Since natural streams tend to attain the state of dynamic equilirbium or quasi-equilibrium as postulated by Gilbert (1877), the question is one of quantifying this state. Langbein's (1964) hypothesis of minimum variance is one way of quantitatively expressing this state. There are several variables that describe channel hydraulic geometry. It is often assumed that these variables are related by single-valued functions, as for example in regime theory, ignoring the interactions and different modes of interactions among variables. Three variables that interact in two different ways can have the same effect on the channel geometry. The range of interaction modes can be theoretically analyzed using the concept of minimum variance.

When applying the concept of minimum variance in practice, the question arises as to which variables to include for variance minimization. Leopold and Maddock (1953) discussed adjustments of width, depth, velocity, slope, velocity's n friction factor, and shear stress and expressed them as power functions of discharge. However, all of the variables may not have the same potential for adjustment, because some variables adjust themselves more than others and some variables could be even regarded as constant. Then, the question is: what weights to assign to individual variables to account for their potential for adjustment. There is no apriori means for determining which variables to include and which not to. For example, Maddock (1969) minimized shear stress and frictional resistance. Dozier (1976) investigated the variability of longitudinal bed shear stress. Langbein (1964) listed five aspects of stream power for use and developed three guidelines for applying the concept of minimum variance on a power basis (Riley, 1978): (1) A change in stream power is equally distributed over each component of power, (2) the total variance of the components of stream power tends toward a minimum, and (3) the total work to be done by streamflow tends toward a minimum.

Hydraulic controls may dominate the cross-sectional geometry of streams. If the bed and banks of a stream were noncohesive, then the change in flow and hence the change in energy would be accommodated by changes in width, depth, and velocity, which, in turn, would also accommodate changes in friction and shear stress. On the other hand, the bed and banks of many streams, especially those of distributaries, are of cohesive materials where the roughness elements are independent of short-term flow regime. This means that the friction and shear stress are independnt variables and should be included in the minimization. The energy spent on the bed will be a function of bed form and discharge. For alluvial streams the slope at any cross-section is not constrained. This suggests that the change in energy can be accommodated by the change in slope. Further, the slope of the water surface profile corresponding to various stages of flow will also affect the way energy is distributed among width, depth, velocity, friction, and shear stress. The sediment load of the stream may also influence the energy distribution, as an increase in the sediment load increases the energy expended per unit length and per unit cross-sectional area.

For at-a-station hydraulic gemeotry, an increase in discharge would be accommodated by changes in width, depth, and velocity alone. The theory of minimum variance invokes certain simplifying assumptions: (1) Discharge is the dominant independent variable whose variation is accommodated by the adjustment of dependent hydraulic and channel variables; (2) dependent variables are related to discharge in simple power form; (3) not all variables have the same potential for adjustment and the three most important dependent variables are width, depth, and velocity; and (4) the joint probability of adjustment of dependent variables tends to a maximum.

The theory of minimum variance for deriving hydraulic geometry relations has been formulated in different ways, including: (1) minimum sum of variances of independent variables, (2) minimum sum of variances of stream power components, (3) maximum joint probability of adjustment of hydraulic geometry, and (4) minimum variance of stream power. This chapter discusses each of these ways.

The theory of minimum variance has been applied to a number of hydraulic problems. Langbein and Leopold (1966) showed that meandering was the most probable planimetric geometry for channels with alternating pools and riffles. Maddock (1969) characterized the behavior of straight channels subject to bed sediment movement. Emmett (1970) derived hillslope profiles for given conditions of initial state, flow type, and runoff rate. However, the minimum variance theory has come under criticism. Kennedy et al. (1964) argued that the effect of sediment transport was not properly accounted for and some of the probabilistic and mathematical reasoning was untenable. For example, the most probable state among alternatives usually may not hold and at any given time, a river network will not likely be in its most probable state. Chorley (1962) showed that a quasiequilibrium state adjusted to prevailing conditions. Richards (1973) argued that log-quadratic functions rather than log-linear would be more appropriate for describing depth–discharge and velocity–discharge relations because of the nonlinearity of resistance variation. The theory does not accommodate geomorphic forms that are controlled by geologic structure and history. It explains only general trends in hydrulic characteristics over a broad range that influence river morphology but do not explain discontinuities and nonlinearities. Dozier (1976) examined the variance minimization tendencies in a supraglacial stream and discussed the lack of physical mechanisms in the development of the theory of minimum variance. Riley (1978) argued that the theory cannot be used to predict with any degree of certainty a hydraulic condition and geometry directly relevant to the stream to be investigated but does allow for the probable conditions to be selected. For a range of hydraulic conditions, the theory can predict a range of possible shapes.

5.2 Minimum Sum of Variances of Independent Variables

A channel system is characterized by a number of variables both geometric (depth, width, and slope) and hydraulic (velocity, roughness, and shear stress). Although these variables are not independent, for simplicity they can be assumed to be independent and this is a key assumtpion, which is difficult to justify. Assuming discharge as the independent variable, it is further assumed that each of the aforementioned independent variables is related to discharge as a single valued function. This assumption is also tenuous.

The theory of minimum variance assumes that the independent variables mutually adjust, subject to constraints, such that the sum of their variances is minimum. Riley (1978) has given a good discussion on the role of this theory that is followed in this section. If a particular variable is considered for an ensemble of streams then this variable has as many values as the number of streams. One can compute the mean and variance of these values. It can be assumed that these values of the variable will have a Gaussian or normal probability distribution. Now consider that each of the variables that characterize the channel system has a mean and variance and a normal probability distribution. Let $X = \{x_i, I = 1, 2, \ldots, n\}$ be the set of n variables x_i, $i = 1, 2, \ldots, n$, having mean $\bar{x}_i$ and variance σ_i^2. Then, the probability, p_i, of variable x_i taking on the value x_{ij} (jth value of variable x_i) can be expressed as:

$$p_i \propto \exp\left(-\sigma_i^2\right). \tag{5.1}$$

The theory of minimum variance says that in order for the ensemble of streams to be in their most probable state, the joint probability, $P = \{p_i\, p_2 \cdots p_n\}$, of the variable set $X = \{x_i, I = 1, 2, \ldots, n\}$ is to be maximized. The joint probability under the assumption of each variable following the normal distribution can be expressed as

$$P = \{p_1\, p_2 \ldots p_i\} \propto \exp\left(-\sum_{i=1}^n \sigma_i^2\right). \tag{5.2}$$

The maximization of the joint probability means that the sum of variances

$$\text{Sum of variances} = \sum_{i=1}^n \sigma_i^2 \tag{5.3}$$

must be minimum.

Since each variable is a unique function of discharge Q raised to some power m, one can write

$$x_i \propto Q_i^{m_i} \quad \text{or} \quad \ln x_i = a_i + m_i \ln Q_i, \tag{5.4}$$

where m_i is the exponent of discharge Q_i corresponding to variable x_i. Taking the variance of Equation (5.4), one finds that the square of the exponent is proportional to the variance and one can then write

$$m_i^2 \propto \sigma_i^2 = \left(\frac{\sigma_{\ln x_i}}{\sigma_{\ln Q_i}}\right)^2, \tag{5.5a}$$

where $\sigma_{\ln x_i}$ and $\sigma_{\ln Q_i}$ are the standard deviations of log x_i and log Q_i, respectively. Using Equation (5.4), it can be shown that the proportionality in Equation (5.5a) can be replaced by the coefficient of determination r^2:

$$m_i^2 = r^2 \left(\frac{\sigma_{\ln x_i}}{\sigma_{\ln Q_i}}\right)^2. \tag{5.5b}$$

The exponent is also proportional to the coefficient of variation of the variable, i.e.,

$$m_i^2 \propto \frac{\sigma_{x_i}^2}{(\bar{x}_i)^2}. \tag{5.6}$$

Either the individual hydraulic variables, for example, velocity, width, and depth, or the components of stream power may be minimized. The variables to be included in the minimization procedure must be chosen, subject to the constraints to be applied, and the variance contributed by each variable must be computed. Then, the resulting function is solved for its minimum value.

It is assumed that velocity (v), depth (d), width (w), friction (f_c), and shear stress (τ) are the variables that determine the river state and that these variables operate free of constraints. Each variable is assumed to contribute equally to the river system, that is, the variables have a weight of unity. At a particular station or cross-section, the following is assumed:

$$v \propto Q^m \tag{5.7}$$

$$d \propto Q^f \tag{5.8}$$

$$w \propto Q^b. \tag{5.9}$$

Then, with perfect correlation

$$\sigma_v^2 = m^2 \tag{5.10}$$

$$\sigma_d^2 = f^2 \tag{5.11}$$

$$\sigma_w^2 = b^2. \tag{5.12}$$

The variance contributed by the shear stress (τ) is

$$\sigma_\tau^2 = f^2 \tag{5.13}$$

because

$$\tau \propto dS, \tag{5.14}$$

where S is the slope of the energy line that can be approximated by the bed slope, which is considered constant.

Using the Darcy–Weisbach friction factor (f_c), the variance contributed by friction

$$f_c \propto \frac{dS}{v^2} \tag{5.15}$$

can be written as

$$\sigma_{f_c}^2 = (f - 2m)^2. \tag{5.16}$$

The sum of variances of variables is

$$var = m^2 + f^2 + b^2 + f^2 + (f - 2m)^2. \tag{5.17a}$$

Eliminating b from Equation (5.17a) by utilizing $m + f + b = 1$,

$$var = m^2 + f^2 + (1 - m - f)^2 + f^2 + (f - 2m)^2$$
$$= 1 + 4f^2 - 2m - 2f - 2fm + 6m^2. \tag{5.17b}$$

Differentiating Equation (5.17b) with respect to m, one gets

$$\frac{\partial var}{\partial m} = -2f - 2 + 12m = 0, \tag{5.18}$$

and differentiating Equation (5.17b) with respect to f, one obtains

$$\frac{\partial var}{\partial f} = 8f - 2 - 2m = 0. \tag{5.19}$$

Solving Equations (5.18) and (5.19) simultaneously, one obtains

$$m = 0.22; f = 0.30; b = 0.48; \frac{b}{f} = 1.57; \frac{b}{f+b} = 0.61; \frac{f}{b+f} = 0.40. \tag{5.20}$$

Thus, a cross-section in which v, w, d, τ, and f_c are free to accommodate changes in discharge and react among themselves will develop a cross-section with the following characteristics:

$$v \propto Q^{0.22} \tag{5.21}$$

$$d \propto Q^{0.30} \tag{5.22}$$

$$w \propto Q^{0.48}. \tag{5.23}$$

Using the regime concept, Lindley (1919) derived a relation for width

$$w = 3.8d^{1.61}, \tag{5.24}$$

which is similar to Equation (5.23) when Q is replaced by d using Equation (5.22).

At cross-sections on many rivers, canals, and laboratory flumes, the variables, velocity, depth, width, slopes, and friction factor (f_c), all considered dependent, often change in proportion to some power of water discharge, considered independent. However, the bed and banks of many streams are of cohesive material and their roughness elements are independent of short-term flow regime. In such cases the energy spent on the bed depends on bed forms and discharge. This then suggests that to Equations (5.7)–(5.9) one can also add

$$S \propto Q^z \tag{5.25}$$

$$f_c \propto Q^J, \tag{5.26}$$

where the z and J exponents, called hydraulic exponents, represent the rate of change of the dependent variables S and f_c, respectively, with the change of independent variable Q.

To illustrate the effect of choice of variables here, consider the case where the river reach is straight and uniform with constant slope and its banks are vertical. This means that the discharge exponent for slope is 0 and the discharge exponent for width, b, is also 0. Then, the depth exponent is f, the velocity exponent m is $1 - f$, the shear stress exponent is f, and the friction factor exponent is $f - 2m = f - 2(1 - f) = 3f - 2$. From the minimum variance hypothesis,

$$f^2 + (1 - f)^2 + (3f - 2)^2 + f^2 \rightarrow \text{minimum.} \qquad (5.27)$$

Equation (5.27) can be simplified as

$$12f^2 - 14f + 5 \rightarrow \text{minimum.} \qquad (5.28)$$

Differentiating Equation (5.28) with respect to f and equating the derivative to 0, the result is

$$24f - 14 = 0 \quad \text{or} \quad f = 0.58. \qquad (5.29)$$

Thus, the exponents become: $b = 0, f = 0.58, m = 0.42$, friction exponent $= -0.26$, and shear stress exponent $= 0.58$. These values are different from those derived previously. However, only small differences occur in the predicted exponents for various combinations of variables and constraints (Richards, 1982).

For the determination of exponents in practice, one can fit, using the least squares method, any two relations, such as depth and width with discharge, and then determine the exponent of velocity from the sum of the three exponents equal to one. For the Lower Namoi-Gwydir drainage basin, Australia, Riley (1978) noted that the depth–discharge relation was more precise than the velocity–discharge relation because the variation in velocity was larger than that in depth or width (Brush, 1961). Also, the sediment load was not a critical constraint for the channel system, but the boundary sediment was a significant constraint, which partly explains the variation in exponents.

It may be noted that the minimum variance theory assumes that the hydraulic geometry is invariant with change in discharge and its adjustment occurs within a stable section. If the primary variables for adjustment are only width, depth, and velocity and there are no constraints then the theory will yield $b = f = m = 0.33$. In practice, the exponents will depend on sedimentological constraint in individual sections. Knighton (1974) discussed mechanisms that link the bank silt-clay content to bank slope, which will relate to the width exponent. Using regression analysis, Williams (1978) expressed the width exponent in terms of the exponent of the width-area relation and the depth exponent in terms of this exponent and the median grain size diameter.

5.3 Minimum Total Variance of Components of Stream Power

Power relations do not necessarily hold for the complete range of flow at a given cross-section (Richards, 1976). More than one power relation can exist for within-bank flows. For instance, very low flows may follow one power law, and flows approaching bankfull flow may follow another. Richards (1973) pointed out that for some cross-sections, simple power functions did not apply. He suggested quadratic or higher order curves for such sections. Hydraulic relations change drastically when a river overflows its banks.

Langbein (1964) reasoned that the mean channel form (equilibrium or quasi-equilibrium) fulfilled necessary hydraulic laws and postulated this most probable condition in the following terms: (1) In accommodating a change in stream power, the channel geometry changes such that each component of power changes as equally as is possible, and (2) the total variance of the components of power tends toward a minimum, or as a corollary, their co-variance tends toward a maximum. A further generalization is that the total work tends toward a minimum. The first term is compatible with the principle of equal action along a river in the development of stream morphology referred to by Gilbert (1877).

From the continuity equation or the definition of discharge,

$$Q = vdw. \tag{5.30}$$

This continuity equation yields $Q \propto Q^m Q^f Q^b$, which leads to $m + f + b = 1.0$.

Consider a river where the rate of fluid friction is measured by the energy slope, and the rate of work (= power) per unit length along a river equals γQS. The variance of power tends toward zero as the product QS tends toward a constant along the river. However, as QS tends toward uniformity, total work, as expressed by the integral $\int \gamma QS \Delta x$, tends toward a minimum. This seems to correspond to the principle of least action or work, which was suggested by Rubey (1953) in stream morphology.

The Langbein hypothesis applies this concept of minimization of effects. In other words, a river system, once having reached an equilibrium condition, changes as little as possible from then on. That is, the system tends to react to an imposed stress in order to minimize the disturbance, or to restore or keep the equilibrium condition. For a cross-section of a river, the hypothesis would suggest that all variables strive to resist any imposed change (maintain equilibrium conditions), with the net result being that all of them change equally insofar as possible. In other words, the dependent variables adjust by an equal percentage of

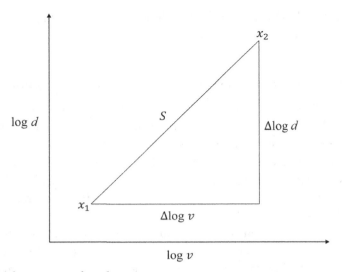

Figure 5.1 log v versus log d

their former values, subject to the restrictions imposed thereon. A typical restriction might be steep cohesive banks that prevent the water surface width from changing significantly as discharge increases.

The problem is to find those particular hydraulic exponent values that, subject to any local physical restrictions, represent a minimum and equal adjustment to a change in the independent variable, usually Q. For these exponent values, the sum of squares of the exponents is a minimum. Consider a system with two dependent variables, say v and d. The question is how these variables change in response to a new discharge.

For a selected discharge, one can plot the associated values of log v and log d on a graph, obtaining point x_1, as shown in Figure 5.1. If it rains upstream, then soon there will be a new discharge at the station. This will force log v and log d to change, by Δlog v and Δlog d, respectively, producing a new point x_2. The line joining x_1 and x_2 forms s (hypotenuse). Assuming that neither depth nor velocity will decrease with the increase in discharge, point x_2 will be somewhere in the first quadrant. Referring to Figure 5.1,

$$s^2 = (\Delta \log v)^2 + (\Delta \log d)^2. \tag{5.31}$$

The change in the independent variable, Δlog Q, is the same for both Δlog v and Δlog d. Dividing all terms in Equation (5.31) by $(\Delta \log Q)^2$,

$$\frac{s^2}{(\Delta \log Q)^2} = \frac{(\Delta \log v)^2}{(\Delta \log Q)^2} + \frac{(\Delta \log d)^2}{(\Delta \log Q)^2} = \left(\frac{\Delta \log v}{\Delta \log Q}\right)^2 + \left(\frac{\Delta \log d}{\Delta \log Q}\right)^2. \tag{5.32}$$

Equation (5.32) expresses the amount of adjustment by the channel system. The terms on the right side can be expressed in terms of hydraulic exponents. For example, for $v \propto Q^m$, one has

$$m = \frac{\Delta \log v}{\Delta \log Q} \tag{5.33a}$$

$$f = \frac{\Delta \log d}{\Delta \log Q}. \tag{5.33b}$$

Thus, Equation (5.32) can be written, using Equations (5.33a) and (5.33b), as

$$\frac{s^2}{(\Delta \log Q)^2} = m^2 + f^2. \tag{5.34}$$

The left side of Equation (5.34) includes only the distance s and the constant Δlog Q and therefore reflects the amount of adjustment s. Thus, the magnitude of adjustment is proportional to the sum of squares of hydraulic exponents. When s is minimum or when the sum of squares of hydraulic exponents is the lowest value, the least adjustment occurs. This type of relation can be extended to three or more variables.

Furthermore, minimizing the sum of squares of exponents corresponds to an equal division of any change in an independent variable. Consider the previous example, wherein the entire increase in discharge was absorbed by v and d. The total adjustment (length of hypotenuse of the triangle whose other sides are $\Delta \log v$ and $\Delta \log d$) is the greatest when the hypotenuse becomes equal to either $\Delta \log d$ or $\Delta \log v$, such that the other is zero. As $\Delta \log d$ becomes more equal than $\Delta \log v$, the hypotenuse decreases, and it reaches its minimum value when $\Delta \log d = \Delta \log v$. In other words, the smallest net adjustment occurs when the dependent variables divide the imposed change equally among themselves.

Another example of Langbein's hypothesis involves the many possible ways in which velocity, depth, and width can adjust to an increase in discharge. Consider the basic relation of exponents, $m + f + b = 1$. For a hypothetical situation involving these three variables without any restrictions, the minimum sum of squares for all possible combinations occurs when $f = b = m = 1/3$. Mathematically, this sum of squares is $(0.33)^2 + (0.33)^2 + (0.33)^2 = 0.3267$. Trial values show that any other combination of values of m, f, and b, where $f + m + b = 1.0$, produces a sum of squares greater than this minimum. For example, if $m = 0.72, f = 0.21$, and $b = 0.07$, then the sum of squares is 0.5674, which is greater than 0.3267. The sum of squares of exponents m, f, and b is a maximum (1.0) when the change is wholly concentrated in one of the three dependent variables, such that the other two remain constant and their exponents are zero. This, of course, is least likely.

The concept of minimizing the sum of squares of changes or deviations is similar to the least squares method for finding the best-fit relation. Let us now consider the meaning of variance. Consider velocity observation at a point. If we compute the variance of logarithmically transformed velocity observations, then such a variance will be proportional to the square of the hydraulic exponent m (exponent of velocity). The same applies to other variables.

Let $\sigma_{\log v}$ be the standard deviation of the log-velocity values. The hydraulic exponent m (for velocity)

$$m = r\left(\frac{\sigma_{\log v}}{\sigma_{\log Q}}\right), \tag{5.35}$$

where r is the coefficient of correlation. Suppose there are 10 different discharge measurements and a mean velocity for each discharge. It is specified that these two variables (velocity and discharge) have a power relation, such as $v \propto Q^m$ or $\log v \propto \log Q$. Let m be 0.5. The ratio of the standard deviations, $\sigma_{\log v}/\sigma_{\log Q}$ should be equal to 0.5. The same applies to each of the other dependent variables, that is,

$$f \propto \frac{\sigma_{\log d}}{\sigma_{\log Q}}; \quad b \propto \frac{\sigma_{\log w}}{\sigma_{\log Q}}; \text{ and so forth.} \tag{5.36}$$

All of the latter relations contain $\sigma_{\log Q}$ as a common factor. The proportionalities, therefore, are still valid if $\sigma_{\log Q}$ is deleted. This leaves $f \propto \sigma_{\log d}$, $b \propto \sigma_{\log w}$, and so on. In other words, each hydraulic exponent is proportional to the standard deviation of the logarithm of its hydraulic variable. This is why variance is used as a term for the "square of hydraulic exponent." The following assumptions are made for these statistical definitions, which are somewhat questionable.

(1) The logarithms of hydraulic variables are treated as random. In actuality, hydraulic variables, rather than being randomly distributed, follow finite laws of fluid mechanics and should be predictable on the basis of these laws. Due to insufficient knowledge, the net result of many predictable actions may often be the same as if the whole channel flow process were random.

(2) The variables are mutually independent. This is, of course, not true, for a change in one variable is usually associated with changes in other variables. Further, some of the variables are interlinked. For example, velocity and depth are not independent but are related.

(3) The variables are approximately normally distributed. This may not necessarily be the case.

Returning to the meaning of variance, the square of a hydraulic exponent is proportional to the variance of the log of the associated dependent variable. Thus, from $d \propto Q^f$ the square of exponent (f^2) is called the variance of depth. The process of finding those exponents whose squares add up to the lowest possible value is designated as "minimizing the variance."

In calculating the most probable exponents in the hydraulic geometry relations, simple laws of exponents apply. Consider shear stress in a wide channel (τ). Because $\tau \propto dS \propto Q^f Q^z \propto Q^{f+z}$, the variance of shear stress equals $(f+z)^2$. The Darcy–Weisbach friction factor (f_c) in wide channels equals

$$f_c = \frac{8gdS}{v^2} \propto \frac{ds}{v^2} \propto \frac{Q^f Q^z}{Q^{2m}}. \tag{5.37}$$

The variance of f_c is

$$var_{f_c} = (f + z - 2m)^2. \tag{5.38}$$

Because $m + f + b = 1.0$, the variance of width is

$$var_w = (1 - m - f)^2. \tag{5.39}$$

Similarly, $f = 1 - m - b$, and $m = 1 - b - f$.

Consider velocity, depth, shear stress, width, friction, and slope as dependent variables.

Then,

$$v \propto Q^m, d \propto Q^f, w \propto Q^b \propto Q^{1-m-f}, S \propto Q^z, \tau \propto dS \propto Q^{f+z}, f_c \propto \frac{dS}{v^2} \propto Q^{f+z-2m}. \tag{5.40}$$

Then, the sum of variances of v, d, w, τ (through depth and slope), and f_c is

$$\text{sum of variances} = m^2 + f^2 + (1 - m - f)^2 + z^2 + (f + z - 2m)^2. \tag{5.41}$$

Squaring the compound variances and collecting the terms,

$$\text{Sum of variances} = 6m^2 + 4f^2 + 2z^2 + 1 - 2m - 2f - 2mf + 4fz - 4mz. \tag{5.42}$$

Taking the first derivative of Equation (5.42) with respect to m and setting it to zero gives

$$12m - 2f - 4z - 2 = 0. \tag{5.43}$$

Taking the first derivative of Equation (5.42) with respect to f and setting it to zero yields

$$-2m + 8f + 4z - 2 = 0. \tag{5.44}$$

Finally, setting the first derivative of Equation (5.42) with respect to z to zero produces

$$-4m + 4f + 4z = 0. \qquad (5.45)$$

Solving Equations (5.43)–(5.45) for m, f, and z, one gets

$$m = 0.14, f = 0.43, z = 0.29, b = 0.43. \qquad (5.46)$$

Thus, the variance minimization yields the hydraulic geometry relations as

$$w \propto 0.43 \qquad (5.47)$$

$$d \propto Q^{0.43} \qquad (5.48)$$

$$v \propto Q^{0.14} \qquad (5.49)$$

$$S \propto Q^{0.29}. \qquad (5.50)$$

Exponents in Equations (5.47)–(5.49) agree reasonably well with field data.

Langbein (1964) dealt with three cases. First, consider the case of a flume with fixed walls on a sand bed with a fixed slope. Then, a change in discharge will be accommodated by a change in depth or velocity. Second, in the case of a river cross-section, the slope is not changing so the change in stream power is accommodated by changes in width, depth, and velcity. Third, in a river channel a change in stream power is accomodated equally by changes in width, depth, velocity, and slope.

5.4 Knight's Formulation of Minimum Variance Hypothesis

The main hypothesis proposed by Knighton (1977) is that stream geometry will approach a quasi-equilibrium state through a sequence of adjustments. This hypothesis leads to a more general formulation of which the minimum variance hypothesis becomes a special case. As before, discharge is regarded as the dominant independent variable to which the dependent hydraulic and channel variables adjust, all other independent variables being controlled to a greater or lesser extent. Simple power functions can be expressed as

$$w = aQ^b \qquad (5.51)$$

$$d = cQ^f \qquad (5.52)$$

$$v = kQ^m, \qquad (5.53)$$

where w is the width; d is the mean depth; v is the mean velocity; Q is the discharge; a, c, and k are constants of proportionality; and b, f, and m are the exponents, with $b + f + m = 1$ from the continuity relation.

For at-a-station hydraulic geometry, an increase in discharge at a cross-section would be accommodated by changes in width, depth, and velocity. It is reasoned that adjustment among those factors that are allowed to vary takes place in such a way that their joint probability of adjustment is a maximum, a condition that is satisfied when all individual probabilities are equal. The condition of maximum probability is considered to be equivalent to the statement that the total variance of the adjustable factors tends toward a minimum. Following Langbein (1964, 1965), the variance of a dependent variable is equal to the square of the corresponding hydraulic exponent of discharge. Thus, the variance of width is b^2, that of depth is f^2, and that of velocity is m^2. Following Knighton (1977), this is based on the following reasoning. Given an independent x and a dependent variable y, the regression coefficient β is

$$\beta = \frac{S_{xy}}{S_x^2},$$

(5.54)

and the correlation coefficient ρ is

$$\rho = \frac{S_{xy}}{\left(S_x^2 S_y^2\right)^{\frac{1}{2}}},$$

(5.55)

where

$$S_{xy} = \text{Covariance of } X \text{ and } Y = \sum_{i=1}^{n}(x_i - \bar{x})(y_i - \bar{y})$$

(5.56)

and

$$S_x^2 = \sum_{i=1}^{n}(x_i - \bar{x})^2$$

(5.57)

$$S_y^2 = \sum_{i=1}^{n}(y_i - \bar{y})^2,$$

(5.58)

where n is the number of values. Substitution of $\sigma_x^2 = S_x^2/n$, where σ_x^2 is the variance of X, into Equations (5.54) and (5.55) yields

$$\beta = \frac{nS_{xy}}{\sigma_x^2}$$

(5.59)

and

$$\rho = \frac{nS_{xy}}{\sigma_x \sigma_y}.$$

(5.60)

Therefore,

$$\beta = \rho \frac{\sigma_y}{\sigma_x}.$$

(5.61)

In the case of hydraulic geometry, the standard deviation of the logarithm of discharge is constant. If a perfect correlation ($\rho = 1$) is assumed, then

$$\beta \propto \sigma_y. \tag{5.62}$$

Since b, m, and f are regression coefficients in the relationships of width, depth, and velocity with discharge, the squares of expressions represent the variances of dependent variables. If width, depth, and velocity are the only variable factors at a channel cross-section, a minimum variance adjustment would be characterized by the minimization of $b^2 + m^2 + f^2$. Langbein (1965) considered that the tendency for the total variance to approach a minimum is equivalent to Le Chatelier's principle of least work. This provides a means of identifying the most probable or quasi-equilibrium state in river channels.

The state of stream behavior of a channel cross-section over a given time period is described by the ordered triple (b, f, m), the components of which are, respectively, the rates of change of width, depth, and velocity with discharge. Now (b, f, m) can be regarded as a vector α belonging to the three-dimensional Euclidean space R^3. For a given time period t_i, the channel is assumed to have a state denoted by $\alpha_i = (b_i, f_i, m_i)$ but by time period t_{i+1}, the channel has changed to the state $\alpha_{i+1} = (b_{i+1}, f_{i+1}, m_{i+1})$, where $b_{i+1} = b_i + b'_i$; $f_{i+1} = f_i + f'_i$; $m_{i+1} = m_i + m'_i$, where b'_i, f'_i, and m'_i are the changes in b_i, f_i, and m_i, respectively. Since $b_i + f_i + m_i = 1$ at any time, it follows that $b'_i + f'_i + m'_i = 0$. Knighton (1972, 1975) reasoned that a change in channel state can justifiably be defined in terms of hydraulic exponents. At-a-station hydraulic geometry can change markedly over a short period of time so segmentation into temporal changes may be necessary.

If x and y are two vectors in R^3 such that $x = (x_1, x_2, x_3)$ and $y = (y_1, y_2, y_3)$ then the scalar or inner product $x \cdot y$ is given by

$$x \cdot y = (x_1, x_2, x_3) \cdot (y_1, y_2, y_3) = x_1 y_1 + x_2 y_2 + x_3 y_3, \tag{5.63}$$

and the length of x, denoted as $\|x\|$ is given by

$$\|x\| = (x \cdot x)^{\frac{1}{2}}, \tag{5.64}$$

and the distance, $d(x, y)$, between the two vectors is

$$d(x, y) = \|x - y\| = [(x - y) \cdot (x - y)]^{\frac{1}{2}}. \tag{5.65}$$

With these ideas in mind, the following is assumed. First, the channel is assumed to be in an unstable state, with the result that its form and its hydraulic geometry are subject to modification. Second, change occurs in distinct phases, each of which has a characteristic hydraulic geometry. Finally, the adjustment occurs in such a way that the channel approaches a kind of quasi-equilibrium, regarded here as a limit state $\alpha = (b, f, m)$. From the definition of the vector convergence, one can write

$$\lim_{k \to \infty} \{\alpha_k\} = \alpha, \tag{5.66}$$

where $\{\alpha_k\}$ denotes the sequence of vector states k produced during channel modification. If the quasi-equilibrium is attained over a finite, rather than an infinite, sequence of changes, then

$$\lim_{k \to n} \{\alpha_k\} = \alpha_n, \tag{5.67}$$

where $\alpha_n = (b_n, f_n, m_n)$ is now the limit state.

The property of vector convergence of Equation (5.66) can be expressed as

$$\lim_{k \to \infty} \|\alpha_k - \alpha\| = 0. \tag{5.68}$$

If limit α exists, then for some real number $\epsilon > 0$, an integer K can be found such that

$$\|\alpha_k - \alpha\| < \epsilon \tag{5.69}$$

for all $k > K$, where K depends on ϵ. A decrease in the distance between them is characterized by the approach to the limit state. Consequently, if the decrease in distance is regular for increasing k, then the $(i + 1)$th state will be closer to the limit vector than to the ith state. Thus, one can express

$$d(\alpha_i, \alpha) \geq d(\alpha_{i+1}, \alpha). \tag{5.70}$$

Neglecting the square root sign since $d(x, y) \geq 0$ for all $x, y \in R^3$ and $d(x, y) = 0$ if and only if $x = y$, Equation (5.70) yields

$$(\alpha_i - \alpha) \cdot (\alpha_i - \alpha) \geq (\alpha_{i+1} - \alpha) \cdot (\alpha_{i+1} - \alpha), \tag{5.71}$$

where

$$(b_i - b)^2 + (f_i - f)^2 + (m_i - m)^2 \geq (b_{i+1} - b)^2 + (f_{i+1} - f)^2 + (m_{i+1} - m)^2. \tag{5.72}$$

Multiplying and substituting $b_{i+1} = b_i + b_i'$; $f_{i+1} = f_i + f_i'$; $m_{i+1} = m_i + m_i'$ in Equation (5.72), one gets

$$b_i^2 + f_i^2 + m_i^2 + 2(b_i'b + f_i'f + m_i'm) \geq b_{i+1}^2 + f_{i+1}^2 + m_{i+1}^2. \tag{5.73}$$

If

$$\alpha_i' \cdot \alpha = b_i' \cdot b + f_i' \cdot f + m_i' \cdot m \leq 0, \tag{5.74}$$

which requires

$$f'_i \leq -\lambda m'_i, \tag{5.75}$$

where $\lambda = (m - b)/(f - b)$ can be regarded as positive, since the rate of change of depth or velocity is usually greater than that of width. Then, Equation (5.73) reduces to

$$b_i^2 + f_i^2 + m_i^2 \geq b_{i+1}^2 + f_{i+1}^2 + m_{i+1}^2. \tag{5.76}$$

Equation (5.76) can be obtained directly from Equation (5.73) without making the assumption expressed by Equation (5.74) if the alternative distance measure defined by

$$\begin{aligned} d^2(\alpha_i, \alpha) = \|\alpha_i - \alpha\|^2 &= (b_i - b)^2 + (f_i - f)^2 + (m_i - m)^2 \\ &= b_i^2 + b^2 + f_i^2 + f^2 + m_i^2 + m^2 \end{aligned} \tag{5.77}$$

is used. Equation (5.76) notes that the limit state is characterized by

$$b^2 + f^2 + m^2 \rightarrow \text{minimum.} \tag{5.78}$$

This is what the Langbein hypothesis yields.

Consider now Langbein's hypothesis involving width, depth, velocity, shear stress, and friction as variable factors. Since shear (τ) equals the product of specific weight of water, depth, and slope, and since specific weight and slope are regarded as constant, shear stress varies only with mean depth. Therefore, one can then write

$$\tau \propto Q^f. \tag{5.79}$$

At constant slope, the Darcy–Weisbach friction factor (f_c) is proportional to the mean depth divided by the square of velocity. Thus, we can write

$$f_c \propto Q^{f-2m}. \tag{5.80}$$

Then, the channel state is defined by $\alpha = (b, f, m, f, (f - 2m))$, where the Euclidean space is now five dimensional. Using the same argument as before and adopting the distance measure of Equation (5.77), the expression corresponding to Equation (5.76) is

$$b_i^2 + f_i^2 + m_i^2 + f_i^2 + (f_i - 2m_i)^2 \geq b_{i+1}^2 + f_{i+1}^2 + m_{i+1}^2 + f_{i+1}^2 + \left(f_{i+1} - 2m_{i+1}\right)^2, \tag{5.81}$$

which implies that the approach to the limit state is characterized by

$$b^2 + f^2 + m^2 + f^2 + (f - 2m)^2 \rightarrow \text{minimum.} \tag{5.82}$$

Thus, this has the same form as the extended version of the minimum variance hypothesis.

Knighton (1977) furher refined his general formulation as follows. The initial assumption that a change in the channel geometry expressed in the form $\alpha_{i+1} = \alpha_i + \alpha_i'$ can be in a more general form of the relation between consecutive channel states as

$$\alpha_{i+1} = P\alpha_i, \tag{5.83}$$

where

$$P = \begin{bmatrix} p_{11} & p_{12} & p_{13} \\ p_{21} & p_{22} & p_{23} \\ p_{31} & p_{32} & p_{33} \end{bmatrix}. \tag{5.84}$$

Thus, the original form of the relation $\alpha_{i+1} = \alpha_i + \alpha_i'$ can be regarded as a spceial case of the more general form expressed by Equation (5.83), in which $p_{i1} = 1$ and

$$\alpha_i' = (p_{12}f_i + p_{13}m_i, p_{21}b_i + p_{23}m_i, p_{31}b_i + p_{32}f_i). \tag{5.85}$$

Equation (5.85) shows that in terms of the original form, the amount of change in any one hydraulic exponent depends linearly on the previous values of the other two hydraulic exponents.

5.5 Minimum Variance of Stream Power

Stream power (*SP*) is the rate of energy dissipation, expressed as

$$SP = \gamma QSL, \tag{5.86}$$

in which *SP* is the stream power, Q is the discharge, L is the length of the channel reach assumed as unity, S is the slope of the energy line but can be assumed to be equal to bed slope, and γ is the weight density of water assumed constant. For unit length $(L = 1)$, Equation (5.86) can be expressed in terms of velocity as

$$SP = \gamma v dw S. \tag{5.87}$$

Using the power relation of each variable on the right side, one can write

$$SP \propto Q^m Q^f Q^b Q^z \propto Q^{m+f+b+z}. \tag{5.88}$$

Defining the variance of *SP* as a sum of variances of individual variables and designating the sum to be a minimum, one can express:

Variance of *SP* $\propto$ variance of velocity + variance of depth + variance of width + variance of slope $\rightarrow$ minimum.

Then,

$$m^2 + f^2 + b^2 + z^2 \rightarrow \text{minimum.} \tag{5.89}$$

Replacing b by $1 - f - m$ in Equation (5.89), one can write

$$m^2 + f^2 + (1 - f - m)^2 + z^2 \rightarrow \text{minimum.} \tag{5.90}$$

Equation (5.90) can be written as

$$1 + 2m^2 + 2f^2 + z^2 + 2mf - 2f - 2m \rightarrow \text{minimum.} \tag{5.91}$$

Differentiating partially with respect to m and setting the derivative to 0 yields

$$2m + f = 1. \tag{5.92}$$

Differentiating partially with respect to f and setting the derivative to 0 yields

$$m + 2f = 1. \tag{5.93}$$

Differentiating partially with respect to z and setting the derivative to 0 yields

$$2z = 0; z = 0. \tag{5.94}$$

Solution of Equations (5.92) and (5.93) and then of b yields

$$m = \frac{1}{3}; \quad f = \frac{1}{3}; \quad b = \frac{1}{3}. \tag{5.95}$$

This states that if the variance of stream power is minimized then the variance of velocity, depth, and width will be equal and that of slope will vanish. This means that any change in stream power will be equally accommodated by the adjustments of velocity, depth, and width, and slope will make no contribution.

5.6 Influence of Choice of Variables

The inclusion or exclusion of variables in the minimization mix has a strong influence on the predicted values of exponents of discharge. Suppose, for example, only three independent variables, width, depth, and velocity, are considered. If width is constant, then a change in discharge will be absorbed by adjustments of velocity and depth, which will lead to certain values of discharge exponents. If width is not constant then some of the change in discharge will be absorbed by the change in width and the changes in depth and velocity will be smaller. This means that the exponents will change. The same arguments will extend to friction and shear stress. To illustrate the effect of the choice of variables on the predicted exponents, recall velocity $\propto Q^m \rightarrow$ variance of velocity is represented by m^2 or $(1-f)^2$

depth $\propto Q^f \rightarrow$ variance of depth is represented by f^2 or $(1-m)^2$

shear stress $\propto Q^f Q^0 \propto Q^f \rightarrow$ variance of shear stress is represented by f^2 or $(1-m)^2$

friction $\propto \dfrac{Q^f Q^0}{Q^{2m}} \propto Q^{f-2m} \rightarrow$ variance of friction is represented by $(f-2m)^2$

$$= 9m^2 - 6m + 1 \text{ or } (3f-2)^2.$$

These expressions provide variances of dependent variables. Thus, in the case of these four dependent variables with width fixed the possible combinations are (Williams, 1978): (1) velocity and depth; (2) velocity and shear stress; (3) velocity and friction factor; (4) depth and shear stress; (5) depth and friction factor; (6) shear stress and friction; (7) velocity, depth, and shear stress; (8) velocity, depth, and friction factor; (9) velocity, shear stress, and friction factor; (10) depth, shear stress, and friction factor; and (11) velocity, depth, shear stress, and friction factor. Some of these combinations may not necessarily be based on hydrologic reasoning. The values of exponents will vary, depending on the combination and may have considerable ranges determined theoretically by variance minimization, as shown in Table 5.1. For example, if the change in discharge is accommodated equally by depth and velocity then $f = 0.5$. This means that 50% of a change in discharge is accommodated by change in depth and 50% by change in velocity. If the shear and

Table 5.1. *Rate of change of dependent variables with increase in discharge for different combinations of variables (after Williams, 1978)*

		Values of exponents			
Combination	Dependent variables	Velocity, v (m)	Depth, d (f)	Shear stress, τ (f)	Friction factor, f_c $(f-2m)$
1	v, d	0.50	0.50	0.50	−0.50
2	v, τ	0.50	0.50	0.50	−0.50
3	v, f_c	0.30	0.70	0.70	0.10
4	d, τ	1.00	0.00	0.00	−2.00
5	d, f_c	0.40	0.60	0.60	−0.20
6	τ, f_c	0.40	0.60	0.60	−0.20
7	v, d, τ	0.67	0.33	0.33	−1.00
8	v, d, f_c	0.36	0.64	0.64	−0.08
9	v, τ, f_c	0.36	0.64	0.64	−0.08
10	d, τ, f_c	0.45	0.55	0.55	−0.35
11	v, d, τ, f_c	0.42	0.58	0.58	−0.26

friction also accommodate part of the change in discharge then $f = 0.58$. That means that 58% of a change in discharge is accommodated by change in depth and 42% by change in velocity.

One can compare actual field data to the exponents of various combinations and select the combination that best matches field observations. Consider the case of a flume of given width and slope, which receives discharge from a pipe considered by Langbein (1965). An increase in discharge will be accomodated by changes in depth, velocity, friction, and shear stress.

Langbein (1964, 1965) discussed four cases and Williams (1978) discussed these cases plus one more but added unit stream power (USP) (velocity times slope) in the minimization mix. These cases are briefly discussed here.

Case 1: The simplest case is where width and slope are approximately constant. The change in discharge will be accommodated by changes in depth and velocity. Table 5.2 lists exponents obtained theoretically by minimizing the variances of various combinations of variables. There, *SP* denotes the unit stream power. Williams (1978) found the average field values of $m = 0.47$, and $f = 0.52$ for the case of constant width and slope.

Case 2: The river is able to change its width along with depth and velocity, but the slope remains constant. Besides width, depth, and velocity, bed shear and friction are also dependent variables. The variance minimization yields $m = 0.22, f = 0.30$, and $b = 0.48$, indicating that 22% of change in discharge is accommodated by change in velocity, 30% by change in depth, and 48% by change in width. Langbein (1964) reported that Rio Galisteo at Domingo, New Mexico, which has constant slope and medium-sand banks, had values of $m = 0.24, f = 0.27$, and $b = 0.49$.

Table 5.2. *Theoretical exponents (rates of change) of dependent variables with increase in discharge for different combinations (after Williams, 1978)*

Number	Dependnt variables	Values of exponents				
		Velocity, v (m)	Depth, d (f)	Shear stress, τ (f)	Friction factor, f_c ($f - 2m$)	Power per unit weight, SP (m)
1	v, d	0.50	0.50	0.50	-0.50	0.50
2	v, d, τ	0.67	0.33	0.33	-1.00	0.67
3	v, d, f_c	0.36	0.64	0.64	-0.08	0.36
4	v, d, SP	0.33	0.67	0.67	0.00	0.33
5	v, d, τ, f_c	0.43	0.57	0.57	-0.20	0.43
6	v, d, τ, SP	0.50	0.50	0.50	-0.50	0.50
7	v, d, f_c, SP	0.33	0.67	0.67	0.00	0.33
8	v, d, τ, f_c, SP	0.38	0.62	0.62	-0.14	0.38

Case 3: The river has a constant slope and sediment tranport is proportional to the rate of discharge. The sediment load per unit width (Colby, 1964) is proportional to the cube of mean velocity. Since load per unit volume is constant, the depth must vary as the square of velocity, or $m = 0.5$. In this case, the respective variances are $f/2$ for velocity, f for depth, $b = 1 - 1.5f$ for width, f for shear, and 0 for friction. Thus, the minimum variance is reached when $f = 1/3$, $m = 1/6$, and $b = 0.50$.

Case 4: This case belongs to straight canals where slope also changes, meaning a longitudinal profile will develop such that the variation in shear and friction is equally divided. The variance of shear stress is $(f + z)^2$ and the variance of friction is $(f + z - 2m)^2 = z^2$. Because of uniform load, $m = 0.5f$, which leads to $z = -m$. However, stream power will also accommodate part of the change in discharge. Various aspects of stream power and their variances can be expressed as:

Power per unit length per unit discharge $\frac{Q\Delta H}{Q\Delta L} \sim S$ variance z^2

Power per unit volume $\frac{Q\Delta H}{Qw\Delta L} \sim vS$ variance $(m + z)^2$

Power per unit bed area $\frac{Q\Delta H}{w\Delta L} \sim vdS$ variance $(m + f + z)^2$

Power per unit length $\frac{Q\Delta H}{\Delta L} \sim QS$ variance $(1 + z)^2$

Power per unit time of stream travel $\frac{Q\Delta H}{\Delta T} \sim SQv$ variance $(1 + m + z)^2$, where H is the head loss.

The total variance therefore is the sum of squares of exponents:

$$z^2 + (m + z)^2 + (m + f + z)^2 + (1 + z)^2 + (1 + m + z)^2$$

The sum of squares is a minimum when $z = -1/6$, and $m = 1/6$, $f = 0.33$, and $b = 0.50$, under the condition that $m = 0.5f$, and $z = -m$.

5.7 Entropy Maximizing

Let the hydraulic geometry be expressed as a function of flow depth d, channel width w, velocity v, slope S, and Manning's roughness coefficient n as

$$G = g(d, w, v, S, n). \tag{5.96}$$

Equation (5.96) can be experssed using Taylor series with respect to reference values (which can be regarded as mean values) of d_0, w_0, v_0, S_0, n_0 and as

$$G(w, d, v, S, n) = G(w_0, d_0, v_0, S_0, n_0) + \frac{\partial G}{\partial w}(w - w_0) + \frac{\partial G}{\partial d} + \frac{\partial G}{\partial v}(v - v_0)$$
$$+ \frac{\partial G}{\partial S}(S - S_0) + \frac{\partial G}{\partial n}(n - n_0) \ldots, \tag{5.97}$$

in which the derivatives are computed at the respective reference values. Taking the variance (*var*) of G considering only the first-order terms in Equation (5.97),

$$var\,(G) = E[G - G_0]^2$$

$$= E\left[\frac{\partial G}{\partial w}(w - w_0) + \frac{\partial G}{\partial d}(d - d_0) + \frac{\partial G}{\partial v}(v - v_0) + \frac{\partial G}{\partial S}(S - S_0) + \frac{\partial G}{\partial n}(n - n_0)\right]^2.$$

$$(5.98)$$

Assuming that the variables are independent, Equation (5.98) can be written as

$$var(G) = \left(\frac{\partial G}{\partial w}\right)^2 var(w) + \left(\frac{\partial G}{\partial d}\right)^2 var(d) + \left(\frac{\partial G}{\partial v}\right)^2 var(v)$$

$$+ \left(\frac{\partial G}{\partial S}\right)^2 var(S) + \left(\frac{\partial G}{\partial n}\right)^2 var(n) + .$$

$$(5.99)$$

Dividing Equation (5.99) by $var(G)$ throughout,

$$1 = \left(\frac{\partial G}{\partial w}\right)^2 \frac{var(w)}{var(G)} + \left(\frac{\partial G}{\partial d}\right)^2 \frac{var(d)}{var(G)} + \left(\frac{\partial G}{\partial v}\right)^2 \frac{var(v)}{var(G)}$$

$$+ \left(\frac{\partial G}{\partial S}\right)^2 \frac{var(S)}{var(G)} + \left(\frac{\partial G}{\partial n}\right)^2 \frac{var(n)}{var(G)}.$$

$$(5.100)$$

Equation (5.100) can be expressed as

$$1 = P_1 + P_2 + P_3 + P_4 + P_5,$$

$$(5.101)$$

where

$$P_1 = \left(\frac{\partial G}{\partial w}\right)^2 \frac{var(w)}{var(G)}; P_2 = \left(\frac{\partial G}{\partial d}\right)^2 \frac{var(d)}{var(G)}; P_3 = \left(\frac{\partial G}{\partial v}\right)^2 \frac{var(v)}{var(G)};$$

$$P_4 = \left(\frac{\partial G}{\partial S}\right)^2 \frac{var(S)}{var(G)}; P_5 = \left(\frac{\partial G}{\partial n}\right)^2 \frac{var(n)}{var(G)}.$$

$$(5.102)$$

Equation (5.101) states that the varaince of G is the sum of these five components. If these components $P_i, I = 1, 2, 3, 4,$ and 5, can be interpreted as probabilities with no constraints imposed on the river, then the principle of maximum entropy will lead these probabilities to be equal, that is, $P_1 = P_2 = P_3 = P_4 = P_5$. This gives rise to 10 possibilities that can minimize the variance in order for the river to achieve stable hydraulic geometry. These possibilities include: $P_1 = P_2; P_1 = P_3;$ $P_1 = P_4; P_1 = P_5; P_2 = P_3; P_2 = P_4; P_2 = P_5; P_3 = P_4; P_3 = P_5; P_4 = P_5.$

In practice, however, this is not the case, and the probabilities will be unequal, depending on the individual contributions of hydraulic variables. Further, some variables dominate in the channel geometry adjustment. In order to account for unequal probabilities, one can introduce weighting factors, such as a_i, $i = 1, 2, 3,$ 4, and 5, such that $a_1 + a_2 + a_3 + a_4 + a_5 = 1$. Then,

$$1 = a_1 P_1 + a_2 P_2 + a_3 P_3 + a_4 P_4 + a_5 P_5. \tag{5.103}$$

For illustrative pupsoes only one possibility (the first) will be considered for deriving the hydraulic geometry, and similar procedures can be followed for other possibilities.

$$a_1 \left(\frac{\partial G}{\partial w}\right)^2 \frac{var(w)}{var(G)} = a_2 \left(\frac{\partial G}{\partial d}\right)^2 \frac{var(d)}{var(G)}. \tag{5.104}$$

Solution of Equation (5.104) leads to

$$w = \left(\frac{a_2}{a_1}\right)^{0.5} \left(\frac{S_d}{S_w}\right) d, \tag{5.105}$$

where S_w is the standard deviation of w and S_d is the standard deviation of d. Substituting Equation (5.105) in Manning's equation to eliminate w,

$$Q = \frac{1}{n} w d^{\frac{5}{3}} S^{\frac{1}{2}} \tag{5.106}$$

leads to

$$Q = \frac{1}{n} \left(\frac{a_2}{a_1}\right)^{0.5} \left(\frac{S_d}{S_w}\right) d^{\frac{8}{3}} S^{\frac{1}{2}}. \tag{5.107}$$

In this possibility, it is assumed that only depth and width are varying for channel adjustment but slope and roughness coefficient are not. Therefore, they can be regarded as constant. Equation (5.107) yields

$$d = b_0 Q^{0.375}; b_0 = \frac{n^{\frac{3}{8}}}{S^{\frac{3}{16}}} \left(\frac{a_1}{a_2}\right)^{\frac{3}{16}} \left(\frac{S_w}{S_d}\right)^{\frac{3}{8}}. \tag{5.108}$$

Eliminating d between Equations (5.105) and (5.106),

$$w = a Q^{0.375}; a = \frac{n^{\frac{3}{8}}}{S^{\frac{3}{16}}} \left(\frac{a_1}{a_2}\right)^{\frac{3}{16}} \left(\frac{S_w}{S_d}\right)^{\frac{3}{8}}. \tag{5.109}$$

Inserting Equations (108) and (109) in the continuity equation

$$Q = wdv, \tag{5.110}$$

the expression for velocity is

$$v = \frac{1}{ab_0} Q^{0.25}. \tag{5.111}$$

Likewise, the expression for slope by inserting Equations (108) and (109) in Manning's Equation (5.106) can be obtained as

$$S = \frac{n^2}{a^2} \left(\frac{1}{b_0}\right)^{\frac{10}{3}} Q^{-1}. \tag{5.112}$$

The exponents for w, d, v, and S are not the same but not too distant either.

References

Brush, L. M. (1961). *Drainage basins, channels, and flow characteristics of selected streams in central Pennsylvania*. U.S. Geological Survey Professional Paper 2 82-F, Washington, DC.

Chorley, R. J. (1962). Geomorphology and general systems theory. Geological Professional Paper 500-B, U.S. Geological Survey, Washington, DC.

Colby, B. R. (1964). *Discharge of sands and mean velocity relationships in sand-bed streams*. Professional Paper 462 A, U.S. geological Survey, Washington, DC..

Dozier, J. (1976). An examination of the variance minimization tendencies of a supraglacial stream. *Journal of Hydrology*, Vol. 31, pp. 359–380.

Emmett, W. W. (1970). The hydraulics of overland flow on hillslopes. *U.S. Geological Survey Professional Paper*, 642-A, p. 68.

Gilbert, G. K. (1877). *Report on the geology of the Henry Mountains*. U.S. Geological Survey Rocky Mountain Region Report, pp. 160.

Kennedy, J. F., Richardson, P. O., and Sutera, S. P. (1964). Discussion of "Geometry of river channels" by W. B Langbein. *Proceedings of the Hydraulics Division, ASCE*, Vol. 90, No. HY6, pp. 332–341.

Knighton, A. D. (1972). Changes in a braided reach. *Geological Society of America Bulletin*, Vol. 83, pp. 3813–3822.

Knighton, A. D. (1974). Variation in width-discharge relation and some implications for hydraulic geometry. *Geological Society of America Bulletin*, Vol. 85, pp. 1069–1076.

Knighton, A. D. (1975). Variations in at-a-station hydraulic geometry. *American Journal of Science*, Vol. 275, pp. 186–218.

Knighton, A. D. (1977). Alternative derivation of the minimum variance hypothesis. *Geological Society of America Bulletin*, Vol. 88, pp. 364–366.

Langbein, W. B. (1964). Geometry of river channels. *Journal of the Hydraulics Division, ASCE*, Vol. 90, No. HY2, pp. 301–311.

Langbein, W. B. (1965). Closure to "Geometry of river channels." *Proceedings of the Hydraulics Division, ASCE*, Vol. 91, No. HY3, pp. 297–313.

Langbein, W. B. and Leopold, L. B. (1966). *River meanders-theory of minimum variance*. U.S. Geological Survey Professional Paper 422-H, 15 pp., Washington, DC.

Leopold, L. B. and Maddock, T. J. (1953). Hydraulic geometry of stream channels and some physiographic implications. *U. S. Geological Survey Professional Paper*, 252, p. 55.

Lindley, E. S. (1919). *Regime Channels. Minutes and Proceedings, Punjab Engineering Congress*, Lahore, 7.

Maddock, T. (1969). The behavior of straight open channels with movable beds. *U.S. Geological Professional Paper*, 622-A, pp. 70.

Richards, K. S. (1973). Hydraulic geometry and channel roughness: A nonlinear system. *American Journal of Science*, Vol. 273, pp. 877–896.

Richards, K. S. (1976). Complex width-discharge relations in natural river sections. *Geological Society of America Bulletin*, Vol. 87, pp. 199–206.

Richards, K. S. (1982). *Rivers: Form and Process in Alluvial Channels*. Metheun, London.

Riley, S. J. (1978). The role of minimum variance theory in defining the regime characteritics of the Lower Namgoi-Gwydir drainage basin. *Water Resources Bulletin*, Vol. 14, No. 1, pp. 1–11.

Rubey, W. W. (1953). *Geology and mineral resources of the Hardin and Brussels quadrangles*. Professional Paper 218, pp. 175, U.S. Geological Survey, Washington, DC.

Williams, G. P. (1978). *Hydraulic geometry of river cross-sections-Theory of minimum variance*. U.S. Geological Survey Professional Paper 1029, Washington, DC.

6

Dimensional Principles

Using dimensional principles, three dimensionless variables can be defined for designing a regime channel. These variables contain six characteristic parameters that reflect fluid, sediment, and geometric characteristics of a channel. This chapter discusses the hydraulic geometry of regime channels using these dimensional principles and illustrates the application of these principles to channel design.

Notation

b = exponent of Q for w

C = Chezy's friction factor or flow resistance factor

C_b = bed form friction factor

C_s = skin friction factor

d = depth of uniform flow

D_s = sediment grain size

d_* = dimensionless flow depth

d_1 = distance from the apex of the triangle to the top of the sediment bed

d_2 = distance from the apex to the water surface

E_* = energy-related quantity

$(E)_{min}$ = minimum energy

F_Q = dimensionless flow rate

F_r = Froude number

f_{cb} = some function

f_{cs} = some function

f_E = some function

f_p = some function

f_Q = function of Q

f_{Qs} = function of Q_s

$f_{\frac{w}{d}}$ = some function

---

f_v = some function

f_λ = some function

g = acceleration due to gravity

g_w = partial gravity submergence force acting on the fluid

M_n = material number

n = Manning's roughness factor

p = net cross-sediment transport rate

p_1 = lateral sediment transport rate from the bed

p_2 = lateral sediment transport rate from the banks

Q = discharge

Q_s = sediment discharge

q_s = longitudinal specific volumetric total sediment transport rate averaged over the bed

R_f = relative flow intensity

S = bed slope

T_R = duration of regime channel formation

v = velocity of flow

v_* = shear velocity

v_{*cr} = critical shear velocity

V_E = dimensionless counterpart of energy equation

w = flow width

X = grain size Reynolds number

Y = mobility number

Y_{cr} = mobility number (Y) corresponding to the critical shear stress

y = a section

γ_s = specific weight of sediment grains or sediment weight density and the cube of sediment size

λ = dimensionless friction

ϕ = some function

υ = kinematic viscosity

μ = viscosity

ρ = fluid density

6.1 Introduction

The hydraulic geometry of a regime channel is specified by flow width (w), depth of uniform flow (d), velocity of flow (v), bed slope (S), and flow resistance factor, such as dimensionless Chezy's friction factor, or flow resistance factor (C), or Manning's roughness factor (n). Let T_R be the duration of regime channel formation. It is assumed that at any time less than T_R, the flow is tranquil or subcritical having the Froude number less than one, the ratio w/d is large (that is, greater than 10), the granular material is cohesionless, and the bed of the symmetric channel cross-section is horizontal and covered by bed forms, such as ripples and/or dunes. The discussion in this chapter closely follows the work of Yalin (1992) and Yalin and Ferreira da Silva (1997).

It is hypothesized that the hydraulic geometry parameters can be determined by three fundamental equations: (1) resistance equation, (2) minimum energy equation, and (3) sediment transport equation. The resistance equation expresses discharge Q as a function of flow width, flow depth, bed slope, and flow resistance factor, and can be written as

$$Q = f_Q(w, d, S, C),   \tag{6.1}$$

where f_Q is a function of w, d, S, and C, which, however, is not uniquely expressed.

The minimum energy equation can be written as

$$E_* = (E)_{min},   \tag{6.2}$$

where E_* denotes the energy-related quantity, and $(E)_{min}$ denotes the minimum energy. However, the energy-related quantity E_* is not fully defined.

Like the flow resistance equation, the sediment transport equation expresses sediment discharge (Q_s) also as a function of flow width, depth of uniform flow, bed slope, and flow resistance factor, and can be written as

$$Q_s = f_{Qs}(w, d, S, C),   \tag{6.3}$$

where f_{Qs} is a function of w, d, S, and C, which is not uniquely expressed. Equations (6.1)–(6.3) are sufficient to solve for w, d, and S. It should be noted that all these quantities in Equations (6.1)–(6.3) correspond to the regime state of the channel, that is, its time invariance.

In order to determine the hydraulic geometry, Yalin (1992) employed six characteristic parameters, which are fluid density, sediment density, kinematic viscosity, sediment size, shear velocity, and friction factor or coefficient. These parameters reflect fluid, hydraulic, and sediment characteristics. Yalin and Ferreira da Silva (1997) utilized these parameters to define three dimensionless variables, in contrast with four dimensionless variables that are used for analogous alluvial channels. The three dimensionless variables are as follows:

(1) Material number (M_n), which is defined as the ratio of the product of the specific weight of sediment grains (γ_s) or sediment weight density and the cube of sediment size or sediment grain size (D_s) to the product of fluid density (ρ) and the square of kinematic viscosity (v):

$$M_n^2 = \frac{\gamma_s D_s^3}{\rho v^2}.   \tag{6.4}$$

(2) Relative flow intensity (R_f), which is defined by the ratio of the product of acceleration due to gravity (g), flow depth (d), and slope of uniform flow (S) to the square of shear velocity (v_*: Equation 6.5a) at the critical stage or critical shear velocity (v_{*cr}: Equation 6.5b):

$$v_* = \sqrt{gSd},\tag{6.5a}$$

where g is acceleration due to gravity.

$$v_{*cr} = \sqrt{\frac{\tau_{cr}}{\rho}},\tag{6.5b}$$

where $\tau_{cr} = 0.0125 + 0.019\,D_s$. D_s is particle size in mm and shear stress is in pounds per square foot.

$$R_f = \frac{gSd}{v_{*cr}^2}.\tag{6.5c}$$

(3) Dimensionless flow depth (d_*), which is defined as the ratio of flow depth to the sediment grain size:

$$d_* = \frac{d}{D_s}.\tag{6.6}$$

For characterizing a regime channel, Yalin and Ferreira da Silva (1997) also defined three related dimensionless variables which are expressed as:

1. Grain size Reynolds number (X), which is defined as the ratio of the product of shear velocity and sediment size (diameter) to the kinematic viscosity:

$$X = \frac{v_* D_s}{v}.\tag{6.7}$$

2. Mobility number (Y), which is defined as the ratio of the product of fluid density and the square of shear velocity to the product of sediment size and specific weight of sediment:

$$Y = \frac{\rho v_*^2}{\gamma_s D_s}.\tag{6.8}$$

3. Dimensionless flow rate (F_Q), which is the ratio of discharge or flow rate to the product of flow width, sediment grain size, and critical shear velocity:

$$F_Q = \frac{Q}{w D_s v_{*cr}}.\tag{6.9}$$

It can be noted that Equation (6.4) for mobility number can be written, using Equations (6.7) and (6.8), as

$$M_n^3 = \frac{X^2}{Y}. \tag{6.10}$$

Likewise, Equation (6.5c) for relative flow intensity can be written using Equation (6.8) as

$$R_f = \frac{Y}{Y_{cr}}, \tag{6.11a}$$

where Y_{cr} is the mobility number (Y) corresponding to the critical shear stress. Y_{cr} can be calculated following Da Silva and Yalin (2017) as

$$Y_{cr} = 0.13 M_n^{-0.392} e^{-0.015 M_n^2} + 0.045 \left[1 - e^{-0.068 M_n} \right]. \tag{6.11b}$$

Thus, there are three independent dimensionless parameters M_n, R_f, and d_* or X, Y, and F_Q in terms of which the three governing equations and, consequently, hydraulic geometry can be expressed.

6.2 Derivation of Hydraulic Geometry

For deriving hydraulic geometry relations, it is assumed that flow in the channel is tranquil, that is, the Froude number is less than one, the bed of the channel cross-section is horizontal and is covered with bed forms-ripples and/or dunes, and the granular material is cohesionless. Since w/d is large in regime channels, a dimensionless counterpart of energy Equation (6.2), V_E, can be defined as an additional variable

$$V_E = f_E \left(\frac{w}{d}, M_n, R_f, d_* \right), \tag{6.12}$$

where f_E is some function of w/d, M_n, R_f, and d_*. It is assumed here that in wide and cohesionless regime channels the angle of repose has a small variation and is therefore ignored.

Now consider sediment transport in a channel, as shown in Figure 6.1. For a regime channel, the cross-sediment transport rate vanishes. In a channel cross-section, the sediment transport process involves (1) the lifting of grains from the bed that diffuse toward the banks due to turbulence, and (2) the detachment of grains from the banks that tend to move toward the bed due to the inclination of banks and turbulence. At any section y, let p_1 define the (lateral) sediment transport rate from the bed, and let p_2 define the (lateral) sediment transport rate from the banks. Then, the net cross-sediment transport rate p can be expressed as

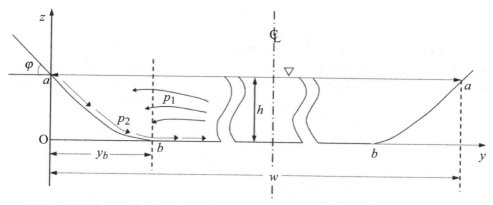

Figure 6.1 Dispersion of sediment sideways during transport in a channel

$$p = p_1 - p_2. \tag{6.13}$$

For the specified cross-section y, the dimensionless cross-transport rate can be expressed as p/q_s, where q_s is the longitudinal specific volumetric total sediment transport rate averaged over the bed. Then, following Equation (6.12), one can write the dimensionless cross-transport rate as

$$\frac{p}{q_s} = f_p \left(\frac{w}{d}, M_n, R_f, d_* \right), \tag{6.14}$$

where f_p is some function of $\frac{w}{d}, M_n, R_f, d_*$.

For a self-forming channel to achieve regime state with no boundary deformation, the cross-sediment rate must necessarily be zero (Parker, 1978). Therefore, Equation (6.13) becomes

$$0 = p_1 - p_2, \quad p_1 = p_2 \quad \text{for} \quad y \in (0, w). \tag{6.15}$$

Equation (6.15) holds for both sand and gravel bed channels. For sand bed channels $p_1 = p_2$, and for gravel bed channels $p_1 = p_2 = 0$. With the insertion of Equation (6.15), Equation (6.14) can be written as

$$0 = f_p \left(\frac{w}{d}, M_n, R_f, d_* \right). \tag{6.16}$$

From Equation (6.16), one can write

$$\frac{w}{d} = f_{\frac{w}{d}} \left(M_n, R_f, d_* \right), \tag{6.17}$$

where $f_{\frac{w}{d}}$ is some function of M_n, R_f, d_*. Equation (6.17) shows that function $f_{\frac{w}{d}}$ is determined by three dimensionless variables $M_n, R_f,$ and d_*. Then, from Equations (6.17) and (6.12) for regime channels, one can write

$$V_E = f_E\left(M_n, R_f, d_*\right). \tag{6.18}$$

The aspect ratio w/d is affected by the granular skin friction of flow boundaries and bed forms at $t = T_R$. Therefore, it would be appropriate to incorporate the friction factor in the w/d expression. Let C_s be the skin friction factor and C_b be the bed form friction factor. Then, these friction factors can be expressed as

$$C_s = f_{cs}\left(M_n, R_f, d_*\right) \tag{6.19}$$

$$C_b = f_{cb}\left(M_n, R_f, d_*\right), \tag{6.20}$$

where f_{cs} and f_{cb} are some functions M_n, R_f, d_*.

By dividing the bed form friction given by Equation (6.20) by the skin friction given by Equation (6.19), one can write the dimensionless friction λ as

$$\frac{C_b}{C_s} = \lambda = \frac{f_{cb}}{f_{cs}} = f_\lambda\left(M_n, R_f, d_*\right), \tag{6.21}$$

where f_λ is some function of M_n, R_f, d_*. Another equivalent relation can also be written as

$$C_b\sqrt{R_f} = \sqrt{R_f}f_{cs} = f_v\left(M_n, R_f, d_*\right) \quad \left(\simeq \frac{v}{v_*}\right), \tag{6.22}$$

where f_v is some function of M_n, R_f, d_*. Eliminating R_f and d_* from Equations (6.22), (6.21), and (6.17), the aspect ratio can be written as

$$\frac{w}{d} = f_{\frac{w}{d}}\left(M_n, \lambda, C_b\sqrt{R_f}\right), \tag{6.23}$$

where $f_{\frac{w}{d}}$ is some function of $M_n, \lambda, C_b\sqrt{R_f}$. Equation (6.23) is an expression for the aspect ratio in terms of dimensionless quantities.

6.3 Derivation of Width and Depth

Equation (6.23) is fundamentally dependent on the cross-transport being zero, and is valid for any granular material. However, the numerical values of M_n, λ, and $C_b\sqrt{R_f}$ will be different from one granular material to another. For gravel bed channels, the effect of viscosity and hence of M_n is negligible, and the bed forms are insignificant ($\lambda \simeq 1$) and the gravel boundary of the channel is at its critical stage $R_f \simeq 1$. Equation (6.23) thus becomes

$$\frac{w}{d} \simeq f_{\frac{w}{d}}\left(1, C_b\sqrt{1}\right), \tag{6.24}$$

in which the friction factor can be expressed as

$$C_b \simeq C_s \simeq 6.82 \left(\frac{d}{D_s}\right)^{\frac{1}{6}},$$ (6.25)

where C_s is the flat bed friction or skin friction.

For sand and gravel bed regime channels, usually (Yalin, 1992)

$$w \sim Q^{\frac{1}{2}},$$ (6.26)

and for gravel bed channels (Yalin, 1992)

$$d \sim Q^{\frac{3}{7}}.$$ (6.27)

With the use of Equations (6.26) and (6.27) in Equations (6.24) and (6.25), one obtains

$$\frac{w}{d} \sim \frac{Q^{\frac{1}{2}}}{Q^{\frac{3}{7}}} \sim Q^{\frac{1}{14}}$$ (6.28)

and

$$C_b \sim d^{\frac{1}{6}} \sim \left(Q^{\frac{3}{7}}\right)^{\frac{1}{6}} \sim Q^{\frac{1}{14}}.$$ (6.29)

Equations (6.28) and (6.29) show that both C_b and w/d at the regime state are proportional to discharge having the same exponent. Thus, one can express

$$\frac{w}{d} \sim C_b.$$ (6.30)

Now consider Equation (6.23) where C_b is in association with R_f. Therefore, Equation (6.30) yields $C_b\sqrt{R_f} \simeq C_b$, $R_f = 1$, or

$$\frac{w}{d} \sim C_b\sqrt{R_f}, \quad \text{where } R_f = 1.$$ (6.31)

Since Equation (6.23) holds for channels of all materials, Equations (6.31) should be valid for sand bed channels as well. This then suggests that

$$\frac{w}{d} = C_b\sqrt{R_f}[\phi(M_n, \lambda)]^2 \left(= \frac{V_w}{V_d}\right),$$ (6.32)

where ϕ is some function.

The resistance equation for wide channels (with hydraulic radius replaced by depth) can be written as

$$Q = (wd)C_b\sqrt{gSd}.$$ (6.33)

Equation (6.33) can identically be expressed as

$$\left(\frac{\frac{Q}{v_{*cr}}}{w^2}\right)\left(\frac{\frac{Q}{v_{*cr}}}{d^2}\right) = C_b^2 \frac{gSd}{v_{*cr}^2} \tag{6.34}$$

or

$$V_w^{-1} V_d^{-1} = C_b \sqrt{R_f}. \tag{6.35}$$

Equations (6.32) and (6.35) can be solved to yield

$$V_w = w\sqrt{\frac{v_{*cr}}{Q}} = \phi(M_n, \lambda) \tag{6.36}$$

and

$$V_d = d\sqrt{\frac{v_{*cr}}{Q}} = \left[C_b \sqrt{R_f} \phi(M_n, \lambda)\right]^{-1}. \tag{6.37}$$

Fundamental to the usefulness of Equations (6.32), (6.36), and (6.37) is the expression of $\phi(M_n, \lambda)$ whose explicit form is not yet theoretically known, but using empirical data Yalin (1992) has shown its form. Lei (1992) showed that M_n is influenced by V_w and for all practical purposes one can express:
If $M_n \leq 15$

$$\phi(M_n) = 0.45 M_n^{0.3} \tag{6.38}$$

and if $M_n > 15$,

$$\phi(M_n) = 1. \tag{6.39}$$

Now the role of λ can be enumerated. One can deduce that $\phi(M_n, \lambda) = \phi_{M_n}(M_n)\phi(\lambda)$.
For gravel-bed channels $\phi_{M_n}(M_n) = 1$ and $\lambda = 1$, Yalin (1992) showed that

$$V_B = 1 \ \phi_\lambda(1) \simeq 1.42. \tag{6.40}$$

For sand bed regime channels,

$$\phi(M_n, \lambda) = 1.42\lambda\phi_{M_n}(M_n). \tag{6.41}$$

Equation (6.41) can be used to evaluate Equations (6.32), (6.36), and (6.37), where $\phi(M_n)$ is given by Equations (6.38) and (6.39).
The flow width w can be expressed from Equation (6.36) as

$$w = \phi(M_n, \lambda)\sqrt{\frac{Q}{v_{*cr}}} \tag{6.42}$$

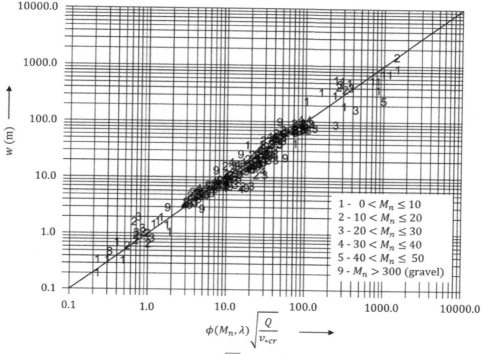

Figure 6.2 Plot of w versus $\phi(M_n, \lambda)\sqrt{\frac{Q}{v_{*cr}}}$ (after Yalin and Ferreira Da Silva, 1997)

and flow depth d can be written from Equation (6.37) as

$$d = \left[C_b\sqrt{R_f}\phi(M_n, \lambda)\right]^{-1}\sqrt{\frac{Q}{v_{*cr}}}. \tag{6.43}$$

Using data from alluvial streams in regime state, Yalin and Ferreira Da Silva (1997) plotted Equations (6.42) and (6.43) on log-log paper, as shown in Figures 6.2 and 6.3. Despite the range of data from fine sand to gravel and some data not corresponding to regime state ($R_f = 1$ for gravel channels and $R_f = 10$ for sand-bed channels), plots were strongly straight lines.

6.4 Computation of Regime Channels

Equation (6.36) and (6.37) contain three unknowns: w, d, and S. Jia (1990) and Yalin (1992) have shown that

$$E_* = F_r = \frac{v^2}{gd} = C^2 S = \frac{CR_f v_{*cr}^2}{gd}. \tag{6.44}$$

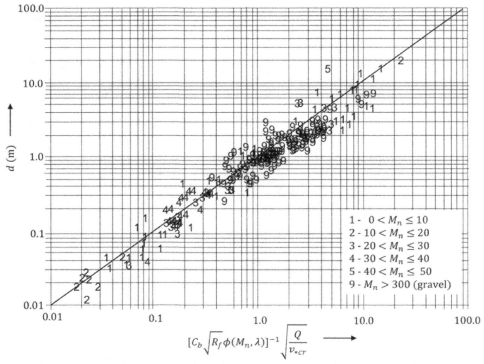

Figure 6.3 Plot of d versus $\left[C_b\sqrt{R_f}\phi(M_n,\lambda)\right]^{-1}\sqrt{\frac{Q}{v_{*cr}}}$ (after Yalin and Ferreira Da Silva, 1997)

Here, w and d satisfy Equations (6.36) and (6.37) for the smallest value of the Froude number. For friction factors C_b and C_s (Engelund, 1966; Yalin, 1977, 1992), one can express

$$C_s = \frac{1}{\kappa}\ln\left(\psi(X)\frac{d_*}{2}\right), \quad \kappa \simeq 0.4 \tag{6.45}$$

$$\frac{1}{C_b^2} = \frac{1}{C_s^2} + \sum_{j=d_u,r_i} a_j\left(\frac{\Delta}{\Lambda}\right)_j^{b_j}\frac{\Lambda_j}{d} \quad (d_u = \text{dunes}, r_i = \text{ripples}), \tag{6.46}$$

where $\wedge$ is the developed bed form length, and $\psi(X) = \exp(\kappa w_s^{-1})$ with w_s given by Yalin (1992) as:

$$w_s = 8.5 + [2.5\ln(2X) - 3]\exp\left(-0.217[(\ln(2X)]^2\right). \tag{6.47}$$

The bed form properties $\left(\frac{\Delta}{\Lambda}\right)_j$ and Λ_j/d in Equation (6.46) can be evaluated with the aid of the following equations (Yalin, 1992) as:

$$\left(\frac{\Delta}{\Lambda}\right)_{dunes} = 0.013\psi_{Xdune}(X)(R_f - 1)\exp\left[-\frac{(R_f - 1)}{13(1 - \exp(-0.0082))}\right] \quad (6.48)$$

$$\left(\frac{\Delta}{\Lambda}\right)_{ripples} = 0.035\psi_{Xripples}(X)(R_f - 1)\exp\left[-0.1\,(R_f - 1)\right] \quad (6.49)$$

and

$$\frac{\Lambda_{dunes}}{d} = 6\left[1 + 0.01\frac{(d_* - 40)\,(d_* - 400)}{d_*}\exp(-0.055\sqrt{d_*} - 0.04X)\right],$$
$$\frac{\Lambda_{ripples}}{d} = \frac{1{,}000}{d_*}, \quad (6.50)$$

in which

$$\psi_{Xdunes}(X) = 1 - \exp\left[-\left(\frac{X}{10}\right)^2\right] \quad (6.51)$$

$$\psi_{Xripples}(X) = \exp\left[-((X - 2.5)/14)^2\right],\text{ where } X > 2.5 \quad (6.52a)$$

and

$$\psi_{Xripples}(X) = 1 \text{ when } X < 2.5. \quad (6.52b)$$

In the case of ripples $a_{ripples} = 1/2$ and $b_{ripples} = 2$; in the case of dunes $a_{dunes} = 1/2$ and $b_{dunes} = 2$, except if $D_s \in [0.1\text{ mm}, 1\text{ mm}]$, then, following Yalin (1992),

$$w_{dunes} = 2.0 - 2.4(D_s - 0.1)(D_s - 1.1)[0.25\log_{10}N - 0.75]. \quad (6.53)$$

The friction factor C_b is a function of X, R_f, and d_*. Because a regime is characterized by the type of granular material and Q, the dimensionless characteristic $f_E = \phi_E(X, R_f, d_*)$ is transformed into $\overline{\phi}_E(M_n, R_f, Q_d)$, which can be accomplished with the use of

$$d_* = \frac{N}{C\sqrt{R_f}} \quad (6.54)$$

$$X = \left[R_f\psi(M_n)M_n^3\right]^{\frac{1}{2}}. \quad (6.55)$$

Equation (6.54) is derived by dividing the resistance equation by v_{*cr} and Equation (6.55) from the definition of M_n, viz,

$$M_n^3 = \frac{X^2}{Y} = \frac{X^2}{R_f\psi(M_n)}. \quad (6.56)$$

It can be deduced (Yalin, 1992) that

$$F_r^2 = \frac{v^2}{gd} = C_b^2 S \qquad (6.57)$$

and C_b are intertwined as

$$F_r = \left(\frac{\alpha}{F_Q}\right)(C_b^2 R_f)^{\frac{3}{2}}, \qquad (6.58)$$

where

$$\alpha = \frac{\gamma_s}{\gamma} Y_{cr} = \frac{v_{*cr}^2}{gD_s}. \qquad (6.59)$$

By plotting the Froude number again R_f for different values of M_n, Yalin and Ferreira da Silva (1997) noted that the regime data tended to correspond to the smallest F_r.

6.5 Computational Procedure

For given Q and D_s, the following steps can be employed for computing the regime channel geometry:

1. Adopt $(w)_i$, $(d)_i$, and $(S)_i$, where i denotes the ith trial.
2. Compute $(R_f)_i$.

   - Compute v_*, X, Y, M_n, d_*, and Y_{cr} using Equations (6.5a), (6.7), (6.8), (6.10), (6.6), and (6.11b), respectively.
   - Compute $(R_f)_i$ using Equation (6.11a) as $R_f = \frac{Y}{Y_{cr}}$.

3. Compute $(C_b)_i$.

   In practice, calculation of C_b cannot rely on the functional expression as $(C_b)_i = \phi_c(M_n, R_f, (F_Q)_i)$, where $(F_Q)_i = \frac{Q}{(w)_i D_s v_{*cr}}$. Therefore, C_b is calculated numerically following Section 6.4:

   - Compute Λ_{dunes}/d and $\Lambda_{ripples}/d$ using Equation (6.50).
   - Compute $\left(\frac{\Delta}{\Lambda}\right)_{dunes}$ and $\left(\frac{\Delta}{\Lambda}\right)_{ripples}$ using Equations (6.48) and (6.49), respectively.
   - Compute C_s using Equation (6.45).
   - Compute $(C_b)_i$ using Equation (6.46).

4. Compute flow discharge $(Q)_i$ using Equation (6.33) as

$$(Q)_i = (w)_i(d)_i(C_b)_i\sqrt{g(S)_i(d)_i}$$

5. Compare $(Q)_i$ to the measured Q. If $(Q)_i$ is not equal to Q, repeat steps 1 to 4 until $(Q)_{i+1} = Qi$ by varying w, d, and S for the $i+1$th trial. The minimum optimization programming is recommended.

Three examples for computation of regime characteristics are given for various sediment grain sizes (sand to gravel) and discharges (low to high). Follow the aforementioned computational procedure.

Example 6.1: Compute w, d, and A, if $D_s = 0.18$ mm and $Q = 1669.7$ m³/s for Bhagirathi River, India (Chitale, 1970). Assume the ratio of weight density of sediment to weight density of water as 1.65 and kinematic viscosity of water as 10^{-6} m² /s. Compare the computed values with those reported by Chitale (1970).

Solution:

Step 1

Adopt w, d, and S at the ith trial as $(w)_i = 215$ m, $(d)_i = 4.5$ m, and $(S)_i = 0.00002$.

Step 2

Compute v_*, X, Y, M_n, d_*, and Y_{cr} as follows:

- $v_* = \sqrt{gSd} = \sqrt{9.81 \times 0.00002 \times 4.5} = 0.03 \frac{\text{m}}{\text{s}}$
- $X = \frac{v_* D_s}{v} = \frac{0.03 \times 0.18 \times 10^{-3}}{10^{-6}} = 5.34$
- $Y = \frac{\rho v_*^2}{\gamma_s D_s} = \frac{1{,}000 \times 0.03^2}{16186.5 \times 0.18 \times 10^{-3}} = 0.3$
- $M_n = \left(\frac{X^2}{Y}\right)^{\frac{1}{3}} = \left(\frac{5.34^2}{0.3}\right)^{\frac{1}{3}} = 4.69$
- $d_* = \frac{d}{D_s} = \frac{4.5}{0.18 \times 10^{-3}} = 25{,}000$
- $Y_{cr} = 0.13 M_n^{-0.392} e^{-0.015 M_n^2} + 0.045 \left[1 - e^{-0.068 M_n}\right]$

$\quad = 0.13 \times 4.69^{-0.392} e^{-0.015 \times 4.69^2} + 0.045 \left[1 - e^{-0.068 \times 4.69}\right] = 0.0645$

Compute $(R_f)_i$ as

$$(R_f)_i = \frac{Y}{Y_{cr}} = \frac{0.3}{0.0645} = 4.69.$$

Step 3

Compute $(C_b)_i$.

- $\frac{\Lambda_{dunes}}{d} = 6\left[1 + 0.01 \frac{(d_* - 40)(d_* - 400)}{d_*} \exp\left(-0.055\sqrt{d_*} - 0.04X\right)\right]$

$\quad = 6\left[1 + 0.01 \frac{(25{,}000 - 40)(25{,}000 - 400)}{25{,}000} \exp\left(-0.055\sqrt{25{,}000} - 0.04 \times 5.34\right)\right]$

$\quad = 6.19$

- $\frac{\Lambda_{ripples}}{d} = \frac{1{,}000}{d_*} = \frac{1{,}000}{25{,}000} = 0.04$

$- \left(\frac{\Delta}{\Lambda}\right)_{dunes} = 0.013\psi_{Xdune}(X)\left(R_f - 1\right)\exp\left[-\frac{(R_f-1)}{13(1-\exp(-0.0082))}\right]$

where $\psi_{Xdunes}(X) = 1 - \exp\left[-\left(\frac{X}{10}\right)^2\right] = 1 - \exp\left[-\left(\frac{5.34}{10}\right)^2\right] = 0.25.$

Therefore,

$$\left(\frac{\Delta}{\Lambda}\right)_{dunes} = 0.013 \times 0.25 \times (4.69 - 1)\exp\left[-\frac{(4.69 - 1)}{13(1 - \exp(-0.0082))}\right] = \sim 0$$

$- \left(\frac{\Delta}{\Lambda}\right)_{ripples} = 0.035\psi_{Xripples}(X)\left(R_f - 1\right)\exp\left[-0.1\left(R_f - 1\right)\right]$

where, since $X > 2.5$, $\psi_{Xripples}(X) = \exp\left[-((X - 2.5)/14)^2\right]$
$$= \exp\left[-((5.34 - 2.5)/14)^2\right] = 0.96.$$

Therefore,

$$\left(\frac{\Delta}{\Lambda}\right)_{ripples} = 0.035 \times 0.96 \times (4.69 - 1)\exp[-0.1(4.69 - 1)] = 0.086$$

$- C_s = \frac{1}{\kappa}\ln\left(\psi(X)\frac{d_*}{2}\right)$

where κ is 0.4 and $\psi(X) = \exp(\kappa w_s^{-1})$.
where w_s is calculated following Equation (6.47) as

$$w_s = 8.5 + [2.5\ln(2X) - 3]\exp\left(-0.217[(\ln(2X)]^2\right)$$
$$= 8.5 + [2.5\ln(2 \times 5.34) - 3]\exp\left(-0.217[(\ln(2 \times 5.34)]^2\right) = 8.03.$$

Therefore, $\psi(X) = \exp(0.4 \times 8.03^{-1}) = 1.05.$
Finally, $C_s = \frac{1}{\kappa}\ln\left(\psi(X)\frac{d_*}{2}\right) = \frac{1}{0.4} \times \ln\left(1.05 \times \frac{25,000}{2}\right) = 23.7$

$-$ Compute $(C_b)_i$ as

$$\frac{1}{C_b^2} = \frac{1}{C_s^2} + \sum_{j=d_u, r_i} a_j\left(\frac{\Delta}{\Lambda}\right)_j^{b_j}\frac{\Lambda_j}{d},$$

where $a_j = \frac{1}{2}$ and $b_j = 2$.
Therefore,

$$\frac{1}{C_b^2} = \frac{1}{23.7^2} + \frac{1}{2}\left(0.0^2 \times 6.19 + 0.086^2 \times 0.04\right)$$

Finally, $(C_b)_i = 22.8.$

Step 4
Compute $(Q)_i$.

$$(Q)_i = (w)_i(d)_i(C_b)_i\sqrt{g(S)_i(d)_i}$$

$$= 215 \times 4.5 \times 22.8\sqrt{9.81 \times 0.00002 \times 4.5} = 655.5\frac{m^3}{s}$$

Step 5
Compare the computed $(Q)_i$ and the measured Q.

Since $(Q)_i$ of $655.5\frac{m^3}{s}$ is different from the measured Q of $1669.7\frac{m^3}{s}$, iterate steps 1 to 4 by inputting 1% increase in w, 10% increase in d, and/or 1–5% decrease in S until $(Q)_{i+1} = Q$.

When $(Q)_{i+1} = Q$, the regime w and d were estimated as

$$w = 219.9 \text{ m}, d = 5.57 \text{ m, and } A = 1224.8 \text{ m}^2$$

For comparison the reported values by Chitale (1970) are listed as

$$w = 218 \text{ m}, d = 6 \text{ m, and } A = 1308 \text{ m}^2$$

Example 6.2: Compute w, d, and A if $D_s = 0.80$ mm and $Q = 848$ m³/s for Savarmati River, India (Chitale, 1970). Compare the computed values with those reported by Chitale. Assume the ratio of weight density of sediment to weight density of water as 1.65 and kinematic viscosity of water as 10^{-6} m² /s.

Solution:

Step 1.
Adopt w, d, and S at the ith trial as $(w)_i = 105$ m, $(d)_i = 4.0$ m, and $(S)_i = 0.0001$.

Step 2.
Compute v_*, X, Y, M_n, d_*, and Y_{cr} as follows:

$$- \ v_* = \sqrt{gSd} = \sqrt{9.81 \times 0.0001 \times 4.0} = 0.062\frac{m}{s}$$

$$- \ X = \frac{v_*D_s}{v} = \frac{0.062 \times 0.8 \times 10^{-3}}{10^{-6}} = 50.11$$

$$- \ Y = \frac{\rho v_*^2}{\gamma_s D_s} = \frac{1,000 \times 0.062^2}{16186.5 \times 0.8 \times 10^{-3}} = 0.3$$

$$- \ M_n = \left(\frac{X^2}{Y}\right)^{\frac{1}{3}} = \left(\frac{50.11^2}{0.3}\right)^{\frac{1}{3}} = 20.24$$

$$- \ d_* = \frac{d}{D_s} = \frac{4.0}{0.8 \times 10^{-3}} = 5,000$$

$$- \ Y_{cr} = 0.13M_n^{-0.392}e^{-0.015M_n^2} + 0.045\left[1 - e^{-0.068M_n}\right]$$

$$= 0.13 \times 20.24^{-0.392}e^{-0.015 \times 20.24^2} + 0.045\left[1 - e^{-0.068 \times 20.24}\right] = 0.034$$

Compute $(R_f)_i$ as

$$(R_f)_i = \frac{Y}{Y_{cr}} = \frac{0.3}{0.034} = 8.99.$$

Step 3.
Compute $(C_b)_i$.

$$-\frac{\Delta_{dunes}}{d} = 6\left[1 + 0.01\frac{(d_* - 40)(d_* - 400)}{d_*}\exp\left(-0.055\sqrt{d_*} - 0.04X\right)\right]$$

$$= 6\left[1 + 0.01\frac{(5{,}000 - 40)(5{,}000 - 400)}{5{,}000}\exp\left(-0.055\sqrt{5{,}000} - 0.04 \times 50.11\right)\right]$$

$$= 6.75$$

$$-\frac{\Delta_{ripples}}{d} = \frac{1{,}000}{d_*} = \frac{1{,}000}{5{,}000} = 0.2$$

$$-\left(\frac{\Delta}{\Lambda}\right)_{dunes} = 0.013\psi_{Xdune}(X)(R_f - 1)\exp\left[-\frac{(R_f - 1)}{13(1 - \exp(-0.0082))}\right]$$

where $\psi_{Xdunes}(X) = 1 - \exp\left[-\left(\frac{X}{10}\right)^2\right] = 1 - \exp\left[-\left(\frac{50.11}{10}\right)^2\right] = 0.999.$

Therefore,

$$\left(\frac{\Delta}{\Lambda}\right)_{dunes} = 0.013 \times 0.999 \times (8.99 - 1)\exp\left[-\frac{(8.99 - 1)}{13(1 - \exp(-0.0082))}\right] = \sim 0$$

$$-\left(\frac{\Delta}{\Lambda}\right)_{ripples} = 0.035\psi_{Xripples}(X)(R_f - 1)\exp\left[-0.1\left(R_f - 1\right)\right]$$

where, since $X > 2.5$, $\psi_{Xripples}(X) = \exp\left[-((X - 2.5)/14)^2\right]$
$$= \exp\left[-((50.11 - 2.5)/14)^2\right] = \sim 0.$$

Therefore,

$$\left(\frac{\Delta}{\Lambda}\right)_{ripples} = 0.035 \times 0.0 \times (8.99 - 1)\exp[-0.1(8.99 - 1)] = \sim 0$$

$$- C_s = \frac{1}{\kappa}\ln\left(\psi(X)\frac{d_*}{2}\right)$$

where κ is 0.4 and $\psi(X) = \exp(\kappa w_s^{-1})$.
where w_s is calculated following Equation (6.47) as

$$w_s = 8.5 + [2.5\ln(2X) - 3]\exp\left(-0.217[(\ln(2X)]^2\right)$$

$$= 8.5 + [2.5\ln(2 \times 50.11) - 3]\exp\left(-0.217[(\ln(2 \times 50.11)]^2\right) = 8.54.$$

Therefore, $\psi(X) = \exp(0.4 \times 8.54^{-1}) = 1.05$

Finally, $C_s = \frac{1}{\kappa} \ln \left(\psi(X) \frac{d_*}{2} \right) = \frac{1}{0.4} \times \ln \left(1.05 \times \frac{5,000}{2} \right) = 19.68$

– Compute $(C_b)_i$ as

$$\frac{1}{C_b^2} = \frac{1}{C_s^2} + \sum_{j=d_u, r_i} a_j \left(\frac{\Delta}{\Lambda} \right)_j^{b_j} \frac{\Lambda_j}{d}$$

where $a_j = \frac{1}{2}$ and $b_j = 2$.
Therefore,

$$\frac{1}{C_b^2} = \frac{1}{19.68^2} + \frac{1}{2} \left(0.0^2 \times 6.75 + 0.0^2 \times 0.2 \right)$$

Finally, $(C_b)_i = 19.68$

Step 4.
Compute $(Q)_i$.

$$(Q)_i = (w)_i (d)_i (C_b)_i \sqrt{g(S)_i (d)_i}$$

$$= 105 \times 4.0 \times 19.68 \sqrt{9.81 \times 0.0001 \times 4.0} = 517.8 \frac{m^3}{s}$$

Step 5.
Compare the computed $(Q)_i$ and the measured Q.

Since $(Q)_i$ of $517.8 \frac{m^3}{s}$ is different from the measured Q of $848 \frac{m^3}{s}$, iterate steps 1 to 4 by inputting 1% increase in w and 10% increase in d until $(Q)_{i+1} = Q$.
When $(Q)_{i+1} = Q$, the regime w and d were estimated as

$$w = 106.7 \text{ m}, d = 4.62 \text{ m}, \text{ and } A = 492.95 \text{ m}^2$$

For comparison, the reported values by Chitale (1970) are listed as

$$w = 107 \text{ m}, d = 5.2 \text{ m}, \text{ and } A = 556.4 \text{ m}^2$$

Example 6.3: Compute w, d, and A if $D_s = 31$ mm and $Q = 4,386$ m³/s for North Saskatchewan River, Canada (Neill, 1973). Assume the ratio of weight density of sediment to weight density of water as 1.65 and kinematic viscosity of water as 10^{-6} m² /s. Compare the computed values with those reported by Neill (1973).

Solution:
Step 1.
Adopt w, d, and S at the ith trial as $(w)_i = 235$ m, $(d)_i = 5.0$ m, and $(S)_i = 0.0007$.

Step 2.
Compute v_*, X, Y, M_n, d_*, and Y_{cr} as follows:

– $v_* = \sqrt{gSd} = \sqrt{9.81 \times 0.0007 \times 5.0} = 0.185 \frac{m}{s}$

– $X = \frac{v_* D_s}{v} = \frac{0.185 \times 31 \times 10^{-3}}{10^{-6}} = 5744.2$

$- \ Y = \frac{\rho v_*^2}{\gamma_s D_s} = \frac{1{,}000 \times 0.185^2}{16186.5 \times 31 \times 10^{-3}} = 0.068$

$- \ M_n = \left(\frac{X^2}{Y}\right)^{\frac{1}{3}} = \left(\frac{5744.2^2}{0.068}\right)^{\frac{1}{3}} = 784.17$

$- \ d_* = \frac{d}{D_s} = \frac{5.0}{31 \times 10^{-3}} = 161.29$

$- \ Y_{cr} = 0.13 M_n^{-0.392} e^{-0.015 M_n^2} + 0.045 \left[1 - e^{-0.068 M_n}\right]$

$\qquad = 0.13 \times 784.17^{-0.392} e^{-0.015 \times 784.17^2} + 0.045 \left[1 - e^{-0.068 \times 784.17}\right] = 0.045$

Compute $\left(R_f\right)_i$ as

$$\left(R_f\right)_i = \frac{Y}{Y_{cr}} = \frac{0.068}{0.045} = 1.52$$

Step 3.
Compute $\left(C_b\right)_i$.

$- \ \frac{\Lambda_{dunes}}{d} = 6 \left[1 + 0.01 \frac{(d_* - 40)(d_* - 400)}{d_*} \exp\left(-0.055\sqrt{d_*} - 0.04X\right)\right]$

$\qquad = 6 \left[1 + 0.01 \frac{(161.29 - 40)(161.29 - 400)}{161.29} \exp\left(-0.055\sqrt{161.29} - 0.04 \times 5744.2\right)\right]$

$\qquad = 6.0$

$- \ \frac{\Lambda_{ripples}}{d} = \frac{1{,}000}{d_*} = \frac{1{,}000}{161.29} = 6.2$

$- \ \left(\frac{\Delta}{\Lambda}\right)_{dunes} = 0.013 \psi_{Xdune}(X)\left(R_f - 1\right) \exp\left[-\frac{\left(R_f - 1\right)}{13(1 - \exp(-0.0082))}\right]$

where $\psi_{Xdunes}(X) = 1 - \exp\left[-\left(\frac{X}{10}\right)^2\right] = 1 - \exp\left[-\left(\frac{5744.2}{10}\right)^2\right] = 1.0$.
Therefore,

$$\left(\frac{\Delta}{\Lambda}\right)_{dunes} = 0.013 \times 1.0 \times (1.52 - 1) \exp\left[-\frac{(1.52 - 1)}{13(1 - \exp(-0.0082))}\right] = \sim 0$$

$- \ \left(\frac{\Delta}{\Lambda}\right)_{ripples} = 0.035 \psi_{Xripples}(X)\left(R_f - 1\right) \exp\left[-0.1\left(R_f - 1\right)\right],$

where, since $X > 2.5, \ \psi_{Xripples}(X) = \exp\left[-((X - 2.5)/14)^2\right]$

$\qquad\qquad\qquad\qquad\qquad\qquad = \exp\left[-((5744.2 - 2.5)/14)^2\right] = \sim 0.$

Therefore,

$$\left(\frac{\Delta}{\Lambda}\right)_{ripples} = 0.035 \times 0.0 \times (1.52 - 1) \exp[-0.1(1.52 - 1)] = \sim 0$$

$- \ C_s = \frac{1}{\kappa} \ln\left(\psi(X) \frac{d_*}{2}\right),$

where κ is 0.4 and $\psi(X) = \exp(\kappa w_s^{-1})$,
where w_s is calculated following Equation (6.47) as

$$
\begin{aligned}
w_s &= 8.5 + [2.5 \ln (2X) - 3] \exp(-0.217[(\ln (2X)]^2) \\
&= 8.5 + [2.5 \ln (2 \times 5744.2) - 3] \exp(-0.217[(\ln (2 \times 5744.2)]^2) = 8.5.
\end{aligned}
$$

Therefore, $\psi(X) = \exp(0.4 \times 8.5^{-1}) = 1.05$
Finally, $C_s = \frac{1}{\kappa} \ln \left(\psi(X) \frac{d_*}{2} \right) = \frac{1}{0.4} \times \ln \left(1.05 \times \frac{161.29}{2} \right) = 11.09$

– Compute $(C_b)_i$ as

$$
\frac{1}{C_b^2} = \frac{1}{C_s^2} + \sum_{j=d_u, r_i} a_j \left(\frac{\Delta}{\Lambda} \right)_j^{b_j} \frac{\Lambda_j}{d}
$$

where $a_j = \frac{1}{2}$ and $b_j = 2$.

Therefore,

$$
\frac{1}{C_b^2} = \frac{1}{11.09^2} + \frac{1}{2} (0.0^2 \times 6.0 + 0.0^2 \times 6.2)
$$

Finally, $(C_b)_i = 11.09$

Step 4.
Compute $(Q)_i$.

$$
\begin{aligned}
(Q)_i &= (w)_i(d)_i(C_b)_i \sqrt{g(S)_i(d)_i} \\
&= 235 \times 5.0 \times 11.09 \sqrt{9.81 \times 0.0007 \times 5.0} = 2414.55 \frac{\text{m}^3}{\text{s}}
\end{aligned}
$$

Step 5.
Compare the computed $(Q)_i$ and the measured Q.
  Since $(Q)_i$ of 2414.55 $\frac{\text{m}^3}{\text{s}}$ is different from the measured Q of 4,386 $\frac{\text{m}^3}{\text{s}}$, iterate steps
1 to 4 by inputting 1% increase in w and 10% increase in d until $(Q)_{i+1} = Q$.
When $(Q)_{i+1} = Q$, the regime w and d were estimated as

$$
w = 242.2 \text{ m}, d = 6.92 \text{ m, and } A = 1,676 \text{ m}^2
$$

For comparison, the reported values by Neill (1973) are listed as

$$
w = 244 \text{ m}, d = 7 \text{ m, and } A = 1,708 \text{ m}^2
$$

6.6 Method of Synthesis

Barr and Herbertson (1968) presented the method of synthesis for regime theory.
They clarified two aspects of the theory. First, regime principles apply to all

quasi-uniform flow in channels of any form or shape over a bed of non-cohesive material in active transport. Second, regime concept is restricted to the design of stable irrigation canals in alluvium. The side slopes may be steep, comprising cohesive but erodible material, and the bed is usually erodible.

The key aspect of the method of synthesis lies in the formulation of n-term nondimensional functional equations from $n + 1$ term dimensionally homogeneous equations whose terms are supposed to be linear. In the context of regime theory, the active forces affecting a channel would be gravity represented by acceleration due to gravity (g) and viscosity represented by kinematic viscosity ($v = \mu/\rho$), assuming the density of water and that of sediment are constant. To enumerate the effect of active forces, velocity measures can be defined as follows.

The velocity measures or ratios can be defined as

$$\frac{Q^{\frac{2}{5}}}{g^{\frac{1}{5}}}; \quad \frac{Q}{v}; \quad \frac{v^{\frac{2}{3}}}{g^{\frac{1}{3}}}.$$

Denoting a linear dimension by l a linear function equation can be formed as

$$\phi\left(l, \frac{\frac{Q^{\frac{2}{5}}}{g^{\frac{1}{5}}}, \frac{Q}{v}, \frac{v^{\frac{2}{3}}}{g^{\frac{1}{3}}}}{2^*}\right) = 0, \tag{6.60}$$

where 2^* signifies any two measures in order to include each active force at least once. In Equation (6.60), Q is a dependent variable that can be replaced by sediment discharge Q_s. One can also include other forms of dynamic velocities, depending on whether slope is constant or variable. For example,

$$\frac{Q_s^{\frac{2}{3}}}{g^{\frac{1}{3}}} \text{ or } \frac{Q^{\frac{2}{3}}}{g^{\frac{1}{3}}}; \quad \frac{v^{\frac{2}{3}}}{(g_w)^{\frac{1}{3}}}; \quad \frac{v^{\frac{2}{3}}}{(Sg)^{\frac{1}{3}}},$$

in which g_w refers to the partial gravity submergence force acting on the fluid, $g_w = \frac{\rho_s - \rho}{\rho}$; and S is slope. Then, using linear dimensions of sediment size and flow depth, the nondimensional function equation can be written as

$$\phi\left(D, d, \frac{Q^3}{g^{\frac{1}{3}}} \text{ or } \frac{Q^{\frac{2}{3}}}{g^{\frac{1}{3}}}, \frac{v^{\frac{2}{3}}}{g^{\frac{1}{3}}}, \frac{v^{\frac{2}{3}}}{(g_w)^{\frac{1}{3}}}, \frac{v^{\frac{2}{3}}}{(Sg)^{\frac{1}{3}}}\right) = 0. \tag{6.61}$$

By eliminating $\frac{v^{\frac{2}{3}}}{(Sg)^{\frac{1}{3}}}$ or d, one can obtain the nondimensional function equations as

$$\phi\left(\frac{Q_s}{Q}, \frac{Q_s^3}{Dg^{\frac{1}{3}}}, \frac{Dg^{\frac{1}{3}}}{v^{\frac{2}{3}}}, \frac{d}{D}, \frac{\rho}{\rho_s}\right) = 0 \tag{6.62}$$

$$\phi\left(\frac{Q_s}{Q}, \frac{Q_s^3}{Dg^{\frac{1}{3}}}, \frac{Dg^{\frac{1}{3}}}{v^{\frac{2}{3}}}, S, \frac{\rho}{\rho_s}\right) = 0. \tag{6.63}$$

However, Equation (6.61) is based on the assumption that the channel is wide, which may not always be valid. Relaxing this assumption,

$$\phi\left(D, d, w, \frac{Q_s^3}{g^{\frac{1}{3}}} \text{ or } \frac{Q^{\frac{2}{3}}}{g^{\frac{1}{3}}}, \frac{v^{\frac{2}{3}}}{g^{\frac{1}{3}}}, \frac{v^{\frac{2}{3}}}{(g_w)^{\frac{1}{3}}}, \frac{v^{\frac{2}{3}}}{(Sg)^{\frac{1}{3}}}\right) = 0. \tag{6.64}$$

Equation (6.62) yields alternatives to Equations (6.62) and (6.63) as

$$\phi\left(\frac{Q_s}{Q}, \frac{Q_s^3}{Dg^{\frac{1}{3}}}, \frac{Dg^{\frac{1}{3}}}{v^{\frac{2}{3}}}, \frac{d}{D}, \frac{w}{D}, \frac{\rho}{\rho_s}\right) = 0 \tag{6.65}$$

$$\phi\left(\frac{Q_s}{Q}, \frac{Q_s^3}{Dg^{\frac{1}{3}}}, \frac{Dg^{\frac{1}{3}}}{v^{\frac{2}{3}}}, S, \frac{w}{D}, \frac{\rho}{\rho_s}\right) = 0. \tag{6.66}$$

Now consider two degrees of freedom. Then, the functional Equations (6.63) and (6.65) reduce to

$$\phi\left(\frac{Q_s}{Q}, Q, d, S\right) = 0, \tag{6.67}$$

in which $Q^{2/3}$ has been replaced by Q for simplicity. If Q_s/Q is constant, then Equation (6.67) becomes

$$\phi(Q, d \text{ or } S) = 0. \tag{6.68}$$

Equation (6.68) shows that discharge varies with flow depth and slope.

Now consider three degrees of freedom, as in rigid-sided channels shown in Figure 6.4, in which side slopes are defined by 1:Z. Both d_1, the distance from

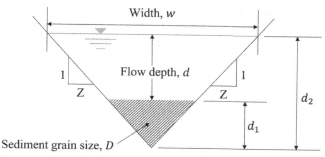

Figure 6.4 A rigid sided channel

the apex of the triangle to the top of the sediment bed, and d_2, the distance from the apex to the water surface are free to adjust. The quantity d_2 leads to defining w, and if d_1 is known then d_2 also d. Thus, w and d can alternatively be defined as dependent variables. Then, the nondimensional function equation can be written as

$$\phi\left(\frac{Q_s}{Q}, \frac{Q_s^{2/3}}{Dg^{\frac{1}{3}}}, \frac{Dg^{\frac{1}{3}}}{v^{\frac{2}{3}}}, S \ or \ \frac{w}{D} \ or \ \frac{d}{D}, Z, \frac{\rho}{\rho_s}\right) = 0. \tag{6.69}$$

Assuming constant sediment size and density and constant side slope, Equation (6.67) reduces to

$$\phi(Q, \ w \ or \ d \ or \ S) = 0. \tag{6.70}$$

If data on Q and Q_s are available, then w, d, and S can be plotted as functions of Q, regardless of the groups of variables being held constant between Equations (6.69) and (6.70).

In the regime equations, width, depth, and slope are plotted against discharge raised to some power (called exponents). Although it is assumed that the exponents of discharge remain constant, Barr and Herbertson (1968) argue that as discharge changes the exponents may change, and all geometric terms will vary with b, which is the exponent of Q for w. Between two dynamically similar channels the ratios of sediment size, channel depth, and width vary in the same proportion. Then, for constant sediment density, Equation (6.69) can be simplified to

$$\phi\left(\frac{Q_s}{Q}, \frac{Q^{\frac{2}{5}}}{bg^{\frac{1}{5}}}, \frac{bg^{\frac{1}{3}}}{v^{\frac{2}{3}}}\right) = 0. \tag{6.71}$$

If Q and Q_s are constant between the two channels then the term $\frac{Q^{\frac{2}{5}}}{bg^{\frac{1}{5}}}$ states that the linear dimensions vary with 0.4 power of the ratios of discharges and the term $\frac{bg^{\frac{1}{3}}}{v^{\frac{2}{3}}}$ states that the ratios of the kinematic viscosity values vary with 0.6 power of the discharge ratios. These are essentially the conditions for dynamic similarity between the two channels. Equation (6.71) specifically states that

$$w \propto Q^{0.4} \tag{6.72}$$

$$d \propto Q^{0.4} \tag{6.73}$$

$$D \propto Q^{0.4} \tag{6.74}$$

$$v \propto Q^{0.2} \tag{6.75}$$

$$S \propto Q^{0.0}. \tag{6.76}$$

Equation (6.76) states that slope is constant. The frequently quoted regime equations with constant D and v are

$$w \propto Q^{0.5} \tag{6.77}$$

$$d \propto Q^{0.33} \tag{6.78}$$

$$S \propto Q^{-0.167}. \tag{6.79}$$

Contrasting Equation (6.77) with Equations (6.72), (6.73), and (6.76) the question arises whether or not the assumption of constant sediment diameter and kinematic viscosity takes into consideration the variations of the w, d, and S relationships.

References

Barr, D. I. H. and Herbertson, J. G. (1968). A similitude framework of regime theory. *Proceedings, Institution of Civil Engineers*, Vol. 41, pp. 761–781.

Chitale, S. Y. (1970). River channel patterns. *Journal of Hydraulics Division, ASCE*, Vol. 96, No. HY1, pp. 201–221.

Da Silva, A. M. F. and Yalin, M. S. (2017). *Fluvial processes*. CRC Press, Boca Raton, FL.

Engelund, F. (1966). Hydraulic resistance of alluvial rivers. *Journal of Hydraulics Division, ASCE*, Vol. 92, No. HY2, pp. 315–326.

Jia, Y. (1990). Minimum Froude number and the equilibrium of alluvial sand rivers. *Earth Surface Processes and Landforms*, Vol. 15, No. 3, pp. 199–209.

Lei, S. (1992). A regime theory based on the minimization of Froude number. Ph.D. Thesis, Department of Civil Engineering, Queen's University, Kingston, Canada.

Neill, C. R. (1973). *Hydraulic geometry of sand rivers in Alberta*. Proceedings of the Hydrology Symposium, Alberta, May.

Parker, G. (1978). Self-formed straight rivers with equilibrium banks and mobile bed: Part I. The sand-silt river. *Journal of Fluid Mechanics*, Vol. 89, No. 1, pp. 109–125.

Yalin, M. S. (1977). On the determination of ripple length. *Journal of Hydraulics Division, ASCE*, Vol. 103, No. HY4, pp. 439–442.

Yalin, M. S. (1992). *River Mechanics*. Pergamon Press, Oxford.

Yalin, M. S. and Ferreira da Silva, A. M. (1997). On the computation of equilibrium channels in cohesionless alluvium. *Journal of Hydroscience and Hydraulic Engineering*, Vol. 16, No. 2, pp. 1–13.

7

Hydrodynamic Theory

Several hydrodynamic theories have been employed for deriving downstream hydraulic geometry relations of width, depth, velocity, and slope in terms of flow discharge. Five theories – the Smith theory, the Julien-Wargadalam (JW) theory, the Parker theory, the Griffiths theory, and the Ackers theory – are discussed in this chapter. These theories employ different forms of the continuity equation, friction equation, and transport equations. The Smith hydrodynamic theory also uses a morphological relation, whereas the JW theory uses an angle between transversal and downstream shear stress components, and the Parker theory uses a depth function. With these equations, the downstream hydraulic geometry relations are derived.

Notation

a = constant
A = cross-sectional area
A_c = constant
a_0 = exponent of the resistance equation
b_e = constant of the resistance equation
b_r = coefficient
b_0 = exponent
C = Chezy's coefficient
C_F = dimensionless parameter defined by the centrifugal force
C_s = suspended sediment concentration
d = flow depth (approximating the hydraulic radius, for wide channels)
D = sediment grain size
D_a = coefficient of deviation angle equation
D_{50} = median grain diameter of the bed material in meters

---

d_* = dimensionless flow depth
d_m = depth corresponding to mean annual flow discharge
f = Darcy–Weisbach friction factor
g = acceleration due to gravity
k = ripple height
K = parameter equal to one for very wide channels
k_1 = a constant (equal to $1/n_m$ for Manning's equation)
k_2 = a constant
k_3 = a constant
k_4 = constant
L = constant
m = exponent
M = average percentage of silt-clay in the channel boundary
n = exponent
n_m = Manning's roughness coefficient
P = wetted perimeter
q = discharge per unit width or specific discharge
Q = discharge
Q_s = total downstream flux of sediment in the channel
Q_* = dimensionless discharge
q_{sa} = a function of the total sediment moving in the x-direction
q_{sd} = x component of sediment discharge (approximate bed load transport)
R = hydraulic radius
R_c = radius of curvature
R_w = ratio of radius of curvature to channel width
S = bed slope
S_b = slope at bankfull
S_s = specific gravity of the sediment
S_w = slope of the water surafce (approximated as bed slope)
t = time
v = flow velcoity
v_f = fall velocity
v_* = dimensionless flow velocity
w = width
w_m = width corresponding to mean annual flow discharge
w_* = dimensionless width
x = longitudinal direction
y = transverse direction
y_1 = edge of the channel
y_2 = edge of the channel
$z = z(x, y, t)$ = surface of the whole landscape at time t

Continued

Z = elevation of the channel bed
$\overline{q}_s$ = two-dimensional vector describing the magnitude and direction of sediment flux
ξ = exponent
μ = exponent
τ = exponent
τ_0 = downstream bed shear force
τ_r = radial shear stress
τ_* = Shields parameter
τ_{*c} = critical value of τ_*
γ_s = weight density of sediment
γ = weight density of water
ρ = mass density of water
ρ_s = mass density of sediment particles
ρ^* = dimensionless submerged mass density of the sediment parameter
υ = kinemtic viscosity

7.1 Introduction

Hey (1979) discussed feedback mechanisms that continually operate between channel sections and as a result hydraulic variables change in time and space. These mechanisms operate upstream through drawdown and backwater and downstream through sediment discharge, and govern the channel evolution. As erosion occurs on a land surface, sediment originating in the headwaters is transported, deposited, and reworked. The transport, aggradation, and degradation vary in space and time, to which the channel and valley respond. Consideration of these mechanisms and channel processes can lead to a dynamic process-based framework for defining hydraulic geometry. However, when the channel achieves a quasi-equilibrium state, a separate consideration of each mechanism and channel process is not warranted. Among the attempts to derive hydraulic geometry from a partial theoretical basis, the work of Langbein (1964), who assumed that a river channel tended toward an equal distribution among velocity, depth, width, and slope in accommodating a change in stream power, is notable. This assumption has not been fully justified on physical grounds.

The downstream hydraulic geometry has been shown to exhibit similarity over a large class of channels (Leopold and Maddock, 1953). Of particular interest is the relation of certain parameters of gross channel form to the total channel discharge, Q. This suggests that it may be possible to derive this geometry using the universally applicable laws of hydrodynamics. The hydrodynamic theory for hydraulic geometry is based on the principles of conservation of mass (expressed by continuity equation), momentum for water (expressed by a resistance law), and sediment transport (expressed by a sediment transport law) (Einstein, 1942; Kalinske, 1947; Bagnold, 1956). Five variations or simplified forms of this theory are discussed in this chapter.

In practice, the continuity equation of water, a resistance equation, a sediment transport equation, and a morphological relation are invoked for deriving hydraulic geometry. These equations have four unknowns: width, depth, velocity, and slope. The morphological relation is derived in different ways, depending on the hypothesis to be employed, as for example in the hypotheses of least mobility (Dou, 1964), minimum stream power (Chang, 1980, 1988; Yang, 1976, 1984), minimum variance (Langbein, 1964; Williams, 1978), or maximum efficiency (Davies and Sutherland, 1983; White et al., 1986).

Ackers (1964) summarized the results of experiments conducted at the Hydraulics Research Station at Wallingford, England, on small streams that achieved stability at discharges between 0.4 and 0.5 cubic feet per second (cfs). Empirical correlations of stream geometry with discharge were found to be consistent with those deduced by the combination of three physical relationships: (1) the resistance formula, (2) sediment transport formula, and (3) the ratio of width to depth. The usefulness of these relationships depends on the quality of data and should be applied in situations similar to those for which the data were collected. This chapter discusses five hydrodynamic theories – Smith, Julien-Wargadalam, Parker, Griffiths, and Ackers – for the derivation of hydraulic geometry.

7.2 Smith Theory

Applying the Manning equation and a well-known empirical formula for the transport of total sediment load, Smith (1974) derived downstream hydraulic geometry for steady-state channels, subject to three conditions: (1) Sediment mass is conserved during transport; (2) the channel has a form just sufficient to transport its total water discharge, and (3) the channel has a form just sufficient to transport its total sediment load. Three assumptions were made for deriving the hydraulic geometry or form parameters of width, depth, velocity, and slope: (a) The channel, as shown in Figure 7.1, has a finite width, which is essential for specifying boundary conditions; (b) the channel is carved in noncohesive materials, with most of the sediment transport occurring close to the channel bed, which is approximately valid for a large class of channels, especially for gravel and alluvial ones; and (c) there is freedom to choose a time scale for which the channel has an essentially steady-state form (Schumm and Litchy, 1964). Smith's hydraulic geometry relations corresponded well with the empirical relations of Leopold and Maddock (1953). The analytical approach developed by Smith (1974) is promising and insightful, and is discussed here. He derived the geometry relations as: channel width $\sim Q^{0.6}$; channel depth $\sim Q^{0.3}$; stream

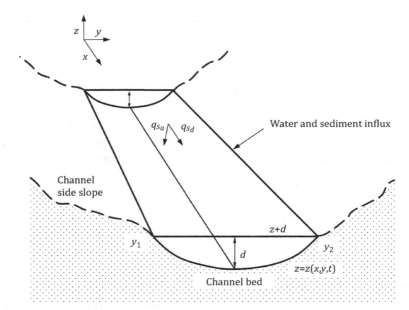

Figure 7.1 A diagrammatic representation of the channel model

velocity $\sim Q^{0.1}$; channel slope $\sim Q^{-0.2}$; where Q is the total stream discharge. The mathematical framework developed by Smith (1974) has considerable flexibility and, in modified form, should prove to be of use in channel morphological studies.

7.2.1 Hydrodynamic Formulation

Let $z = z(x, y, t)$ represent the surface of the whole landscape at time t, where x and y represent the longitudinal and transverse directions. We may assume a time scale for which dz/dt is close to zero, and for which the channel has a steady-state form. Let there be a channel with a straight center line (thalweg) aligned along the x axis. It is assumed that: (1) the channel is symmetrical about the center line; (2) any curves in the channel are not major controlling factors of the hydraulic geometry; (3) the channel has a finite width; (4) the channel is in noncohesive materials, with most of the sediment transport occurring close to the channel bed; (5) the channel is in steady-state condition; (6) there is an appropriate time scale; and (7) changes in channel geometry occur at a very short time scale when compared with the time scale at which significant landscape lowering occurs, implying that at that time scale $\partial z/\partial t$ is close to zero and the channel has a steady-state form.

It is hypothesized that sediment mass is conserved during transport; the channel has a form, just sufficient to carry its total discharge and its total sediment discharge. The finite width concerns the flux of water in the channel that moves in the direction of slope of the water surface, which is a reasonable approximation to the more general hypothesis that water moves down the energy gradient. For the symmetry of the channel, the water surface has no slope component in the y direction perpendicular to the direction of flow. In other words, the water moves only in the x direction.

It is assumed that flow in the x direction is adequately given by Manning's equation:

$$v = k_1 d^{\frac{2}{3}} S_W^{\frac{1}{2}}, \tag{7.1}$$

where v is the flow velcoity, S_w is the slope of the water surafce (approximated as bed slope), d is the flow depth (approximating the hydraulic radius, for wide channels), and k_1 is a constant (equal to $1/n_m$ for Manning's equation in MKS system and $1.487/n_m$ in the FPS system of units, n_m is the Manning roughness coefficient). The discharge per unit width (Q/w) or specific discahrge, q, is given as

$$q = vd = k_1 d^{\frac{5}{3}} S_W^{\frac{1}{2}} \quad (k_1 = 1/n_m). \tag{7.2}$$

Implicit in this equation is that Manning's n_m is constant along the channel.

Since the channel has a finite width, at every point downstream there are two values of y (y_1 and y_2) representing the edges of the channel where the flux of water is zero. Since q is the magnitude of the flux in the x direction, the total flux Q is obtained by integrating q across the channel in the y direction from y_1 to y_2:

$$Q = \int_{y_1}^{y_2} q \, dy = \int_{y_1}^{y_2} k_1 d^{\frac{5}{3}} S_W^{\frac{1}{2}} \, dy, \tag{7.3}$$

in which Q is determined by the contributing drainage area. This means that Equation (7.3) states that the channel must have a form sufficient to transport this discharge. For simplicity, it is assumed that Q increases linearly with the downstream distance:

$$\int_{y_1}^{y_2} k_1 d^{\frac{5}{3}} S_W^{\frac{1}{2}} \, dy = A_c x \quad A_c = \text{constant}. \tag{7.4}$$

Now, consider the flux of sediment mass governed by the law of conservation of mass. It is assumed that sediment moves close to the channel bed. This will mean that there is a continual interchange of particles between the channel bed and water. Then, by the conservation law it can be stated that the sediment moving out of a small section of channel bed minus the sediment moving into the small section of channel bed equals the net erosion (or deposition) in that section of the bed. Mathematically, this can be approximated as

$$\frac{\partial Z}{\partial t} = \nabla \cdot \overline{q_s}, \tag{7.5}$$

where Z is the elevation of the channel bed, and $\overline{q_s}$ is the two-dimensional vector describing the magnitude and direction of sediment flux.

Now expressions for the two components of sediment flux vector are sought, based on physical reasoning. For noncohesive materials, two major sets of forces control the sediment motion. The first is the drag force of water and the second is gravitational. It is reasonable to assume that a component of sediment moves in the direction of drag force, or in other words, in the direction of flow. Let this component be denoted as q_{sd}. An approximate bed load transport equation can be expressed as (Smith and Bretherton, 1972):

$$q_{sd} = k_2 q^n S_w^m = k_1^n k_2 d^{\frac{5n}{3}} S_w^{m+\frac{n}{2}}, \tag{7.6}$$

where k_2 is a constant, and m and n are exponents $\in (1,4)$.

The other component of sediment discharge is controlled by gravity. The channel bed, unlike the water surface, must have a component of slope in the y direction (otherwise the channel would not be finite in width). Let the sediment moving in the x direction be denoted as q_{sd}, as a continual exchange of particles controlled by turbulence. It is then reasonable to assume that if a net cross-channel slope exists, then there should be a net diffusion of particles in the direction of cross-channel gradient under the influence of gravity. Using this argument, it is assumed that the quantity of sediment thus moving, q_{sa}, is a function of the total sediment moving in the x-direction, q_{sd}, and the cross-channel gradient. The cross-channel gradient may be written as $\partial d / \partial y$. Then,

$$q_{sa} = k_3 q_{sd} \frac{\partial d}{\partial y} = k_1^n k_2 k_3 d^{\frac{5n}{3}} S_W^{m+\frac{n}{2}} \frac{\partial d}{\partial y}, \tag{7.7}$$

where k_3 is constant. Hence,

$$\overline{q_s} = \left(k_1^n k_2 d^{\frac{5n}{3}} S_W^{m+\frac{n}{2}}, \quad k_1^n k_2 k_3 d^{\frac{5n}{3}} S_W^{m+\frac{n}{2}} \frac{\partial d}{\partial y} \right). \tag{7.8}$$

Thus,

$$\frac{\partial z}{\partial t} = -\frac{\partial}{\partial x} \left(k_1^n k_2 d^{\frac{5n}{3}} S_W^{m+\frac{n}{2}} \right) - \frac{\partial}{\partial y} \left(k_1^n k_2 k_3 d^{\frac{5n}{3}} S_W^{m+\frac{n}{2}} \frac{\partial d}{\partial y} \right). \tag{7.9}$$

The total downstream flux of sediment in the channel Q_s may be obtained by integrating the x component of sediment discharge, q_{sd}, across the channel:

$$\int_{y_1}^{y_2} q_{sd} \, dy = \int_{y_1}^{y_2} k_1^n k_2 d^{\frac{5n}{3}} S_w^{m+\frac{n}{2}} \, dy = Q_s. \tag{7.10}$$

The total discharge of water Q is proportional to the area, and by assumption it increases linearly downstream. Since, by the choice of time scale, the whole landscape is everywhere eroding downward at the same rate, it follows that the sediment discharge must also increase linearly with distance downstream:

$$\int_{y_1}^{y_2} q_{sd}\, dy = Q_s = Lx, \quad L = \text{constant}. \tag{7.11}$$

Equation (7.11) states that the channel must have a form sufficient to carry the total volume of sediment, given the sediment transport law. Equation (7.11) can be now expressed with the use of Equation (7.10) as

$$\int_{y_1}^{y_2} k_1^n k_2 d^{\frac{5n}{3}} S_w^{m+\frac{n}{2}}\, dy = Lx, \quad L = \text{constant}. \tag{7.12}$$

Now the mathematical equations for deriving the channel geometry comprise Equations (7.4), (7.9), and (7.12). For steady state, $dz/dt = 0$. Therefore, Equation (7.9) becomes

$$\frac{\partial}{\partial x}\left(k_1^n k_2 d^{\frac{5n}{3}} S_w^{m+\frac{n}{2}}\right) + \frac{\partial}{\partial y}\left(k_1^n k_2 k_3 d^{\frac{5n}{3}} S_w^{m+\frac{n}{2}} \frac{\partial d}{\partial y}\right) = 0. \tag{7.13}$$

7.2.2 Derivation of Hydraulic Geometry

Smith (1974) proposed a similarity solution of the form

$$S_w = k_4 x^{\mu} \qquad k_4 = \text{constant} \tag{7.14}$$

$$d = x^{\xi} f\left(\frac{y}{x^{\tau}}\right), \quad \eta = \frac{y}{x^{\tau}} \qquad \mu, \xi, \tau = \text{exponents}, \tag{7.15}$$

where $f(.)$ is some function that is to be determined. First, introducing Equations (7.14) and (7.15) in Equation (7.13), one can write

$$\frac{\partial}{\partial x}\left[k_1^n k_2 \left(x^{\xi} f\left(\frac{y}{x^{\tau}}\right)\right)^{\frac{5n}{3}} (k_4 x^{\mu})^{m+\frac{n}{2}}\right] + \frac{\partial}{\partial y}\left(k_1^n k_2 k_3 \left(x^{\xi} f\left(\frac{y}{x^{\tau}}\right)\right)^{\frac{5n}{3}} (k_4 x^{\mu})^{m+\frac{n}{2}} \cdot \frac{\partial}{\partial y}\left(x^{\xi} f\left(\frac{y}{x^{\tau}}\right)\right)\right) = 0. \tag{7.16}$$

Equation (7.16) can be cast as

$$\frac{\partial}{\partial x}\left[k_1^n k_2 x^{\frac{5n\xi}{3}} f(\eta)^{\frac{5n}{3}} k_4^{m+\frac{n}{2}} x^{\mu\left(m+\frac{n}{2}\right)}\right] + \frac{\partial}{\partial y}\left[k_1^n k_2 k_3 x^{\frac{5n\xi}{3}} f(\eta)^{\frac{5n}{3}} k_4^{m+\frac{n}{2}} x^{\mu\left(m+\frac{n}{2}\right)} \cdot x^{\xi} \frac{\partial}{\partial y} f(\eta)\right] = 0. \tag{7.17}$$

Note that

$$\frac{\partial f(\eta)}{\partial y} = \frac{\partial f(\eta)}{\partial \eta} \cdot \frac{\partial \eta}{\partial y} = \frac{\partial f(\eta)}{\partial \eta} \cdot \frac{1}{x^{\tau}}. \tag{7.18}$$

Inserting Equation (7.18) in Equation (7.17), one gets

$$\frac{\partial}{\partial x}\left[k_1^n k_2 f(\eta)^{\frac{5n}{3}} x^{\frac{5n\xi}{3}+\mu\left(m+\frac{n}{2}\right)}\right] + \frac{\partial}{\partial y}\left[k_1^n k_2 k_3 f(\eta)^{\frac{5n}{3}} x^{\frac{5n\xi}{3}+\mu\left(m+\frac{n}{2}\right)+\xi-\tau} \cdot \frac{\partial f(\eta)}{\partial \eta}\right] = 0. \quad (7.19)$$

Equation (7.19) can be simplified as

$$k_1^n k_2 \left[\frac{5n\xi}{3} + \mu\left(m+\frac{n}{2}\right)\right] x^{\frac{5n\xi}{3}+\mu\left(m+\frac{n}{2}\right)-1} \frac{\partial}{\partial x}\left[f(\eta)^{\frac{5n}{3}}\right]$$

$$+ k_1^n k_2 k_3 x^{\frac{5n\xi}{3}+\mu\left(m+\frac{n}{2}\right)+\xi-\tau} \frac{\partial}{\partial y}\left[f(\eta)^{\frac{5n}{3}} \frac{\partial f(\eta)}{\partial \eta}\right] = 0. \quad (7.20)$$

Further simplification of Equation (7.20) leads to

$$k_1^n k_2 \left[\frac{5n\xi}{3} + \mu\left(m+\frac{n}{2}\right)\right] x^{\frac{5n\xi}{3}+\mu\left(m+\frac{n}{2}\right)-1} \frac{\partial}{\partial x}\left[f(\eta)^{\frac{5n}{3}}\right]$$

$$+ k_1^n k_2 k_3 x^{\frac{5n\xi}{3}+\mu\left(m+\frac{n}{2}\right)+\xi-2\tau} \frac{\partial}{\partial \eta}\left[f(\eta)^{\frac{5n}{3}} \frac{\partial f(\eta)}{\partial \eta}\right] = 0. \quad (7.21)$$

Now consider Equation (7.4). With the use of Equation (7.14) and (7.15), and noting that $\eta = \frac{y}{x^\tau}$; $\partial \eta = \frac{\partial y}{x^\tau}$; $\partial y = x^\tau \partial \eta$, one can write

$$\int_{y_1}^{y_2} k_1 \left(x^\xi f(\eta)\right)^{\frac{5}{3}} (k_4 x^\mu)^{\frac{1}{2}} \, dy = A_c x = \int_{y_1}^{y_2} k_1 x^{\frac{5\xi}{3}+\frac{\mu}{2}+\tau} \cdot k_4 f(\eta)^{\frac{5}{3}} \, d\eta$$

$$= k_1 x^{\frac{5\xi}{3}+\frac{\mu}{2}+\tau} k_4 \int_{y_1}^{y_2} f(\eta)^{\frac{5}{3}} \, d\eta = A_c x. \quad (7.22)$$

Now consider Equation (7.12). Using Equations (7.14) and (7.15), one can write

$$\int_{y_1}^{y_2} k_1^n k_2 \{x^\xi f(\eta)\}^{\frac{5n}{3}} (k_4 x^\mu)^{m+\frac{n}{2}} \, dy = Lx$$

$$\int_{y_1}^{y_2} k_1^n k_2 k_4^{m+\frac{n}{2}} x^{\frac{5n\xi}{3}+\mu\left(m+\frac{n}{2}\right)+\tau} f(\eta)^{\frac{5n}{3}} \, d\eta = k_1^n k_2 k_4^{m+\frac{n}{2}} \int_{y_1}^{y_2} x^{\frac{5n\xi}{3}+\mu\left(m+\frac{n}{2}\right)+\tau} f(\eta)^{\frac{5n}{3}} \, d\eta = Lx. \quad (7.23)$$

From Equation (7.23), one may choose the exponents of x such that it remains only a function of η. This yields

$$\frac{5n\xi}{3} + \mu\left(m+\frac{n}{2}\right) - 1 = \frac{5n\xi}{3} + \mu\left(m+\frac{n}{2}\right) + \xi - 2\tau. \quad (7.24)$$

Equation (7.24) simplifies to

$$\xi - 2\tau = -1. \tag{7.25}$$

From Equation (7.22), one obtains for the exponents of x:

$$\frac{5\xi}{3} + \frac{\mu}{2} + \tau = 1. \tag{7.26}$$

From Equation (7.23), one obtains for the exponents of x:

$$\frac{5n\xi}{3} + \mu\left(m + \frac{n}{2}\right) + \tau = 1. \tag{7.27}$$

Now Equations (7.25)–(7.27) are solved for μ, ξ, and τ. From Equation (7.25),

$$\tau = \frac{\xi}{2} + \frac{1}{2}. \tag{7.28}$$

Substituting Equation (7.28) into Equation (7.26),

$$5\frac{\xi}{3} + \frac{\mu}{2} + \frac{\xi}{2} + \frac{1}{2} = 1 \tag{7.29}$$
$$10\xi + 3\xi + 3\mu = \xi(10 + 3) + 3\mu = 3.$$

Substituting Equation (7.28) into Equation (7.27),

$$5\frac{n\xi}{3} + \mu\left(m + \frac{n}{2}\right) + \frac{\xi}{2} + \frac{1}{2} = 1$$

or

$$10n\xi + 3\mu(2m + n) + 3\xi = \xi(10n + 3) + 3\mu(2m + n) = 3. \tag{7.30}$$

After multiplying Equation (7.29) by $(2m + n)$,

$$13\xi(2m + n) + 3\mu(2m + n) = 3(2m + n). \tag{7.31}$$

On subtraction of Equation (7.30) from Equation (7.31),

$$\xi[26m + 3n - 3] = 3(2m + n - 1). \tag{7.32}$$

Then,

$$\xi = \frac{3(2m + n - 1)}{26m + 3n - 3} = \frac{2m + n - 1}{\frac{26}{3}m + n - 1}. \tag{7.33}$$

Substituting Equation (7.33) in Equation (7.28), one obtains

$$\tau = \frac{1}{2} + \frac{3(2m + n - 1)}{2(26m + 3n - 3)} = \frac{16m + 3n - 3}{26m + 3n - 3} = \frac{\frac{16}{3}m + n - 1}{\frac{26}{3}m + n - 1}. \tag{7.34}$$

Substituting Equations (7.33) and (7.34) in Equation (7.26), one obtains μ:

$$\frac{5}{3}\left[\frac{2m+n-1}{\frac{26}{3}m+n-1}\right] + \frac{\mu}{2} + \frac{\frac{16}{3}m+n-1}{\frac{26}{3}m+n-1} = 1 \tag{7.35}$$

$$\frac{\mu}{2} = 1 - \frac{\frac{16}{3}m+n-1}{\frac{26}{3}m+n-1} - \frac{5}{3}\frac{(2m+n-1)}{\left(\frac{26}{3}m+n-1\right)}$$

$$= \frac{\frac{26}{3}m+n-1-\frac{16}{3}m-n+1-\frac{5}{3}(2m)-\frac{5}{3}n+\frac{5}{3}}{\frac{26}{3}m+n-1}. \tag{7.36}$$

Then,

$$\frac{\mu}{2} = \frac{-\frac{5}{3}n+\frac{5}{3}}{\frac{26}{3}m+n-1} \tag{7.37}$$

$$\mu = \frac{-\frac{10}{3}n+\frac{10}{3}}{\frac{26}{3}m+n-1}. \tag{7.38}$$

Equations (7.33), (7.34), and (7.38) give the values of ξ, τ, and μ. If $m = n = 2$ (Smith and Bretherton, 1972), then these equations yield:

$$\xi = \frac{5}{\frac{55}{3}} = \frac{15}{55} = \frac{3}{11} \tag{7.39}$$

$$\tau = \frac{35}{55} = \frac{7}{11} \tag{7.40}$$

$$\mu = -\frac{2}{11}. \tag{7.41}$$

Now consider the hypotheses assumed in the beginning.

$$Q \propto x \quad \text{(distance downstream)} \tag{7.42}$$

$$\text{Slope,} \quad S_w = k_4 x^\mu \propto Q^\mu \propto Q^{-\frac{2}{11}} \tag{7.43}$$

$$\text{Depth of flow,} \quad d = x^\xi f(\eta) \propto Q^\xi \propto Q^{\frac{3}{11}} \tag{7.44}$$

$$\text{Width,} \quad w = y_2 - y_1 = x^\tau(\eta_2 - \eta_2) \propto Q^\tau \propto Q^{\frac{7}{11}}. \tag{7.45}$$

Finally, recalling Manning's equation,

$$v \propto d^{\frac{2}{3}}s_w^{\frac{1}{2}} \propto Q^{\frac{2\xi}{3}}Q^{\frac{\mu}{2}} \propto Q^{\frac{2\xi}{3}+\frac{\mu}{2}} \propto Q^{\frac{1}{11}}. \tag{7.46}$$

The exponents from Equation (7.45) for w, Equation (7.44) for d, Equation (7.46) for v, and Equation (7.43) for slope, respectively, are: $b = 7/11$, $f = 3/11$, $m = 1/11$, and $y = -2/11$, which are close to $b = 0.6$, $f = 0.3$, and $m = 0.1$ obtained by Leopold and Maddock (1953).

Note that $\frac{5\xi}{3} + \frac{\mu}{2} + \tau = 1$ holds from Equation (7.26). Therefore,

$$Q = wdv \propto Q^\tau Q^\xi Q^{\frac{\mu}{2}+\frac{2\xi}{3}} = Q^{\tau+\frac{\mu}{2}+\frac{5\xi}{3}}. \tag{7.47}$$

Thus, $\tau + [5\xi/3] + [\mu/2] = 1$ holds, as before. Usually, m and n lie between 1 and 4. Then, the values of τ and ξ are fairly independent of n, or the exponents for width and depth in the downstream hydraulic geometry depend largely on the exponent of slope in the transport law. On the other hand, exponent μ largely depends on the value of n so that the exponent for slope in the downstream hydraulic geometry depends on how quickly the sediment transporting capacity varies with the discharge per unit width. Furthermore, the exponents for width and depth τ and ξ are not very sensitive to changes in n and m if n and m take on values between 1 and 4. Also, none of the exponents is very sensitive to a change in the rate of increase of total discharge downstream (e.g. if $Q \propto x^{\frac{1}{2}}$ rather than $Q \propto x$).

7.3 Julien–Wargadalam (JW) Theory

Julien (1988) analytically derived the downstream hydraulic geometry of noncohesive alluvial channels using four relationships: (1) steady-state continuity equation, (2) resistance equation, (3) rate of sediemnt transport linked to shear force, and (4) the angle between transversal and downstream shear stress components. These relationships are outlinbed in this section. Julien (2014) reviewed a three-level approach to the downstream hydraulic geometry of alluvial rivers: empirical concepts, theoretical developments, and equivalent channel width.

7.3.1 Continuity Equation

For steady-state uniform bankfull flow condition, the continuity equation can be experessed as

$$Q = wdv, \tag{7.48}$$

where Q is the dominant discharge. For wide channels with large width–depth ratios, the flow depth d is approximately equal to the hydraulic radius $R = A/P$, where A is the flow cross-sectional area and P is the wetted perimeter.

7.3.2 Resistance Equation

The flow resistance equation was expressed in power form equivalent to the Keulegan (1938) equation:

$$\frac{1}{\sqrt{f}} = b_e \ln \left(\frac{d}{D}\right)^{a_0}, \tag{7.49}$$

where f is the Darcy–Weisbach friction factor, D is the sediment grain size, a_0 is the exponent of the resistance equation, and b_e is the resistance constant. The value of a_0 that defines the slope of the tangent to the semi-logarithmic resistance Equation (7.49) was expressed as a function of the relative submergence (d/D):

$$a_0 = \frac{1}{\ln \left(\frac{12.2d}{D}\right)}. \tag{7.50}$$

Equation (7.50) shows that a_0 tends to 0 as d/D tends to infinity. Then, the Darcy–Weisbach friction factor and Chezy's coefficient C remain constant as seen from $f = 8g/C^2$. For $d/D = 200$ (intermediate values of relative submergence), $a_0 = 1/6$ (Manning–Strickler relationship) is comparable to the Keulegan equation. For coarse bed material channels, exponent a_0 increases quickly at small values of submergence. Literature shows that $a_0 = 0.5$ for $0.7 < \frac{d}{D} < 10$ (Leopold and Wolman, 1957); $a_0 = 0.25$ for $3 < \frac{d}{D} < 13$ (Ackers, 1964); $a_0 = 0.25$ (Kellerhals, 1967); $a_0 = 0.44$ for $2 < \frac{d}{D} < 10$ (Charlton et al., 1978); $a_0 = 0.281$ for gravel-bed stream (Bray, 1973); and $a_0 = 0.46$ for $0.2 < \frac{d}{D} < 4$ for very steep cobble and boulder bed stream (Mussetter, 1989).

The average flow velocity was expressed as a power form (Einstein and Chien, 1954):

$$v = a\sqrt{8g}\left(\frac{d}{D}\right)^{a_0} d^{\frac{1}{2}} S^{\frac{1}{2}}, \tag{7.51}$$

where a is a constant. The downstream bed shear force τ_0 was expressed as a function of bed slope S, mass density of water ρ, and hydraulic radius $R = Kd$, where K is a parameter equal to one for very wide channels (large width–depth ratio):

$$\tau_0 = K\rho g d S. \tag{7.52}$$

Equation (7.51) emphasizes the influence of relative submergence of bed sediment on flow resistance. However, the flow resistance in alluvial channels is affected by channel irregularity, channel boundary sediment composition, vegetation, and conditions of sediment transport. Manning's equation may therefore be preferable as a resistance equation.

7.3.3 Sediment Transport Equation

The sediment transport was connected to the Shields parameter τ_* expressed as the ratio of the relative magnitude of downstream shear force and the weight of the sediment particles, also called Shields number:

$$\tau_* = \frac{\tau_0}{(\rho_s - \rho)gD}, \tag{7.53}$$

where ρ_s is the mass density of sediment particles. For the incipient motion of sediment particles in turbulent flows over rough boundaries, the critical value of τ_*, τ_{*c}, is about 0.047. The sediment transport rate increases with increasing τ_*, which is associated with the vertical processes of aggradation and degradation in alluvial channels.

7.3.4 Angle between Transversal and Downstream Shear Stress Components

Channels have curves that lead to secondary circulation. Streamlines near the bed deviate toward the inner bank, and streamlines near the surface are deflected toward the outer bank. The drag on the bed sediment particles and tangential bed shear stress will be pointed toward the inner bank. The ratio of radial shear stress τ_r to the downstream bed shear stress τ_0 expresses the deviation angle θ of the streamlines near the bed. For flow in bends with the radius of curvature, R_c, is proportional to channel width w. Therefore,

$$\tan \theta = \frac{\tau_r}{\tau_0} = D_a \frac{d}{R_c}, \tag{7.54}$$

in which D_a is the coefficient of deviation angle equation expressed as

$$D_a = \frac{8a^2}{KC_F} \left(\frac{d}{D}\right)^{2a_0}, \tag{7.55}$$

where C_F is the dimensionless parameter defined by the centrifugal force generating secondary motion to the shear force abating the motion and dissipating energy. For $D_a = 11$, D varies with relative submergence d/D. Rozovskii (1961) approximated the angle of deviation for $D_a = 11$ as

$$\tan \theta = b_r \left(\frac{d}{D}\right)^{b_0} \frac{d}{R_c} \tag{7.56}$$

with $b_0 = 0$, $b_r = 11$.

7.3.5 Hydraulic Geometry Relations

The governing equations are (7.48), (7.49), (7.52), and (7.55), which contain 13 variables, including w, d, v, S, Q, D, τ_*, $\tan\theta$, g, a, b_r, γ_{s*}, and b_0, where γ_{s*} is the specific density of sediment. Julien (1988) showed that the primary independent variables were Q, D, and S, and other parameters could be considered nearly constant. These equations are combined and solved for channel width, flow depth, average flow velocity, and slope and can be expressed as power functions of discharge, sediment size, Shields parameter, and deviation angle. Using the ratio of radius of curvature (R_c) to channel width (w) (R_w), $\rho^* = (\rho_s - \rho)/\rho K = $ dimensionless submerged mass density of the sediment parameter α defined as $\alpha = R_w/b_r\sqrt{8g\rho^*}$, the hydraulic geometry relations were expressed as

$$d = \alpha^{\frac{1}{2+a_0+b_0}} Q^{\frac{1}{2+a_0+b_0}} D^{\frac{-1+2a_0+2b_0}{4+2a_0+2b_0}} \tau_*^{\frac{1}{4+2a_0+2b_0}} (\tan\theta)^{\frac{1}{2+a_0+b_0}} \tag{7.57}$$

$$w = \frac{b_r}{R_w} \alpha^{\frac{1+b_0}{2+a_0+b_0}} Q^{\frac{1+b_0}{2+a_0+b_0}} D^{\frac{-1+2a_0-3b_0}{4+2a_0+2b_0}} \tau_*^{\frac{-1-b_0}{4+2a_0+2b_0}} (\tan\theta)^{\frac{-1-a_0}{2+a_0+b_0}} \tag{7.58}$$

$$v = \frac{R_w}{b_r} \alpha^{\frac{-2-b_0}{2+a_0+b_0}} Q^{\frac{a_0}{2+a_0+b_0}} D^{\frac{2-4a_0+2b_0}{4+2a_0+2b_0}} \tau_*^{\frac{2+b_0}{4+2a_0+2b_0}} (\tan\theta)^{\frac{-1-a_0}{2+a_0+b_0}} \tag{7.59}$$

$$S = \rho_* \alpha^{\frac{1}{2+a_0+b_0}} Q^{\frac{-1}{2+a_0+b_0}} D^{\frac{5}{4+2a_0+2b_0}} \tau_*^{\frac{5+2a_0+2b_0}{4+2a_0+2b_0}} (\tan\theta)^{\frac{1}{2+a_0+b_0}} . \tag{7.60}$$

Four coefficients C_d, C_w, C_v, and C_S were defined to group R_w, b_r, ρ_*, α, and θ considering their small variability as

$$C_d = \frac{d}{Q^{\frac{1}{2+a_0+b_0}} D^{\frac{-1+2a_0+2b_0}{4+2a_0+2b_0}} \tau_*^{\frac{1}{4+2a_0+2b_0}} (\tan\theta)^{\frac{1}{2+a_0+b_0}}} = \alpha^{\frac{1}{2+a_0+b_0}} (\tan\theta)^{\frac{1}{2+a_0+b_0}} \tag{7.61}$$

$$C_w = \frac{w}{Q^{\frac{1+b_0}{2+a_0+b_0}} D^{\frac{-1+2a_0-3b_0}{4+2a_0+2b_0}} \tau_*^{\frac{-1-b_0}{4+2a_0+2b_0}}} = \frac{b_r}{R_w} \alpha^{\frac{1+b_0}{2+a_0+b_0}} (\tan\theta)^{\frac{-1-a_0}{2+a_0+b_0}} \tag{7.62}$$

$$C_v = \frac{v}{Q^{\frac{a_0}{2+a_0+b_0}} D^{\frac{2-4a_0+2b_0}{4+2a_0+2b_0}} \tau_*^{\frac{2+b_0}{4+2a_0+2b_0}}} = \frac{R_w}{b_r} \alpha^{\frac{-2-b_0}{2+a_0+b_0}} (\tan\theta)^{\frac{-1-a_0}{2+a_0+b_0}} \tag{7.63}$$

$$C_S = \frac{S}{Q^{\frac{-1}{2+a_0+b_0}} D^{\frac{5}{4+2a_0+2b_0}} \tau_*^{\frac{5+2a_0+2b_0}{4+2a_0+2b_0}}} = \rho_* \alpha^{\frac{1}{2+a_0+b_0}} (\tan\theta)^{\frac{1}{2+a_0+b_0}} . \tag{7.64}$$

Wargadalam (1993) found $b_0 = 2a_0$ to minimize the variability of these four coefficients. Using 382 measurements, the coefficients were found to be:

$C_d = 0.133$, $C_w = 0.512$, $C_v = 14.7$, and $C_S = 12.4$. Then, the downstream hydraulic geometry equations for average depth, width, velocity, and slope were simplified as:

$$d = 0.133 Q^{\frac{1}{3a_0+2}} D^{\frac{6a_0-1}{6a_0+4}} \tau_*^{\frac{-1}{6a_0+4}} \tag{7.65}$$

$$w = 0.512 Q^{\frac{1+2a_0}{2+3a_0}} D^{\frac{-1-4a_0}{4+6a_0}} \tau_*^{\frac{-1-2a_0}{4+2a_0+2b_0}} \tag{7.66}$$

$$v = 14.7 Q^{\frac{a_0}{2+3a_0}} D^{\frac{2-2a_0}{4+6a_0}} \tau_*^{\frac{2+2a_0}{6a_0+4}} \tag{7.67}$$

$$S = 12.4 Q^{\frac{-1}{2+3a_0}} D^{\frac{5}{4+6a_0}} \tau_*^{\frac{5+6a_0}{4+6a_0}}. \tag{7.68}$$

For known channel slope, the hydraulic geometry was expressed as

$$d = 0.2 Q^{\frac{2}{6a_0+5}} D^{\frac{6a_0}{6a_0+5}} S^{\frac{-1}{6a_0+5}} \tag{7.69}$$

$$w = 1.33 Q^{\frac{2+4a_0}{5+6a_0}} D^{\frac{-4a_0}{5+6a_0}} S^{\frac{-(1+2a_0)}{5+6a_0}} \tag{7.70}$$

$$v = 3.7 Q^{\frac{1+2a_0}{5+6a_0}} D^{\frac{-2a_0}{5+6a_0}} S^{\frac{2+2a_0}{6a_0+5}} \tag{7.71}$$

$$\tau_* = 0.12 Q^{\frac{2}{5+6a_0}} D^{\frac{-5}{5+6a_0}} S^{\frac{4+6a_0}{5+6a_0}}. \tag{7.72}$$

Using large field data sets from 835 rivers and canals using Manning's resistance equation $(a_0 = 1/6)$, Julien and Wargadalam (1995) simplified downstream hydraulic geometry relationships as:

$$d = 0.2 Q^{0.33} D_{50}^{0.17} S^{-0.17} \tag{7.73}$$

$$w = 1.33 Q^{0.44} D_{50}^{-0.11} S^{-0.22} \tag{7.74}$$

$$v = 3.76 Q^{0.22} D_{50}^{-0.05} S^{0.39} \tag{7.75}$$

$$\tau_* = 0.121 Q^{0.33} D_{50}^{-0.83} S^{0.83} \tag{7.76}$$

$$S = 12.41 Q^{-0.4} D_{50} \tau_*^{1.2}, \tag{7.77}$$

in which Q is the bankfull discharge in cubic meters per second, D_{50} is the median grain diameter of the bed material in meters, and S is channel slope.

Using 1,485 measurements covering a wide range of flow conditions for gravel-bed and cobble-bed streams with meandering to braided planform geometry, Lee and Julien (2006) obtained similar equations with regression analysis:

$$d = 0.201Q^{0.336}D_{50}^{-0.025}S^{-0.06} \tag{7.78}$$

$$w = 3.004Q^{0.436}D_{50}^{-0.002}S^{-0.153} \tag{7.79}$$

$$v = 2.996Q^{0.198}D_{50}^{-0.007}S^{0.242} \tag{7.80}$$

$$\tau_* = 0.090Q^{0.423}D_{50}^{-0.995}S^{0.912} \tag{7.81}$$

$$S = 4.981Q^{-0.346}\cdot D_{50}^{0.955}\tau_*^{0.966}. \tag{7.82}$$

Shin and Julien (2010) applied the JW theory to evaluate the changes in hydraulic geometry of a 45 kilometer long reach from the Hapcheon Re-regulation Dam to the confluence of the Hwang River with the Nakdong River. The non-vegetated active channel width decreased by an average of 152 m (47% width reduction) since 1982. The average median bed sediment size increased from 1.07 mm in 1983 to 5.72 mm in 2003, and the bed slope decreased from 94 to 85 cm/km from 1983 to 2003.

7.3.6 Calculation Procedure

The given information on noncohesive alluvial channels is the dominant discharge Q (m³/s), median grain size D_{50} (m), and dimensionless Shield parameter τ_*. The calculation procedure involves five steps:

1. Approximately estimate the flow depth.
2. Compute parameter a_0 from the above depth and grain size.
3. Calculate the flow depth using the values of Q, D_{50}, and τ_*.
4. Repeat steps 2 and 3 with the computed flow depth until convergence is achieved.
5. Compute channel width, velocity, and slope using the last value of a_0.

> **Example 7.1:** Consider a gravel-bed channel for which the bankfull discharge is 100 m³/s, the median sediment grain size diameter is 0.045 m near incipient motion τ_*. Compute the equilibrium hydraulic geometry-depth, width, velocity, and slope.
>
> (1) **Solution:** When τ_* is unknown and bed material is a gravel, the incipient motion τ_* can be 0.047. Then, the equilibrium hydraulic geometry-depth, width, velocity, and slope are computed as follows: The initial flow depth d is set to 1 m for the iterative procedure.
>
> (2) a_0 is calculated using Equation (7.50) as
>
> $$a_0 = \frac{1}{\ln\left(\frac{12.2 \times 1}{0.045}\right)} = 0.18.$$

(3) Calculate d using Equation (7.65) as

$$d = 0.133 \times 100^{\frac{1}{3 \times 0.18 + 2}} \times 0.045^{\frac{6 \times 0.18 - 1}{6 \times 0.18 + 4}} \times 0.047^{\frac{-1}{6 \times 0.18 + 4}} = 1.43 \text{ m}.$$

(4) Repeat steps 2 and 3 until flow depth d converges. The converged d is 1.56 m and the corresponding a_0 is 0.165.

(5) Calculate width (Equation (7.66)), velocity (Equation (7.67)), and slope (Equation (7.68)) using the a_0 of step 4 as

$$w = 0.512 \times 100^{\frac{1+2 \times 0.165}{2+3 \times 0.165}} \times 0.045^{\frac{-1-4 \times 0.165}{4+6 \times 0.165}} \times 0.047^{\frac{-1-2 \times 0.165}{4+6 \times 0.165}} = 37.8 \text{ m}$$

$$v = 14.7 \times 100^{\frac{0.165}{2+3 \times 0.165}} \times 0.045^{\frac{2-2 \times 0.165}{4+6 \times 0.165}} \times 0.047^{\frac{2+2 \times 0.165}{4+6 \times 0.165}} = 1.70 \text{ m/s}$$

$$S = 12.4 \times 100^{\frac{-1}{2+3 \times 0.165}} \times 0.045^{\frac{5}{4+6 \times 0.165}} \times 0.047^{\frac{5+6 \times 0.165}{4+6 \times 0.165}} = 0.0022.$$

Therefore, d is 1.56 m, w is 37.8 m, v is 1.7 m/s, and S is 0.0022.

Example 7.2: Consider a gravel-bed channel for which the bankfull discharge is 100 m^3/s, the median sediment grain size diameter is 0.045 m, and $\tau_* = 0.05$. Compute the equilibrium hydraulic geometry-depth, width, velocity, and slope.

Solution: Since τ_* is known, the equilibrium hydraulic geometry-depth, width, velocity, and slope- are computed as follows:

(1) The initial flow depth d is set to 1 m for the iterative procedure.
(2) a_0 is calculated using Equation (7.50) as

$$a_0 = \frac{1}{\ln \left(\frac{12.2 \times 1}{0.045} \right)} = 0.18.$$

(3) Calculate d using Equation (7.65) as

$$d = 0.133 \times 100^{\frac{1}{3 \times 0.18 + 2}} \times 0.045^{\frac{6 \times 0.18 - 1}{6 \times 0.18 + 4}} \times 0.05^{\frac{-1}{6 \times 0.18 + 4}} = 1.41 \text{ m}.$$

(4) Repeat steps 2 and 3 until flow depth d converges. The converged d is 1.54 m and the corresponding a_0 is 0.166.

(5) Calculate width (Equation (7.66)), velocity (Equation (7.67)), and slope (Equation (7.68)) using the a_0 value of step 4 as

$$w = 0.512 \times 100^{\frac{1+2 \times 0.166}{2+3 \times 0.166}} \times 0.045^{\frac{-1-4 \times 0.166}{4+6 \times 0.166}} \times 0.05^{\frac{-1-2 \times 0.166}{4+6 \times 0.166}} = 37.2 \text{ m}$$

$$v = 14.7 \times 100^{\frac{0.166}{2+3 \times 0.166}} \times 0.045^{\frac{2-2 \times 0.166}{4+6 \times 0.166}} \times 0.05^{\frac{2+2 \times 0.166}{4+6 \times 0.166}} = 1.75 \text{ m/s}$$

$$S = 12.4 \times 100^{\frac{-1}{2+3 \times 0.166}} \times 0.045^{\frac{5}{4+6 \times 0.166}} \times 0.05^{\frac{5+6 \times 0.166}{4+6 \times 0.166}} = 0.0024.$$

Therefore, d is 1.54 m, w is 37.2 m, v is 1.75 m/s, and S is 0.0024.

Example 7.3: Consider a gravel-bed channel for which the bankfull discharge is 100 m^3/s, the median sediment grain size diameter is 0.045 m, and $S = 3 \times 10^{-3}$. Compute the equilibrium hydraulic geometry-depth, width, velocity, and Shields parameter.

Solution:
Since S is known, the equilibrium hydraulic geometry-depth, width, velocity, and Shields parameter- are computed as follows:

(1) The initial flow depth d is set to 1 m for the iterative procedure.
(2) a_0 is calculated using Equation (7.50) as

$$a_0 = \frac{1}{\ln\left(\frac{12.2 \times 1}{0.045}\right)} = 0.18.$$

(3) Calculate d using Equation (7.69) as

$$d = 0.2 \times 100^{\frac{2}{6 \times 0.18 + 5}} \times 0.045^{\frac{6 \times 0.18}{6 \times 0.18 + 5}} \times 0.003^{\frac{-1}{6 \times 0.18 + 5}} = 1.37 \text{ m}.$$

(4) Repeat steps 2 and 3 until flow depth d converges. The converged d is 1.45 m and the corresponding a_0 is 0.167.
(5) Calculate width (Equation (7.70)), velocity (Equation (7.71)), and Shields parameter (Equation (7.72)) using the a_0 value of step 4 as

$$w = 1.33 \times 100^{\frac{2 + 4 \times 0.167}{5 + 6 \times 0.167}} \times 0.045^{\frac{-4 \times 0.167}{5 + 6 \times 0.167}} \times 0.003^{\frac{-1 - 2 \times 0.167}{5 + 6 \times 0.167}} = 52.97 \text{ m}$$

$$v = 3.76 \times 100^{\frac{1 + 2 \times 0.167}{5 + 6 \times 0.167}} \times 0.045^{\frac{-2 \times 0.167}{5 + 6 \times 0.167}} \times 0.003^{\frac{2 + 2 \times 0.167}{5 + 6 \times 0.167}} = 1.3 \text{ m/s}$$

$$\tau_* = 0.121 \times 100^{\frac{2}{5 + 6 \times 0.167}} \times 0.045^{\frac{-5}{5 + 6 \times 0.167}} \times 0.003^{\frac{4 + 6 \times 0.167}{5 + 6 \times 0.167}} = 0.06.$$

Therefore, d is 1.45 m, w is 52.97 m, v is 1.3 m/s, and τ_* is 0.06.

7.3.7 Further Discussion

Hager (1996) attempted to generalize Equations (7.69)–(7.72) and provide an explicit approach for the flow depth. In these equations, the common denominator is $\beta = 5 + a_0$. Then, using the square of the sediment Froude number

$$q^2 = \frac{Q^2}{gSD^5}, \tag{7.83}$$

it can be shown that

$$\frac{1}{g}\left(5\frac{d}{D}\right)^{\beta} = q^2 \tag{7.84}$$

$$\frac{1}{g}\left(\frac{3}{4}\frac{w}{D}\right)^{\frac{\beta}{1+2a_0}} = q^2 \tag{7.85}$$

$$\frac{g^{\frac{\beta}{1+2a_0}}}{g}\left(\frac{v^2}{3.76^2 gSD}\right)^{\frac{\beta}{1+2a_0}} = q^2 \tag{7.86}$$

$$\frac{1}{g}\left(\frac{\tau_*}{0.121S}\right)^{\beta} = q^2. \tag{7.87}$$

Hager (1996) noted that the scaling length of both flow depth and width is the mean sediment diameter. For velocity the dimensionless coefficient $g^{\frac{\beta}{1+2a_0}-1}$ should vanish in a correct system of equations, that is, $a_0 = -1$. The dimensionless Shields parameter needs to be scaled with the bottom slope so that the dominant parameter $\tau_* = Kd\rho/[(\rho_s - \rho)D_{50}]$.

Equation (7.84) can be simplified because a_0 is a function of d/D only. Using Equation (7.50),

$$\beta = 5 + \frac{6}{\ln\left(12.2\frac{d}{D}\right)}. \tag{7.88}$$

With $g = 9.81$ m/s^2, one can write $q = f_q\left(\frac{d}{D}\right)$. Hager (1996) computed this function and noted that q increases rapidly with increasing d/D as seen below:

d/D	q
10^0	1.23×10^2
10^1	6.50×10^4
10^2	2.46×10^7
10^3	8.53×10^9
10^4	2.85×10^{12}.

Hager approximated the function to 20% of the above values and to 5% for $1 < d/D < 10^2$ as

$$\frac{d}{D} = \left(\frac{q}{123}\right)^{0.375}. \tag{7.89}$$

Huang (1996) showed that these equations were similar to the downstream geometry relations established by Huang and Warner (1995). Noting the range of a_0 as 0.0–0.5, Equations (7.69)–(7.71) can be expressed as

$$d = K_d Q^{0.4-0.25} D^{0.0-0.375} S^{-(0.2-0.125)} \tag{7.90}$$

$$w = K_w Q^{0.4-0.5} D^{-(0.0-0.25)} S^{-(0.2-0.25)} \tag{7.91}$$

$$v = K_v Q^{0.20-0.25} D^{-(0.0-0.15)} S^{0.4-0.375}. \tag{7.92}$$

The average values of the coefficients found by Julien and Wargadalam (1995) are: $K_d = 0.2$, $K_w = 1.33$, and $K_v = 3.76$. Huang and Warner (1995) found

$$d = K_d Q^{0.3} n_m^{0.383} S^{-0.206} \tag{7.93}$$

$$w = K_w Q^{0.5} n_m^{0.355} S^{-0.156} \tag{7.94}$$

$$v = K_v Q^{0.20} n_m^{-0.738} S^{0.362}, \tag{7.95}$$

where n_m is Manning's roughness coefficient and $K_d = 0.427$, $K_w = 4.059$, and $K_v = 0.576$. Equations (7.90) are consistent with Equations (7.93)–(7.95).

Considering the relatively small ranges of the coefficients C_d, C_w, and C_v defined by Equations (7.61)–(7.63) and accounting for the effects of only major variables, Huang (1996) showed that

$$C_d \propto \left[\left(\frac{d}{D} \right)^{b_0} \frac{d}{w} \right]^{\frac{1}{2+a_0+b_0}} \tag{7.96}$$

$$C_w \propto C_h^{-(1+a_0)} \tag{7.97}$$

$$C_v \propto C_h^{a_0}. \tag{7.98}$$

Because of the limited range, the quantity

$$\xi = \left[\left(\frac{d}{D} \right)^{b_0} \frac{d}{w} \right] \tag{7.99}$$

can also be expressed as

$$\frac{w^{\frac{1}{b_0}}}{d} = \xi^{-\frac{1}{b_0}} D^{-\frac{b_0}{1+b_0}} = \xi_*. \tag{7.100}$$

For $b_0 = 2a_0$ and $0 < a_0 < 0.5$, Equation (7.100) can be written as

$$\frac{w^{1.0-0.5}}{d} = \xi_*. \tag{7.101}$$

The quantity ξ_* can be related to channel sediment composition, bank vegetation, and bank strength. Schumm (1960) showed the influence of sediment composition on channel geometry as

$$w = 2.3 \frac{Q^{0.38}}{M^{0.39}} \tag{7.102}$$

$$d = 0.6 Q^{0.29} M^{0.34}, \tag{7.103}$$

where M is the average percentage of silt-clay in the channel boundary.

Regarding the influence of bank vegetation on channel geometry, Hey and Thorne (1986) showed

$$w = K_{veg} Q^{0.50} \tag{7.104}$$

$$d = 0.22 Q^{0.37} D_{50}^{-0.11}, \tag{7.105}$$

where K_{veg} is 4.33, 3.33, 2.73, and 2.34 for grassy banks with no trees or bushes and the banks were covered with 1–5%, 5–50%, and >50% trees/shrub, respectively.

Discharge Q can be eliminated between Equations (7.102) and (7.103),

$$\frac{w^{0.763}}{d} = 31.147 \, M^{-0.638}. \tag{7.106}$$

Similarly, eliminating Q between Equations (7.104) and (7.105),

$$\frac{w^{0.74}}{d} = \frac{K_{veg}^{0.74}}{0.22} D_{50}^{0.11}. \tag{7.107}$$

Thus, from Equations (7.96)–(7.101) and (7.106)–(7.107), it can be interpreted that

$$C_d(C_v) \sim \left(M, K_{veg} \right)^+ \tag{7.108}$$

$$C_w \sim \left(M, K_{veg} \right)^-. \tag{7.109}$$

The coefficients C_d, C_w, and C_v can be related as

$$K_d \propto C_d^{\frac{6a_0}{5+6a_0}} \tag{7.110}$$

$$K_w \propto C_d^{-\frac{(2a_0+1)(3a_0+1)}{5+6a_0}} \tag{7.111}$$

$$K_v \propto C_d^{\frac{a_0((a_0-1)+1}{5+6a_0}}. \tag{7.112}$$

Hence, for $0 < a_0 < 0.5$, from Equations (7.111)–(5.112),

$$K_d(K_v) \sim \left(M, K_{veg} \right)^+ \tag{7.113}$$

$$K_w \sim \left(M, K_{veg} \right)^-. \tag{7.114}$$

Equations (7.113) and (7.114) show that channels that have a high silt-clay content or have banks with high strength because of coverage by trees or brush have a smaller K_w but larger K_d and vice versa.

Using data from 85 cross-sections on gravel-bed rivers in Iran, Afzalimehr et al. (2010) determined at-a-station hydraulic geometry in terms of discharge (Q), median sediment size (D_{50}), and Shields parameter (τ_*) as

$$w = 4.92 \, Q^{0.726} D_{50}^{-0.035} \tau_*^{-0.018} \qquad (7.115)$$

$$d = 0.346 \, Q^{0.363} D_{50}^{0.073} \tau_*^{0.048} \qquad (7.116)$$

$$S = 1.441 \, Q^{0.033} D_{50}^{0.814} \tau_*^{0.846}. \qquad (7.117)$$

Comparing Equations (7.115)–(7.117) with the Julien–Wargadalam equations, it was found that for determining width and depth for at-a-station geometry it was not necessary to take into account the grain size and the Shields parameter. The difference between the two sets of equations would be significant if discharge is used instead of dominant discharge. Thus, a new set of equations were obtained as

$$w = 5.87 \, Q^{0.743} \qquad (7.118)$$

$$d = 0.226 \, Q^{0.345} \qquad (7.119)$$

$$S = 1.565 \, D_{50}^{0.821} \tau_*^{0.851}. \qquad (7.120)$$

Equation (7.120) shows that for determining slope, discharge is not an effective parameter.

7.4 Parker Theory

Parker (1978a, b) developed a mathematical theory of hydraulic geometry for both sand-bed and gravel-bed rivers. He argued that in a sand-bed channel, bank stability was achieved by the balance between two items. First, there is the lateral diffusion of momentum, which is accompanied by the backward diffusion of suspended bed material that settles in the weaker flow near the banks. Second, the sediment entrained from the bed there is replaced and migrates toward the centerline. Combining the flow depth function in terms of sediment size and slope with appropriate forms of friction and transport laws, he was able to determine three of the five variables (width, depth, slope, discharge, and sediment discharge) given the other two variables as well as grain size and fall velocity. The velocity was determined from the continuity equation for pre-determined flow depth and width.

For gravel-bed rivers, there is a backward fall in the shear stress within the central strip. This fall has a moderate rate of bedload transport at the centerline, which implies that the bed is mobile. The transport reduces to zero at the junction of mobile bed and curving threshold banks. Parker (1978b) reasoned that this is how bank stability is achieved. The shear stress at the center line was taken to be 20% higher than the threshold. He derived three equations connecting width, flow depth, discharge, sediment discharge, and particle size, and expressed explicit hydraulic geometry relations as:

$$w \propto Q^{0.5} D^{-0.35} \quad \text{or} \quad w = 4.4 \left(\frac{Q}{g \Delta D^5} \right)^{0.5} \tag{7.121}$$

$$d \propto Q^{0.42} D^{-0.04}, \tag{7.122}$$

with $d = 0.95$ to $0.97d$ according to discharge and grain size.

$$S \propto Q^{-0.41} D^{0.16}. \tag{7.123}$$

Ferguson (1986) deduced from Parker's equations:

$$w = 27.5 Q S^{1.22} \left(g \Delta D^3 \right)^{-0.4} \cong 6.8 S^{1.22} D^{-1.4}. \tag{7.124}$$

Parker et al. (2007) derived bankfull hydraulic geometry of single thread gravel-bed rivers involving bankfull width, bankfull depth, and down-channel slope as functions of bankfull discharge and bed surface median sediment size:

$$\frac{g^{\frac{1}{5}} w}{Q^{\frac{2}{5}}} = 4.63 \left(\frac{Q}{\sqrt{g D_{50}} D_{50}^2} \right)^{0.0667} \tag{7.125}$$

$$\frac{g^{\frac{1}{5}} d}{Q^{\frac{2}{5}}} = 0.382 \left(\frac{Q}{\sqrt{g D_{50}} D_{50}^2} \right)^{-0.0004} \tag{7.126}$$

$$S = 0.101 \left(\frac{Q}{\sqrt{g D_{50}} D_{50}^2} \right)^{-0.344}. \tag{7.127}$$

Parker et al. (2007) verified Equations (7.125)–(7.127) with four baseline data sets and three independednt data sets.

7.5 Griffiths Theory

Griffiths (1980) theoretically derived hydraulic geometry relations for wide ($w/d > 15$) gravel bed rivers having relatively steep slopes. His theory comprises (1) uniform flow hydraulic equation, (2) empirical definition for friction, (3) sediment transport equation, and (4) continuity equation of water discharge. He discussed the application of his theory to stable channel design. Before presenting his theory, it may be convenient to introduce the dimensionless variables:

$$w_* = \frac{w}{D_{50}}; d_* = \frac{d}{D_{50}}; v_* = \frac{v}{[g(S_s - 1)D_{50}]^{0.5}}; Q_* = \frac{Q}{[g(S_s - 1)]^{0.5} D_{50}^{2.50}}, \quad (7.128)$$

where S_s is the specific gravity of the sediment, that is, $\left(\frac{\gamma_s}{\gamma}\right)$, γ_s is the weight density of sediment, and γ is the weight density of water. It may be noted that these variables correspond to the bankfull condition.

7.5.1 Derivation of Downstream Hydraulic Geometry

The hydraulic equation for uniform flow can be expressed as

$$f \propto d_* \frac{S}{v_*^2}, \quad (7.129)$$

in which f is the friction factor, and S is the slope. Of course, Manning's roughness can also be employed, but Hey (1979) argued that the use of f was more rigorous.

Using the data from gravel bed rivers in New Zealand and elsewhere, Griffiths (1981) empirically derived a relation for f as

$$\frac{1}{\sqrt{f}} = 0.76 + 1.981 \log_{10}\left(\frac{R}{D_{50}}\right). \quad (7.130)$$

Equation (7.130) was approximated for $5 \leq R/D_{50} \leq 200$ as

$$f \propto \left(\frac{D_{50}}{d}\right)^{0.52}, \quad (7.131)$$

where $R \simeq d$. Thus, Equation (7.130) reduces to

$$f \propto \left(\frac{1}{d_*}\right)^{0.52}. \quad (7.132)$$

Bray (1973) developed a similar equation but with an exponent of 0.56.

The next equation is one of sediment transport. At the bankfull discharge, the threshold conditions of incipient motion of bed particles of size D_{50}, bed surface of a stable channel being armoured, and bed load transport of only fine particles, and bank erosion, if any, with an equilibrium between bank erosion and deposition, Shields' equation can be assumed to occur:

$$D_{50} \propto dS_b \quad \text{or} \quad d_* \propto \frac{1}{S_b}, \quad (7.133)$$

where S_b is the slope at bankfull condition or energy gradient equal to water surface slope and bed slope.

The continuity of water discharge can be expressed as

$$Q_* = v_* w_* d_*. \tag{7.134}$$

Assuming the aspect ratio to be constant (Li et al. 1976; Griffiths, 1980), one can write

$$d_* \propto w_*. \tag{7.135}$$

For deriving hydraulic geometry relations, Equations (7.129), (7.132), (7.133), (7.134), and (7.135) are utilized. In order to derive an expression for w, Equation (7.134) is written as

$$w_* = \frac{Q_*}{v_* d_*}. \tag{7.136}$$

The variable v_* is eliminated using Equations (7.129), (7.133), and (7.134). To that end, from Equations (7.129) and (7.132) one can write

$$\left(\frac{1}{d_*}\right)^{0.52} \propto \frac{d_* S}{v_*^2} \quad \text{or} \quad v_*^2 \propto d_*^{1.52} S \quad \text{or} \quad v_* \propto d_*^{0.76} S^{0.5}. \tag{7.137}$$

Using Equation (7.133), Equation (7.137) can be written as

$$v_* \propto d_*^{0.26}. \tag{7.138}$$

Inserting Equations (7.135) and (7.138) in Equation (7.136), the result is

$$w_* \propto Q_*^{0.44}. \tag{7.139}$$

Because of Equation (7.135), one can also write

$$d_* \propto Q_*^{0.44}. \tag{7.140}$$

Because of the continuity Equation (7.134), one can express velocity as

$$v_* \propto Q_*^{0.12}. \tag{7.141}$$

Using Equation (7.132) in Equation (7.140), the equation for friction factor follows:

$$f \propto Q_*^{-0.23}. \tag{7.142}$$

Similarly, substituting Equation (7.140) in Equation (7.133), the equation for slope results as

$$S \propto Q_*^{-0.44}. \tag{7.143}$$

The next equation is for suspended sediment concentration C_s. It is noted that the bed material load is regarded as the sum of bed load and suspended load. In gravel bed rivers at bankfull discharge, the total sediment load is the sum of washload and bed material load; the bed load may be around 10% and the washload may be around 60% of the total load (Griffiths, 1979). It may therefore be assumed that the slope of the curve relating the wash load concentration to discharge is similar to that of the suspended component of bed material load for mean annual to bankfull discharge, as shown by Griffiths (1979) for New Zealand rivers. From the Engelund and Hansen (1967) equation, the bed material load concentration can be expressed as

$$C_s \propto v_* \frac{(d_* S)^{1.5}}{d_*}. \tag{7.144}$$

Inserting Equations (7.140), (7.141), and (7.143) in Equation (7.144),

$$C_s \propto Q_*^{-0.33}. \tag{7.145}$$

Equation (7.145) can also be obtained if the Ackers and White formula for sediment transport (White et al. 1973) is employed. Griffiths (1980) showed that the exponents of Equations (7.139)–(7.144) were similar to those obtained using data from gravel bed rivers in New Zealand as well as to those derived by Langbein (1964), Engelund and Hansen (1967), Lane et al. (1959), and Parker (1979).

7.5.2 Stable Channel Design

In design of stable channels, Equations (7.129), (7.132), (7.133), and (7.134) are given, for which the discharge is obtained from upstream and the slope is controlled by the landscape and D_{50}, which depends on the existing gravel surface. The discharge must be confined within the channel banks, and the designer has the freedom to choose width expressed as

$$w_* \propto Q_* S^{1.26}. \tag{7.146}$$

Inserting Equation (7.133) in Equation (7.146), one can express

$$w_* \propto \frac{Q_*}{d_*^{1.26}}. \tag{7.147}$$

Based on field observations, Griffiths (1980) expressed depth (d_m) corresponding to mean annual flow discharge as

$$d_{m*} \propto d_* \propto Q_*^{0.43}. \tag{7.148}$$

Substituting Equation (7.148) in Equation (7.147), the result is

$$w_{m*} \propto w_* \propto Q_*^{0.48}. \tag{7.149}$$

It has been suggested that the constant of proportionality and exponent in the width–discharge relation should be empirically determined for each watershed (Parker, 1979) and then other hydraulic geometry relations can be determined theoretically. Griffiths (1981) rewrote Equation (7.146) as

$$\frac{wD_{50}g^{0.5}}{QS^{1.26}} = \text{constant} \tag{7.150}$$

and suggested determining the value of constant for a given river by taking the mean value from the values of variables at several cross-sections. For the design reach, the width can be determined using Equation (7.150) if other variables are known.

It may be noted that Henderson (1966), Kellerhals (1967), and Parker (1979) expressed equations similar to Equation (7.146) but with slightly different exponents. For example, Henderson (1966) wrote

$$w_* \propto Q_* S^{1.17}, \tag{7.151}$$

which is widely employed in New Zealand. Then, Equation (7.151) leads to

$$w_* \propto Q_*^{0.50}, \tag{7.152}$$

which is similar to Equation (7.149).

7.6 Ackers Theory

Ackers (1964) conducted extensive laboratory investigations at the Hydraulics Research Station at Wallingford into channels that achieved stability for discharges between 0.4 and 5.4 cfs. He developed hydraulic geometry relations using three equations: channel resistance equation, sediment transport equation, and width to depth ratio (or aspect ratio). Since stream geometry may be defined by water surface width, mean depth, and slope, these three equations should suffice. From his experiments, he observed that the ratio (w/d) was almost unchanged from an average value of 12.4 (for finer sands) for channels with eroded banks, and was about 10 for channels with uneroded banks, but ranged up to 45 for meandered channels.

7.6.1 Governing Equations

The resistance equation was expressed as

$$v = a_2 d^{0.75} S^{0.50}, \tag{7.153}$$

where

$$a_2 = \frac{7.13\sqrt{g}}{k^{0.25}}, \tag{7.154}$$

in which k is the ripple height, which varied little over the range of tests conducted. Introducing a parameter X defined as

$$X = \frac{\gamma_s}{\gamma_w}\frac{Q_s}{Q} = sG; s = \frac{\gamma_s}{\gamma_w}; G = \frac{Q_s}{Q}, \tag{7.155}$$

he expressed the sediment transport equation using Einstein's parameters in terms of discharge as

$$\frac{X}{w}\frac{Q}{s} = 2{,}400v_fD^{-7}\left[\frac{dS}{\left(\frac{\gamma_s}{\gamma_w}-1\right)}\right]^8, \tag{7.156}$$

where v_f is the fall velocity. Equation (7.156) can be expressed in terms of flow velocity as

$$v = a_1 d^7 S^8, \tag{7.157}$$

where

$$a_1 = \frac{2{,}400v_fD^{-7}}{\frac{X\,(s-1)^8}{s}}. \tag{7.158}$$

The width–depth ratio was exprssed as

$$\frac{w}{d} = a_3. \tag{7.159}$$

For small streams, the aspect ratio was regarded as constant, although it may depend on the cohesive strength for the bank material, that is, clay content (Schumm, 1960).

7.6.2 Hydraulic Geometry Equations

Equations (7.153), (7.156), and (7.159) can be combined to yield hydraulic geometry relations as follows:

$$d = a_d Q^{0.43} \tag{7.160}$$

$$w = a_w Q^{0.43} \tag{7.161}$$

$$A = a_A Q^{0.86} \tag{7.162}$$

$$v = a_v Q^{0.14} \tag{7.163}$$

$$S = a_S Q^{-0.36}, \tag{7.164}$$

where

$$a_d = a_1^{0.03} a_2^{-0.47} a_3^{0.43} \tag{7.165}$$

$$a_w = a_1^{0.03} a_2^{-0.47} a_3^{-0.57} \tag{7.166}$$

$$a_A = a_1^{0.06} a_2^{-0.94} a_3^{-0.14} \tag{7.167}$$

$$a_v = a_1^{-0.06} a_2^{0.94} a_3^{0.14} \tag{7.168}$$

$$a_S = a_1^{-0.16} a_2^{0.56} a_3^{-0.36}. \tag{7.169}$$

7.6.3 Significance of Width–Depth Ratio

In irrigation canals, the width–depth ratio depends on design discharge in a manner that

$$\frac{w}{d} \propto d^{0.5}. \tag{7.170}$$

This dependence of bank slope on flow depth is a result of the characteristics of soil in the bank. Equation (7.170) can be expressed as

$$\frac{w}{d} = a_3 = b_3 d^{0.5}. \tag{7.171}$$

Using Equation (7.171), Equations (7.160)–(7.164) can be modified as

$$d \propto Q^{0.35} \tag{7.172}$$

$$w \propto Q^{0.53} \tag{7.173}$$

$$A \propto Q^{0.88} \tag{7.174}$$

$$v \propto Q^{0.12} \tag{7.175}$$

$$S \propto Q^{-0.29}. \tag{7.176}$$

Comparing the exponents in Equations (7.172)–(7.176) with those of the traditional regime equations, which, resepctively, are 0.33, 0.50, 0.83, 0.17, and −0.17, or with those of Simons and Albertson (1960), which for d, w, A, and v, respectively are 0.36, 0.51, 0.87, and 0.13, the Ackers theory yielded comparable exponents.

7.6.4 Comparison with Regime Relations

For purposes of comparing with the regime theory of Inglis (1946) and Lacey (1957–1958), the Inglis–Lacey equations are often expressed as

$$v \propto \frac{g^{0.5} dS}{D^{0.5}} \tag{7.177}$$

$$\frac{v^3}{w} \propto g^{1.5} D^{0.5} \tag{7.178}$$

$$\frac{v^2}{gd} \propto \left(Xv_f\right)^{0.5} (vg)^{-\frac{1}{6}}, \tag{7.179}$$

where v is the kinematic viscosity.

Combining Equations (7.153) and (7.157) and assuming an arbitrary value of exponent n, the result is

$$v^{1+n} = a_1 a_2 d^{7+0.75n} S^{8+0.5n}. \tag{7.180}$$

To compare with Equation (7.179), Equation (7.180) with $n = -26$ can be expressed as

$$\frac{v^2}{d} = a_1^{-0.08} a_2^{2.08} S^{0.4}. \tag{7.181}$$

If the ripple height is assumed constant then Equation (7.181) can be cast as

$$\frac{v^2}{d} \propto X^{0.08} D^{0.5} S^{0.4}. \tag{7.182}$$

If $n = -16$, then Equation (7.180) can be writtena as

$$\frac{v^3}{d} = a_1^{-0.2} a_2^{3.2}. \tag{7.183}$$

Equation (7.183) can be compared with Equation (7.178). The width can be introduced through Equation (7.159), yielding

$$\frac{v^3}{w} \propto X^{0.2} D^{1.2} a_3. \tag{7.184}$$

Now for comparison with Equation (7.177), let $n = 4$, the equation becomes

$$v \propto a_1^{0.2} a_2^{0.8} (dS)^2. \tag{7.185}$$

Neglecting the effect of ripple height, Equation (7.185) can be written as

$$v \propto X^{0.2} D^{-0.2} \left(\frac{dS}{\sqrt{D}} \right)^2 . \tag{7.186}$$

Ackers (1964) also derived an equation similar to the Blench equation (Blench, 1957) from the algebraic manipulation of the governing equations:

$$\frac{v^2}{dS} = a_1^{0.025} a_2^{1.16} a_3^{0.375} (vw)^{0.375} . \tag{7.187}$$

If the ripple height is regarded as constant then Equation (7.187) can be written as

$$\frac{v^2}{dS} \propto X^{-0.025} D^{-0.15} a_3^{0.375} (vw)^{0.375} . \tag{7.188}$$

Another regime equation, often quoted, is

$$v = 16 d_h^{\frac{2}{3}} S^{\frac{1}{3}} , \tag{7.189}$$

where d_h is the hydraulic mean depth. Taking $n = -46$, Equation (7.180) becomes

$$v = a_1^{-0.02} a_2^{1.02} d^{0.61} S^{0.33} . \tag{7.190}$$

With $n = -36$, an alternative equation can be written as

$$v = a_1^{-0.026} a_2^{0.71} \left(d^2 S \right)^{0.285} . \tag{7.191}$$

The exponent of 0.285 is consistent with 0.286 of Simons and Albertson (1960) for coarse noncohesive materials.

References

Ackers, P. (1964). Experiments on small streams in alluvium. *Journal of the Hydraulics Division, Proc. ASCE*, Vol. 90, No. HY4, pp. 1-37.

Afzalimehr, H., Abdolhosseini, M., and Singh, V. P. (2010). Hydraulic geometry relations for stable channel design. *Journal of Hydrologic Engineering*, Vol. 15, No. 10, pp. 859–864.

Bagnold, R. A. (1956). The flow of cohesionless grains in fluids. *Philosophical Transactions, Royal Society of London, Series A*, Vol. 249, No. 964, pp. 235–297.

Blench, T. (1957). *Regime Behavior of Canals and Rivers*. Butterworth, London.

Bray, D. L. (1973). Regime relations for Alberta gravel bed river. In: *Fluvial Processes and Sedimentation, Proceedings of Hydrology Symposium held at University for Alberta, Edmonton*, Thorn Press, Limited, Ottawa.

Chang, H. H. (1980). Geometry of gravel stream. *Journal of the Hydraulics Division, ASCE*, Vol. 106, No HY9, pp. 1443–1456.

Chang, H. H. (1988). *Fluvial Processes in River Engineering*. John Wiley & Sons, New York.

Charlton, F. G., Brown, P. M., and Benson, R. W. (1978). *The hydraulic geometry of some gravel rivers in Britain*. Report IT 180, Hydraulic Research Station, Wallingford.

Davies, T. R. H. and Sutherland, A. J. (1983). Extremal hypotheses for river behavior. *Water Resources Research*, Vol. 19, No.1, pp. 141–148.

Dou, G. R. (1964). Hydraulic geometry of plain alluvial rivers and tidal river mouth. *Journal of Hydraulic Engineering (in Chinese)*, No. 2, pp. 1–13.

Einstein, H. A. (1942). Formulas for transportation of bed-load. *Transactions, ASCE*, Vol. 107, pp. 561–575.

Einstein, H. A. and Chien, N. (1954). Similarity of distorted river models with unstable beds. *Proceedings, ASCE*, Vol. 80, No. 566, p. 21.

Engelund, F. and Hansen, E. (1967). A Monograph of Sediment Transport in Alluvial Streams. *Copenhagen, Teknisk Forlag*, 62 p. 62.

Ferguson, R. I. (1986). Hydraulic and hydraulic geometry. *Progress in Physical Geography*, Vol. 0, pp. 1–31.

Griffiths, G. A. (1979). Recent sedimentation history of the Waimakariri River. *Journal of Hydrology (New Zealand)*, Vol. 18, No. 1, pp. 6–28.

Griffiths, G. A. (1980). Hydraulic geometry relationships of some New Zealand gravel bed rivers. *Journal of Hydrology (New Zealand)*, Vol. 19, No. 2, pp. 106–118.

Griffiths, G. A. (1981). Stable channel design in gravel bed rivers. *Journal of Hydrology*, Vol. 52, pp. 291–396.

Hager, W. H. (1996). Discussion of "Alluvial channel geometry: theory and applications," by P. Y. Julien and J. Wargadalam. *Journal of Hydraulic Engineering*, Vol. 122, No. 12, p. 750. A

Henderson, F. M. (1966). *Open Channel Flow*. The Macmillan Company, New York, pp. 522.

Hey, R. D. (1979). Dynamic process-response Julien del of river channel development. *Earth Surface Processes*, Vol. 4, pp. 59–72.

Hey, R. D. and Thorne, C. R. (1986). Stable channels with gravel beds. *Journal of Hydraulic Engineering*, Vol. 112, No. 6, pp. 671–689.

Huang, H. Q. (1996). Discussion of "Alluvial channel geometry: theory and applications," by P. Y. Julien and J. Wargadalam. *Journal of Hydraulic Engineering*, Vol. 122, No. 12, pp. 750–751.

Huang, H. Q. and Warner, R. F. (1995). The multivariate control of hydraulic geometry: a causal investigation in terms of boundary shear distribution. *Earth Surface Processes and Landforms*, Vol. 20, No. 2, pp. 115–130.

Inglis, C. C. (1946). *Meanders and their bearing in river training*. Maritime Paper No. 7, Institution of Civil Engineers, London.

Julien, P. Y. (1988). *Downstream Hydraulic Geometry of Noncohesive Alluvial Channels. International Conference on River Regime*, John Wiley & Sons, New York, pp. 9–18.

Julien, P. Y. (2014). Downstream hydraulic geometry of alluvial rivers. Proceedings of a Symposium on Sediment Dynamics from the Summit to the Sea, held in New Orleans, December 11–14, IAHS Publication 367, pp. 3–11.

Julien, P. Y. and Wargadalam, J. (1995). Alluvial channel geometry: Theory and applications. *Journal of Hydraulic Engineering*, Vol. 121, No. 4, pp. 312–325.

Kalinske, A. A. (1947). Movement of sediment as bed load in rivers. *Transactions, American Geophysical Union*, Vol. 28, pp. 615–620.

Kellerhals, R. (1967). Stable channels with gravel-paved beds. *Journal of the Waterways and Harbors Division, ASCE*, Vol. 93, No. WW1, pp. 63–84.

Keulegan, G. H. (1938). Laws of turbulent flow in open channels. *Gaithersburg, MD: National Bureau of Standards*, Vol. 21, pp. 707–741

Lacey, G. (1957–1958). *Flow in alluvial channels with sandy mobile beds*. Proceedings, Institution of Civil Engineers, Vol. 9, pp. 145–164, London.

Lane, E. W., Lin, P. N. and Liu, H. K. (1959). *The most efficient stable channel for comparatively clean water in non-cohesive material*. Report CER59HKLS, Colorado State University, Fort Collins.

Langbein, W. B. (1964). Geometry of river channels. *Journal of the Hydraulics Division, ASCE*, Vol. 90, No. HY2, pp. 301–311.

Lee, J.-S. and Julien, P. Y. (2006). Downstream hydraulic geometry of alluvial channels. *Journal of Hydraulic Engineering*, Vol. 132, No. 12, pp. 1347–1352.

Leopold, L. B. and Maddock, T. J. (1953). Hydraulic geometry of stream channels and some physiographic implications. *U. S. Geological Survey Professional Paper* 252, p. 55.

Leopold, L. B. and Wolman, M. G. (1957). *River channel patterns: braided, meandering and straight*. Geological Survey Professional Paper 282-B, pp. 38–85, U.S. Geological Survey, Washington, DC.

Li, R. M., Simons, D. B., and Stevens, M. A. (1976). Morphology of stable streams in small watersheds. *Journal of the Hydraulics Division, ASCE*, Vol. 102, No. HY8, pp. 1101–1117.

Mussetter, R. A. (1989). *Dynamics of mountain streams*. Ph.D. dissertation, Department of Civil Engineering, Colorado State University, Fort Collins.

Parker, G. (1978a). Self-formed rivers with stable banks and mobile bed: Part 1. The sand-silt river. *Journal of Fluid Mechanics*, Vol. 89, pp. 109–125.

Parker, G. (1978b). Self-formed rivers with stable banks and mobile bed: Part 2. The gravel river. *Journal of Fluid Mechanics*, Vol. 89, pp. 127–146.

Parker, G. (1979). Hydraulic geometry of active gravel rivers. *Journal of the Hydraulics Division, ASCE*, Vol. 105, pp. 1185–1201.

Parker, G., Wilcock, P. R., Paola, C., Dietrich, W. E., and Pitlick, J. (2007). Physical basis for quasi-universal relations describing bankfull hydraulic geometry of single-thread gravel rivers. *Journal of Geophysical Research*, Vol. 112, F04005, pp. 1–21.

Rozovskii, I. L. (1961). *Flow of Water in Bends of Open Channels*. Translated by Y. Prushansky, Israel Program for Scientific Translation, Jerusalem, Israel.

Schumm, S. A. (1960). *The shape of alluvial channels in relation to sediment type*. Professional Paper 352-B, Geological survey, U.S. Department of Interior, Washington, DC.

Schumm, S. A. and Litchy, R. W. (1964). Time, space and causality in geomorphology. *American Journal of Science*, Vol. 263, pp. 110–119.

Shin, Y. H. and Julien, P. Y. (2010). Changes in hydraulic geometry of the Hwang River below the Hapheon re-regulation dam, South Korea. *International Journal of River Basin Management*, Vol. 8, No. 2, pp. 139–150.

Simons, D. B. and Albertson, M. L. (1960). Uniform water conveyance in alluvial material. Journal of the Hydraulics Division, ASCE, Vol. 86, No. HY5, *Proc. Paper* 2484, pp. 33–71.

Smith, T. E. (1974). A derivation of the hydraulic geometry of steady-state channels from conservation principles and sediment transport laws. *Journal of Geology*, Vol. 82, pp. 98–104.

Smith, T. E. and Bretherton, F. P. (1972). Stability and the conservation of mass in drainage basin evolutaion. *Water Resources Research*, Vol. 8, No. 6, pp. 1506–1524.

Wargadalam, J. (1993). *Hydraulic geometry equations of alluvial channels*. Ph.D. dissertation, Colorado State University, Fort Collins.

White, W. R., Bettes, R., and Wang, S. Q. (1986). *A study on river regime*. Hydraulics Research Station, Wallingford.
White, W. R., Milli, H., and Crabbe, A. D. (1973). *Sediment transport: An appraisal of available methods*. Hydraulics Research Station, Wallingford.
Williams, G. P. (1978). *Hydraulic geometry of river cross-sections- Theory of minimum variance*. Professional paper 1029, U.S. Geological Survey, Washington, DC.
Yang, C. T. (1976). Minimum unit stream power and fluvial hydraulics. *Journal of the Hydraulic Division, ASCE*, Vol. 102, No. HY7, pp. 919–934.
Yang, C. T. (1984). Unit stream power for gravel. *Journal of Hydraulic Engineering, ASCE*, Vol. 110, No. 12, pp. 1783–1797.

8

Scaling Theory

Empirical equations of downstream hydraulic geometry, entailing width, depth, velocity, and bed slope, can be derived using the scaling theory. The theory employs the momentum equation, a flow resistance formula, and continuity equation for gradually varied open channel flow. The scaling equations are expressed as power functions of water discharge and bed sediment size, and are applicable to alluvial, ice, and bedrock channels. These equations are valid for any value of water discharge as opposed to just mean or bankfull values that are used in empirical equations. This chapter discusses the use of scaling theory for the derivation of downstream hydraulic geometry. The scaling theory-based hydraulic geometry equations are also compared with those derived using the regime theory, threshold theory, and stability index theory, and the equations are found to be consistent.

Notation

c = a constant
d = flow depth
D = sediment size
D_{go} = median size of the initial sediment distribution
D_r = sediment size ratio
D_{50} = medium pavement size
d_m = flow depth in the model channel
d_p = flow depth in the prototype channel
d_r = ratio of flow depths
f = friction factor
f_r = ratio of friction factor

Continued

f_s = silt factor

F_s = Shields entrainment function

F_r = Froude number

F_{rp} = Froude number of flow in the prototype channel

F_{rm} = Froude number of flow in the model channel

F_{rr} = ratio of Froude numbers

g = acceleration due to gravity

g_w = bed load transport rate (weight per unit width per unit time)

G_w = bed load transport rate

G_{wr} = scaled bed load transport rate

I_S = a dimensionless stable-section index

k_s = characteristic surface roughness length parameter

k_{sr} = characteristic surface roughness length parameter ratio

m = subscript referring to model

n = roughness

p = subscript associated with any quantity referring to prototype

P = wetted perimeter

P_r = ratio of wetted perimeter

Q = discharge

Q_r = discharge ratio

Q_s = sediment discharge

r = subscript referring to ratio

R = hydraulic radius

S = bed slope

S_f = friction slope

S_{fm} = friction or energy slope of the model channel

S_{fp} = friction or energy slope of the prototype channel

S_{fr} = ratio of friction or energy slopes

S_m = model channel bed slope

S_p = prototype channel bed slope

S_r = ratio of channel bed slopes

S_s = specific gravity of sediment particles

v = mean cross-sectional velocity

v_r = ratio of velocity

v_s = grain shear velocity

v_* = shear velocity

w = width

w_r = horizontal scale

x_m = downstream distance in the model channel

x_p = downstream distance in the prototype channel

x_r = ratio of downstream distances or horizontal scale

$X_i, i = 1, 2, 3, 4,$ and 5 = five dimensionless variables

α = constant energy coefficient
μ = dynamic viscosity
ρ = fluid density
σ_{go} = geometric standard deviation of the distribution
ρ_s = sediment density
γ_s = specific sediment weight

8.1 Introduction

The downstream hydraulic geometry usually entails relations for geometric parameters of width (w), depth (d), and bed slope (S); and hydraulic parameters of velocity (v) and roughness (n). The empirical relations express these parameters as power functions of discharge (Q) corresponding to mean annual flow or bankfull discharge. These relations are based on the assumption of homogenous alluvial rivers, where incoming water discharge and sediment load are constant and of invariable channel geomorphic history. This means that these relations may not apply if there is a change in the frequency of bankfull flow or if the channel has transitions, such as tributary junctions, changes in channel patterns, and crossings of major geologic boundaries.

A physically based approach is to solve for width, depth, velocity, and slope for given Q, sediment discharge Q_s, and sediment size D, as done by Laursen (1958) for the design of stable channels. In this approach, equations of continuity, resistance, and sediment transport are employed. However, there are four unknowns and only three equations. For closure, another relation is employed, such as a constant w/d ratio (Griffiths, 1980), a stability index (Griffiths, 1981), or something else.

Griffiths (2003) developed a scaling theory that provides a theoretical basis for the hydraulic geometry relations using the continuity and momentum equations and a resistance relation for gradually varied open channel flow. In this theory, hydraulic geometry relations are treated as scaling relations between dynamically similar systems (Ashmore, 1991, 2001). There are two major advantages of the scaling theory. First, the scaling relations apply to alluvial, ice, and bedrock channels. Second, they hold for any discharge and are not restricted to mean annual or bankfull discharge as assumed by empirical relations developed by Leopold and Maddock (1953). The discussion in this chapter follows the work of Griffiths (2003).

8.2 Scaling Theory

The scaling theory employs the equation of continuity, gradually varied open channel flow equation, a resistance relation, and the principle of similitude. It is assumed that the channel shape at a given cross-section remains constant but may vary with distance downstream. That is, the channel can be non-prismatic with alluvial, ice, or rock sediment boundary. Both mobile and immobile boundary channels are considered.

Since scaling theory uses prototype and model concepts, the prototype channel refers to the upstream reach or reference channel, whereas the model channel refers to the downstream channel reach. Subscript p associated with any quantity will refer to prototype, subscript m to model, and subscript r to ratio.

8.2.1 Immobile Boundary Channels

Here both channel bed and banks are immobile. The flow equation for the prototype channel can be expressed as

$$\frac{dd_p}{dx_p}\left(1 - \alpha F_{rp}^2\right) = S_p - S_{fp}, \tag{8.1}$$

in which d_p is the flow depth in the prototype at the downstream distance x_p, F_{rp} is the Froude number of flow in the prototype, α is the constant energy coefficient, S_p is the prototype channel bed slope, and S_{fp} is the friction or energy slope of the prototype channel. Here subscript p corresponds to prototype.

Similar to Equation (8.1), the flow equation for the model channel can be expressed as

$$\frac{dd_m}{dx_m}\left(1 - \alpha F_{rm}^2\right) = S_m - S_{fm}, \tag{8.2}$$

where all quantities have the same meaning as just shown for the prototype but are for the model channel, that is, subscript m corresponds to model.

The Froude number F_r can be defined as

$$F_r = \frac{v}{\sqrt{gR}}, \tag{8.3}$$

in which v is the mean cross-sectional flow velocity, R is the hydraulic radius, and g is the acceleration due to gravity. Here the channel slope is considered small, and the channel is considered wide, so $R \sim d$.

From similitude, one can write (subscript r corresponds to the ratio of a quantity for model to that for prototype)

$$d_r = \frac{d_m}{d_p} \tag{8.4}$$

$$F_{rr} = \frac{F_{rm}}{F_{rp}} \tag{8.5}$$

$$S_r = \frac{S_m}{S_p}$$ (8.6)

$$S_{fr} = \frac{S_{fm}}{S_{fp}}$$ (8.7)

$$x_r = \frac{x_m}{x_p}.$$ (8.8)

Substitution of Equations (8.4)–(8.8) in Equation (8.2) leads to

$$\frac{1}{S_p} \frac{dd_p}{dx_p} \left(\frac{d_r}{S_r x_r} \right) = \frac{1 - \frac{S_{fp}}{S_p} \left(\frac{S_{fr}}{S_r} \right)}{1 - \propto F_{rp}^2 \left(F_{rr}^2 \right)}.$$ (8.9)

Now the concept of dynamic similarity is invoked. Accordingly, Equations (8.1) and (8.2) must be identical. Then, from Equation (8.9), one must have

$$\frac{d_r}{S_r x_r} = 1$$ (8.10)

$$\frac{S_{fr}}{S_r} = 1$$ (8.11)

$$F_{rr}^2 = 1.$$ (8.12)

The next step is to specify an expression for the friction factor so the friction slope S_f can be explicitly expressed. Griffiths (2003) used the Darcy–Weisbach equation, which can be written as

$$S_f = \frac{fv^2}{8gR} = \frac{fF_r^2}{8},$$ (8.13)

where f is the friction factor. The effect of roughness or using a different friction relation will be discussed in Section 8.2.2. Equation (8.13) can be expressed in scaling terms as

$$S_{fr} = f_r F_{rr}^2.$$ (8.14)

Now an expression for the friction factor needs to be specified. For fully turbulent flow, following Henderson (1966), the friction factor formula can be written as

$$f = 0.13 \left(\frac{D}{d} \right)^{\frac{1}{3}},$$ (8.15)

where D is a characteristic bed material size, typically D_{50}. The grain shear will dominate the flow resistance for immobile stream beds. In scaling terms, Equation (8.15) can be expressed as

$$f_r = \left(\frac{k_{sr}}{d_r}\right)^{0.33},\qquad(8.16)$$

where k_s is the characteristic surface roughness length parameter. Equation (8.16) is similar to the Manning–Strickler expression.

The horizontal scale x_r can be replaced by w_r. Then, Equation (8.10) yields

$$\frac{d_r}{w_r} = S_r.\qquad(8.17)$$

It may be noted that S_r is not unity, suggesting that the channel is distorted. Using Equation (8.12) in Equation (8.14) one can write

$$S_{fr} = f_r.\qquad(8.18)$$

Then, Equation (8.16) can be written with the use of Equations (8.17) and (8.18) as

$$f_r = \left(\frac{k_{sr}}{d_r}\right)^{0.33} = \frac{d_r}{w_r}.\qquad(8.19)$$

Equation (8.19) yields

$$v_r = d_r^{0.5}.\qquad(8.20)$$

Recalling the continuity equation in scaling terms,

$$Q_r = w_r d_r v_r.\qquad(8.21)$$

With the use of $k_{sr} = D_r$ applicable in alluvial channels, Equation (8.19) can be written as

$$f_r = \left(\frac{D_r}{d_r}\right)^{0.33} = \frac{d_r}{w_r} = S_r.\qquad(8.22)$$

With the use of Equations (8.20) and (8.22) in Equation (8.21), the following can be obtained:

$$w_r = D_r^{-0.18} Q_r^{0.47}\qquad(8.23)$$

$$d_r = D_r^{0.12} Q_r^{0.35}\qquad(8.24)$$

$$v_r = D_r^{0.06} Q_r^{0.18}.\qquad(8.25)$$

Inserting Equations (8.23) and (8.24) in Equation (8.17), the slope can be expressed as

$$S_r = D_r^{0.30} Q_r^{-0.12}.\qquad(8.26)$$

Table 8.1. *Exponents of hydraulic geometry relations [JW = Julien–Wargadalam]*

Equation	Scaling theory		Regime theory		Empirical theory		JW Theory[*]	
	j	k	j	k	j	k	j	k
$w_r = D_r^k Q_r^j$	0.47	−0.18	–	0	0.50	0	0.44	−0.11
$d_r = D_r^k Q_r^j$	0.35	0.12	–	0	0.40	0	0.33	0.17
$v_r = D_r^k Q_r^j$	0.18	0.06	–	0	0.01	0	0.22	−0.05
$S_r = D_r^k Q_r^j$	−0.12	0.30	–	0	−0.10 to −0.44	0	−0.4	1
$G_{wr} = D_r^k Q_r^j$	0.82	0.45	–	0	0.80	0	–	–

[*] In the JW equations, slope is also included in the equations.

It may be interesting to compare the exponents of discharge in Equations (8.23)–(8.26) with those obtained empirically by Leopold and Maddock (1953), as shown in Table 8.1. The two sets of exponents are comparable and the differences between them may be attributed to the inclusion of sediment size in the scaling equations. These equations may not necessarily be for stable geometry, indicating that they may not provide a simple solution for stable channel design.

8.2.2 Effect of Roughness

For the Darcy–Weisbach equation, Henderson (1966) provided the friction factor as given by Equation (8.15). If one uses another appropriate friction formula then that may result in different discharge exponents. Examples of such formulae can be cited here. Griffiths (1979) derived a Keulegan type relation expressed as

$$\frac{1}{\sqrt{f}} = 0.76 + 1.98 \log \left(\frac{R}{D_{50}} \right), \tag{8.27}$$

which was approximated within the region $(4 < R/D_{50} < 200)$ as

$$f = 0.50 \left(\frac{D_{50}}{R} \right)^{0.52}. \tag{8.28}$$

Using field data, Bray (1979) obtained

$$f = 0.54 \left(\frac{D_{50}}{R} \right)^{0.56}. \tag{8.29}$$

Following Bennett (1977), Bennett and Nordin (1978) found that the grain shear velocity (v_s) can be expressed as

$$v_s = \frac{v}{2.5 \, \ln\left(\frac{11 \, R}{k_s}\right)}, \tag{8.30}$$

where k_s is the equivalent roughness related to grains. Using data from Guy et al. (1966), Yang (1996) approximated k_s as $2D_{50}$. The friction slope due to grain roughness can then be expressed as

$$S_f = \frac{F_r^2}{2.5 \ln\left(\frac{11R}{2D_{50}}\right) + 6.0}. \tag{8.31}$$

The logarithmic term in Equation (8.31) can be approximated for $2 < R/D_{50} < 30$ as

$$2.5 \ln\left(\frac{11 \, R}{2D_{50}}\right) = cD_{50}^{\frac{1}{6}}, \tag{8.32}$$

where c is a constant. Equation (8.31) can be expressed as

$$S_f = \frac{F_r^2}{cD_{50}^{\frac{1}{6}} + 6.0}. \tag{8.33}$$

In terms of scaling, Equation (8.33) can be written as

$$S_{fr} = \left(\frac{D_r}{d_r}\right) F_{rr}^2, \tag{8.34}$$

where D_r denotes the mean particle diameter ratio.

Using Equation (8.11), one gets

$$S_{fr} = \frac{d_r}{w_r} = \left(\frac{D_r}{d_r}\right)^{\frac{1}{3}} \frac{v_r^2}{d_r} \tag{8.35}$$

$$w_r = \frac{d_r^{\frac{4}{3}}}{D_r^{\frac{1}{3}}}. \tag{8.36}$$

For eliminating w_r, inserting Equations (8.35) and (8.36) in Equation (8.21), one obtains

$$Q_r = d_r^{\frac{1}{2}} \left(\frac{d_r^{\frac{1}{3}}}{D_r^{\frac{1}{3}}}\right) d_r \tag{8.37}$$

$$d_r = D_r^{\frac{2}{11}} Q_r^{\frac{6}{11}}. \tag{8.38}$$

Substituting Equation (8.38) in Equation (8.36), the result is

$$w_r = D_r^{-\frac{1}{11}} Q_r^{\frac{8}{11}}.$$ (8.39)

Using Equations (8.38) and (8.39) in Equation (8.21),

$$v_r = D_r^{\frac{-1}{11}} Q_r^{\frac{-3}{11}}.$$ (8.40)

Inserting Equations (8.38) and (8.39) in Equation (8.17),

$$S_r = D_r^{\frac{3}{11}} Q_r^{-\frac{2}{11}}.$$ (8.41)

It is seen that the discharge exponents of Equations (8.38)–(8.41) are slightly different from those derived earlier, showing that each friction equation may lead to a unique set of exponents.

8.2.3 Mobile Bed

In channels with fixed banks but mobile bed, aggradation and degradation of the channel bed will occur, resulting in the transport of sediment along the channel. This means a scaling relation of sediment transport is needed, which can be derived using dimensional analysis as did Yalin (1971). It is assumed that the flow in the channel is steady and uniform and the bed is made up of cohesionless uniform grains. Then, sediment generation can be described by seven parameters, including dynamic viscosity (μ), fluid density (ρ), sediment particle size (D), sediment density (ρ_s), shear velocity ($v_* = \sqrt{gdS}$), flow depth (d), and specific sediment weight (γ_s). Since ρ, v_*, and D are recurring quantities, four dimensionless variables denoted as X_i, $i = 1, 2, 3,$ and 4, can be defined as:

$$X_1 = \frac{\rho v_* D}{\mu}$$ (8.42)

$$X_2 = \frac{\rho v_*^2}{\gamma_s D}$$ (8.43)

$$X_3 = \frac{d}{D}$$ (8.44)

$$X_4 = \frac{\rho_s}{\rho}.$$ (8.45)

Assuming the negligible role of dynamic viscosity in fully turbulent flow and of X_3 and X_4 in sediment transport near the bed, the bed load transport rate (weight per unit width per unit time) denoted by g_w can be expressed as

$$g_w = g_w(X_2).$$ (8.46)

If the specific gravity of sediment particles is denoted by S_s then the specific weight of sediment can be expressed as

$$\gamma_s = \rho g (S_s - 1).$$ (8.47)

Inserting Equation (8.47) in Equation (8.43) and recalling the definition of shear velocity, the result is

$$X_2 = \frac{dS}{(S_s - 1)D}.$$ (8.48)

Equation (8.48) is the Shields entrainment function. With the use of Equations (8.24) and (8.26), Equation (8.48) can be written in scaling terms as

$$(S_s - 1)_r = \frac{d_r S_r}{D_r} = D_r^{-0.58} Q^{0.23}.$$ (8.49)

Now for bed load transport the dimensionless quantity can be expressed as

$$X_5 = \frac{g_w}{\rho v_*^3}.$$ (8.50)

The bed load transport rate can be written as $G_w = w g_w$. Therefore, in scaling terms the bed load transport rate can be written as

$$G_{wr} = w_r (d_r S_r)^{1.5}.$$ (8.51)

Substituting Equations (8.23), (8.24), and (8.26) in Equation (8.51), the bed load transport rate can be written as

$$G_{wr} = D_r^{0.45} Q_r^{0.83}.$$ (8.52)

8.3 Comparison with Stable Channel Design Theories

The scaling relations can now be compared with the relations derived using other theories for stable channel design.

8.3.1 Threshold Theory

Using the Manning–Strickler resistance equation and the Shields entrainment function, the stable channel width of wide rivers composed of coarse sediment with bankfull discharge at the threshold of motion has been predicted by several investigators (Henderson, 1966; Kellerhals, 1967; Parker, 1979; Griffiths, 1979, 1981). Examples of width prediction equations (in FPS units) are:

$$\text{Griffiths (1981) equation: } w = 5.28S^{1.26}D_{50}^{-1.50} \tag{8.53}$$

$$\text{Henderson (1966) equation: } w = 2.06S^{1.25}D_{50}^{-1.50} \tag{8.54}$$

$$\text{Kellerhals (1967) equation: } w = 21.18S^{1.25}D_{50}^{-1.44} \tag{8.55}$$

$$\text{Parker (1979) equation: } w = 6.84S^{1.22}D_{50}^{-1.50}. \tag{8.56}$$

For illustration, consider the Henderson equation, in which S_s was taken as 2.65 or $S_s - 1 = 1.68$. Equation (8.54) can be written as a scaling relation as

$$w_r = \frac{Q_r d_r^{1.17}}{(S_s - 1)_r D_r^{1.5}}. \tag{8.57}$$

Then, using Equations (8.26) and (8.49), Equation (8.54) can be expressed in scaling terms as

$$w_r = D_r^{-0.16} Q_r^{0.48}. \tag{8.58}$$

Equation (8.58) is similar to Equation (8.23), demonstrating that Equation (8.54) is consistent with similitude requirements. One can show likewise for the other three Equations (8.53), (8.55), and (8.56).

8.3.2 Stability Index Theory

For gravel-bed rivers, Griffiths (1981) introduced a dimensionless stable-section index (I_S) as

$$I_S = \frac{wD_{50}^{1.5}g^{0.5}}{S^{1.26}Q} = \text{constant}, \tag{8.59}$$

where

$$I_S = \frac{1}{4[F_s(\gamma_s - 1)]^{1.76}}, \tag{8.60}$$

in which

$$F_s = \frac{RS}{(\gamma_s - 1)D_{50}}. \tag{8.61}$$

Equation (8.61) is the Shields entrainment function, which is constant for $v_* D_{50}/v \gtrsim 400$. Shields (1936) reported $F_s = 0.056$ but Little and Mayer (1976) reported the values in the range of 0.02 and 0.1. For threshold conditions, Equation (8.59) can be expressed as

$$I_S = \frac{w D_{50}^{1.5} g^{0.5} (\gamma_s - 1)^{1.76}}{S^{1.26} Q}. \tag{8.62}$$

In scaling terms, the stable channel width at the threshold condition obtained from Equation (8.62) is the same as Equation (8.57). Griffiths (1981) evaluated Equation (8.59) for a number of sites in the Buller catchment in South Island, New Zealand. He found that the stability indices were remarkably constant for stable reaches of a given river or watershed but would considerably vary from river to river.

8.3.3 Regime Theory

The regime equations derived by Lacey (1930) for stable canals can be written as

$$v \sim (f_s R)^{0.5} \tag{8.63}$$

$$f_s \sim D^{0.5} \tag{8.64}$$

$$v \sim R^{0.67} S^{0.33} \tag{8.65}$$

$$P \sim Q^{0.5}, \tag{8.66}$$

where P is the wetted perimeter, and f_s is the silt factor. From the continuity equation,

$$Q = vRP. \tag{8.67}$$

Using Equation (8.67), the scaling versions of Lacey Equations (8.63)–(8.66) can be expressed as

$$P_r = Q_r^{0.50} \tag{8.68}$$

$$R_r = D_r^{-0.17} Q_r^{0.33} \tag{8.69}$$

$$v_r = D_r^{0.17} Q_r^{0.17} \tag{8.70}$$

$$S_r = D_r^{0.83} Q_r^{-0.17}. \tag{8.71}$$

For wide channels, $P_r \sim w$ and $R_r \sim d_r$. Then, Equations (8.68)–(8.71) are similar to Equations (8.23)–(8.26). Equation (8.68) does not contain sediment size, but its effect on P_r will be minor in stable channels.

It may now be interesting to compare the threshold theory with the regime theory in the context of hydraulic similitude. By eliminating $(S_s - 1)_r$ from Equation (8.57), one obtains

$$w_r = \frac{Q_r^{0.62} S_r^{1.17}}{D_r^{0.51}}. \tag{8.72}$$

Now one can write Equation (8.71) as

$$\frac{S_r Q_r^{0.17}}{D_r^{0.83}} = 1.$$ (8.73)

Multiplying Equation (8.68) with Equation (8.73),

$$P_r = Q_r^{0.50} \left(\frac{S_r Q_r^{0.17}}{D_r^{0.83}} \right) = \frac{S_r Q^{0.67}}{D_r^{0.83}}.$$ (8.74)

Equation (8.74) is the equivalent regime equation.

8.4 Application to Channel Design

A stable gravel-bed channel is often designed for river engineering works, such as highway encroachment, bridge waterway section, diversion, meander cutoff, and low-flow channel flanked by berms bounded by flood-confining stop-banks or levees. When designing such a channel, there are restrictions imposed by the site and the only degree of freedom available is the channel width w. For example, the bankfull discharge Q is determined from the upstream conditions, so it cannot be chosen. The water surface slope S is assumed to be equal to the bed slope, which is controlled by local landscape features. Thus, the channel width can be computed using Equation (8.59), wherein the average value of the stability index can be obtained from neighboring stable reaches. In this procedure, the medium pavement size, D_{50}, can be determined by the locally available surface gravels. However, it is estimated using a rule of thumb as Henderson (1966) suggested:

$$D_{50} \text{ (surface bed material)} \simeq D_{75}\text{(whole bed material)}.$$ (8.75)

Using laboratory and field data, Davies (1974) obtained an empirical equation for the prediction of D_{50} of the armoured surface particle distribution owing to a given flow over a sediment bed:

$$D_{50} = 1.839 \times 10^{-3} \left\{ \left[\log \left(1{,}000 D_{go}\sigma_{go}^2 \right) \right] \left[\frac{v_*^3}{v(\gamma_s - 1)g} \right]^{0.389} \right\},$$ (8.76)

where D_{go} is the median size of the initial sediment distribution, and σ_{go} is the geometric standard deviation of the distribution. The condition for the bed to armour was defined as

$$D_5(\text{initial}) < D_{50} < D_{95}(\text{initial}).$$ (8.77)

For Equation (8.76) the data had the following characteristics:

$$0.001 \leq D_{go} \leq 0.0182 \text{ m}$$

$$1.89 \leq \sigma_{go} \leq 15.14$$

$$0.03 \leq v_* \leq 0.17 \text{ m/s}.$$

Equations (8.61) and (8.76) can be combined (Little and Meyer, 1976) as

$$D_{50} = 2.706 \times 10^{-7} \left\{ \left[\log \left(1{,}000 D_{go} \sigma_{go}^2 \right) \right] [(\gamma_s - 1)g]^{0.195} \left[\frac{F_s^{0.584}}{v^{0.389}} \right] \right\}^{2.401}. \quad (8.78)$$

In design, field data are utilized to determine γ_s, v, D_{go}, and σ_{go}. Equations (8.59), (8.60), and (8.78) can be employed for computing design channel width for bank-full discharge Q and slope S.

Following Griffiths (1981), a design example is now given. He discussed that a river channel was diverted to avoid flooding of a town, but active channel bars and bank erosion indicated that the diversion was unstable. The instability was also evidenced by the stability index being much smaller than the stability index of the stable upstream river sections. The maximum limit of local bank erosion corresponded to a channel width of 37 m and seemed to be waning. Therefore, a stable channel width was designed for this straight uniform diversion.

From field survey and hydraulic relations and sediment analysis, the design cross-section was $\gamma_s = 2.65$, $v = 1.31 \times 10^{-6}$, $S = 0.005$, $Q = 50 \text{ m}^3/\text{s}$, and for the initial bed-sediment size distribution:

$D_5 = 0.0009$, $D_{95} = 0.04$, $D_{go} = 0.011$, and $\sigma_{go} = 2.30$. Using five stable upstream reaches, Griffiths (1981) computed the mean stability index of 10.48 with a standard deviation of 1.91. Inserting γ_s and the stability index in Equation (8.60), the value of F_s was computed as

$$10.48 = \frac{1}{4[F_s(2.65 - 1)]^{1.76}},$$

which yielded $F_s = 0.073$.

Inserting the values of $F_s = 0.073$, $D_{go} = 0.011$, $\sigma_{go} = 2.30$, $\gamma_s = 2.65$, $v = 1.31 \times 10^{-6}$, and $g = 9.81 \text{ m}^2/\text{s}$ in Equation (8.78),

$$D_{50} = 2.706$$
$$\times 10^{-7} \left\{ [\log(1{,}000) \times 0.011 \times 2.30][(2.65 - 1)9.81]^{0.195} \left[\frac{0.073^{0.584}}{(1.31 \times 10^{-6})^{0.389}} \right] \right\}^{2.401}.$$

The median value of the bed material is found to be: $D_{50} = 0.031$ m.

Now, check if the bed will armour. Because if $D_5 = 0.0009 < D_{50} = 0.031 < D_{95} = 0.04$, the bed will armour. The design width of the channel is computed from Equation (8.59) as

$$10.48 = \frac{w0.031^{1.5}9.81^{0.5}}{0.005^{1.26}50.00}$$

$W = 38.7$ m, which is 64% higher than the original value of 23 m. Griffiths (1981) indicated that the design width using Equation (8.54) of Henderson (1966) yielded a value of 38.3 m.

References

Ashmore, P. E. (1991). Channel morphology and bed load pulses in braided, gravel-bed streams. *Geographical Annals*, Vol. 73A, pp. 37–52.

Ashmore, P. E. (2001). Braiding phenomena: Statics and kinetics. In: *Gravel-Bed Rivers V*, edited by M. P. Mosley, pp. 95–114, New Zealand Hydrological Society, Wellington.

Bennett, J. P. (1978). *One-dimensional surface water transport modeling*. U.S. Geological Survey National Training Center, Video Series No. 2500, tapes 10–14, Denver, CO.

Bennett, J. P. and Nordin, C. F. (1977). Simulation of sediment transport and armoring. *Hydrological Sciences Bulletin*, Vol. 22, No. 4, pp. 555–560.

Bray, D. I. (1979). Estimating average velocity in gravel-bed rivers. *Journal of Hydraulics Division, ASCE*, Vol. 105, pp. 1103–1122.

Davies, L. H. (1974). *Problems posed by new town development with particular to Milton Keynes*. Paper no. 2 in: Proceedings of Research Colloquium on Rainfall, Runoff and Surface Water Drainage of Urban Catchments, Department of Civil Engineering, Bristol University.

Griffiths, G. A. (1979). Rigid boundary flow resistance of gravel rivers. *Ministry of Works Development, Christchurch, Water and Soil Division, Report WS* 127, pp. 20.

Griffiths, G. A. (1980). Downstream hydraulic geometry of some New Zealand gravel bed rivers. *Journal of Hydrology (N.Z.)*, Vol. 18., pp. 106–108.

Griffiths, G. A. (1981). Stable-channel design in gravel bed rivers. *Journal of Hydrology*, Vol. 52, pp. 291–308.

Griffiths, G. A. (2003). Downstream hydraulic geometry and similitude. *Water Resources Research*, Vol. 39, No. 4, 1094, doi:10.1029/2002WR001488.

Guy, H. P., Simons, D. B., and Richardson, E. V. (1966). *Summary of Alluvial Channel Data from Flume Experiments, 1956–61*. US Government Printing Office.

Henderson, F. M. (1966). *Open Channel Flow*. Macmillan, New York.

Kellerhals, H. (1967). Stable channel with gravel paved beds, *Journal of Waterways and Harbors Division, ASCE*, Vol. 93, pp. 63–84.

Lacey, G. (1930). Stable channels in alluvium. *Proceedings, Institution of Civil Engineers*, Vol. 229, No. 1930, pp. 259–292.

Laursen, E. M. (1958). Sediment-transport mechanics in stable-channel design. *Transactions of American Society of Civil Engineers*, Vol. 123, pp. 195–203.

Leopold, L. B., and Maddock, T. (1953). The hydraulic geometry of stream channels and some physiographic implications. *U.S. Geological Survey Professional Paper 252*, p. 57.

Little, W. C. and Mayer, P. G. (1976). Stability of channel beds by armoring. *Journal of Hydraulics Division, ASCE*, Vol. 102, pp. 1647–1661.

Parker, G. (1979). Hydraulic geometry of active gravel rivers. *Journal of Hydraulics Division, ASCE*, Vol. 105, pp. 1185–1201.

Shields, A. (1936). *Application of Similarity Principles and Turbulence Research to Bed-Load Movement*. (Translated from German), California Institute of Technology, Pasadena.

Yalin, M. S. (1971). *Theory of Hydraulic Models*. pp. 266, Macmillan, New York.

Yang, C. T. (1996). *Sediment Transport: Theory and Practice*. McGraw-Hill Book Company, New York.

9

Tractive Force Theory

This chapter derives the at-a-station as well as the downstream hydraulic geometry using the tractive force theory. Here the threshold discharge does not necessarily correspond to the bankfull discharge for at-a-station hydraulic geometry. At a given section, there can be a threshold discharge where the channel is flowing partially. In the case of downstream hydraulic geometry, the channel is supposed to be flowing full for threshold discharge. The fundamental concept in the tractive force theory is the balance of forces at a station.

Notation

A_0 = cross-sectional area at the threshold discharge

C = constant

C_a = an area coefficient

C_0 = constant of integration

C_p = wetted perimeter coefficient

C_t = width–depth ratio coefficient

C_1 = constant of integration

d_0 = hydraulic depth

D = sediment size

D_s = sediment particle size on the channel boundary

$E(k, \alpha)$ = elliptical integral of the second kind with modulus k

F_d = resultant drag force

F_l = lift force on the particles

f_l = silt factor

G = constant

h = maximum depth of the partially full channel

Continued

k = modulus

k_1, k_2, k_3, k_4, and k_5 = constants

n = Manning's roughness coefficient

P_0 = wetted perimeter

Q_0 = threshold discharge

Q_s = sediment discharge in the channel

q_s = sediment discharge per unit width of the overland flow area

r = $\beta \tan \phi$

R_0 = hydraulic radius

S_0 = channel bed slope

t = time

T = top width of the partially full channel

T_0 = top width of the channel

v = mean velocity

W = submerged weight of the particles

x = downslope distance

y = flow depth at any point away from the center

y_0 = depth of flow at the center of the cross-section or maximum flow depth

z = bed elevation

β = ratio of lift force to drag force

ξ = a distance away from the centerline where the flow depth is y

η = porosity of the bed material

γ = unit weight of water

γ_s = specific sediment weight

ϕ = angle of repose, which is the same as the friction angle

σ = constant of proportionality

τ = local bed shear stress or tractive force

τ_c = critical shear stress

τ_0 = maximum shear stress

τ_{oc} = critical shear stress at the center

θ = local side slope angle

ψ = Shields entrainment function

9.1 Introduction

The geomorphic processes of a channel are dependent on the water and sediment the channel receives from its watershed, and these processes result in the channel shape and cross-section. For simplicity it is assumed that the watershed over its space is uniformly subjected to precipitation events, meaning the watershed is small, say 10 (about 25 square km) to 20 square miles (about 50 square km). When the watershed is subjected to a number of large precipitation events, they tend to lead to threshold conditions in the channel. In other words, the sediment particles on the channel bed and banks are subjected to flow conditions that are just sufficient to initiate the movement of these particles. Such a channel is called the threshold channel, and the discharges that collectively formed the threshold channel are called the threshold discharge. The watershed will remain in equilibrium until it is subjected to a precipitation event that is greater than the events that generated discharges that caused threshold channels. That is, channel morphology will not change and the particles on the bed and banks will remain stable for discharges less than the threshold discharge.

Inside a watershed, channels in equilibrium form cross-sections and shapes according to the maximum threshold discharge that has occurred. Particles on the periphery of the channel cross-section are at the "threshold" of movement under the corresponding flow conditions. Threshold channels develop over a long period of time, say 100 to 200 years, due to a large number of precipitation events. Defining the threshold condition and threshold discharge, this chapter discusses the derivation of hydraulic geometry using the tractive force theory, also called the threshold theory, and closely follows the work of Lane (1955), Henderson (1960), and Li (1975).

9.2 Threshold Condition

The tractive force theory assumes that the channel will be in equilibrium until subjected to a discharge or flow just sufficient to initiate the movement of particles. Under equilibrium the particles remain on the bed and banks and do not move. The threshold condition is a limiting condition defined by discharge below which there is no movement of sediment particles and above which the movement of particles is initiated. Thus, threshold discharge can be defined as the discharge that is just sufficient to initiate the movement of sediment particles on the channel bed and banks. Bankfull discharge can be regarded as an acceptable measure of threshold discharge, because for discharges smaller than this value the sediment particles on the channel bed and banks do not move appreciably and channel morphology will remain more or less unchanged. Likewise, a threshold channel can be defined as the channel whose bank and bed material movement is just initiated by the threshold discharge and before which the channel would be in the equilibrium condition.

The threshold condition for channel bed and banks can be derived from the sediment continuity equation. Consider a channel cross-section, as shown in Figure 9.1. Let z define the bed elevation, Q_s sediment discharge in the channel, η the porosity of the bed material, q_s sediment discharge per unit width of the overland flow area, which is the lateral inflow of sediment into the channel, x the

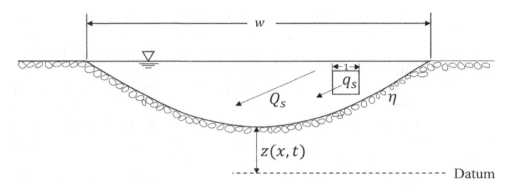

Figure 9.1 Channel cross-section subject to lateral sediment supply

downslope distance, and t the time. The sediment continuity equation can be expressed as

$$\frac{\partial Q_s}{\partial x} + (1 - \eta)\frac{\partial z}{\partial t} = q_s. \tag{9.1}$$

For the equilibrium condition, the second term on the left side of Equation (9.1), $\frac{\partial z}{\partial t} = 0$, for all x. Equation (9.1) then reduces to

$$\frac{\partial Q_s}{\partial x} = q_s. \tag{9.2}$$

If it is assumed that the value of q_s supplied by overland flow is constant, then Equation (9.2) integrates to

$$Q_s = q_s x + C_0, \tag{9.3}$$

where C_0 is the constant of integration, depending on the boundary condition. If the watershed is in the equilibrium condition, meaning that there is no appreciable movement of sediment particles in the overland flow area, then $q_s = 0$. Then, at $x = 0$, $Q_s = 0$, and $Q_s = 0$ for all x. This means at the threshold condition the channel is in equilibrium.

9.3 Tractive Force Theory

The tractive force theory for deriving hydraulic geometry is based on the balance of forces acting on the channel alluvial material. Lane (1955) and Lane et al. (1959) developed the tractive force theory for defining the shape of the threshold channel in homogeneous coarse material, which is followed here.

9.3.1 Assumptions

The tractive force theory makes the following assumptions:

(1) At and above the water surface, the channel side slope is at the angle of repose of the alluvial channel material.
(2) Particles at all points of the channel periphery are at the condition of incipient motion.
(3) The lift and drag forces of the fluid and the downslope component of the gravity force acting on the particles are balanced by the friction force developed between particles.
(4) The lift and drag forces are directly proportional to the tractive force.
(5) If the side slope is zero, then the flow-wise tractive force suffices to cause incipient motion.
(6) The particles are held together against the channel bed by the component of submerged weight of the particles acting normal to the bed.
(7) The tractive force acts in the direction of flow and equals the component of the weight of the water above the area on which the force acts.

9.3.2 Forces Acting on Threshold Channel Shape

Now consider the threshold channel shape as shown in Figure 9.2. Let ϕ define the angle of repose that is the same as the friction angle, θ define the local side slope angle, W define the submerged weight of the particles, F_l define the lift force on the particles, F_d define the resultant drag force, T_0 define the top width of the

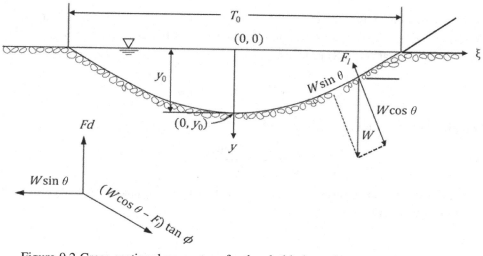

Figure 9.2 Cross-sectional geometry of a threshold channel

channel, y_0 define the depth of flow at the center of the cross-section, and y define the flow depth at any point away from the center.

Friction is the product of normal force and the tangent of the friction angle. The normal (to the side slope) component of the submerged weight can be defined as $W \cos \theta$. Then, the normal force is the difference between this normal component and the lift force: $W \cos \theta - F_l$. The downslope component of the submerged weight of the particles is $W \sin \theta$. Under the incipient motion on the periphery of the channel, it can be stated that F_d and $W \sin \theta$ are balanced by $(W \cos \theta - F_l) \tan \phi$. Thus, from force balance,

$$W^2 \sin^2\theta + F_d^2 = (W \cos \theta - F_l)^2 \tan^2\phi. \tag{9.4}$$

By assumption,

$$F_d = \sigma\tau, \tag{9.5}$$

where τ is the local bed shear stress or tractive force, and σ is a constant of proportionality. Likewise,

$$F_l = \beta F_d, \tag{9.6}$$

where β is the ratio of lift force to drag force. Inserting Equations (9.5) and (9.6) in Equation (9.4), the result is

$$W^2 \sin^2\theta + (\sigma\tau)^2 = (W \cos \theta - \sigma\beta\tau)^2 \tan^2\phi. \tag{9.7}$$

Equation (9.7) can be cast as

$$\tau^2 + \left(\frac{W}{\sigma}\right)^2 \sin^2\theta = \left(\frac{W}{\sigma} \cos \theta - \beta\tau\right)^2 \tan^2\phi. \tag{9.8}$$

Now, recall that at the center of the channel when $y = y_0$ (the maximum flow depth), the side slope angle θ is 0, and the shear stress becomes maximum denoted as τ_0. Then, Equation (9.8) reduces to

$$\tau_0^2 = \left(\frac{W}{\sigma} - \beta\tau_0\right)^2 \tan^2\phi. \tag{9.9}$$

Equation (9.9) can be rewritten as

$$\frac{W}{\sigma} = \frac{1 + \beta \tan \phi}{\tan \phi}\tau_0. \tag{9.10}$$

The maximum shear stress occurs at the center of the channel cross-section and is often expressed as

$$\tau_0 = \gamma y_0 S_0, \tag{9.11}$$

where γ is the unit weight of water and S_0 is the channel bed slope.

By assumption, the tractive force varies with the weight of the fluid above the area. Consider a distance ξ away from the centerline where the flow depth is y. The weight of the fluid in a column of unit area and depth y is γy. The normal component of this fluid weight is $\gamma y \cos \theta$. Thus, the tractive force at that distance (or point) is

$$\tau = \gamma y S_0 \cos \theta. \tag{9.12}$$

Equation (9.12) reduces to Equation (9.11) at the centerline where y becomes y_0. Inserting Equations (9.10)–(9.12) in Equation (9.8), one obtains

$$(\gamma y S_0 \cos \theta)^2 + \left(\frac{1 + \beta \tan \phi}{\tan \phi} \gamma y_0 S_0 \right)^2 \sin^2 \theta$$

$$= \left(\frac{1 + \beta \tan \phi}{\tan \phi} \gamma y_0 S_0 \cos \theta - \beta \gamma y S_0 \cos \theta \right)^2 \tan^2 \phi. \tag{9.13}$$

Equation (9.13) can be simplified to

$$\left(\frac{y}{y_0} \right)^2 + \left(\frac{1 + \beta \tan \phi}{\tan \phi} \right)^2 \tan^2 \theta = \left(\frac{1 + \beta \tan \phi}{\tan \phi} - \beta \frac{y}{y_0} \right)^2 \tan^2 \phi. \tag{9.14}$$

The quantity $\tan \theta$ can be expressed as $-dy/d\xi$. Let $\beta \tan \phi$ be defined as r. With the substitution of these quantities, Equation (9.14) can be written as

$$\left(\frac{y}{y_0} \right)^2 + \left(\frac{1 + r}{\tan \phi} \right)^2 \left(\frac{-dy}{d\xi} \right)^2 = \left(\frac{1 + r}{\tan \phi} - \frac{r}{\tan \phi} \frac{y}{y_0} \right)^2 \tan^2 \phi. \tag{9.15}$$

Equation (9.15) can be simplified as

$$\left(\frac{-dy}{d\xi} \right)^2 = \tan^2 \phi \left[\left(1 - \frac{r}{1 + r} \frac{y}{y_0} \right)^2 - \left(\frac{1}{1 + r} \frac{y}{y_0} \right)^2 \right]. \tag{9.16}$$

Equation (9.16) can be cast as

$$\frac{-d \left(\frac{y}{y_0} \right)}{\sqrt{1 - \frac{2r}{1+r} \left(\frac{y}{y_0} \right) - \frac{1-r}{1+r} \left(\frac{y}{y_0} \right)^2}} = \tan \phi \, d \left(\frac{\xi}{y_0} \right). \tag{9.17}$$

Equation (9.17) can be integrated as follows. It may be recalled from the table of integrals that

$$\int \frac{dx}{\sqrt{ax^2 + bx + c}} = \frac{1}{\sqrt{-a}} \sin^{-1} \frac{-2ax - b}{\sqrt{b^2 - 4ac}}, \qquad a < 0. \tag{9.18}$$

In Equation (9.18), $a = -\frac{1-r}{1+r}$. Since r is a positive quantity, $a < 0$; $b = -\frac{2r}{1+r}$; $c = 1$; $x = y/y_0$; and $dx = dy/y_0$. Therefore, integration of Equation (9.18) yields

$$-\sqrt{\frac{1+r}{1-r}}\,\sin^{-1}\left[(1-r)\frac{y}{y_0} + r\right] = \frac{\xi}{y_0}\tan\phi + C_1, \qquad (9.19)$$

where C_1 is the constant of integration, which can be evaluated from the boundary condition $y/y_0 = 1$ when $\xi/y_0 = 0$. Substituting this boundary condition in Equation (9.19), one obtains

$$C_1 = -\frac{\pi}{2}\sqrt{\frac{1+r}{1-r}}. \qquad (9.20)$$

Inserting Equation (9.20) in Equation (9.19), the result is

$$\frac{y}{y_0} = \frac{1}{1-r}\left[\cos\left(\tan\phi\sqrt{\frac{1-r}{1+r}}\,\frac{\xi}{y_0}\right) - r\right]. \qquad (9.21)$$

Lane et al. (1959) derived Equation (9.21) for defining the shape of a channel in a noncohesive material, in which the cosine function produces the shape as shown in Figure 9.2, taking $\beta = 0.85$ and $\phi = 35°$.

9.3.3 Channel Geometry

The threshold channel geometry corresponding to the threshold discharge can be derived from Equation (9.21).

9.3.3.1 Top Width

At $y = 0$, $\xi = T_0/2$, as shown in Figure 9.2. Then, Equation (9.21) yields

$$0 = \frac{1}{1-r}\left[\cos\left(\tan\phi\sqrt{\frac{1-r}{1+r}}\,\frac{T_0}{2y_0}\right) - r\right]. \qquad (9.22)$$

Equation (9.22) can be cast as

$$\frac{T_0}{y_0} = \frac{2}{\tan\phi}\left[\sqrt{\frac{1+r}{1-r}}\cos^{-1}r\right]. \qquad (9.23)$$

Equation (9.23) can be expressed as

$$\frac{T_0}{y_0} = C_t(\beta,\phi). \qquad (9.24)$$

The right side of Equation (9.24) defines the width–depth ratio coefficient, which is a function of β and ϕ.

9.3.3.2 Cross-Section

The cross-section area at the threshold discharge can be expressed as

$$A_0 = 2 \int_0^{T_0/2} y \, d\xi. \tag{9.25}$$

Inserting Equation (9.21) in Equation (9.25) one can express

$$A_0 = 2y_0 \int_0^{T_0/2} \frac{1}{1-r} \left[\cos\left(\tan\phi \sqrt{\frac{1-r}{1+r}} \frac{\xi}{y_0} \right) - r \right] d\xi. \tag{9.26}$$

Let $a = \tan\phi \sqrt{\frac{1-r}{1+r}} \frac{1}{y_0}$. Then, Equation (9.26) can written as

$$A_0 = 2y_0 \int_0^{T_0/2} \frac{1}{1-r} [\cos(a\xi) - r] d\xi. \tag{9.27}$$

Integrating Equation (9.27),

$$\frac{A_0(1-r)}{2y_0} = \frac{\sin(aT_0/2)}{a} - r\frac{T_0}{2}. \tag{9.28}$$

Substituting Equation (9.23) and the definition of a in Equation (9.28), the result is

$$\frac{A_0(1-r)}{2y_0} = \frac{\sin\left(\tan\phi \sqrt{\frac{1-r}{1+r}} \frac{1}{y_0} \frac{y_0}{\tan\phi} \sqrt{\frac{1+r}{1-r}} \cos^{-1} r \right)}{\tan\phi \sqrt{\frac{1-r}{1+r}} \frac{1}{y_0}} - r\frac{y_0}{\tan\phi} \sqrt{\frac{1+r}{1-r}} \cos^{-1} r. \tag{9.29}$$

Equation (9.29) simplifies to

$$\frac{A_0(1-r)}{y_0} = \sin\left(\cos^{-1} r \right) T_0 \frac{1}{\cos^{-1} r} - rT_0. \tag{9.30}$$

It may be recalled that $\sin(\cos^{-1} r) = \sqrt{1-r^2}$. Then, Equation (9.30) leads to

$$\frac{A_0}{y_0 T_0} = \frac{1}{1-r} \left(\frac{\sqrt{1-r^2}}{\cos^{-1} r} - r \right). \tag{9.31}$$

Equation (9.31) can be expressed as

$$\frac{A_0}{y_0 T_0} = C_a(r) = C_a(\beta, \phi), \tag{9.32}$$

where C_a is an area coefficient that depends only on r. The area coefficient is the ratio of actual cross-sectional area to the area of a rectangular section with the same top width and maximum depth as the depth of actual cross-section and is always smaller than 1.0.

9.3.3.3 Wetted Perimeter

The wetted perimeter can be expressed as

$$P_0 = 2 \int_0^{T_0/2} \sqrt{1 + \left(\frac{dy}{d\xi}\right)^2} \, d\xi. \tag{9.33}$$

The quantity $dy/d\xi$ can be obtained from Equation (9.21) as

$$\frac{dy}{d\xi} \frac{1}{y_0} = -\frac{a}{1-r} \sin(a\xi), \tag{9.34}$$

where a is defined as before, $a = \tan\phi\sqrt{\frac{1-r}{1+r}\frac{1}{y_0}}$. Then, inserting Equation (9.34) in Equation (9.33) one can express

$$P_0 = 2 \int_0^{T_0/2} \sqrt{1 + \left(\frac{ay_0}{1-r} \sin(a\xi)\right)^2} \, d\xi. \tag{9.35}$$

Inserting the expression for a, Equation (9.35) can be written as

$$P_0 = 2 \int_0^{T_0/2} \sqrt{1 + \left(\frac{\tan\phi}{1-r}\sqrt{\frac{1-r}{1+r}} \sin\left(\tan\phi\sqrt{\frac{1-r}{1+r}}\,\xi/y_0\right)\right)^2} \, d\xi. \tag{9.36}$$

With the use of Equation (9.23), the wetted perimeter from Equation (9.36) can be written as

$$P_0 = \frac{2y_0}{(1-r)k} \int_{\frac{\pi}{2}-\cos^{-1}r}^{\frac{\pi}{2}} \sqrt{1 - k^2 \sin^2\alpha} \, d\alpha, \tag{9.37}$$

in which

$$k = \frac{\tan\phi}{\sqrt{1 + \tan^2\phi - r^2}}. \tag{9.38}$$

The wetted perimeter can now be expressed as

$$\frac{P_0}{y_0} = \frac{2}{(1-r)k} \left[E\left(k, \frac{\pi}{2}\right) - E\left(k, \frac{\pi}{2} - \cos^{-1}r\right) \right], \tag{9.39}$$

where $E(k, \alpha)$ is an elliptical integral of the second kind with modulus k. Equation (9.39) can be expressed as

$$\frac{P_0}{y_0} = C_p(\beta, \phi),$$

(9.40)

in which C_p is the wetted perimeter coefficient.

9.3.3.4 Hydraulic Depth

By definition, hydraulic depth (d_0) is the ratio of cross-sectional area to the top width. Using Equations (9.24) and (9.32),

$$d_0 = \frac{A_0}{T_0} = \frac{C_a y_0 T_0}{C_t y_0}.$$

(9.41)

Equation (9.41) can be written as

$$\frac{d_0}{T_0} = \frac{C_a}{C_t}.$$

(9.42)

9.3.3.5 Hydraulic Radius

The hydraulic radius is the ratio of cross-sectional area to the wetted perimeter. Using Equations (9.32) and (9.40), it can be expressed as

$$R_0 = \frac{A_0}{P_0} = \frac{C_a y_0 T_0}{C_p y_0}.$$

(9.43)

Equation (9.43) can be written as

$$\frac{R_0}{T_0} = \frac{C_a}{C_p}.$$

(9.44)

Coefficients C_t, C_a, and C_p are the three basic dimensionless coefficients for describing the threshold channel geometry. Although they have complicated forms, they can be approximated by simple power functions, as shown in Figures 9.3–9.5. To that end, the lift-to-drag ratio must be specified. Bhowmik (1968) found an approximate value of $\beta = 0.85$. Then, r can be expressed as

$$r = 0.85 \, \tan \phi.$$

(9.45)

Using Equation (9.45), coefficients C_t, C_a, and C_p can be computed and plotted for various values ϕ, as shown in Figures 9.3–9.5. The plots in the figures can be expressed as power functions (Li, 1975) as

Tractive Force Theory

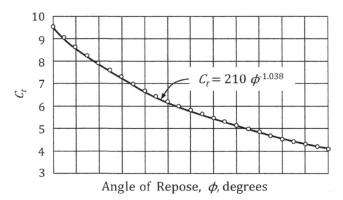

Figure 9.3 Dimensionless geometric coefficient C_t for threshold channel

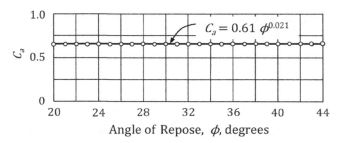

Figure 9.4 Dimensionless geometric coefficient C_a for threshold channel

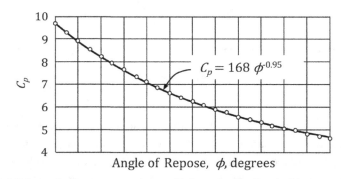

Figure 9.5 Dimensionless geometric coefficient C_p for threshold channel

$$C_t = 210\phi^{-1.038} \qquad (9.46)$$

$$C_a = 0.61\phi^{0.021} \qquad (9.47)$$

$$C_p = 168\phi^{-0.95}. \qquad (9.48)$$

If the lift force is assumed negligible, that is, $\beta = 0$, then $r = 0$ and

$$C_t = \frac{\pi}{\tan \phi} \tag{9.49}$$

$$C_a = \frac{2}{\pi} \tag{9.50}$$

$$C_p = \frac{2}{\pi} E\left(k, \frac{\pi}{2}\right). \tag{9.51}$$

The threshold channel geometry is only slightly influenced by the lift-to-drag ratio. Equations (9.49)–(9.51) were derived by Lane et al. (1959).

9.3.4 Hydraulic Geometry

Under threshold conditions, if the channel is flowing partially, as shown in Figure 9.6, the stream morphology does not change, and sediment particles on the bed and banks do not move. Thus, the channel shape and slope do not change either.

Equation (9.21) specifies the channel shape. The maximum depth of the partially full channel can be written as

$$h = y_0 - y\left(\xi = \frac{T_0}{2}\right), \tag{9.52}$$

where T is the top width of the partially full channel. With the use of Equation (9.21),

$$\frac{h}{y_0} = 1 - \frac{1}{1-r}\left[\cos\left(\tan\phi\sqrt{\frac{1-r}{1+r}}\frac{T}{2y_0}\right) - r\right] = \eta = f(r). \tag{9.53a}$$

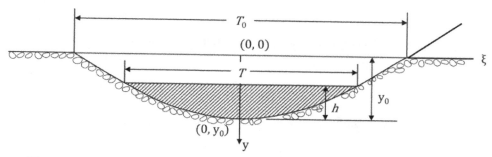

Figure 9.6 Geometry of a partially-full threshold channel

One can define $\frac{h}{y_0} = \eta$. Then, with its substitution, Equation (9.53a) can be written as

$$\eta = 1 - \frac{1}{1-r}\left[\cos\left(\tan\phi\sqrt{\frac{1-r}{1+r}}\frac{T}{2y_0}\right) - r\right] = f(r). \qquad (9.53b)$$

The top width can be written from Equation (9.53b) as

$$\frac{T}{y_0} = \frac{2}{\tan\phi}\sqrt{\frac{1+r}{1-r}}\cos^{-1}\varphi = f(\eta, r), \qquad (9.54)$$

where

$$\varphi = 1 - \frac{h}{y_0} + r\frac{h}{y_0} = 1 - \eta + r\eta. \qquad (9.55)$$

Equation (9.54) can be written with the use of Equation (9.23) as

$$\frac{T}{T_0} = \frac{\cos^{-1}\varphi}{\cos^{-1}r}. \qquad (9.56a)$$

For $\phi = 35°$ and $\beta = 0.85$,

$$\frac{T}{T_0} = \eta^{0.517}. \qquad (9.56b)$$

The cross-sectional area for the partially full channel can be derived in a like manner, yielding

$$\frac{A}{A_0} = \frac{\sqrt{1-\varphi^2} - \varphi\cos^{-1}\varphi}{\sqrt{1-r^2} - r\cos^{-1}r}. \qquad (9.57)$$

Likewise for the wetted perimeter, k,

$$\frac{P}{P_0} = \frac{E\left(k, \frac{\pi}{2}\right) - E\left(k, \frac{\pi}{2} - \cos^{-1}\varphi\right)}{E\left(k, \frac{\pi}{2}\right) - E\left(k, \frac{\pi}{2} - \cos^{-1}r\right)}. \qquad (9.58)$$

The hydraulic depth for the partially full channel can be written as

$$\frac{d}{d_0} = \frac{\frac{A}{A_0}}{\frac{T}{T_0}}. \qquad (9.59a)$$

For $\phi = 35°$ and $\beta = 0.85$,

$$\frac{d}{d_0} = \frac{A}{A_0} \neq \frac{T}{T_0} = \eta^{0.993}, \qquad (9.59b)$$

in which A/A_0 is given by Equation (9.57) and T/T_0 is given by Equation (9.56). Likewise, the hydraulic radius for partially full channel is given as

$$\frac{R}{R_0} = \frac{\frac{A}{A_0}}{\frac{P}{P_0}},$$

(9.60)

in which A/A_0 is given by Equation (9.57) and P/P_0 is given by Equation (9.58).

Assuming bed slope and Manning's coefficient as constant, the discharge in the partially full channel can be expressed as

$$\frac{Q}{Q_0} = \frac{\left(\frac{A}{A_0}\right)^{5/3}}{\left(\frac{P}{P_0}\right)^{\frac{2}{3}}}.$$

(9.61a)

For $\phi = 35°$ and $\beta = 0.85$,

$$\frac{Q}{Q_0} = \eta^{2.148}.$$

(9.61b)

The dimensionless ratio T/T_0 in Equation (9.56), d/d_0 in Equation (9.59), and Q/Q_0 in Equation (9.61) can be evaluated for specific values of β and ϕ. If $\phi = 35°$ and $\beta = 0.85$, these ratios can be expressed by simple power functions as

$$\frac{T}{T_0} = \left(\frac{h}{y_0}\right)^{0.517} = \eta^{0.517}$$

(9.62)

$$\frac{d}{d_0} = \left(\frac{h}{y_0}\right)^{0.993} = \eta^{0.993}$$

(9.63)

$$\frac{Q}{Q_0} = \left(\frac{h}{y_0}\right)^{2.148} = \eta^{2.148}.$$

(9.64)

The top width ratio, depth ratio, and discharge ratio, given, respectively, by Equations (9.56), (9.59), and (9.61) can be evaluated. For $\phi = 35°$ and $\beta = 0.85$, these ratios are graphed in Figure 9.7. The graphs in the figure can be approximated as power functions. From Equations (9.62) and (9.64),

$$\frac{T}{T_0} = \left(\frac{Q}{Q_0}\right)^{0.24}.$$

(9.65)

Equation (9.65) can be written as

$$T \propto Q^{0.24},$$

(9.66)

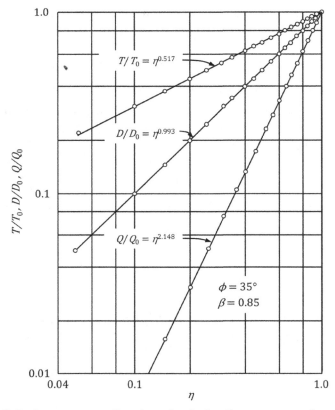

Figure 9.7 Ratios as power functions for hydraulic geometry of partially full threshold channel

since Q_0 and T_0 are constant. Similarly, from Equations (9.63) and (9.64),

$$d \propto Q^{0.46}. \tag{9.67}$$

Since the bed slope is constant,

$$S_0 \propto Q^{0.00}. \tag{9.68}$$

Using Equations (9.66) and (9.67), the mean velocity in the partially full channel can be written as

$$v \propto Q^{0.30}. \tag{9.69}$$

Equations (9.66)–(9.69) are the at-a-station hydraulic geometry relations. Figure 9.8 shows the variation of exponents of Q with β for $\phi = 35°$ and Figure 9.9 shows the variation of exponents of Q with ϕ for $\beta = 0$. These figures

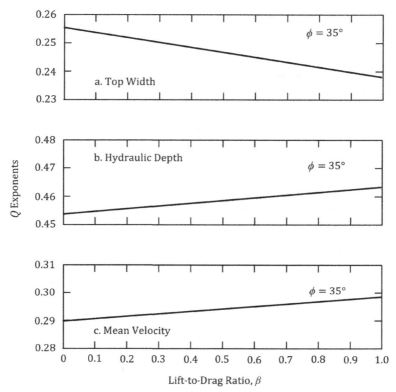

Figure 9.8 Variation of exponents of Q with β for $\phi = 35°$

illustrate that the values of exponents are only slightly dependent on β and ϕ, meaning the particle size exercises little influence.

The variations of exponents of Q with β for $\phi = 35°$ are illustrated in Figure 9.8. The variations of exponents with ϕ for $\beta = 0.0$ and $\beta = 0.85$ are illustrated in Figure 9.9. These figures show that the values of the exponents with respect to Equations (9.66), (9.67), and (9.69) are not sensitive to the variations in ϕ and β, meaning they are approximately independent of sediment particle size.

Judd and Peterson (1969) developed at-a-station hydraulic geometry relations. For two sites 70 and 71 at Boulder Creek, Colorado, the Q exponents were given for T, d, S_0, and V, respectively, as 0.18, 0.51, 0.00 (estimated), and 0.31 against the theoretical values of 0.24, 0.46, 0.00, and 0.30.

9.3.5 Downstream Hydraulic Geometry

Li (1975) and Li et al. (1976) extended the tractive force theory (Lane, 1955) to threshold channel concept and derived downstream hydraulic geometry relations.

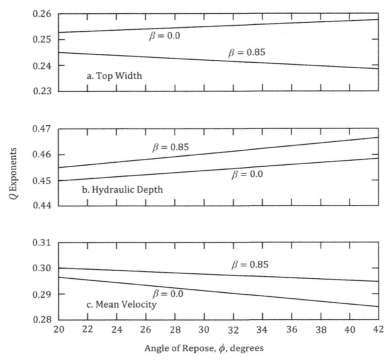

Figure 9.9 Variation of exponents of Q with ϕ for $\beta = 0$.

It may be recalled that the channel-forming discharge is discharge that sculpts the threshold channel. It is assumed that the sediment materials that make up the channel are uniform over its length. The three dimensionless coefficients defined previously can now be recast for a fixed ϕ as:

$$\frac{d_0}{y_0} = \frac{d_0 T_0}{T_0 y_0} = C_a = k_1 \tag{9.70}$$

$$\frac{T_0}{d_0} = \frac{C_t}{C_a} = k_2 \tag{9.71}$$

$$\frac{R_0}{d_0} = \frac{R_0}{T_0} \frac{T_0}{d_0} = \frac{C_t}{C_p} = k_3, \tag{9.72}$$

in which k_1, k_2, and k_3 are the constants.

It is also assumed that in small channels the friction slope is approximately equal to the bed slope. Then, for a threshold channel the threshold discharge Q_0 can be expressed using Manning's equation as:

$$Q_0 = \frac{1}{n} A_0 R_0^{\frac{2}{3}} S_0^{\frac{1}{2}} \tag{9.73}$$

or

$$Q_0 = \frac{1}{n} T_0 d_0 R_0^{\frac{2}{3}} S_0^{1/2},$$ (9.74)

where S_0 is the slope of the channel bed, A_0 is the cross-sectional area at the threshold discharge, T_0 is the topwidth of the threshold channel, d_0 is the hydraulic depth $= A_0/T_0$, $R_0 = A_0/P_0$ is the hydraulic radius, P_0 is the wetted perimeter, and n is Manning's roughness coefficient.

Following Shield's criterion for incipient motion in turbulent flow, one can write

$$D_s = \gamma y_0 S_0,$$ (9.75)

where D_s is the sediment particle size on the channel boundary, γ is the unit weight of water over a reach length, and γS_0 is constant. Then,

$$y_0 S_0 = k_4.$$ (9.76)

Thus, from Equations (9.70) and (9.76), one obtains

$$d_0 S_0 = k_5 = \text{constant}.$$ (9.77)

Using the Strickler formula, Manning's roughness coefficient n can be related to the particle size D_s as

$$n = C D_s^{\frac{1}{6}},$$ (9.78)

where C is a constant. Since D_s is constant, n is also constant. Inserting Equations (9.71), (9.72), (9.77), and (9.78) into Manning's Equation (9.74), the result is

$$Q_0 \propto d_0^{\frac{13}{6}}$$ (9.79)

or

$$d_0 \propto Q_0^{0.46}.$$ (9.80)

Similarly, using Equations (9.71) and (9.80), one obtains

$$T_0 \propto Q_0^{0.46}.$$ (9.81)

Likewise, using Equations (9.77) and (9.80), one gets

$$S_0 \propto Q_0^{-0.46}.$$ (9.82)

Furthermore,

$$v_0 = \frac{Q_0}{A_0} = \frac{Q_0}{T_0 d_0}.$$ (9.83)

Inserting Equations (9.80) and (9.81) in Equation (9.83), the result is

$$v_0 \propto Q_0^{0.08}. \tag{9.84}$$

Equations (9.80), (9.81), (9.82), and (9.84) are the hydraulic geometry relations based on the threshold channel concept and tractive force theory.

For coarse material with an angle of repose of $30°$ the hydraulic geometry equations can be expressed (Hey, 1978) as

$$v = 2.91\, D^{0.30} Q^{0.08} \tag{9.85}$$

$$S = 0.087\, D^{1.15} Q^{-0.46} \tag{9.86}$$

$$P = 1.853 D^{-01.5} Q^{0.46} \tag{9.87}$$

$$R = 0.185 D^{-0.15} Q^{0.46}, \tag{9.88}$$

in which v is m/s, D is in mm, and P and R are in m. These equations are derived using the continuity equation, flow resistance equation

$$v = 15.46\, R^{0.5} S^{0.38}. \tag{9.89}$$

The bed threshold equation

$$R = \frac{5.79D}{S} \tag{9.90}$$

and bank threshold equation

$$P = 10R \tag{9.91}$$

simplify to Equation (9.87).

9.3.6 Validity of Assumptions

It has been assumed that

$$\frac{T_0}{d_0} = K_2 \tag{9.92}$$

and

$$d_0 S_0 = K_5, \tag{9.93}$$

where K_2 and K_5 are constants. It follows that these are independent of threshold discharge. Using the data due to Brush (1961), the ratio T_0/d_0 is plotted against drainage area for various values of discharge and are found to be independent of drainage area. Likewise, the ratio $d_0 S_0$ is plotted against drainage area for various

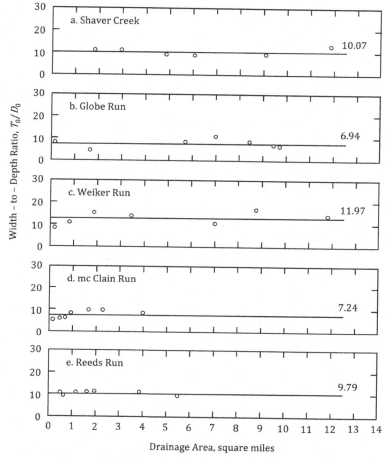

Figure 9.10 Variation of T_0/d_0 with drainage area for various values of threshold discharge

values of discharge and is found to be independent of drainage area, as seen from Figure 9.10. Since discharge is related to drainage area, it follows that T_0/d_0 and $d_0 S_0$ are independent of threshold discharge, as seen from Figure 9.11. For the Brush data, the theoretical values of exponents of Q are compared with those for the datum given for T_0, P_0, S_0, v_0, respectively, as 0.52, 0.43, −0.48 (estimated), and 0.05 against theoretical value of 0.46, 0.46, −0.46, and 0.08.

9.4 Henderson Theory

Henderson (1960) developed a theory to derive hydraulic geometry equations similar to those of Lacey's regime theory. The theory consists of an extension of

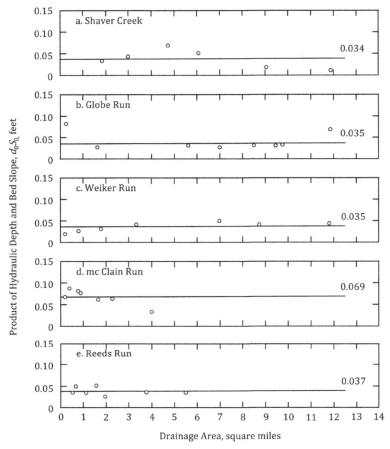

Figure 9.11 Variation of $d_0 S_0$ with drainage area for various values of threshold discharge

the tractive force theory of stable channel design and Strickler's formula. It may be noted that the tractive force theory assumes that the channel bed is on the condition of threshold of motion, which requires that the shear stress, and hence the product of hydraulic radius (R) and slope (S), is constant at all points along the channel. On the other hand, the regime theory assumes that the bed is live, which implies that the product of the square root of hydraulic radius and slope is constant along the channel. Henderson (1960) showed, using the Einstein bed load function, that this condition was satisfied by a stable channel with live bed. He argued that the tractive force theory should be more suitably called the threshold theory. The discussion here follows closely his work, which attempted to rationalize the threshold and live bed theories.

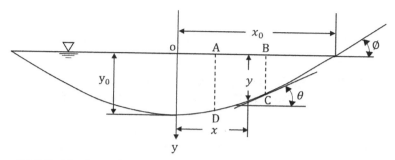

Figure 9.12 Stable channel cross-section

9.4.1 Governing Equations of Threshold Theory

When the channel is flowing full, the bed material would be on the threshold of motion at every point on the cross-section. The shear stress is smaller at the center than at side slopes, but both lateral slope and shear stress combine to dislodge the bed material. The critical shear stress on a side slope is related to the critical shear stress on the bed. Referring to Figure 9.12, the ratio of shear stresses can be expressed, following Lane and Carlson (1958), as

$$\frac{\tau}{\tau_{oc}} = \cos\theta \sqrt{1 - \frac{\tan^2\theta}{\tan^2\phi}}, \tag{9.94}$$

in which τ is the local shear stress, τ_{oc} is the critical shear stress at the center, θ is the side slope (angle), and ϕ is the angle of repose. Assuming that the local shear stress varies with the local flow depth and ignoring the lateral transfer of shear stress across vertical faces such as AD and BC (Figure 9.12), then Equation (9.94) yields the $x - y$ relation as

$$\frac{y}{y_o} = \cos\frac{x\tan\phi}{y_o}. \tag{9.95}$$

Equation (9.95) yields a sine profile. To determine the cross-section area (A), Equation (9.95) can be integrated to yield

$$A = \frac{2y_o^2}{\tan\phi}. \tag{9.96}$$

Extending the threshold theory, Henderson (1960) expressed the wetted perimeter (P) as

$$P = \frac{2y_0}{\sin\phi} \int_0^{\frac{\pi}{2}} \sqrt{1 - \sin^2\phi \sin^2\alpha}\, d\alpha, \tag{9.97}$$

where

$$\alpha = \frac{x \tan \phi}{y_0}. \tag{9.98}$$

The integral in Equation (9.97) is the complete elliptical integral of the second kind, often denoted by E. Hence, Equation (9.97) can be written as

$$P = \frac{2y_o E}{\sin \phi}. \tag{9.99}$$

The hydraulic radius can be expressed as

$$R = \frac{A}{P} = \frac{2y_o^2}{\tan \phi} \frac{\sin \phi}{2y_o E} = \frac{y_o \cos \phi}{E}. \tag{9.100}$$

A typical value of the angle of repose is $\phi = 35°$, for which the cross-section parameters can be written as

$$\text{Surface width } w = 2x_o = \frac{\pi y_o}{\tan \phi} = 4.49\, y_o \tag{9.101}$$

$$\text{Area } A = \frac{2y_o^2}{\tan \phi} = \frac{2y_o^2}{0.698} = 2.86 y_o^2 \tag{9.102}$$

$$\text{Perimeter } P = \frac{2y_o E}{\sin \phi} = \frac{2y_o}{0.574}(1.432) = 4.99 y_o \tag{9.103}$$

$$\text{Hydraulic radius } R = \frac{y_o \cos \phi}{E} = \frac{0.818 y_0}{1.432} = 0.572 y_o. \tag{9.104}$$

The y–θ relation used in Equation (9.95) does not lead to uniquely defined channel cross-sections. Three alternative channel cross-sections are shown in Figure 9.13.

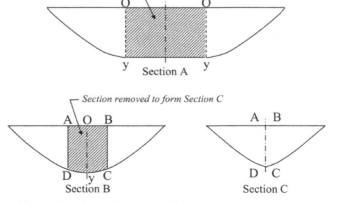

Figure 9.13 Alternative channel cross-sections

For section A, the y–θ relation will be satisfied if a section of constant depth is inserted in the middle between the two curved banks; for section C it will be satisfied by removing a section in the middle.

For particle Reynolds number above 400, Henderson (1960) introduced the particle diameter through the Shields relation or entrainment function as

$$\frac{\tau_c}{\gamma\left(\frac{\gamma_s}{\gamma}-\right)D} = 0.056, \tag{9.105}$$

in which τ_c is the critical shear stress, γ is the fluid specific weight, γ_s is the sediment specific weight, and D is the sediment size. For a typical value of $\gamma_s/\gamma = 2.6$, and $\tau = \gamma RS$, Equation (9.105) for the threshold motion becomes

$$D = 11RS. \tag{9.106}$$

Equation (9.106) holds only when $D > 0.25$ in.

For narrow channel cross-sections, of more interest is maximum stress instead of mean stress. The maximum stress corresponds to maximum depth y_o and can be expressed as $\tau = \gamma y_o S$. Thus, Equation (9.106) can be written as

$$D = 11y_oS. \tag{9.107}$$

For section B, the tractive force theory gives $R = 0.572y_o$, which will lead to

$$D = \frac{11RS}{0.572} = 19RS. \tag{9.108}$$

Recalling Strickler's formula for Manning's n to bed material size D, which is median size,

$$n = 0.034D^{\frac{1}{6}}, \tag{9.109}$$

the velocity from Manning's equation can be written as

$$v = \frac{1.49R^{\frac{2}{3}}S^{\frac{1}{2}}}{0.034D^{\frac{1}{6}}}. \tag{9.110}$$

Now D in Equation (9.110) can be replaced by Equation (9.107) for wide channels, resulting in

$$v = 29R^{\frac{1}{2}}S^{\frac{1}{3}} \tag{9.111}$$

and by Equation (9.108) for section B in Lane's theory, yielding

$$v = 27R^{\frac{1}{2}}S^{\frac{1}{3}}. \tag{9.112}$$

Both Equations (9.111) and (9.112) are similar to Lacey's equation:

$$v = 16R^{\frac{2}{3}}S^{\frac{1}{3}}. \tag{9.113}$$

In Equations (9.111) and (9.112) the bed material is uniform and is the dominant size, around D_{80} to D_{90}, which may be two to three times bigger than the median size. Therefore, allowance must be made for this difference in size. In that case the coefficient 0.034 in Strickler's formula will become 0.3 and the coefficients in Equations (9.111) and (9.112) will increase by about 10%.

For section B, the discharge can be expressed (with 27 replaced by 30) as

$$Q = 30PR^{\frac{3}{2}}S^{\frac{1}{2}}; \quad R = \frac{D}{19S}; \quad P = 8.75R. \tag{9.114}$$

By eliminating P and R, Equation (9.114) can be recast for S as

$$S = 0.44D^{1.15}Q^{-0.45}. \tag{9.115}$$

Equation (9.115) gives the limiting slope at which the channel having section B will be stable. For slope greater than this value, a wide channel of section A with less scouring capacity will be needed. If slope is less, then a channel of section C will be needed.

Now the surface width can be written as

$$w = 4.49y_o = \frac{4.49D}{11S} = \frac{0.41D}{S}. \tag{9.116}$$

Inserting Equation (9.115), Equation (9.116) yields

$$w = 0.93D^{-0.15}Q^{0.45}. \tag{9.117}$$

Likewise, the wetted perimeter can be expressed as

$$P = 1.03D^{-0.15}Q^{0.45}. \tag{9.118}$$

Equation (9.118) is similar to Lacey's equation:

$$P = 2.67 \, Q^{0.50}. \tag{9.119}$$

The velocity equation can also be expressed in terms of D and R by combining Equations (9.108) and (9.112) as

$$v = \frac{30R^{\frac{1}{2}}D^{\frac{1}{2}}}{(19R)^{\frac{1}{3}}} = 11D^{\frac{1}{2}}R^{\frac{1}{6}}. \tag{9.120}$$

The Lacey equation is

$$v = 1.17\sqrt{f_l R}, \tag{9.121}$$

where f_l is the silt factor, which is about 1.1 for silt. Equations (9.121) and (9.120) are quite different.

In these derivations, a value of $r = \gamma_s/\gamma = 2.6$ was assumed. For the general case, the geometric parameters are now given:

$$D = \frac{18RS}{r} = \frac{18y_oS}{r} \tag{9.122}$$

$$v = 30.6r^{\frac{1}{6}}R^{\frac{1}{2}}S^{\frac{1}{3}} \tag{9.123}$$

$$v = 30.6R^{\frac{1}{2}}S^{\frac{1}{3}}\left(\frac{r\cos\phi}{E}\right)^{\frac{1}{6}} \tag{9.124}$$

$$S = 0.24\,r^{1.15}\left(\cot\phi\sqrt{\frac{\cos\phi}{E}}\right)^{0.46}D^{1.15}Q^{-0.46} \tag{9.125}$$

$$w = 0.73\frac{\cos\phi\cot\phi}{E}\left(\tan\phi\sqrt{\frac{E}{\cos\phi}}\right)^{0.46}r^{-0.15}D^{-0.15}Q^{0.46} \tag{9.126}$$

$$P = 0.47\cot\phi\left(\tan\phi\sqrt{\frac{E}{\cos\phi}}\right)^{0.46}r^{-0.15}D^{-0.15}Q^{0.46} \tag{9.127}$$

$$v = 11.5D^{\frac{1}{3}}R^{\frac{1}{6}}\left(\frac{r\cos\phi}{E}\right)^{\frac{1}{2}}. \tag{9.128}$$

If the sediment size is less than $D < 0.25$ in. and particle Reynolds number is less than 400, and the entrainment function is less than 0.056, for example, $0.056k$, then the coefficient 11 in Equation (9.106) should be increased by $11/k$, and the coefficients in other equations should be altered. That is, the coefficient in Equations (9.106) and (9.122) should be multiplied by $1/k$, that in Equations (9.112) and (9.124) by $k^{1/6}$, that in Equations (9.115) and (9.125) by $k^{1.23}$, that in Equations (9.118) and (9.127) by $k^{-0.23}$, and that in Equations (9.120) and (9.128) by $k^{1/2}$.

9.4.2 Bed Load Equation and Lacey's Equations

Henderson (1960) discussed the relation between the live bed theory and the threshold theory. The regime theory – live bed theory – can be interpreted through Shields entrainment function ψ:

$$\frac{1}{\Psi} = \frac{\tau}{(r-1)\gamma D} = \frac{5RS}{8D} \quad \text{(with } r = 2.6\text{)}. \tag{9.129}$$

For canal data used by Lacey, this function is much greater than the Shields criterion for threshold of motion, which seems to fall on the straight-line portion of the Einstein bed load function and explains the basic form of the regime theory.

In the regime theory,

$$R^{\frac{1}{2}}S = \text{constant}. \tag{9.130}$$

The value of constant will vary from one canal system to another. In the threshold theory the shear stress that is proportional to RS is constant along the canal, that is,

$$RS = \text{constant}. \tag{9.131}$$

Equations (9.130) and (9.131) are quite different. However, Equation (9.131) can be deduced from the Einstein bed load function.

9.4.3 Application to Design of Artificial Channels

If the size of the bed material is coarse (>0.25 in.), it is preferable to design the channel using the threshold theory, and for finer material (<0.25 in.) it is better to use the live bed theory. Plots of velocity versus hydraulic radius raised to some power multiplied by slope raised to some power have been constructed for different types of data. For example, Simons and Albertson (1960) plotted v versus R^2S as in the Lacey equation. Henderson (1960) showed that v versus $R^{3/2}S$ would plot equally well. Similarly, for the same data,

$$v = 24R^{\frac{1}{2}}S^{\frac{1}{3}} \tag{9.132}$$

will also be a good plot, which is similar to Equation (9.112). The smaller value of the coefficient can be regarded as a factor of safety against the onset of bed motion.

For the same data, the wetted perimeter can be plotted as

$$P = GQ^{0.512}, \tag{9.133}$$

where G varies from 1.75 for coarse-bed channels to 3.5 for channels with sand bed and banks. The value of G should decrease as the size of the bed material increases.

The Lacey theory shows that

$$R \propto Q^{\frac{1}{3}}. \tag{9.134}$$

For fine and coarse beds,

$$R \propto Q^{0.361}. \tag{9.135}$$

For any type of slope,

$$PR = A \propto Q^{0.873} \tag{9.136}$$

or

$$PR^{1.35} \propto Q. \tag{9.137}$$

In the threshold theory,

$$v \propto D^{\frac{1}{3}} R^{\frac{1}{6}}. \tag{9.138}$$

Then,

$$PR^{\frac{7}{6}} \propto D^{\frac{-1}{3}} Q. \tag{9.139}$$

Equation (9.139) is valid for all channel slopes and shapes. The corresponding equation in the regime theory can be written as

$$PR^{\frac{3}{2}} \propto f_l^{\frac{-1}{2}} Q. \tag{9.140}$$

However, the Simons and Albertson data show

$$PR^{\frac{7}{6}} \propto Q^{0.933} \tag{9.141}$$

and

$$PR^{\frac{3}{2}} \propto Q^{1.053}. \tag{9.142}$$

The data do not discriminate between the Lacey theory and the threshold theory. To differentiate between the theories, one can write from Equations (9.106) and (9.111) by eliminating R as

$$P = \frac{1.14 Q S^{\frac{7}{6}}}{D^{\frac{3}{2}}}. \tag{9.143}$$

If $P = \sqrt{Q}$ holds, then the slope–discharge relation should be of the form

$$S^{\frac{7}{6}} \propto Q^{-\frac{1}{2}}, \tag{9.144}$$

which yields

$$S \propto Q^{-0.46}. \tag{9.145}$$

Similarly, if

$$P \propto D^{-0.15} Q^{\frac{1}{2}} \tag{9.146}$$

holds, then

$$S \propto D^{1.15} Q^{-0.46} \tag{9.147}$$

should also hold. These equations show that the dependence of a P–D–Q relation on an S–D–Q relation for wide channels is similar in form as for narrow channels.

9.4.4 Application to Rivers

For rivers, the first question is one of defining the dominant discharge, which for a stable single thread river can be the bankfull discharge. The mean annual flood with a recurrence interval of about 1.4 years is also used, but the return period value has varied from 0.5 year to 2.2 years. Rivers tend to adjust their slope–discharge relation, that is their longitudinal profiles. The question arises as to the shape of the longitudinal profile. It is known that rivers that braid, meaning they spread into many channels that divide and rejoin, are associated with steep slopes, and rivers that meander are associated with gentler slopes. It is assumed that a river braids because it has excess transporting capacity, which is dissipated by depositing sediment toward the banks, whereas a meandering river does not have excess transporting capacity and has a fairly stable single-thread channel. The meander channel formation itself may not be stable because meander loops tend to move downstream, but the cross-section is relatively stable. For the Wolman–Leopold data (Wolman and Leopold, 1957), the slope–discharge can be expressed as

$$S = 0.06Q^{-0.44}. \tag{9.148}$$

In the plot, the data above this curve represent braided channels and those below the meandering channels. With the sediment size included,

$$S = 0.64D^{1.14}Q^{-0.44}, \tag{9.149}$$

which holds for straight and meandering channels. Single-thread channels in coarse bed material tend to adjust their slopes.

References

Bhowmik, N. (1968). *The mechanics of flow and stability of alluvial channels formed in coarse materials.* Unpublished Ph.D. dissertation, Colorado State University, Fort Collins.

Brush, L. M. (1961). Drainage basins, channels, and flow characteristics of selected streams in central Pennsylvania. *U. S. Geological Survey Professional Paper* 282-F, pp. 145–175.

Henderson, F. M. (1960). Stability of alluvial channels. *Journal of the Hydraulics Division, Proceedings of the ASCE*, Vol. 87, No. HY6, pp. 109–138.

Hey, R. D. (1978). Determinate hydraulic geometry of river channels. *Journal of the Hydraulic Division, ASCE*, Vol. 104, No. HY6, pp. 869–885.

Judd, H. E. and Peterson, D. F. (1969). *Hydraulics of Large Bed Element Channels.* Report No. PRWG 17-6, pp. 115, Utah Water Resources Research Laboratory, Logan.

Lane, E. W. (1955). Design of stable canals. *Transactions, ASCE*, Vol. 120, pp. 1234–1260.

Lane, E. W. and Carlson, E. J. (1958). *Some factor affecting the stability of canals constructed in coarse granular materials.* Proceedings, Minnesota International

Hydraulic Convention, International Association of Hydraulic Research, Minneapolis, MN.

Lane, E. W., Lin, P. N., and Liu, H. K. (1959). *The most efficient stable channel for comparatively clean water in non-cohesive material.* Colorado State University Report CER59HKLS, April, Fort Collins.

Li, R. M. (1975). *Mathematical modeling of response from small watershed.* Unpublished Ph. D. dissertation, pp. 212, Colorado State University, Fort Collins.

Li, R. M., Simons, D. B., and Stevens, M. A. (1976). Morphology of cobble streams in small watersheds. *Journal of Hydraulics Division, ASCE,* Vol. 102, No. HY8, pp. 1101–1117.

Simons, D. B. and Albertson, M. L. (1960). Uniform water conveyance channels in alluvial material. *Journal of the Hydraulic Division, Proc. ASCE,* Vol. 86, No. 5, pp. 33–71.

Wolman, M. G. and Leopold, L. B. (1957). *River flood plains: some observations on their formations.* U.S. Geological Survey Professional Paper 282-C, Washington, DC.

10

Thermodynamic Theory

A regime channel geometry can be computed using the second law of thermodynamics and the Gibbs equation, which constitue the foundation of the thermodynamic method. With the use of a regime width relation, the need for a sediment transport rate relation can be obviated. This chapter discusses the thermodynamic methdology for deriving the hydraulic geometry of regime channels.

Notation

A_R = a geometric variable

A_* = energy-related characteristics

A_1 = cross-sectional area at section 1

A_2 = cross-sectional area at section 2

a_R = parameter

B_s = a known function of the grain size Reynolds number $X = \frac{\rho v_* D}{\mu}$

b_R = parameter

C = dimensionless Chezy coefficient for beds with bed forms

C_R = regime value of friction factor C

CS = surface confining the volume V between sections 1 and 2

CS_* = surface contacting the alluvium and air

CV = control volume

c_R = total friction factor

$\overline{c_R}$ = component of total friction

$\bar{c}$ = granular roughness of the (flat) bed surface

d = flow depth

D = representative sediment grain size or diameter

d_0 = intial flow depth at time equal to 0

d_R = flow depth at regime state

---

e = total fluid energy per unit fluid volume (specific total energy) at space point m

E = total energy of the fluid system

E_{Sys} = energy content of the fluid in Sys

E_{CV} = energy cotent of the fluid in CV

E_k = kinetic energy component of E_{CV}

E_p = potential energy component of E_{CV}

$\dot{E}_1$ = energy flux passing through the "in" cross-section A_1

$\dot{E}_2$ = energy flux passing through the "out" cross-section A_2

F_r = Froude number

F_1 = energy flux (or energy transfer) through section A_1 plus the displacement work at the section per unit time

F_2 = energy flux (or energy transfer) through section A_2 plus the displacement work at the section per unit time.

g = acceleration due to gravity

$°k$ = absolute temperature

L = channel length

N = dimensionless specific flow rate

m_1 and m_2 = space points

p = pressure at space point m

Q = flow rate

Q_s = sediment transport rate

$\dot{Q}_*$ = net heat time rate

R = corresponds to the regime state

S = slope

S_e = entropy of CV

S_0 = initial slope at time equal to 0

S_R = slope at regime state

Sys = alluvial stream system

t = time

T = time to equilibrium (stable) state

T_w = time the regime value is nearly acquired by w

u = internal energy

u_m = space average of u

U = internal energy component of E_{CV}

v = fluid velocity at space point m

V = volume confined between sections 1 and 2 or volume of CV and the coincidental Sys

v_m = mean velocity

v_{*cr} = critical shear velocity

v_* = shear velocity

w = flow width

w_0 = initial flow width at time equal to 0

Continued

w_R = width at regime state

$\dot{W}_*$ = time rate of work exchanged between the stream and its surroundings (grannular medium and air)

X = grain size Reynolds number

Y = Mobility number

Y_{c_r} = modified transport function

Z = relative depth (d/D)

Z_c = elevation of the centroid of CV

z_f = elevation of free surface

z_1 = elevation of any corresponding point of section A_1

z_2 = elevation of any corresponding point of section A_2

Δt = observation time

α = coriolis coefficient

ξ^3 = material number

γ = specific weight of fluid

γ_s = specific weight of sediment grains

μ = dynamic viscosity

η = flow intensity

ρ = fluid density

τ_0 = bed shear stress

τ_{ij} = viscous and turbulent stresses

θ = dimensionless normalized time

ΔH = hydraulic energy loss

10.1 Introduction

For the development of theory based on thermodynamics, it is assumed that the alluvium in which the channel is formed is cohesionless, the flow rate is constant, and the flow is turbulent. This assumption is usually true. Further, most alluvial channels are wide. Based on these considerations, a regime channel is determined using six characteristic parameters, which include: constant flow discharge Q, fluid density ρ, dynamic viscosity μ, specific weight of sediment grains γ_s or critical shear velocity v_{*c_r}, representative grain size or diameter D, and acceleration due to gravity g. The regime relations are of the type

$$A_R = a_R Q^{b_R}, \tag{10.1}$$

where A_R stands for a geometric variable, such as width w_R, flow depth d_R, or slope S_R; and a_R and b_R are the corresponding parameters, often considered constant. Subscript R corresponds to the regime state. Thus, the regime relations can be expressed as

$$A_R = f_{A_R}(Q, \rho, \mu, \gamma_s \text{ or } v_{*cr}, D, g). \tag{10.2}$$

The discussion in this chapter derives from Equation (10.2) and closely follows the work of Yalin and Ferreira da Silva (1999).

10.2 Hypothesis

Consider a wide alluvial channel with w = flow width, d = flow depth, and slope S. The channel tends to acquire its regime state, for one of its energy-related characteristics, say A_*, tends to acquire its minimum value. The three regime characteristics (subscript R denoting regime value), width w_R, flow depth d_R, and slope S_R, can be determined by the solution of three equations:

$$\text{Resistance equation: } Q = f_Q(w_R, d_R, S_R, C_R) \tag{10.3}$$

$$\text{Minimum } A_*: A_* = (A_*) \, min \text{ or } dA_* = 0 \tag{10.4}$$

$$\text{Sediment transport equation: } Q_s = f_{Q_s}(w_R, d_R, S_R, C_R), \tag{10.5}$$

where Q is the flow rate, Q_s is the sediment transport rate, and C_R is the regime value of the friction factor C, which among other things, is a function of d and S, that is, $C_R = f_c(d_R, S_R)$. Equation (10.3) is straightforward, but Equations (10.4) and (10.5) are not.

The regime channel dimensions w_R, d_R, and S_R can be determined using Equation (10.5), provided Q_s is known at the regime state, which usually is not the case. Further, it is not yet settled what exactly constitutes the energy-related characteristic A_*. Different investigators have proposed different expressions. For example, Chang (1979, 1980) used $A_* = S$; Yang (1984) used $A_* = Sv_m$ (v_m = mean velocity); White et al. (1982) used $A_* = 1/Q_s$; Davies and Sutherland (1983) used $A_* = 1/C$; Yang (1987) used $A_* = SL$ (L = length of channel reach); and Yalin (1982) used $A_* = F_r$ (F_r = Froude number). Thus, the energy quantity A_* must be determined first.

10.3 Determination of A_*

Let the flow commence at time $t = 0$ in a channel initially (unstable) characterized by (w_0, d_0, S_0). The channel then progressively deforms in order to become, at time $t = T$, an equilibrium (stable) or regime channel characterized by (w_R, d_R, S_R). For convenience, the channel development time is normalized as $\theta = \frac{t}{T} \in (0, 1)$, where θ is the dimensionless normalized time. It is assumed that the flow in channel is treated steady and uniform at any time θ during the course of its development. This assumption, however, is tenous. Noting that Q is constant, the unsteadiness of the developing stream during $0 < t < T$ can be attributed to the time variation of the channel. Accordingly, Equation (10.3) can be expressed using Chezy's equation as

$$v_m = C\sqrt{gSd} = \frac{Q}{wd},\qquad (10.6)$$

where v_m is the mean velocity, and C is the dimensionless Chezy roughness coefficient for beds with bed forms $= v_m/v_*$, $v_* = $ shear velocity $= \sqrt{\tau_0/\rho}$, $\tau_0 = $ bed shear stress, and $\rho = $ flow density. Equation (10.6) can also be expressed as

$$F_r^2 = \frac{v_m^2}{gd} = \frac{Q^2}{gd^3 w^2} = C^2 S.\qquad (10.7)$$

The flow uniformity is justified by noting that the ratio of channel length L to flow depth d $(L/d) > 10^3$, that is, the stream channel is long. The steadiness of flow is justified as follows. Let Δt be the duration of observations or measurements taken on the stream. The duration of development, T, of a natural stream from any time θ is larger by orders of magnitude than the duration of observations. By time T, the stream boundaries are virtually immobile and the flow charateristics are virtually constant.

If the initial channel slope is S_0 then during the period $(0 < \theta < 1)$ of regime development, the channel slope can only decrease, as borne out by experimental evidence, that is, $S_0 > S$. The decrease in slope can be due to aggradation-degradation in short channels with large D/d or due to meandering in long channels with small D/d.

For regime formation by meandering, consider an alluvial stream system (Sys) and the control volume (CV) at a development stage θ. Here the control volume is taken to be the fluid volume confined between two sections one in-section (say section 1 with cross-sectional area A_1) and the other out-section (say section 2 with cross-sectional area A_2). Let CS be the surface confining the volume V between sections 1 and 2. During the observation time Δt ($\ll T$) the boundaries are virtually stationary, and so is CV. At any t, selected within Δt, the stationary CV coincides with the instantaneous (moving) fluid system Sys. The surface, CS_*, contacting the alluvium and air, is part of CS and can be expressed as $CS_* = CS - (A_1 + A_2)$.

Let $\dot{Q}_*$ be the net heat time rate and $\dot{W}_*$ be the time rate of work exchanged between the stream and its surroundings (granular medium and air). It is assumed that any possible heat exchange between the stream and its surroundings is negligible. Similarly, it is assumed that no work is exchanged between the stream and its environment. The stream system is a specified fluid mass moving with the average flow velocity v_m. Accordingly, the thermodynamics of open channel flow employs:

$$\dot{Q}_* = \frac{dQ_*}{dt} = 0; \quad \dot{W}_* = \frac{dW_*}{dt} = 0. \tag{10.8}$$

Now the variation of flow energy is described.

10.3.1 Variation of Flow Energy

Let E_{Sys} and E_{CV} be the energy contents of the fluids in *Sys* and *CV*, respectively; $\dot{E}_1$ the energy flux passing through the "in" cross-section A_1, and $\dot{E}_2$ the energy flux passing through the "out" cross-section A_2, as shown in Figure 10.1. Let F_1 define the energy flux (or energy transfer) through section A_1 plus the displacement work at the section per unit time, and F_2 define the energy flux (or energy transfer) through section A_2 plus the displacement work at the section per unit time.

Applying the first law of thermodynamics to the stream system,

$$\frac{DE_{sys}}{Dt} = \frac{dE_{CV}}{dt} + (F_2 - F_1) = \dot{Q}_* - \dot{W}_*, \tag{10.9}$$

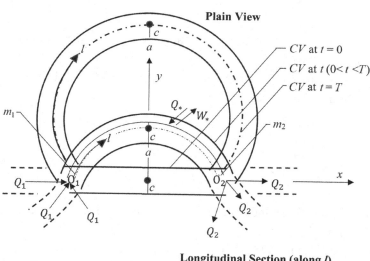

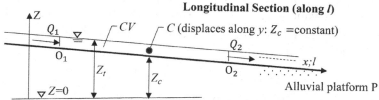

Figure 10.1 Channel section showing inflow and outflow of energy

where

$$E_{sy} = E_{CV} = \rho \int_V e \, dV \tag{10.10}$$

$$F_i = \int_{A_i} e_i \rho v \, dA_i + \int_{A_i} \frac{p}{\rho} \, dA_i = \dot{E}_i + W_i, \quad i = 1, 2, \tag{10.11}$$

where V is the volume of CV and the coincidental Sys, A_i is the cross-sectional area of section i, $\dot{E}_i$ is the energy flux through the cross-sectional area A_i, W_i is the fluid displacement work at A_i, p is the pressure at space point m, v is the fluid velocity at space point m, and e is the total fluid energy per unit fluid volume (specific total energy) at space point m. The total energy E of the fluid system can be written as the sum of kinetic energy (E_k), potenial energy (E_p), and internal energy (U):

$$E = E_k + E_p + U = \rho \int_V (e_k + e_p + u) dV. \tag{10.12}$$

The value of e for the element dV can be expressed as the sum of elemental kinetic energy, elemental potentail energy, and elemental internal energy:

$$e = \frac{\rho}{2} v^2 + \gamma z + u, \tag{10.13}$$

where $\rho v^2 / 2$ is the kinetic energy, γ is the weight density, γz is the potential energy, and u is the internal energy.

Invoking the uniformity of flow and Equation (10.8), and from Equations (10.9), (10.10), and (10.11), one obtains:

$$\dot{E}_2 = \dot{E}_1, \quad \text{i.e.,} \quad e_2 = e_1 \tag{10.14}$$

and

$$\frac{dE_{sys}}{dt} = \frac{dE_{CV}}{dt}; \quad E_{sys} = E_{CV}; \quad i.e, \quad E = \text{constant}. \tag{10.15}$$

Equation (10.15) states that the energy of the stream system can be replaced by the energy of the control volume.

With the use of Equation (10.13) and noting that $(z_2 - z_1) = (z_{f2} - z_{f1})$, where z_1 and z_2 are the elevations of any two corresponding points of sections A_1 and A_2 (points m_1 and m_2 in Figure 10.1), and z_f is the elevation of free surface, and Equation (10.14), one can write

$$u_2 - u_1 = g(z_{f1} - z_{f2}) = (\Delta H)_{1-2}. \tag{10.16}$$

Equation (10.16) states that the hydraulic energy loss is $(\Delta H)_{1-2} = H_1 - H_2 = z_1 - z_2$, but it is not a loss from a thermodynamic point of view; rather it is the amount of energy conversion – the specific potential energy gz_f of the downstream moving fluid (mixture of water and sediment) gets transformed into its (specific) internal energy u, that is, the elevation of free surface decreases, while u increases along the reach. This energy conversion is by means of the Joulean work done by all forces that oppose the fluid motion within the *CV*. These forces include viscous and turbulent stresses (τ_{ij}), hydrodynamic forces interacting between transported grains and transporting fluid, form drag and skin friction forces interacting between the fluid and bed forms, and shear stresses acting on the bank surfaces. The stress on the bank results from the motion of grains on the bank, which leads to the displacement of banks and expansion of meander loops. The work of these forces increases the internal energy of the fluid-sediment mixture, that is, $\frac{\partial U}{dt} > 0$, which compensates for the rate of its decrease $\dot{U}_1 - \dot{U}_2 = \rho Q(u_1 - u_2)(<0)$, thus maintaining the level of internal energy of the *CV*.

10.3.2 Variation of Flow Energy with Dimensionless Time

Equation (10.16) states that the energy-analysis of the present system (corresponding to $\dot{E}_1 = \dot{E}_2$ and $\dot{Q}_* = \dot{W}_* = 0$) can be replaced by that of the (coincidental) control volume. Denoting the internal, kinetic, and potential energy components of E_{CV} by U, E_k, and E_p, respectively, one can write

$$E_{CV} = U + E_k + E_p (= E_{sys}), \tag{10.17}$$

which can be expressed with the aid of Equation (10.12) as

$$E_{CV} = \int_{V_{CV}} e\, dV = \int_{V_{CV}} u\, dV + \frac{\rho}{2} \int_{V_{CV}} v^2\, dV + \gamma \int_{V_{CV}} z\, dV$$

$$= V_{CV}\left(u_m + \frac{\rho}{2}\alpha v_m^2 + \gamma z_c\right)\rho = E_{sys}, \tag{10.18}$$

where u_m is the space average of u, z_C is the elevation of the centroid of *CV*, and α is the coriolis coefficient, which can be ignored.

Using Equation (10.18), and taking into account that θ and t are related to each other by a constant proportion ($\theta = t/T$), one can express Equation (10.17) as

$$\frac{dU}{d\theta} + \frac{dE_k}{d\theta} + \frac{dE_p}{d\theta} = 0. \tag{10.19}$$

In the course of regime development, there is no systematic increment or decrement of the average elevation of a stream, that is, of its centroid the time reduction of the stream slope S is either by comparable amounts of bed erosion in the upper reaches and deposition in the lower reaches or by meandering. Hence, $\frac{dE_p}{d\theta} = \frac{dz_c}{d\theta} = 0$. This reduces Equation (10.19) to

$$\frac{dU}{d\theta} + \frac{dE_k}{d\theta} = 0, \tag{10.20}$$

Since the flow is uniform, the possible "in" and "out" fluxes of heat and/or work through sections A_1 and A_2 are identical. Moreover, no heat and/or work is crossing the part of the control surface that is in contact with the granular medium and air [Equation (10.8)]. Hence, the frictional (and thus irreversible) fluid flow CV is virtually isolated. According to the second law of thermodynamics, the entropy S_e of CV must progressively increase from zero onwards with the passage of time:

$$\frac{dS_e}{dt} \geq 0. \tag{10.21}$$

The time increments of U and S_e are interrelated by Gibbs' TdS_e relation:

$$^\circ k \frac{dS_e}{d\theta} = \frac{dU}{d\theta}, \tag{10.22}$$

where $^\circ k$ is the (always positive) absolute temperature. It is assumed that $^\circ k =$ constant > 0. Eliminating $dU/d\theta$ between Equations (10.20) and (10.22), one obtains

$$^\circ k \frac{dS_e}{d\theta} + \frac{dE_k}{d\theta} = 0. \tag{10.23}$$

This states that the time increment of S_e implied by Equation (10.20) must be necessarily accompanied by the corresponding time decrement of the flow of kinetic energy:

$$E_k = \frac{\rho}{2} V_{CV} v_m^2. \tag{10.24}$$

For a finite system, entropy cannot increase indefinitely. It increases until a certain maximum is reached, which corresponds to the state of the final equilibrium (regime state) at $\theta = 1$. Then $\theta > 1$, $S_e = S_{e_{max}}$.

Note that the present stream conveys a constant flow rate Q and that it satisfies the resistance Equation (10.7):

$$\frac{v_m^2}{gd} = c^2 S = F_r^2 = \textit{Squared Froude number}. \tag{10.25}$$

Using $V_{CV} = Lwd$, $Q = wdv_m =$ constant, and Equation (10.25), Equation (10.24) can be written as

$$E_k = \frac{\rho}{2} Lw F_r^2 g d^2. \tag{10.26}$$

From Equation (10.6),

$$d^2 = \left(\frac{Q^2}{C^2 w^2 g S}\right)^{\frac{2}{3}}.$$ (10.27)

Inserting Equation (10.27) in Equation (10.26),

$$E_k = \frac{\gamma}{2} L w F_r^2 \left(\frac{Q^2}{C^2 w^2 g S}\right)^{\frac{2}{3}}.$$ (10.28)

Inserting Equation (10.25) in Equation (10.28),

$$E_k = \frac{\gamma}{2} L w F_r^2 \left(\frac{Q^2}{F_r w^2 g}\right)^{\frac{2}{3}}.$$ (10.29)

Equation (10.29) can be expressed as

$$E_k = (\text{constant}) \frac{L}{w^{\frac{1}{3}}} F_r^{\frac{2}{3}},$$ (10.30)

where

$$\text{constant} = \frac{\gamma}{2} \frac{Q^{\frac{4}{3}}}{g^{\frac{2}{3}}}.$$ (10.31)

The regime development of w takes place much faster than that of S and d; by the time $T_w (\ll T)$ the regime value is nearly acquired by w. Considering the time internal $T_w < t < T$, and ignoring $dw/d\theta$, one obtains by the substitution of Equation (10.30) in Equation (10.23):

$$\frac{°k dS_e}{d\theta} + \text{constant} \left(\frac{dL}{d\theta} - \frac{1}{3w^{4/3}} \frac{dw}{d\theta} + \frac{2}{3F_r} \frac{dF_r}{d\theta}\right) = 0,$$ (10.32)

with $E_k = (\text{constant}) L \left(F r^2 / w\right)^{1/3}$.

Since $dL/d\theta \geq 0$ (if the channel is straight, if meandering $dL/d\theta > 0$), Equation (10.32) indicates that the time increment of entropy $\left(\frac{dS_e}{d\theta} \geq 0\right)$ must necessarily be accompanied by $dF/d\theta \leq 0$.

It, then, follows that regime-formation, in general, follows the kinetic energy E_k. For an alluvial channel, under these conditions, the minimization of E_k reduces to the minimization of $F_r^2 = c^2 S$ (which is thus identified with A_*):

$$0 < \theta < 1 : F_r \rightarrow (F_r)_{min}; \quad \frac{dF_r}{dt} < 0$$ (10.33)

$$\theta \geq 1 : F_r \to (F_r)_{min}; \quad \frac{dF_r}{d\theta} = 0. \tag{10.34}$$

The Froude number of a sediment transporting open channel flow can be expressed as a function of the following dimensionless variables:

$$\xi^3 = \frac{\rho\gamma_s D^2}{\mu^2}, \eta = \frac{v_*^2}{v_{*c_r}^2}; N = \frac{Q}{wDv_{*c_r}}; F_r = \phi(\xi, \eta, N) \tag{10.35}$$

$$\xi^3 = \frac{X^2}{Y} = \frac{\rho\gamma_s D^3}{\mu^2} = \text{material number}$$

$$X = \frac{\rho v_* D}{\mu} = \text{grain size Reynolds number}$$

$$Y = \frac{\rho v_*^2}{\gamma_s D} = \text{mobility number}$$

$$\eta = \frac{Y}{Y_{c_r}} = \left(\frac{v_*}{v_{*c_r}}\right)^2 = \text{flow intensity},$$

where Y_{cr} can be calculated following Da Silva and Yalin (2017) as

$$Y_{cr} = 0.13\xi^{-0.392}e^{-0.015\xi^2} + 0.045\left[1 - e^{-0.068\xi}\right]$$

$$N = \frac{Q}{wDv_*c_r} = \text{dimensionless specific flow rate}$$

$$v_* = \sqrt{gSd},$$

where g is acceleration due to gravity.

$$v_{*cr} = \sqrt{\frac{\tau_{cr}}{\rho}},$$

where $\tau_{cr} = 0.0125 + 0.019D$. D is particle size in mm and shear stress is in pounds per square foot.

The quantity c_r signifies the value corresponding to the initiation of sediment transport (or to the "critical stage").

10.4 Transport Equation

The cross-section of a regime channel does not vary with time, that is, the net cross-transport rate is zero all over. The regime aspect ratio can be expressed as

$$\frac{w_R}{d_R} = \left[1.42\left(\frac{c_R}{\overline{c}_R}\right)\phi(\xi)\right]^2 c_R\sqrt{\eta_R},\tag{10.36}$$

where $\phi(\xi)$ is determined as

$$\phi(\xi) = 1, \quad \text{if } \xi \geq 1.5; \phi(\xi) = 0.45\xi^{0.3}, \quad \text{if } \xi < 15\tag{10.37}$$

and $\sqrt{\eta_R}$ implies $\sqrt{\frac{gS_R d_R}{v_{*c_r}}}$.

Here c_R and $\overline{c}_R$ are regime values of the total friction factor and its component $\overline{c}_R$. The component $\overline{c}$ reflects the influence of the granular roughness of the (flat) bed surface and it is given by

$$\overline{c} = \frac{1}{k}\ln\left(b_s\frac{Z}{2}\right), \quad k = 0.4,\tag{10.38}$$

where $b_s = \exp\left[kB_s^{-1}\right]$, B_s being a known function of the grain size Reynolds number $X = \frac{\rho v_* D}{\mu}$.

10.5 System of Regime Equations

Using Equation (10.6) as resistance equation, $F_r^2 = c^2 S$ as A_* and replacing the Q_s-equation by Equation (10.36),

$$\frac{Q}{w_R d_R} = c_R\sqrt{gS_R d_R}\tag{10.39}$$

$$\left(F_r^2\right)_{min} = \left(c^2 S\right)_{min} = c_R^2 S_R\tag{10.40}$$

$$\frac{w_R}{d_R} = \left[1.42\left(\frac{c_R}{\overline{c}_R}\right)\phi(\xi)\right]^2 c_R\sqrt{\frac{gS_R d_R}{v_{*c_r}}},\tag{10.41}$$

where v_{*c_r} is determined by the transport inception curve.

Yalin and Ferreira da Silva (1999) expressed w_R as

$$w_R = \alpha_R\sqrt{\frac{Q}{v_{*c_r}}}, \quad \alpha_R = \phi_\xi(\xi)\cdot\phi_c(\overline{c}_R)\cdot c_R,\tag{10.42}$$

in which

$$\phi_\xi(\xi) = 0.639\xi^{0.30}, \quad \text{if } \xi \leq 15\tag{10.43}$$

$$\phi_\xi(\xi) = 1.42, \quad \text{if } \xi > 15\tag{10.44}$$

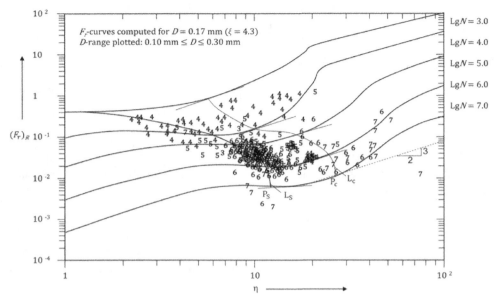

Figure 10.2 Families of F_r-curves and experimental data for sand [Yalin and Ferreira da Silva (1999)]

$$\text{If } \overline{c_R} < 21 \text{ then } \phi_c(\overline{c_R}) = \frac{1}{\overline{c_R}}. \tag{10.45}$$

If $21 \leq \overline{c_R} \leq 33$,

$$\text{then } \phi_c(\overline{c_R}) = \frac{1}{\left[0.00033(\overline{c_R} - 27)^3 - 0.03564(\overline{c_R} - 27) + 0.857\right] \cdot \overline{c_R}}. \tag{10.46}$$

$$\text{If } \overline{c_R} > 33 \text{ then } \phi_c(\overline{c_R}) = \frac{1}{0.714\,\overline{c_R}}. \tag{10.47}$$

The flow depth d_R can be determined by eliminating w_R from Equations (10.42) and (10.3),

$$d_R = \left[\alpha_R^2 (c_R^2 S_R)\right]^{-\frac{1}{3}} \left(\frac{Q v_{*c_r}}{g}\right)^{\frac{1}{3}}. \tag{10.48}$$

Yalin and Ferreira da Silva (1999) tested Equations (10.42)–(10.48) using field and laboratory data on sediment ranging from fine sand to gravel and found excellent agreement, as shown in Figures 10.2–10.3.

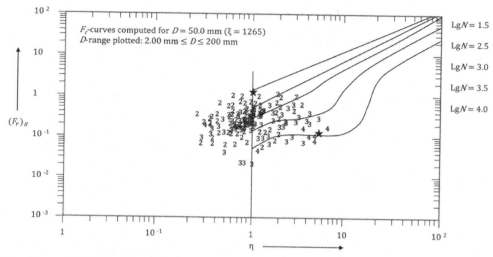

Figure 10.3 Families of F_r-curves and experimental data for gravel [Yalin and Ferreira da Silva (1999)]

10.6 Computation of Hydraulic Geometry

The procedure for computing hydraulic geometry involves three basic steps: (1) choosing an energy quantity and specifying basic equations, (2) determination of roughness factor and Froude number, and (3) computing the geometry. Each step is now presented.

10.6.1 Basic Equations

Yalin and Da Silva Ferriera (1999) chose $A_* = v_m$. Equation (10.4) can be replaced by Equation (10.42). Thus, the basic equations, in lieu of Equations (10.4)–(10.6), can be expressed as

$$\text{Resistance equation } \left(F_r^2\right)_R = \frac{Q^2}{g\left(w_R^2 d_R^3\right)} \tag{10.49}$$

$$w_R = \left[\phi_\xi(\xi) \cdot \phi_c(\overline{c}_R) \cdot c_R\right]\sqrt{\frac{Q}{v_{*c_r}}} \tag{10.50}$$

$$\left(F_r^2\right)_R = c_R^2 S_R \rightarrow min \quad \text{(minimum Froude number)}. \tag{10.51}$$

It may be noted that F_r varies with η and N, since ξ is constant and its minimum value is attained only because of η since it decreases monotonically.

10.6.2 Determination of Friction Coefficient

The total friction factor due to grains, dunes, and ripples was expressed by Yalin and Ferreira da Silva (1999) as

$$\frac{1}{c^2} = \frac{1}{C_s^2} + \sum_{j=du,r} \frac{1}{2}\left(\frac{\Delta}{\Lambda}\right)_j^2 \frac{\Lambda_j}{d} \quad (du = \text{dunes}, r = \text{ripples}) \tag{10.52}$$

$$C_s = \frac{1}{\kappa}\ln\left(\psi(X)\frac{Z}{2}\right), \quad \kappa \simeq 0.4, \tag{10.53a}$$

where $\psi(X) = \exp\left(\kappa w_s^{-1}\right)$ with w_s given by Yalin (1992) as

$$w_s = 8.5 + [2.5\ln(2X) - 3]\exp\left(-0.217[(\ln(2X)]^2\right). \tag{10.53b}$$

The friction factor C_s is due to granular roughness or skin. The bed form characteristics are evaluated as

$$\left(\frac{\Delta}{\Lambda}\right)_{du} = 0.013\psi_{du}(X)(\eta - 1)\exp\left[-\frac{(\eta - 1)}{13(1 - \exp(-0.0082))}\right] \tag{10.54}$$

$$\left(\frac{\Delta}{\Lambda}\right)_r = 0.035\psi_r(X)(\eta - 1)\exp\left[-0.1\,(\eta - 1)\right] \tag{10.55}$$

$$\text{and } \frac{\Lambda_{du}}{d} = 6\left[1 + 0.01\frac{(Z - 40)(Z - 400)}{Z}\exp\left(-0.055\sqrt{Z} - 0.04X\right)\right],$$
$$\frac{\Lambda_r}{d} = \frac{1{,}000}{Z}, \tag{10.56}$$

where $Z = d/D$. Functions $\psi_{du}(X)$ and $\psi_r(X)$ are determined using Equations (10.57) and (10.58) as

$$\psi_{du}(X) = 1 - \exp\left[-\left(\frac{X}{10}\right)^2\right] \tag{10.57}$$

$$\psi_r(X) = \exp\left[-\left(\frac{X - 2.5}{14}\right)^2\right], \text{if } X > 2.5; \psi_r(X) = 1, \text{if } X \leq 2.5. \tag{10.58}$$

It may, however, be more convenient to express the friction factor in terms of $\xi, \eta,$ and N, which can be done by X and Z in Equations (10.50)–(10.58):

$$X^2 = \eta\psi(\xi)\xi^2 \quad \text{and} \quad Z = \frac{N}{c\sqrt{\eta}}N, \tag{10.59}$$

where $\psi(\xi) = Y_{c_r}$ is the modified transport function. Thus, c is determined as

$$c = \phi_c(\xi, \eta, N). \tag{10.60}$$

The Froude number and friction factor have been related as

$$F_r = \frac{\alpha}{N}(c^2 \eta)^{\frac{3}{2}} \quad \text{with} \quad \alpha = \frac{\gamma_s}{\gamma} Y_{c_r} = \frac{v_{*c_r}^2}{gD}. \tag{10.61}$$

Through Equation (10.61), F_r becomes a function of $\xi, \eta,$ and N.

10.6.3 Computation of Geometry

Initially, one obtains some numerical values of the characteristic parameters on the right side of Equation (10.2). Then, the steps involved in obtaining the regime channel geometry (w_R, d_R, and S_R) are as follows:

1. Adopt $(w)_i, (d)_i,$ and $(S)_i$.
2. Compute $(\eta)_i$.

   - Compute $v_{*cr}, \dot{v}_*, X, Y, \xi,$ and Y_{cr} using Equation 10.35. Calculate Z as defined in Equation 10.56. Note that v_{*cr} is a unit of m/s.
   - Compute $(\eta)_i$ using Equation 10.35 as $\eta = \frac{Y}{Y_{cr}}$.

3. Compute $(C)_i$.

   In practice, calculation of C cannot rely on the functional expression as $(C)_i = \phi_c(\xi, \eta, (N)_i)$ where $(N)_i = \frac{Q}{w_i D v_{*c_r}}$. Therefore, C is calculated numerically following Section 10.6.2:

   - Compute Λ_{du}/d and Λ_r/d using Equation (10.56).
   - Compute $\left(\frac{A}{\Lambda}\right)_{du}$ and $\left(\frac{A}{\Lambda}\right)_r$ using Equations (10.54) and (10.55), respectively.
   - Compute C_s using Equation (10.53a).
   - Compute $(C)_i$ using Equation (10.52).

4. Compute flow discharge $(Q)_i$ using Equation (10.39) as

$$(Q)_i = (w)_i(d)_i(C)_i\sqrt{g(S)_i(d)_i}$$

5. Repeat steps 1 to 4 until $(Q)_i$ is equal to the measured Q. The minimum optimization programming is recommended.
6. Calculate N_R and $(F_r)_R$ using Equations (10.35) and (10.61), respectively, with the computed w_R and d_R.

Six examples for computation of regime characteristics are given for various sediment grain sizes (sand to gravel) and discharges (low to high). Follow the aforementioned computational procedure.

Example 10.1: Consider the following for the Bhagirathi River, India: $Q = 1669.7$ m³/s, $D = 0.18$ mm, and $\gamma_s/\gamma = 1.65$. Compute the regime channel geometry. The measure values are: $w_R = 218.1$ m, $d_R = 5.95$ m, $S_R = 0.000058$, $N_R = 10^{6.5}$, $(F_r)_R = 0.028$, $\eta_R = 18.6$, and $c_R = 22.1$.

Solution:

Step 1.

Adopt w, d, and S at the ith trial as $(w)_i = 212.8$ m, $(d)_i = 4.5$ m, and $(S)_i = 0.00002$.

Step 2.

Compute $v_{*cr}, v_*, X, Y, \xi, d_*,$ and Y_{cr} as follows:

- $v_{*cr} = \sqrt{\dfrac{\tau_{cr}}{\rho}}$ where $\tau_{cr} = 0.0125 + 0.019D$.

  $\tau_{cr} = 0.0125 + 0.019 \times 0.18 = 0.016\frac{\text{lb}}{\text{ft}^2}\frac{\text{ft}}{\text{s}^2}$ and $\rho = 62.4\frac{\text{lb}}{\text{ft}^3}$.

  Therefore, $v_{*cr} = \sqrt{\dfrac{0.016}{62.4}\dfrac{\text{ft}}{\text{s}}} \times 0.3\frac{\text{m}}{\text{ft}} = 0.005\frac{\text{m}}{\text{s}}$

- $v_* = \sqrt{gSd} = \sqrt{9.81 \times 0.00002 \times 4.5} = 0.03\frac{\text{m}}{\text{s}}$

- $X = \dfrac{v_* D}{v} = \dfrac{0.03 \times 0.18 \times 10^{-3}}{10^{-6}} = 5.34$

- $Y = \dfrac{\rho v_*^2}{\gamma_s D} = \dfrac{1000 \times 0.03^2}{16186.5 \times 0.18 \times 10^{-3}} = 0.3$

- $\xi = \left(\dfrac{X^2}{Y}\right)^{\frac{1}{3}} = \left(\dfrac{5.34^2}{0.3}\right)^{\frac{1}{3}} = 4.69$

- $d_* = \dfrac{d}{D} = \dfrac{4.5}{0.18 \times 10^{-3}} = 25,000$

- $Y_{cr} = 0.13\xi^{-0.392}e^{-0.015\xi^2} + 0.045\left[1 - e^{-0.068\xi}\right]$

  $= 0.13 \times 4.69^{-0.392}e^{-0.015 \times 4.69^2} + 0.045\left[1 - e^{-0.068 \times 4.69}\right] = 0.0645.$

Compute $(\eta)_i$ as

$$(\eta)_i = \frac{Y}{Y_{cr}} = \frac{0.3}{0.0645} = 4.69.$$

Step 3.

Compute $(C)_i$.

- $\dfrac{\Lambda_{dunes}}{d} = 6\left[1 + 0.01\dfrac{(d_* - 40)(d_* - 400)}{d_*}\exp\left(-0.055\sqrt{d_*} - 0.04X\right)\right]$

  $= 6\left[1 + 0.01\dfrac{(25,000 - 40)(25,000 - 400)}{25,000}\exp\left(-0.055\sqrt{25,000} - 0.04 \times 5.34\right)\right]$

  $= 6.19$

- $\dfrac{\Lambda_{ripples}}{d} = \dfrac{1,000}{d_*} = \dfrac{1,000}{25,000} = 0.04$

$- \left(\frac{\Delta}{\Lambda}\right)_{dunes} = 0.013\psi_{Xdune}(X)(\eta - 1)\exp\left[-\frac{(\eta-1)}{13(1-\exp{(-0.0082)})}\right],$

where $\psi_{Xdunes}(X) = 1 - \exp\left[-\left(\frac{X}{10}\right)^2\right] = 1 - \exp\left[-\left(\frac{5.34}{10}\right)^2\right] = 0.25.$

Therefore,

$$\left(\frac{\Delta}{\Lambda}\right)_{dunes} = 0.013 \times 0.25 \times (4.69 - 1)\exp\left[-\frac{(4.69 - 1)}{13(1 - \exp{(-0.0082)})}\right] = \sim 0$$

$- \left(\frac{\Delta}{\Lambda}\right)_{ripples} = 0.035\psi_{Xripples}(X)(\eta - 1)\exp\left[-0.1\,(\eta - 1)\right]$

where, since $X > 2.5$, $\psi_{Xripples}(X) = \exp\left[-((X - 2.5)/14)^2\right]$
$$= \exp\left[-((5.34 - 2.5)/14)^2\right] = 0.96.$$

Therefore,

$$\left(\frac{\Delta}{\Lambda}\right)_{ripples} = 0.035 \times 0.96 \times (4.69 - 1)\exp\left[-0.1\,(4.69 - 1)\right] = 0.086$$

$- C_s = \frac{1}{\kappa}\ln\left(\psi(X)\,\frac{d_*}{2}\right),$

where κ is 0.4 and $\psi(X) = \exp\left(\kappa w_s^{-1}\right)$,
where w_s is calculated following Equation (10.53b) as

$w_s = 8.5 + [2.5\ln{(2X)} - 3]\exp\left(-0.217[(\ln{(2X)}]^2\right)$
$= 8.5 + [2.5\ln{(2 \times 5.34)} - 3]\exp\left(-0.217[(\ln{(2 \times 5.34)}]^2\right) = 8.03.$

Therefore, $\psi(X) = \exp\left(0.4 \times 8.03^{-1}\right) = 1.05.$

Finally, $C_s = \frac{1}{\kappa}\ln\left(\psi(X)\,\frac{d_*}{2}\right) = \frac{1}{0.4} \times \ln\left(1.05 \times \frac{25,000}{2}\right) = 23.7.$

– Compute $(C)_i$ as

$$\frac{1}{C^2} = \frac{1}{C_s^2} + \sum_{j=d_u, r_i} a_j\left(\frac{\Delta}{\Lambda}\right)_j^{b_j}\frac{\Lambda_j}{d},$$

where $a_j = \frac{1}{2}$ and $b_j = 2.$
Therefore,

$$\frac{1}{C^2} = \frac{1}{23.7^2} + \frac{1}{2}\left(0.0^2 \times 6.19 + 0.086^2 \times 0.04\right).$$

Finally, $(C)_i = 22.8.$

Step 4.

Compute $(Q)_i$.

$$(Q)_i = (w)_i(d)_i(C)_i\sqrt{g(S)_i(d)_i}$$

$$= 212.8 \times 4.5 \times 22.8\sqrt{9.81 \times 0.00002 \times 4.5} = 648.7\,\frac{\text{m}^3}{\text{s}}.$$

Step 5.

Compare the computed $(Q)_i$ and the measured Q.

Since $(Q)_i$ of $648.7\,\frac{\text{m}^3}{\text{s}}$ is different from the measured Q of $1669.7\,\frac{\text{m}^3}{\text{s}}$, repeat steps 1 to 4 by inputting 1% increase in w, d, and/or S for the following iterations.

After going through several iterations, when $(w)_{i+n} = 215$ m, $(d)_{i+n} = 5.0$ m, and $(S)_{i+n} = 0.00006$, the computed $(Q)_{i+n}$ is shown with the computational procedure as follows:

Step 1.

Adopt w, d, and S at the $i + n$th trial as $(w)_{i+n} = 215$ m, $(d)_{i+n} = 5.0$ m, and $(S)_{i+n} = 0.00006$.

Step 2.

Compute $v_{*cr}, v_*, X, Y, \xi, d_*,$ and Y_{cr} as follows:

- $v_{*cr} = \sqrt{\frac{\tau_{cr}}{\rho}}$ where $\tau_{cr} = 0.0125 + 0.019D$.

    $\tau_{cr} = 0.0125 + 0.019 \times 0.18 = 0.016\,\frac{\text{lb}}{\text{ft}^2}\,\frac{\text{ft}}{\text{s}^2}$ and $\rho = 62.4\,\frac{\text{lb}}{\text{ft}^3}$.

    Therefore, $v_{*cr} = \sqrt{\frac{0.016}{62.4}\frac{\text{ft}}{\text{s}}} \times 0.3\,\frac{\text{m}}{\text{ft}} = 0.005\,\frac{\text{m}}{\text{s}}$

- $v_* = \sqrt{gSd} = \sqrt{9.81 \times 0.00006 \times 5} = 0.055\,\frac{\text{m}}{\text{s}}$

- $X = \frac{v_*D}{v} = \frac{0.055 \times 0.18 \times 10^{-3}}{10^{-6}} = 9.76$

- $Y = \frac{\rho v_*^2}{\gamma_s D} = \frac{1000 \times 0.055^2}{16186.5 \times 0.18 \times 10^{-3}} = 1.01$

- $\xi = \left(\frac{X^2}{Y}\right)^{\frac{1}{3}} = \left(\frac{9.76^2}{1.01}\right)^{\frac{1}{3}} = 4.55$

- $d_* = \frac{d}{D} = \frac{5}{0.18 \times 10^{-3}} = 27777.8$

- $Y_{cr} = 0.13\xi^{-0.392}e^{-0.015\xi^2} + 0.045\left[1 - e^{-0.068\xi}\right]$

    $= 0.13 \times 4.55^{-0.392}e^{-0.015 \times 4.55^2} + 0.045\left[1 - e^{-0.068 \times 4.55}\right] = 0.0646.$

Compute $(\eta)_{i+n}$ as

$$(\eta)_{i+n} = \frac{Y}{Y_{cr}} = \frac{1.01}{0.0646} = 15.65.$$

Step 3.

Compute $(C)_{i+n}$.

$$- \frac{\Lambda_{dunes}}{d} = 6\left[1 + 0.01 \frac{(d_* - 40)(d_* - 400)}{d_*} \exp\left(-0.055\sqrt{d_*} - 0.04X\right)\right]$$

$$= 6\left[1 + 0.01 \frac{(27777.8 - 40)(27777.8 - 400)}{27777.8} \exp\left(-0.055\sqrt{27777.8} - 0.04 \times 9.76\right)\right]$$

$$= 6.12$$

$$- \frac{\Lambda_{ripples}}{d} = \frac{1,000}{d_*} = \frac{1,000}{27777.8} = 0.036$$

$$- \left(\frac{\Delta}{\Lambda}\right)_{dunes} = 0.013\psi_{Xdune}(X)(\eta - 1)\exp\left[-\frac{(\eta - 1)}{13(1 - \exp(-0.0082))}\right],$$

where $\psi_{Xdunes}(X) = 1 - \exp\left[-\left(\frac{X}{10}\right)^2\right] = 1 - \exp\left[-\left(\frac{9.76}{10}\right)^2\right] = 0.61.$
Therefore,

$$\left(\frac{\Delta}{\Lambda}\right)_{dunes} = 0.013 \times 0.49 \times (15.65 - 1)\exp\left[-\frac{(15.65 - 1)}{13(1 - \exp(-0.0082))}\right] = \sim 0$$

$$- \left(\frac{\Delta}{\Lambda}\right)_{ripples} = 0.035\psi_{Xripples}(X)(\eta - 1)\exp\left[-0.1\,(\eta - 1)\right]$$

where, since $X > 2.5$, $\psi_{Xripples}(X) = \exp\left[-((X - 2.5)/14)^2\right]$
$$= \exp\left[-((9.76 - 2.5)/14)^2\right] = 0.76.$$

Therefore,

$$\left(\frac{\Delta}{\Lambda}\right)_{ripples} = 0.035 \times 0.76 \times (15.65 - 1)\exp\left[-0.1\,(15.65 - 1)\right] = 0.09$$

$$- C_s = \frac{1}{\kappa}\ln\left(\psi(X)\frac{d_*}{2}\right)$$

where κ is 0.4 and $\psi(X) = \exp\left(\kappa w_s^{-1}\right)$,
where w_s is calculated following Equation (10.53b) as

$$w_s = 8.5 + [2.5\ln(2X) - 3]\exp\left(-0.217[(\ln(2X)]^2\right)$$
$$= 8.5 + [2.5\ln(2 \times 9.76) - 3]\exp\left(-0.217[(\ln(2 \times 9.76)]^2\right) = 8.49.$$

Therefore, $\psi(X) = \exp\left(0.4 \times 8.49^{-1}\right) = 1.048$

Finally, $C_s = \frac{1}{\kappa}\ln\left(\psi(X)\frac{d_*}{2}\right) = \frac{1}{0.4} \times \ln\left(1.048 \times \frac{27777.8}{2}\right) = 23.96$

- Compute $(C)_{i+n}$ as

$$\frac{1}{C^2} = \frac{1}{C_s^2} + \sum_{j=d_u, r_i} a_j\left(\frac{\Delta}{\Lambda}\right)_j^{b_j} \frac{\Lambda_j}{d},$$

where $a_j = \frac{1}{2}$ and $b_j = 2$.
Therefore,

$$\frac{1}{C^2} = \frac{1}{23.96^2} + \frac{1}{2}\left(0.0^2 \times 6.12 + 0.09^2 \times 0.036\right).$$

Finally, $(C)_{i+n} = 23.01$

Step 4.
Compute $(Q)_{i+n}$.

$$(Q)_{i+n} = (w)_{i+n}(d)_{i+n}(C)_{i+n}\sqrt{g(S)_{i+n}(d)_{i+n}}$$

$$= 215 \times 5 \times 23.01\sqrt{9.81 \times 0.00006 \times 5} = 1341.9\frac{m^3}{s}.$$

The computed $(Q)_{i+n}$ is getting close to Q but still far from the observed Q of $1669.7\frac{m^3}{s}$. Therefore, we showed an additional iteration step when $(w)_{i+k} = 217.15$ m, $(d)_{i+k} = 5.25$ m, and $(S)_{i+k} = 0.00007$ as follows:

Step 1.
Adopt w, d, and S at the $i + k$th trial as $(w)_{i+k} = 217.15$ m, $(d)_{i+k} = 5.25$ m, and $(S)_{i+k} = 0.00007$.

Step 2.
Compute $v_{*cr}, v_*, X, Y, \xi, d_*,$ and Y_{cr} as follows:

$-$ $v_{*cr} = \sqrt{\frac{\tau_{cr}}{\rho}},$ where $\tau_{cr} = 0.0125 + 0.019D.$

$\tau_{cr} = 0.0125 + 0.019 \times 0.18 = 0.016\frac{lb}{ft^2}\frac{ft}{s^2}$ and $\rho = 62.4\frac{lb}{ft^3}.$

Therefore, $v_{*cr} = \sqrt{\frac{0.016\frac{ft}{s}}{62.4}} \times 0.3\frac{m}{ft} = 0.005\frac{m}{s}$

$-$ $v_* = \sqrt{gSd} = \sqrt{9.81 \times 0.00007 \times 5.25} = 0.06\frac{m}{s}$

$-$ $X = \frac{v_* D}{v} = \frac{0.06 \times 0.18 \times 10^{-3}}{10^{-6}} = 10.81$

$-$ $Y = \frac{\rho v_*^2}{\gamma_s D} = \frac{1,000 \times 0.06^2}{16186.5 \times 0.18 \times 10^{-3}} = 1.24$

$-$ $\xi = \left(\frac{X^2}{Y}\right)^{\frac{1}{3}} = \left(\frac{10.81^2}{1.24}\right)^{\frac{1}{3}} = 4.55$

$-$ $d_* = \frac{d}{D} = \frac{5.25}{0.18 \times 10^{-3}} = 29166.7$

$-$ $Y_{cr} = 0.13\xi^{-0.392}e^{-0.015\xi^2} + 0.045\left[1 - e^{-0.068\xi}\right]$

$\qquad = 0.13 \times 4.55^{-0.392}e^{-0.015 \times 4.55^2} + 0.045\left[1 - e^{-0.068 \times 4.55}\right] = 0.0646.$

Compute $(\eta)_{i+k}$ as

$$(\eta)_{i+k} = \frac{Y}{Y_{cr}} = \frac{1.24}{0.0646} = 19.17.$$

Step 3.
Compute $(C)_{i+k}$.

$$- \frac{\Lambda_{dunes}}{d} = 6\left[1 + 0.01\frac{(d_* - 40)(d_* - 400)}{d_*}\exp\left(-0.055\sqrt{d_*} - 0.04X\right)\right]$$
$$= 6\left[1 + 0.01\frac{(29166.7 - 40)(29166.7 - 400)}{29166.7}\exp\left(-0.055\sqrt{29166.7} - 0.04 \times 10.81\right)\right]$$
$$= 6.093$$

$$- \frac{\Lambda_{ripples}}{d} = \frac{1,000}{d_*} = \frac{1,000}{29166.7} = 0.034$$

$$- \left(\frac{\Delta}{\Lambda}\right)_{dunes} = 0.013\psi_{Xdune}(X)(\eta - 1)\exp\left[-\frac{(\eta - 1)}{13(1 - \exp(-0.0082))}\right],$$

where $\psi_{Xdunes}(X) = 1 - \exp\left[-\left(\frac{X}{10}\right)^2\right] = 1 - \exp\left[-\left(\frac{10.81}{10}\right)^2\right] = 0.69$.
Therefore,

$$\left(\frac{\Delta}{\Lambda}\right)_{dunes} = 0.013 \times 0.69 \times (19.17 - 1)\exp\left[-\frac{(19.17 - 1)}{13(1 - \exp(-0.0082))}\right] = \sim 0$$

$$- \left(\frac{\Delta}{\Lambda}\right)_{ripples} = 0.035\psi_{Xripples}(X)(\eta - 1)\exp\left[-0.1(\eta - 1)\right],$$

where, since $X > 2.5$, $\psi_{Xripples}(X) = \exp\left[-((X - 2.5)/14)^2\right]$
$$= \exp\left[-((10.81 - 2.5)/14)^2\right] = 0.70.$$

Therefore,

$$\left(\frac{\Delta}{\Lambda}\right)_{ripples} = 0.035 \times 0.7 \times (19.17 - 1)\exp\left[-0.1(19.17 - 1)\right] = 0.073$$

$$- C_s = \frac{1}{\kappa}\ln\left(\psi(X)\frac{d_*}{2}\right),$$

where κ is 0.4 and $\psi(X) = \exp\left(\kappa w_s^{-1}\right)$,
where w_s is calculated following Equation (10.53b) as

$$w_s = 8.5 + [2.5\ln(2X) - 3]\exp\left(-0.217[(\ln(2X)]^2\right)$$
$$= 8.5 + [2.5\ln(2 \times 10.81) - 3]\exp\left(-0.217[(\ln(2 \times 10.81)]^2\right) = 8.52.$$

Therefore, $\psi(X) = \exp\left(0.4 \times 8.52^{-1}\right) = 1.05$.
Finally, $C_s = \frac{1}{\kappa}\ln\left(\psi(X)\frac{d_*}{2}\right) = \frac{1}{0.4} \times \ln\left(1.05 \times \frac{29166.7}{2}\right) = 24.08$

– Compute $(C)_{i+k}$ as

$$\frac{1}{C^2} = \frac{1}{C_s^2} + \sum_{j=d_u, r_i} a_j \left(\frac{\Delta}{\Lambda}\right)_j^{b_j} \frac{\Lambda_j}{d},$$

where $a_j = \frac{1}{2}$ and $b_j = 2$.
Therefore,

$$\frac{1}{C^2} = \frac{1}{24.08^2} + \frac{1}{2}\left(0.0^2 \times 6.093 + 0.073^2 \times 0.034\right).$$

Finally, $(C)_{i+k} = 23.47$.

Step 4.
Compute $(Q)_{i+k}$.

$$(Q)_{i+k} = (w)_{i+k}(d)_{i+k}(C)_{i+k}\sqrt{g(S)_{i+k}(d)_{i+k}}$$

$$= 217.15 \times 5.25 \times 23.47\sqrt{9.81 \times 0.00007 \times 5.25} = 1606.6\frac{m^3}{s}.$$

To obtain $(Q)_{i+R} \approx Q$, additional iterations were undertaken. When $(Q)_{i+R} \approx Q$ of $1669.7\frac{m^3}{s}$, the regime characteristics were estimated as
$w_R = 219.5$ m, $d_R = 5.31$ m, $S_R = 0.00007$, $N_R = 10^{6.9}$, $(F_r)_R = 0.013$, $\eta_R = 18.9$
and $c_R = 23.5$,
where $N_R = \frac{Q}{w_R D v_* c_r} = \frac{1669.7}{219.5 \times 0.18^{-3} \times 0.005} = 8452037.5 = 10^{6.9}$ and $(F_r)_R = \frac{\frac{\gamma_s}{\gamma} Y_{cr}}{N_R}\left(C_R^2 \eta_R\right)^{\frac{3}{2}} =$
$\frac{1.65 \times 0.0645}{8452037.5}\left(23.5^2 \times 18.9\right)^{\frac{3}{2}} = 0.013$.

Example 10.2: Consider the following for a river, India: $Q = 15.6$ m^3/s, $D = 0.33$ mm, and $\gamma_s/\gamma = 1.65$. Compute the regime channel geometry. The measure values are: $w_R = 18.1$ m, $d_R = 1.24$ m, $S_R = 0.0002$, $N_R = 10^{5.3}$, $(F_r)_R = 0.039$, $\eta_R = 12.0$, and $c_R = 14.1$.

Solution:
Step 1.
Adopt w, d, and S at the ith trial as $(w)_i = 16.5$ m, $(d)_i = 0.8$ m, and $(S)_i = 0.00015$.

Step 2.
Compute $v_{*cr}, v_*, X, Y, \xi, d_*$, and Y_{cr} as follows:

– $v_{*cr} = \sqrt{\frac{\tau_{cr}}{\rho}}$, where $\tau_{cr} = 0.0125 + 0.019D$.
$\tau_{cr} = 0.0125 + 0.019 \times 0.33 = 0.019 \frac{lb}{ft^2}\frac{ft}{s^2}$ and $\rho = 62.4\frac{lb}{ft^3}$.

Therefore, $v_{*cr} = \sqrt{\frac{0.019}{62.4}\frac{ft}{s}} \times 0.3\frac{m}{ft} = 0.0052\frac{m}{s}$

- $v_* = \sqrt{gSd} = \sqrt{9.81 \times 0.00015 \times 0.8} = 0.034\frac{m}{s}$
- $X = \frac{v_* D}{v} = \frac{0.034 \times 0.33 \times 10^{-3}}{10^{-6}} = 11.33$
- $Y = \frac{\rho v_*^2}{\gamma_s D} = \frac{1,000 \times 0.034^2}{16186.5 \times 0.33 \times 10^{-3}} = 0.22$
- $\xi = \left(\frac{X^2}{Y}\right)^{\frac{1}{3}} = \left(\frac{11.33^2}{0.22}\right)^{\frac{1}{3}} = 8.35$
- $d_* = \frac{d}{D} = \frac{0.8}{0.33 \times 10^{-3}} = 2424.2$
- $Y_{cr} = 0.13 \xi^{-0.392} e^{-0.015\xi^2} + 0.045 \left[1 - e^{-0.068\xi}\right]$

$= 0.13 \times 8.35^{-0.392} e^{-0.015 \times 8.35^2} + 0.045 \left[1 - e^{-0.068 \times 8.35}\right] = 0.039.$

Compute $(\eta)_i$ as

$$(\eta)_i = \frac{Y}{Y_{cr}} = \frac{0.22}{0.039} = 5.6.$$

Step 3.
Compute $(C)_i$.

- $\frac{\Lambda_{dunes}}{d} = 6\left[1 + 0.01 \frac{(d_* - 40)(d_* - 400)}{d_*} \exp\left(-0.055\sqrt{d_*} - 0.04X\right)\right]$

$= 6\left[1 + 0.01 \frac{(2424.2 - 40)(2424.2 - 400)}{2424.2} \exp\left(-0.055\sqrt{2424.2} - 0.04 \times 11.33\right)\right]$

$= 11.06$

- $\frac{\Lambda_{ripples}}{d} = \frac{1,000}{d_*} = \frac{1,000}{2424.2} = 0.41$
- $\left(\frac{\Delta}{\Lambda}\right)_{dunes} = 0.013 \psi_{Xdune}(X)(\eta - 1) \exp\left[-\frac{(\eta - 1)}{13(1 - \exp(-0.0082))}\right],$

where $\psi_{Xdunes}(X) = 1 - \exp\left[-\left(\frac{X}{10}\right)^2\right] = 1 - \exp\left[-\left(\frac{11.33}{10}\right)^2\right] = 0.72.$
Therefore,

$$\left(\frac{\Delta}{\Lambda}\right)_{dunes} = 0.013 \times 0.72 \times (5.6 - 1) \exp\left[-\frac{(5.6 - 1)}{13(1 - \exp(-0.0082))}\right] = \sim 0$$

- $\left(\frac{\Delta}{\Lambda}\right)_{ripples} = 0.035 \psi_{Xripples}(X)(\eta - 1) \exp\left[-0.1(\eta - 1)\right],$

where, since $X > 2.5$, $\psi_{Xripples}(X) = \exp\left[-((X - 2.5)/14)^2\right]$
$= \exp\left[-((11.33 - 2.5)/14)^2\right] = 0.67.$
Therefore,

$$\left(\frac{\Delta}{\Lambda}\right)_{ripples} = 0.035 \times 0.67 \times (5.6 - 1) \exp\left[-0.1(5.6 - 1)\right] = 0.068$$

$$- C_s = \tfrac{1}{\kappa} \ln \left(\psi(X) \, \tfrac{d_*}{2} \right),$$

where κ is 0.4 and $\psi(X) = \exp \left(\kappa w_s^{-1} \right)$,
where w_s is calculated following Equation (10.53b) as

$$
\begin{aligned}
w_s &= 8.5 + [2.5 \ln (2X) - 3] \exp \left(-0.217[(\ln (2X)]^2 \right) \\
&= 8.5 + [2.5 \ln (2 \times 11.33) - 3] \exp \left(-0.217[(\ln (2 \times 11.33)]^2 \right) = 8.54.
\end{aligned}
$$

Therefore, $\psi(X) = \exp \left(0.4 \times 8.54^{-1} \right) = 1.048$.
Finally, $C_s = \tfrac{1}{\kappa} \ln \left(\psi(X) \, \tfrac{d_*}{2} \right) = \tfrac{1}{0.4} \times \ln \left(1.048 \times \tfrac{2424.2}{2} \right) = 17.87$

- Compute $(C)_i$ as

$$\frac{1}{C^2} = \frac{1}{C_s^2} + \sum_{j=d_u, r_i} a_j \left(\frac{\Delta}{\Lambda} \right)_j^{b_j} \frac{\Lambda_j}{d},$$

where $a_j = \tfrac{1}{2}$ and $b_j = 2$.
Therefore,

$$\frac{1}{C^2} = \frac{1}{17.87^2} + \frac{1}{2} \left(0.0^2 \times 11.06 + 0.068^2 \times 0.41 \right).$$

Finally, $(C)_i = 15.62$.

Step 4.
Compute $(Q)_i$.

$$
\begin{aligned}
(Q)_i &= (w)_i (d)_i (C)_i \sqrt{g(S)_i (d)_i} \\
&= 16.5 \times 0.8 \times 15.62 \sqrt{9.81 \times 0.00015 \times 0.8} = 7.1 \, \frac{\text{m}^3}{\text{s}}.
\end{aligned}
$$

Step 5.
Compare the computed $(Q)_i$ and the measured Q.
   Since $(Q)_i$ of $7.1 \frac{\text{m}^3}{\text{s}}$ is different from the measured Q of $15.6 \frac{\text{m}^3}{\text{s}}$, repeat steps 1 to 4 by inputting 1% increase in w, d, and/or S for the following iterations.
   After going through several iterations, when $(w)_{i+n} = 17.65$ m, $(d)_{i+n} = 0.95$ m, and $(S)_{i+n} = 0.00025$, the computed $(Q)_{i+n}$ is shown with the computational procedure as follows:

Step 1.
Adopt w, d, and S at the $i + n$th trial as $(w)_{i+n} = 17.65$ m, $(d)_{i+n} = 0.95$ m, and $(S)_{i+n} = 0.00025$.

Step 2.
Compute $v_{*cr}, v_*, X, Y, \xi, d_*$, and Y_{cr} as follows:

$$- v_{*cr} = \sqrt{\frac{\tau_{cr}}{\rho}} \quad \text{where } \tau_{cr} = 0.0125 + 0.019D.$$

$\tau_{cr} = 0.0125 + 0.019 \times 0.33 = 0.019 \frac{\text{lb}}{\text{ft}^2} \frac{\text{ft}}{\text{s}^2}$ and $\rho = 62.4 \frac{\text{lb}}{\text{ft}^3}$.

Therefore, $v_{*cr} = \sqrt{\frac{0.019}{62.4} \frac{\text{ft}}{\text{s}}} \times 0.3 \frac{\text{m}}{\text{ft}} = 0.0052 \frac{\text{m}}{\text{s}}$

$- v_* = \sqrt{gSd} = \sqrt{9.81 \times 0.00025 \times 0.95} = 0.048 \frac{\text{m}}{\text{s}}$

$- X = \frac{v_* D}{v} = \frac{0.048 \times 0.33 \times 10^{-3}}{10^{-6}} = 15.92$

$- Y = \frac{\rho v_*^2}{\gamma_s D} = \frac{1{,}000 \times 0.048^2}{16186.5 \times 0.33 \times 10^{-3}} = 0.44$

$- \xi = \left(\frac{X^2}{Y}\right)^{\frac{1}{3}} = \left(\frac{15.92^2}{0.44}\right)^{\frac{1}{3}} = 8.35$

$- d_* = \frac{d}{D} = \frac{0.95}{0.33 \times 10^{-3}} = 2878.8$

$- Y_{cr} = 0.13\xi^{-0.392} e^{-0.015\xi^2} + 0.045\left[1 - e^{-0.068\xi}\right]$

$\qquad = 0.13 \times 8.35^{-0.392} e^{-0.015 \times 8.35^2} + 0.045\left[1 - e^{-0.068 \times 8.35}\right] = 0.04.$

Compute $(\eta)_{i+n}$ as

$$(\eta)_{i+n} = \frac{Y}{Y_{cr}} = \frac{0.44}{0.04} = 11.07.$$

Step 3.
Compute $(C)_{i+n}$.

$- \frac{\Lambda_{dunes}}{d} = 6\left[1 + 0.01 \frac{(d_* - 40)(d_* - 400)}{d_*} \exp\left(-0.055\sqrt{d_*} - 0.04X\right)\right]$

$\qquad = 6\left[1 + 0.01 \frac{(2878.8 - 40)(2878.8 - 400)}{2878.8} \exp\left(-0.055\sqrt{2878.8} - 0.04 \times 15.92\right)\right]$

$\qquad = 10.06$

$- \frac{\Lambda_{ripples}}{d} = \frac{1{,}000}{d_*} = \frac{1{,}000}{2878.8} = 0.035$

$- \left(\frac{\Delta}{\Lambda}\right)_{dunes} = 0.013\psi_{Xdune}(X)(\eta - 1)\exp\left[-\frac{(\eta - 1)}{13(1 - \exp(-0.0082))}\right],$

where $\psi_{Xdunes}(X) = 1 - \exp\left[-\left(\frac{X}{10}\right)^2\right] = 1 - \exp\left[-\left(\frac{15.92}{10}\right)^2\right] = 0.92.$
Therefore,

$$\left(\frac{\Delta}{\Lambda}\right)_{dunes} = 0.013 \times 0.92 \times (11.07 - 1)\exp\left[-\frac{(11.07 - 1)}{13(1 - \exp(-0.0082))}\right] = \sim 0$$

$- \left(\frac{\Delta}{\Lambda}\right)_{ripples} = 0.035\psi_{Xripples}(X)(\eta - 1)\exp\left[-0.1(\eta - 1)\right],$

where, since $X > 2.5$, $\psi_{Xripples}(X) = \exp\left[-((X - 2.5)/14)^2\right]$

$\qquad\qquad\qquad\qquad = \exp\left[-((15.92 - 2.5)/14)^2\right] = 0.40.$

Therefore,

$$\left(\frac{\Delta}{\Lambda}\right)_{ripples} = 0.035 \times 0.40 \times (11.07 - 1) \exp\left[-0.1\,(11.07 - 1)\right] = 0.05$$

$- C_s = \frac{1}{\kappa} \ln\left(\psi(X)\,\frac{d_*}{2}\right),$

where κ is 0.4 and $\psi(X) = \exp\left(\kappa w_s^{-1}\right)$,
where w_s is calculated following Equation (10.53b) as

$$w_s = 8.5 + [2.5 \ln(2X) - 3] \exp\left(-0.217[(\ln(2X)]^2\right)$$
$$= 8.5 + [2.5 \ln(2 \times 15.92) - 3] \exp\left(-0.217[(\ln(2 \times 15.92)]^2\right) = 8.59.$$

Therefore, $\psi(X) = \exp\left(0.4 \times 8.5^{-1}\right) = 1.048.$
Finally, $C_s = \frac{1}{\kappa} \ln\left(\psi(X)\,\frac{d_*}{2}\right) = \frac{1}{0.4} \times \ln\left(1.048 \times \frac{2878.8}{2}\right) = 18.3$

− Compute $(C)_{i+n}$ as

$$\frac{1}{C^2} = \frac{1}{C_s^2} + \sum_{j=d_u,\,r_i} a_j \left(\frac{\Delta}{\Lambda}\right)_j^{b_j} \frac{\Lambda_j}{d},$$

where $a_j = \frac{1}{2}$ and $b_j = 2$.
Therefore,

$$\frac{1}{C^2} = \frac{1}{18.3^2} + \frac{1}{2}\left(0.0^2 \times 10.06 + 0.05^2 \times 0.035\right).$$

Finally, $(C)_{i+n} = 17.04.$

Step 4.
Compute $(Q)_{i+n}$.

$$(Q)_{i+n} = (w)_{i+n}(d)_{i+n}(C)_{i+n}\sqrt{g(S)_{i+n}(d)_{i+n}}$$
$$= 17.65 \times 0.95 \times 17.04\sqrt{9.81 \times 0.00025 \times 0.95} = 13.8\,\frac{\mathrm{m}^3}{\mathrm{s}}.$$

To obtain $(Q)_{i+R} \approx Q$, additional iterations were undertaken. When $(Q)_{i+R} \approx Q$ of $15.6\,\frac{\mathrm{m}^3}{\mathrm{s}}$, the regime characteristics were estimated as
$w_R = 17.85$ m, $d_R = 1.0$ m, $S_R = 0.00027$, $N_R = 10^{5.7}$, $(F_r)_R = 0.028$, $\eta_R = 12.2$ and $c_R = 17.3$,
where $N_R = \frac{Q}{w_R D v_* c_r} = \frac{15.6}{17.85 \times 0.33^{-3} \times 0.0052} = 509294.6 = 10^{5.7}$ and $(F_r)_R = \frac{\frac{\gamma_s}{\gamma}Y_{cr}}{N_R}\left(C_R^2 \eta_R\right)^{\frac{3}{2}} = \frac{1.65 \times 0.039}{509294.6}\left(17.3^2 \times 12.2\right)^{\frac{3}{2}} = 0.028.$

Example 10.3: Consider the following for the Beaver River: $Q = 141.4 \text{ m}^3/\text{s}$, $D = 0.55$ mm, and $\gamma_s/\gamma = 1.65$. Compute the regime channel geometry. The measure values are: $w_R = 54.9$ m, $d_R = 2.74$ m, $S_R = 0.00024$, $N_R = 10^{5.5}$, $(F_r)_R = 0.0329$, $\eta_R = 25.5$, and $c_R = 11.7$.

Solution:

Step 1.

Adopt w, d, and S at the ith trial as $(w)_i = 50$ m, $(d)_i = 2.0$ m, and $(S)_i = 0.00015$.

Step 2.

Compute $v_{*cr}, v_*, X, Y, \xi, d_*,$ and Y_{cr} as follows:

- $v_{*cr} = \sqrt{\dfrac{\tau_{cr}}{\rho}}$ where $\tau_{cr} = 0.0125 + 0.019D$.

  $\tau_{cr} = 0.0125 + 0.019 \times 0.55 = 0.023 \frac{\text{lb}}{\text{ft}^2} \frac{\text{ft}}{\text{s}^2}$ and $\rho = 62.4 \frac{\text{lb}}{\text{ft}^3}$.

  Therefore, $v_{*cr} = \sqrt{\frac{0.023}{62.4} \frac{\text{ft}}{\text{s}}} \times 0.3 \frac{\text{m}}{\text{ft}} = 0.0057 \frac{\text{m}}{\text{s}}$

- $v_* = \sqrt{gSd} = \sqrt{9.81 \times 0.00015 \times 2.0} = 0.054 \frac{\text{m}}{\text{s}}$

- $X = \frac{v_* D}{v} = \frac{0.054 \times 0.55 \times 10^{-3}}{10^{-6}} = 29.84$

- $Y = \frac{\rho v_*^2}{\gamma_s D} = \frac{1{,}000 \times 0.054^2}{16186.5 \times 0.55 \times 10^{-3}} = 0.33$

- $\xi = \left(\frac{X^2}{Y}\right)^{\frac{1}{3}} = \left(\frac{29.84^2}{0.33}\right)^{\frac{1}{3}} = 13.9$

- $d_* = \frac{d}{D} = \frac{2}{0.55 \times 10^{-3}} = 3636.4$

- $Y_{cr} = 0.13 \xi^{-0.392} e^{-0.015 \xi^2} + 0.045 \left[1 - e^{-0.068 \xi}\right]$

  $= 0.13 \times 13.9^{-0.392} e^{-0.015 \times 13.9^2} + 0.045 \left[1 - e^{-0.068 \times 13.9}\right] = 0.03.$

Compute $(\eta)_i$ as

$$(\eta)_i = \frac{Y}{Y_{cr}} = \frac{0.33}{0.03} = 10.99.$$

Step 3.

Compute $(C)_i$.

- $\dfrac{\Lambda_{dunes}}{d} = 6\left[1 + 0.01 \dfrac{(d_* - 40)(d_* - 400)}{d_*} \exp\left(-0.055\sqrt{d_*} - 0.04X\right)\right]$

  $= 6\left[1 + 0.01 \dfrac{(3636.4 - 40)(3636.4 - 400)}{3636.4} \exp\left(-0.055\sqrt{3636.4} - 0.04 \times 29.84\right)\right]$

  $= 8.11$

$$- \frac{\Lambda_{ripples}}{d} = \frac{1{,}000}{d_*} = \frac{1{,}000}{3636.4} = 0.275$$

$$- \left(\frac{\Delta}{\Lambda}\right)_{dunes} = 0.013\psi_{Xdune}(X)(\eta - 1)\exp\left[-\frac{(\eta-1)}{13(1-\exp(-0.0082))}\right],$$

where $\psi_{Xdunes}(X) = 1 - \exp\left[-\left(\frac{X}{10}\right)^2\right] = 1 - \exp\left[-\left(\frac{29.84}{10}\right)^2\right] = 0.99$.

Therefore,

$$\left(\frac{\Delta}{\Lambda}\right)_{dunes} = 0.013 \times 0.99 \times (10.99 - 1)\exp\left[-\frac{(10.99 - 1)}{13(1 - \exp(-0.0082))}\right] = \sim 0$$

$$- \left(\frac{\Delta}{\Lambda}\right)_{ripples} = 0.035\psi_{Xripples}(X)(\eta - 1)\exp\left[-0.1\,(\eta - 1)\right],$$

where, since $X > 2.5$, $\psi_{Xripples}(X) = \exp\left[-((X - 2.5)/14)^2\right]$

$$= \exp\left[-((29.84 - 2.5)/14)^2\right] = 0.022.$$

Therefore,

$$\left(\frac{\Delta}{\Lambda}\right)_{ripples} = 0.035 \times 0.022 \times (10.99 - 1)\exp\left[-0.1\,(10.99 - 1)\right] = 0.003$$

$$- C_s = \frac{1}{\kappa}\ln\left(\psi(X)\frac{d_*}{2}\right),$$

where κ is 0.4 and $\psi(X) = \exp\left(\kappa w_s^{-1}\right)$,
where w_s is calculated following Equation (10.53b) as

$$w_s = 8.5 + [2.5\ln(2X) - 3]\exp\left(-0.217[(\ln(2X)]^2\right)$$
$$= 8.5 + [2.5\ln(2 \times 29.84) - 3]\exp\left(-0.217[(\ln(2 \times 29.84)]^2\right) = 8.57.$$

Therefore, $\psi(X) = \exp\left(0.4 \times 8.57^{-1}\right) = 1.047$.

Finally, $C_s = \frac{1}{\kappa}\ln\left(\psi(X)\frac{d_*}{2}\right) = \frac{1}{0.4} \times \ln\left(1.047 \times \frac{3636.4}{2}\right) = 18.88$

- Compute $(C)_i$ as

$$\frac{1}{C^2} = \frac{1}{C_s^2} + \sum_{j=d_u, r_i} a_j \left(\frac{\Delta}{\Lambda}\right)_j^{b_j} \frac{\Lambda_j}{d},$$

where $a_j = \frac{1}{2}$ and $b_j = 2$.

Therefore,

$$\frac{1}{C^2} = \frac{1}{18.88^2} + \frac{1}{2}\left(0.0^2 \times 8.11 + 0.003^2 \times 0.275\right).$$

Finally, $(C)_i = 18.88$.

Step 4.

Compute $(Q)_i$.

$$(Q)_i = (w)_i (d)_i (C)_i \sqrt{g(S)_i (d)_i}$$

$$= 50 \times 2 \times 18.88 \sqrt{9.81 \times 0.00015 \times 2} = 102.4 \frac{m^3}{s}.$$

Step 5.

Compare the computed $(Q)_i$ and the measured Q.

Since $(Q)_i$ of $102.4 \frac{m^3}{s}$ is different from the measured Q of $141.4 \frac{m^3}{s}$, repeat steps 1 to 4 by inputting 1% increase in w, d, and/or S for the following iterations.

After going through several iterations, when $(w)_{i+n} = 53.2$ m, $(d)_{i+n} = 2.2$ m, and $(S)_{i+n} = 0.00017$, the computed $(Q)_{i+n}$ is shown with the computational procedure as follows:

Step 1.

Adopt w, d, and S at the $i + n$th trial as $(w)_{i+n} = 53.2$ m, $(d)_{i+n} = 2.2$ m, and $(S)_{i+n} = 0.00017$.

Step 2.

Compute $v_{*cr}, v_*, X, Y, \xi, d_*$, and Y_{cr} as follows:

- $v_{*cr} = \sqrt{\frac{\tau_{cr}}{\rho}}$ where $\tau_{cr} = 0.0125 + 0.019D$.

$\tau_{cr} = 0.0125 + 0.019 \times 0.55 = 0.023 \frac{lb}{ft^2} \frac{ft}{s^2}$ and $\rho = 62.4 \frac{lb}{ft^3}$.

Therefore, $v_{*cr} = \sqrt{\frac{0.023 \frac{ft}{s}}{62.4}} \times 0.3 \frac{m}{ft} = 0.0057 \frac{m}{s}$

- $v_* = \sqrt{gSd} = \sqrt{9.81 \times 0.00017 \times 2.2} = 0.06 \frac{m}{s}$
- $X = \frac{v_* D}{v} = \frac{0.06 \times 0.55 \times 10^{-3}}{10^{-6}} = 33.31$
- $Y = \frac{\rho v_*^2}{\gamma_s D} = \frac{1,000 \times 0.06^2}{16186.5 \times 0.55 \times 10^{-3}} = 0.41$
- $\xi = \left(\frac{X^2}{Y}\right)^{\frac{1}{3}} = \left(\frac{33.31^2}{0.41}\right)^{\frac{1}{3}} = 13.91$
- $d_* = \frac{d}{D} = \frac{2.2}{0.55 \times 10^{-3}} = 4,000$
- $Y_{cr} = 0.13 \xi^{-0.392} e^{-0.015 \xi^2} + 0.045 \left[1 - e^{-0.068\xi}\right]$

$$= 0.13 \times 13.91^{-0.392} e^{-0.015 \times 13.91^2} + 0.045 \left[1 - e^{-0.068 \times 13.91}\right] = 0.03.$$

Compute $(\eta)_{i+n}$ as

$$(\eta)_{i+n} = \frac{Y}{Y_{cr}} = \frac{0.41}{0.03} = 13.7.$$

Step 3.

Compute $(C)_{i+n}$.

$$-\frac{\Lambda_{dunes}}{d} = 6\left[1 + 0.01\frac{(d_* - 40)(d_* - 400)}{d_*}\exp\left(-0.055\sqrt{d_*} - 0.04X\right)\right]$$

$$= 6\left[1 + 0.01\frac{(4{,}000 - 40)(4{,}000 - 400)}{4{,}000}\exp\left(-0.055\sqrt{4{,}000} - 0.04 \times 33.31\right)\right]$$

$$= 7.74$$

$$-\frac{\Lambda_{ripples}}{d} = \frac{1{,}000}{d_*} = \frac{1{,}000}{4{,}000} = 0.25$$

$$-\left(\frac{\Delta}{\Lambda}\right)_{dunes} = 0.013\psi_{Xdune}(X)(\eta - 1)\exp\left[-\frac{(\eta - 1)}{13(1 - \exp(-0.0082))}\right],$$

where $\psi_{Xdunes}(X) = 1 - \exp\left[-\left(\frac{X}{10}\right)^2\right] = 1 - \exp\left[-\left(\frac{33.31}{10}\right)^2\right] = 0.99.$

Therefore,

$$\left(\frac{\Delta}{\Lambda}\right)_{dunes} = 0.013 \times 0.99 \times (13.7 - 1)\exp\left[-\frac{(13.7 - 1)}{13(1 - \exp(-0.0082))}\right] = \sim 0$$

$$-\left(\frac{\Delta}{\Lambda}\right)_{ripples} = 0.035\psi_{Xripples}(X)(\eta - 1)\exp\left[-0.1\,(\eta - 1)\right],$$

where, since $X > 2.5$, $\psi_{Xripples}(X) = \exp\left[-((X - 2.5)/14)^2\right]$

$$= \exp\left[-((33.31 - 2.5)/14)^2\right] = 0.008.$$

Therefore,

$$\left(\frac{\Delta}{\Lambda}\right)_{ripples} = 0.035 \times 0.008 \times (13.7 - 1)\exp\left[-0.1\,(13.7 - 1)\right] = 0.001$$

$$-\;C_s = \frac{1}{\kappa}\ln\left(\psi(X)\,\frac{d_*}{2}\right),$$

where κ is 0.4 and $\psi(X) = \exp\left(\kappa w_s^{-1}\right)$,
where w_s is calculated following equation (10.53b) as

$$w_s = 8.5 + [2.5\ln(2X) - 3]\exp\left(-0.217[(\ln(2X)]^2\right)$$

$$= 8.5 + [2.5\ln(2 \times 33.31) - 3]\exp\left(-0.217[(\ln(2 \times 33.31)]^2\right) = 8.57.$$

Therefore, $\psi(X) = \exp\left(0.4 \times 8.57^{-1}\right) = 1.048.$

Finally, $C_s = \frac{1}{\kappa}\ln\left(\psi(X)\,\frac{d_*}{2}\right) = \frac{1}{0.4} \times \ln\left(1.048 \times \frac{4{,}000}{2}\right) = 19.11$

– Compute $(C)_{i+n}$ as

$$\frac{1}{C^2} = \frac{1}{C_s^2} + \sum_{j=d_u, r_i} a_j\left(\frac{\Delta}{\Lambda}\right)_j^{b_j}\frac{\Lambda_j}{d},$$

where $a_j = \frac{1}{2}$ and $b_j = 2$.
Therefore,

$$\frac{1}{C^2} = \frac{1}{19.11^2} + \frac{1}{2}\left(0.0^2 \times 7.74 + 0.001^2 \times 0.25\right).$$

Finally, $(C)_{i+n} = 19.12$.

Step 4.
Compute $(Q)_{i+n}$.

$$(Q)_{i+n} = (w)_{i+n}(d)_{i+n}(C)_{i+n}\sqrt{g(S)_{i+n}(d)_{i+n}}$$

$$= 53.2 \times 2.2 \times 19.12\sqrt{9.81 \times 0.00017 \times 2.2} = 131.0\frac{m^3}{s}.$$

To obtain $(Q)_{i+R} \approx Q$, additional iterations were undertaken. When $(Q)_{i+R} \approx Q$ of $141.4\frac{m^3}{s}$, the regime characteristics were estimated as
$w_R = 53.7$ m, $d_R = 2.3$ m, $S_R = 0.00017$, $N_R = 10^{5.9}$, $(F_r)_R = 0.02$, $\eta_R = 13.6$ and $c_R = 19.2$,
where $N_R = \frac{Q}{w_R D v_* c_r} = \frac{141.4}{54.3 \times 0.55^{-3} \times 0.0057} = 823676.3 = 10^{5.9}$ and $(F_r)_R = \frac{\gamma_s Y_{cr}}{\gamma} \frac{1}{N_R}\left(C_R^2 \eta_R\right)^{\frac{3}{2}} = \frac{1.65 \times 0.039}{823676.3}\left(19.2^2 \times 13.6\right)^{\frac{3}{2}} = 0.02$.

Example 10.4: Consider the following for the Mississippi River, USA: $Q = 13252.0$ m^3/s, $D = 0.656$ mm, and $\gamma_s/\gamma = 1.65$. Compute the regime channel geometry. The measure values are: $w_R = 532.2$ m, $d_R = 15.2$ m, $S_R = 0.000077$, $N_R = 10^{6.3}$, $(F_r)_R = 0.018$, $\eta_R = 31.8$, and $c_R = 15.31$.

Solution:
Step 1.
Adopt w, d, and S at the ith trial as $(w)_i = 525$ m, $(d)_i = 12$ m, and $(S)_i = 0.00003$.

Step 2.
Compute $v_{*cr}, v_*, X, Y, \xi, d_*$, and Y_{cr} as follows:

- $v_{*cr} = \sqrt{\frac{\tau_{cr}}{\rho}}$ where $\tau_{cr} = 0.0125 + 0.019D$.

  $\tau_{cr} = 0.0125 + 0.019 \times 0.656 = 0.025\frac{lb}{ft^2}\frac{ft}{s^2}$ and $\rho = 62.4\frac{lb}{ft^3}$.

  Therefore, $v_{*cr} = \sqrt{\frac{0.025}{62.4}\frac{ft}{s}} \times 0.3\frac{m}{ft} = 0.006\frac{m}{s}$

- $v_* = \sqrt{gSd} = \sqrt{9.81 \times 0.00003 \times 12} = 0.059\frac{m}{s}$

- $X = \frac{v_* D}{v} = \frac{0.059 \times 0.656 \times 10^{-3}}{10^{-6}} = 38.98$

- $Y = \frac{\rho v_*^2}{\gamma_s D} = \frac{1,000 \times 0.059^2}{16186.5 \times 0.656 \times 10^{-3}} = 0.33$

$$- \xi = \left(\frac{X^2}{Y}\right)^{\frac{1}{3}} = \left(\frac{38.98^2}{0.33}\right)^{\frac{1}{3}} = 16.59$$

$$- d_* = \frac{d}{D} = \frac{12}{0.656 \times 10^{-3}} = 18292.7$$

$$- Y_{cr} = 0.13\xi^{-0.392}e^{-0.015\xi^2} + 0.045\left[1 - e^{-0.068\xi}\right]$$

$$= 0.13 \times 16.59^{-0.392}e^{-0.015\times16.59^2} + 0.045\left[1 - e^{-0.068\times16.59}\right] = 0.031.$$

Compute $(\eta)_i$ as

$$(\eta)_i = \frac{Y}{Y_{cr}} = \frac{0.33}{0.031} = 10.68.$$

Step 3.
Compute $(C)_i$.

$$- \frac{\Lambda_{dunes}}{d} = 6\left[1 + 0.01\frac{(d_* - 40)(d_* - 400)}{d_*}\exp\left(-0.055\sqrt{d_*} - 0.04X\right)\right]$$

$$= 6\left[1 + 0.01\frac{(18292.7 - 40)(18292.7 - 400)}{18292.7}\exp\left(-0.055\sqrt{18292.7} - 0.04\times38.98\right)\right]$$

$$= 6.13$$

$$- \frac{\Lambda_{ripples}}{d} = \frac{1{,}000}{d_*} = \frac{1{,}000}{18292.7} = 0.055$$

$$- \left(\frac{\Delta}{\Lambda}\right)_{dunes} = 0.013\psi_{Xdune}(X)(\eta - 1)\exp\left[-\frac{(\eta - 1)}{13(1 - \exp(-0.0082))}\right],$$

where $\psi_{Xdunes}(X) = 1 - \exp\left[-\left(\frac{X}{10}\right)^2\right] = 1 - \exp\left[-\left(\frac{38.98}{10}\right)^2\right] = 0.99.$
Therefore,

$$\left(\frac{\Delta}{\Lambda}\right)_{dunes} = 0.013 \times 0.99 \times (10.68 - 1)\exp\left[-\frac{(10.68 - 1)}{13(1 - \exp(-0.0082))}\right] = \sim 0$$

$$- \left(\frac{\Delta}{\Lambda}\right)_{ripples} = 0.035\psi_{Xripples}(X)(\eta - 1)\exp\left[-0.1\left(\eta - 1\right)\right],$$

where, since $X > 2.5$, $\psi_{Xripples}(X) = \exp\left[-((X - 2.5)/14)^2\right]$
$$= \exp\left[-((38.98 - 2.5)/14)^2\right] = 0.001.$$

Therefore,

$$\left(\frac{\Delta}{\Lambda}\right)_{ripples} = 0.035 \times 0.001 \times (10.68 - 1)\exp\left[-0.1(10.68 - 1)\right] = 0.00014$$

$$- C_s = \frac{1}{\kappa}\ln\left(\psi(X)\frac{d_*}{2}\right),$$

where κ is 0.4 and $\psi(X) = \exp\left(\kappa w_s^{-1}\right)$,
where w_s is calculated following Equation (10.53b) as

$$w_s = 8.5 + [2.5 \ln (2X) - 3] \exp \left(-0.217[(\ln (2X)]^2\right)$$
$$= 8.5 + [2.5 \ln (2 \times 38.98) - 3] \exp \left(-0.217[(\ln (2 \times 38.98)]^2\right) = 8.55.$$

Therefore, $\psi(X) = \exp \left(0.4 \times 8.55^{-1}\right) = 1.048.$

Finally, $C_s = \frac{1}{\kappa} \ln \left(\psi(X) \frac{d_*}{2}\right) = \frac{1}{0.4} \times \ln \left(1.048 \times \frac{18292.7}{2}\right) = 22.9.$

– Compute $(C)_i$ as

$$\frac{1}{C^2} = \frac{1}{C_s^2} + \sum_{j=d_u, r_i} a_j \left(\frac{\Delta}{\Lambda}\right)_j^{b_j} \frac{\Lambda_j}{d},$$

where $a_j = \frac{1}{2}$ and $b_j = 2$.

Therefore,

$$\frac{1}{C^2} = \frac{1}{22.9^2} + \frac{1}{2} \left(0.0^2 \times 6.13 + 0.00014^2 \times 0.055\right).$$

Finally, $(C)_i = 22.9.$

Step 4.
Compute $(Q)_i$.

$$(Q)_i = (w)_i (d)_i (C)_i \sqrt{g(S)_i (d)_i}$$
$$= 525 \times 12 \times 22.9 \sqrt{9.81 \times 0.00003 \times 12} = 8573.57 \frac{m^3}{s}.$$

Step 5.
Compare the computed $(Q)_i$ and the measured Q.

Since $(Q)_i$ of $8,573.57 \frac{m^3}{s}$ is different from the measured Q of $13,252\frac{m^3}{s}$, repeat steps 1 to 4 by inputting 1% increase in w, d, and/or S for the following iterations.

After going through several iterations, when $(w)_{i+n} = 528.5$ m, $(d)_{i+n} = 15.05$ m, and $(S)_{i+n} = 0.000033$, the computed $(Q)_{i+n}$ is shown with the computational procedure as follows:

Step 1.
Adopt w, d, and S at the $i + n$th trial as $(w)_{i+n} = 528.5$ m, $(d)_{i+n} = 15.05$ m, and $(S)_{i+n} = 0.000033$.

Step 2.
Compute $v_{*cr}, v_*, X, Y, \xi, d_*,$ and Y_{cr} as follows:

– $v_{*cr} = \sqrt{\frac{\tau_{cr}}{\rho}}$ where $\tau_{cr} = 0.0125 + 0.019D.$

   $\tau_{cr} = 0.0125 + 0.019 \times 0.656 = 0.025 \frac{\text{lb}}{\text{ft}^2} \frac{\text{ft}}{\text{s}^2}$ and $\rho = 62.4 \frac{\text{lb}}{\text{ft}^3}.$

Therefore, $v_{*cr} = \sqrt{\frac{0.025}{62.4} \frac{\text{ft}}{\text{s}}} \times 0.3 \frac{\text{m}}{\text{ft}} = 0.006 \frac{\text{m}}{\text{s}}$

$- v_* = \sqrt{gSd} = \sqrt{9.81 \times 0.000033 \times 15.05} = 0.07 \frac{\text{m}}{\text{s}}$

$- X = \frac{v_* D}{v} = \frac{0.07 \times 0.656 \times 10^{-3}}{10^{-6}} = 45.79$

$- Y = \frac{\rho v_*^2}{\gamma_s D} = \frac{1,000 \times 0.07^2}{16186.5 \times 0.656 \times 10^{-3}} = 0.46$

$- \xi = \left(\frac{X^2}{Y}\right)^{\frac{1}{3}} = \left(\frac{45.79^2}{0.46}\right)^{\frac{1}{3}} = 16.6$

$- d_* = \frac{d}{D} = \frac{15.05}{0.656 \times 10^{-3}} = 22942.1$

$- Y_{cr} = 0.13\xi^{-0.392}e^{-0.015\xi^2} + 0.045\left[1 - e^{-0.068\xi}\right]$

$\quad = 0.13 \times 16.6^{-0.392}e^{-0.015 \times 16.6^2} + 0.045\left[1 - e^{-0.068 \times 16.6}\right] = 0.031.$

Compute $(\eta)_{i+n}$ as

$$(\eta)_{i+n} = \frac{Y}{Y_{cr}} = \frac{0.46}{0.031} = 14.74.$$

Step 3.
Compute $(C)_{i+n}$.

$- \frac{\Lambda_{dunes}}{d} = 6\left[1 + 0.01\frac{(d_* - 40)(d_* - 400)}{d_*}\exp\left(-0.055\sqrt{d_*} - 0.04X\right)\right]$

$\quad = 6\left[1 + 0.01\frac{(22942.1 - 40)(22942.1 - 400)}{22942.1}\exp\left(-0.055\sqrt{22942.1} - 0.04 \times 45.79\right)\right]$

$\quad = 6.05$

$- \frac{\Lambda_{ripples}}{d} = \frac{1,000}{d_*} = \frac{1,000}{22942.1} = 0.044$

$- \left(\frac{\Delta}{\Lambda}\right)_{dunes} = 0.013\psi_{Xdune}(X)(\eta - 1)\exp\left[-\frac{(\eta - 1)}{13(1 - \exp(-0.0082))}\right],$

where $\psi_{Xdunes}(X) = 1 - \exp\left[-\left(\frac{X}{10}\right)^2\right] = 1 - \exp\left[-\left(\frac{45.79}{10}\right)^2\right] = 0.99.$
Therefore,

$$\left(\frac{\Delta}{\Lambda}\right)_{dunes} = 0.013 \times 0.99 \times (14.74 - 1)\exp\left[-\frac{(14.74 - 1)}{13(1 - \exp(-0.0082))}\right] = \sim 0$$

$- \left(\frac{\Delta}{\Lambda}\right)_{ripples} = 0.035\psi_{Xripples}(X)(\eta - 1)\exp\left[-0.1\,(\eta - 1)\right],$

where, since $X > 2.5$, $\psi_{Xripples}(X) = \exp\left[-((X - 2.5)/14)^2\right]$

$\quad = \exp\left[-((45.79 - 2.5)/14)^2\right] = \sim 0.$

Therefore,

$$\left(\frac{\Delta}{\Lambda}\right)_{ripples} = 0.035 \times 0.0 \times (14.74 - 1) \exp\left[-0.1\,(14.74 - 1)\right] = \sim 0$$

$$- \quad C_s = \frac{1}{\kappa} \ln\left(\psi(X)\,\frac{d_*}{2}\right),$$

where κ is 0.4 and $\psi(X) = \exp\left(\kappa w_s^{-1}\right)$,
where w_s is calculated following Equation (10.53b) as

$$w_s = 8.5 + [2.5 \ln (2X) - 3] \exp\left(-0.217[(\ln (2X)]^2\right)$$
$$= 8.5 + [2.5 \ln (2 \times 45.79) - 3] \exp\left(-0.217[(\ln (2 \times 45.79)]^2\right) = 8.55.$$

Therefore, $\psi(X) = \exp\left(0.4 \times 8.55^{-1}\right) = 1.048$.
Finally, $C_s = \frac{1}{\kappa} \ln\left(\psi(X)\,\frac{d_*}{2}\right) = \frac{1}{0.4} \times \ln\left(1.048 \times \frac{22942.1}{2}\right) = 23.49$.

$-$ Compute $(C)_{i+n}$ as

$$\frac{1}{C^2} = \frac{1}{C_s^2} + \sum_{j=d_u,\,r_i} a_j \left(\frac{\Delta}{\Lambda}\right)_j^{b_j} \frac{\Lambda_j}{d},$$

where $a_j = \frac{1}{2}$ and $b_j = 2$.
Therefore,

$$\frac{1}{C^2} = \frac{1}{23.49^2} + \frac{1}{2}\left(0.0^2 \times 6.05 + 0.0^2 \times 0.044\right).$$

Finally, $(C)_{i+n} = 23.49$.

Step 4.
Compute $(Q)_{i+n}$.

$$(Q)_{i+n} = (w)_{i+n}(d)_{i+n}(C)_{i+n}\sqrt{g(S)_{i+n}(d)_{i+n}}$$
$$= 528.5 \times 15.05 \times 23.49\sqrt{9.81 \times 0.000033 \times 15.05} = 13{,}047\,\frac{m^3}{s}.$$

To obtain $(Q)_{i+R} \approx Q$, additional iterations were undertaken. When $(Q)_{i+R} \approx Q$ of $13{,}252\,\frac{m^3}{s}$, the regime characteristics were estimated as
$w_R = 533.2$ m, $d_R = 15.2$ m, $S_R = 0.000033$, $N_R = 10^{6.8}$, $(F_r)_R = 0.006$, $\eta_R = 14.7$
and $c_R = 23.5$,
where $N_R = \frac{Q}{w_R D v_* c_r} = \frac{13252}{533.2 \times 0.656^{-3} \times 0.006} = 6313992.2 = 10^{6.7}$ and $(F_r)_R = \frac{\frac{\gamma_s}{\gamma}Y_{cr}}{N_R}\left(C_R^2 \eta_R\right)^{\frac{3}{2}} = \frac{1.65 \times 0.031}{6313992.2}\left(23.5^2 \times 14.7\right)^{\frac{3}{2}} = 0.006$.

Example 10.5: Consider the following for the Savannah River, USA: $Q = 848.9$ m^3/s, $D = 0.8$ mm, and $\gamma_s/\gamma = 1.65$. Compute the regime channel geometry. The measure values are: $w_R = 106.7$ m, $d_R = 5.18$ m, $S_R = 0.00011$, $N_R = 10^{5.6}$, $(F_r)_R = 0.050$, $\eta_R = 11.6$, and $c_R = 20.5$.

Solution:

Step 1.

Adopt w, d, and S at the ith trial as $(w)_i = 104$ m, $(d)_i = 4.0$ m, and $(S)_i = 0.0001$.

Step 2.

Compute v_{*cr}, v_*, X, Y, ξ, d_*, and Y_{cr} as follows:

- $v_{*cr} = \sqrt{\frac{\tau_{cr}}{\rho}}$ where $\tau_{cr} = 0.0125 + 0.019D$.

  $\tau_{cr} = 0.0125 + 0.019 \times 0.8 = 0.028 \frac{\text{lb}}{\text{ft}^2} \frac{\text{ft}}{\text{s}^2}$ and $\rho = 62.4 \frac{\text{lb}}{\text{ft}^3}$.

  Therefore, $v_{*cr} = \sqrt{\frac{0.028 \frac{\text{ft}}{\text{s}}}{62.4}} \times 0.3 \frac{\text{m}}{\text{ft}} = 0.006 \frac{\text{m}}{\text{s}}$

- $v_* = \sqrt{gSd} = \sqrt{9.81 \times 0.0001 \times 4.0} = 0.062 \frac{\text{m}}{\text{s}}$

- $X = \frac{v_* D}{v} = \frac{0.062 \times 0.8 \times 10^{-3}}{10^{-6}} = 50.11$

- $Y = \frac{\rho v_*^2}{\gamma_s D} = \frac{1,000 \times 0.062^2}{16186.5 \times 0.8 \times 10^{-3}} = 0.3$

- $\xi = \left(\frac{X^2}{Y}\right)^{\frac{1}{3}} = \left(\frac{50.11^2}{0.3}\right)^{\frac{1}{3}} = 20.24$

- $d_* = \frac{d}{D} = \frac{4.0}{0.8 \times 10^{-3}} = 5,000$

- $Y_{cr} = 0.13\xi^{-0.392}e^{-0.015\xi^2} + 0.045\left[1 - e^{-0.068\xi}\right]$

  $= 0.13 \times 20.24^{-0.392}e^{-0.015 \times 20.24^2} + 0.045\left[1 - e^{-0.068 \times 20.24}\right] = 0.034$.

Compute $(\eta)_i$ as

$$(\eta)_i = \frac{Y}{Y_{cr}} = \frac{0.3}{0.034} = 8.99.$$

Step 3.

Compute $(C)_i$.

- $\frac{\Lambda_{dunes}}{d} = 6\left[1 + 0.01\frac{(d_* - 40)(d_* - 400)}{d_*}\exp\left(-0.055\sqrt{d_*} - 0.04X\right)\right]$

  $= 6\left[1 + 0.01\frac{(5,000 - 40)(5,000 - 400)}{5,000}\exp\left(-0.055\sqrt{5,000} - 0.04 \times 50.11\right)\right]$

  $= 6.75$

- $\frac{\Lambda_{ripples}}{d} = \frac{1,000}{d_*} = \frac{1,000}{5,000} = 0.2$

$$- \left(\tfrac{\Delta}{\Lambda}\right)_{dunes} = 0.013 \psi_{Xdune}(X)(\eta - 1) \exp\left[-\tfrac{(\eta-1)}{13(1-\exp(-0.0082))}\right],$$

where $\psi_{Xdunes}(X) = 1 - \exp\left[-\left(\tfrac{X}{10}\right)^2\right] = 1 - \exp\left[-\left(\tfrac{50.11}{10}\right)^2\right] = 0.999.$

Therefore,

$$\left(\tfrac{\Delta}{\Lambda}\right)_{dunes} = 0.013 \times 0.999 \times (8.99 - 1) \exp\left[-\tfrac{(8.99-1)}{13(1-\exp(-0.0082))}\right] = \sim 0$$

$$- \left(\tfrac{\Delta}{\Lambda}\right)_{ripples} = 0.035 \psi_{Xripples}(X)(\eta - 1) \exp\left[-0.1\,(\eta - 1)\right],$$

where, since $X > 2.5$, $\psi_{Xripples}(X) = \exp\left[-((X - 2.5)/14)^2\right]$
$$= \exp\left[-((50.11 - 2.5)/14)^2\right] = \sim 0.$$

Therefore,

$$\left(\tfrac{\Delta}{\Lambda}\right)_{ripples} = 0.035 \times 0.0 \times (8.99 - 1) \exp\left[-0.1\,(8.99 - 1)\right] = \sim 0$$

$- C_s = \tfrac{1}{\kappa} \ln\left(\psi(X)\,\tfrac{d_*}{2}\right),$

where κ is 0.4 and $\psi(X) = \exp\left(\kappa w_s^{-1}\right)$,
where w_s is calculated following Equation (6.47) as

$$w_s = 8.5 + [2.5 \ln(2X) - 3] \exp\left(-0.217[(\ln(2X)]^2\right)$$
$$= 8.5 + [2.5 \ln(2 \times 50.11) - 3] \exp\left(-0.217[(\ln(2 \times 50.11)]^2\right) = 8.54.$$

Therefore, $\psi(X) = \exp\left(0.4 \times 8.54^{-1}\right) = 1.05.$
Finally, $C_s = \tfrac{1}{\kappa} \ln\left(\psi(X)\,\tfrac{d_*}{2}\right) = \tfrac{1}{0.4} \times \ln\left(1.05 \times \tfrac{5,000}{2}\right) = 19.68.$

$-$ Compute $(C)_i$ as

$$\frac{1}{C^2} = \frac{1}{C_s^2} + \sum_{j=d_u, r_i} a_j \left(\frac{\Delta}{\Lambda}\right)_j^{b_j} \frac{\Lambda_j}{d},$$

where $a_j = \tfrac{1}{2}$ and $b_j = 2$.
Therefore,

$$\frac{1}{C^2} = \frac{1}{19.68^2} + \frac{1}{2}\left(0.0^2 \times 6.75 + 0.0^2 \times 0.2\right).$$

Finally, $(C)_i = 19.68.$

Step 4.
Compute $(Q)_i$.

$$(Q)_i = (w)_i (d)_i (C)_i \sqrt{g(S)_i (d)_i}$$
$$= 104 \times 4.0 \times 19.68 \sqrt{9.81 \times 0.0001 \times 4.0} = 512.84 \frac{\text{m}^3}{\text{s}}$$

Step 5.

Compare the computed $(Q)_i$ and the measured Q.

Since $(Q)_i$ of $512.84\,\frac{m^3}{s}$ is different from the measured Q of $848.9\,\frac{m^3}{s}$, repeat steps 1 to 4 by inputting 1% increase in w, d, and/or S for the following iterations.

After going through several iterations, when $(w)_{i+n} = 105.5$ m, $(d)_{i+n} = 4.56$ m, and $(S)_{i+n} = 0.00016$, the computed $(Q)_{i+n}$ is shown with the computational procedure as follows:

Step 1.

Adopt w, d, and S at the $i + n$th trial as $(w)_{i+n} = 105.5$ m, $(d)_{i+n} = 4.56$ m, and $(S)_{i+n} = 0.00016$.

Step 2.

Compute $v_{*cr}, v_*, X, Y, \xi, d_*$, and Y_{cr} as follows:

- $v_{*cr} = \sqrt{\frac{\tau_{cr}}{\rho}}$ where $\tau_{cr} = 0.0125 + 0.019D$.

  $\tau_{cr} = 0.0125 + 0.019 \times 0.8 = 0.028\,\frac{lb}{ft^2}\,\frac{ft}{s^2}$ and $\rho = 62.4\,\frac{lb}{ft^3}$.

  Therefore, $v_{*cr} = \sqrt{\frac{0.028}{62.4}\frac{ft}{s}} \times 0.3\,\frac{m}{ft} = 0.006\,\frac{m}{s}$

- $v_* = \sqrt{gSd} = \sqrt{9.81 \times 0.00016 \times 4.56} = 0.085\,\frac{m}{s}$

- $X = \frac{v_* D}{v} = \frac{0.085 \times 0.8 \times 10^{-3}}{10^{-6}} = 67.68$

- $Y = \frac{\rho v_*^2}{\gamma_s D} = \frac{1{,}000 \times 0.085^2}{16186.5 \times 0.8 \times 10^{-3}} = 0.55$

- $\xi = \left(\frac{X^2}{Y}\right)^{\frac{1}{3}} = \left(\frac{67.68^2}{0.55}\right)^{\frac{1}{3}} = 20.24$

- $d_* = \frac{d}{D} = \frac{4.56}{0.8 \times 10^{-3}} = 5699.9$

- $Y_{cr} = 0.13\xi^{-0.392}e^{-0.015\xi^2} + 0.045\left[1 - e^{-0.068\xi}\right]$

  $= 0.13 \times 20.24^{-0.392}e^{-0.015 \times 20.24^2} + 0.045\left[1 - e^{-0.068 \times 20.24}\right] = 0.033$.

Compute $(\eta)_{i+n}$ as

$$(\eta)_{i+n} = \frac{Y}{Y_{cr}} = \frac{0.55}{0.033} = 16.4.$$

Step 3.

Compute $(C)_{i+n}$.

- $\frac{\Lambda_{dunes}}{d} = 6\left[1 + 0.01\frac{(d_* - 40)(d_* - 400)}{d_*}\exp\left(-0.055\sqrt{d_*} - 0.04X\right)\right]$

  $= 6\left[1 + 0.01\frac{(5699.9 - 40)(5699.9 - 400)}{5699.9}\exp\left(-0.055\sqrt{5699.9} - 0.04 \times 67.68\right)\right]$

  $= 6.33$

- $\dfrac{\Lambda_{ripples}}{d} = \dfrac{1{,}000}{d_*} = \dfrac{1{,}000}{5699.9} = 0.18$

- $\left(\dfrac{\Delta}{\Lambda}\right)_{dunes} = 0.013\psi_{Xdune}(X)(\eta - 1)\exp\left[-\dfrac{(\eta - 1)}{13(1 - \exp(-0.0082))}\right],$

  where $\psi_{Xdunes}(X) = 1 - \exp\left[-\left(\dfrac{X}{10}\right)^2\right] = 1 - \exp\left[-\left(\dfrac{67.68}{10}\right)^2\right] = 1.0$.
  Therefore,

  $$\left(\dfrac{\Delta}{\Lambda}\right)_{dunes} = 0.013 \times 1.0 \times (16.4 - 1)\exp\left[-\dfrac{(16.4 - 1)}{13(1 - \exp(-0.0082))}\right] = \sim 0$$

- $\left(\dfrac{\Delta}{\Lambda}\right)_{ripples} = 0.035\psi_{Xripples}(X)(\eta - 1)\exp\left[-0.1\,(\eta - 1)\right],$

  where, since $X > 2.5$, $\psi_{Xripples}(X) = \exp\left[-((X - 2.5)/14)^2\right]$
  $$= \exp\left[-((67.68 - 2.5)/14)^2\right] = \sim 0.$$
  Therefore,

  $$\left(\dfrac{\Delta}{\Lambda}\right)_{ripples} = 0.035 \times 0.0 \times (16.4 - 1)\exp\left[-0.1\,(16.4 - 1)\right] = \sim 0$$

- $C_s = \dfrac{1}{\kappa}\ln\left(\psi(X)\,\dfrac{d_*}{2}\right),$

  where κ is 0.4 and $\psi(X) = \exp\left(\kappa w_s^{-1}\right)$,
  where w_s is calculated following Equation (10.53b) as

  $w_s = 8.5 + [2.5\ln(2X) - 3]\exp\left(-0.217[(\ln(2X)]^2\right)$
  $= 8.5 + [2.5\ln(2 \times 67.68) - 3]\exp\left(-0.217[(\ln(2 \times 67.68)]^2\right) = 8.53.$

  Therefore, $\psi(X) = \exp\left(0.4 \times 8.53^{-1}\right) = 1.048$.
  Finally, $C_s = \dfrac{1}{\kappa}\ln\left(\psi(X)\,\dfrac{d_*}{2}\right) = \dfrac{1}{0.4} \times \ln\left(1.048 \times \dfrac{5699.9}{2}\right) = 20.0$.

- Compute $(C)_{i+n}$ as

  $$\dfrac{1}{C^2} = \dfrac{1}{C_s^2} + \sum_{j=d_u, r_i} a_j\left(\dfrac{\Delta}{\Lambda}\right)_j^{b_j}\dfrac{\Lambda_j}{d},$$

  where $a_j = \dfrac{1}{2}$ and $b_j = 2$.
  Therefore,

  $$\dfrac{1}{C^2} = \dfrac{1}{20.0^2} + \dfrac{1}{2}\left(0.0^2 \times 6.33 + 0.0^2 \times 0.18\right).$$

Finally, $(C)_{i+n} = 20.0$.

Step 4.
Compute $(Q)_{i+n}$.

$$(Q)_{i+n} = (w)_{i+n}(d)_{i+n}(C)_{i+n}\sqrt{g(S)_{i+n}(d)_{i+n}}$$

$$= 105.5 \times 4.56 \times 20.0\sqrt{9.81 \times 0.00016 \times 4.56} = 814\frac{m^3}{s}.$$

To obtain $(Q)_{i+R} \approx Q$, additional iterations were undertaken. When $(Q)_{i+R} \approx Q$ of $848.9\frac{m^3}{s}$, the regime characteristics were estimated as
  $w_R = 107.6$ m, $d_R = 4.62$ m, $S_R = 0.00016$, $N_R = 10^{6.2}$, $(F_r)_R = 0.02$, $\eta_R = 16.9$
and $c_R = 20.1$.
  where $N_R = \frac{Q}{w_R D v_* c_r} = \frac{848.9}{107.6 \times 0.8^{-3} \times 0.006} = 1{,}643{,}626 = 10^{6.2}$ and $(F_r)_R = \frac{\frac{\gamma_s}{\gamma}Y_{cr}}{N_R}(C_R^2 \eta_R)^{\frac{3}{2}} = \frac{1.65 \times 0.033}{1643626}(20.1^2 \times 16.9)^{\frac{3}{2}} = 0.02.$

Example 10.6: Consider the following for the North Saskatchewan River, Canada: $Q = 4386.0$ m³/s, $D = 31$ mm, and $\gamma_s/\gamma = 1.65$. Compute the regime channel geometry. The measure values are: $w_R = 244.0$ m, $d_R = 7.62$ m, $S_R = 0.00035$, $N_R = 10^{3.6}$, $(F_r)_R = 0.074$, $\eta_R = 1.1$, and $c_R = 14.6$.

Solution:
Step 1.
Adopt w, d, and S at the ith trial as $(w)_i = 235$ m, $(d)_i = 5.0$ m, and $(S)_i = 0.0007$.

Step 2.
Compute $v_{*cr}, v_*, X, Y, \xi, d_*$, and Y_{cr} as follows:

$-\ v_{*cr} = \sqrt{\frac{\tau_{cr}}{\rho}}$ where $\tau_{cr} = 0.0125 + 0.019D$.

  $\tau_{cr} = 0.0125 + 0.019 \times 31 = 0.6\frac{lb}{ft^2}\frac{ft}{s^2}$ and $\rho = 62.4\frac{lb}{ft^3}$.

  Therefore, $v_{*cr} = \sqrt{\frac{0.6\frac{ft}{s}}{62.4}} \times 0.3\frac{m}{ft} = 0.03\frac{m}{s}$

$-\ v_* = \sqrt{gSd} = \sqrt{9.81 \times 0.0007 \times 5.0} = 0.185\frac{m}{s}$

$-\ X = \frac{v_* D}{v} = \frac{0.185 \times 31 \times 10^{-3}}{10^{-6}} = 5744.2$

$-\ Y = \frac{\rho v_*^2}{\gamma_s D} = \frac{1{,}000 \times 0.185^2}{16186.5 \times 31 \times 10^{-3}} = 0.068$

$-\ \xi = \left(\frac{X^2}{Y}\right)^{\frac{1}{3}} = \left(\frac{5744.2^2}{0.068}\right)^{\frac{1}{3}} = 784.17$

$-\ d_* = \frac{d}{D} = \frac{5.0}{31 \times 10^{-3}} = 161.29$

$-\ Y_{cr} = 0.13\xi^{-0.392}e^{-0.015\xi^2} + 0.045\left[1 - e^{-0.068\xi}\right]$

  $= 0.13 \times 784.17^{-0.392}e^{-0.015 \times 784.17^2} + 0.045\left[1 - e^{-0.068 \times 784.17}\right] = 0.045.$

Compute $(\eta)_i$ as

$$(\eta)_i = \frac{Y}{Y_{cr}} = \frac{0.068}{0.045} = 1.52.$$

Step 3.
Compute $(C)_i$.

$$-\frac{\Lambda_{dunes}}{d} = 6\left[1+0.01\frac{(d_* - 40)(d_* - 400)}{d_*}\exp\left(-0.055\sqrt{d_*} - 0.04X\right)\right]$$

$$= 6\left[1+0.01\frac{(161.29 - 40)(161.29 - 400)}{161.29}\exp\left(-0.055\sqrt{161.29} - 0.04 \times 5744.2\right)\right]$$

$$= 6.0$$

$$-\frac{\Lambda_{ripples}}{d} = \frac{1{,}000}{d_*} = \frac{1{,}000}{161.29} = 6.2$$

$$-\left(\frac{\Delta}{\Lambda}\right)_{dunes} = 0.013\psi_{Xdune}(X)(\eta - 1)\exp\left[-\frac{(\eta - 1)}{13(1 - \exp(-0.0082))}\right],$$

where $\psi_{Xdunes}(X) = 1 - \exp\left[-\left(\frac{X}{10}\right)^2\right] = 1 - \exp\left[-\left(\frac{5744.2}{10}\right)^2\right] = 1.0.$

Therefore,

$$\left(\frac{\Delta}{\Lambda}\right)_{dunes} = 0.013 \times 1.0 \times (1.52 - 1)\exp\left[-\frac{(1.52 - 1)}{13(1 - \exp(-0.0082))}\right] = \sim 0$$

$$-\left(\frac{\Delta}{\Lambda}\right)_{ripples} = 0.035\psi_{Xripples}(X)(\eta - 1)\exp[-0.1(\eta - 1)],$$

where, since $X > 2.5$, $\psi_{Xripples}(X) = \exp\left[-((X - 2.5)/14)^2\right]$
$$= \exp\left[-((5744.2 - 2.5)/14)^2\right] = \sim 0.$$

Therefore,

$$\left(\frac{\Delta}{\Lambda}\right)_{ripples} = 0.035 \times 0.0 \times (1.52 - 1)\exp[-0.1(1.52 - 1)] = \sim 0$$

$$- C_s = \frac{1}{\kappa}\ln\left(\psi(X)\frac{d_*}{2}\right),$$

where κ is 0.4 and $\psi(X) = \exp\left(\kappa w_s^{-1}\right)$,
where w_s is calculated following Equation (6.47) as

$$w_s = 8.5 + [2.5\ln(2X) - 3]\exp\left(-0.217[(\ln(2X)]^2\right)$$
$$= 8.5 + [2.5\ln(2 \times 5744.2) - 3]\exp\left(-0.217[(\ln(2 \times 5744.2)]^2\right) = 8.5.$$

Therefore, $\psi(X) = \exp\left(0.4 \times 8.5^{-1}\right) = 1.05.$

Finally, $C_s = \frac{1}{\kappa}\ln\left(\psi(X)\frac{d_*}{2}\right) = \frac{1}{0.4} \times \ln\left(1.05 \times \frac{161.29}{2}\right) = 11.09$

– Compute $(C)_i$ as

$$\frac{1}{C^2} = \frac{1}{C_s^2} + \sum_{j=d_u, r_i} a_j \left(\frac{\Delta}{\Lambda}\right)_j^{b_j} \frac{\Lambda_j}{d},$$

where $a_j = \frac{1}{2}$ and $b_j = 2$.
Therefore,

$$\frac{1}{C^2} = \frac{1}{11.09^2} + \frac{1}{2}\left(0.0^2 \times 6.0 + 0.0^2 \times 6.2\right).$$

Finally, $(C)_i = 11.09$.

Step 4.
Compute $(Q)_i$.

$$(Q)_i = (w)_i (d)_i (C)_i \sqrt{g(S)_i (d)_i}$$

$$= 235 \times 5.0 \times 11.09\sqrt{9.81 \times 0.0007 \times 5.0} = 2,414.55 \frac{m^3}{s}.$$

Step 5.
Compare the computed $(Q)_i$ and the measured Q.

Since $(Q)_i$ of $2,414.55 \frac{m^3}{s}$ is different from the measured Q of $4,386 \frac{m^3}{s}$, repeat steps 1 to 4 by inputting 1% increase in w, d, and/or 1% decrease in S for the following iterations.

After going through several iterations, when $(w)_{i+n} = 241.5$ m, $(d)_{i+n} = 7.55$ m, and $(S)_{i+n} = 0.00048$, the computed $(Q)_{i+n}$ is shown with the computational procedure as follows:

Step 1.
Adopt w, d, and S at the $i + n$th trial as $(w)_{i+n} = 241.5$ m, $(d)_{i+n} = 7.55$ m, and $(S)_{i+n} = 0.00048$.

Step 2.
Compute $v_{*cr}, v_*, X, Y, \xi, d_*,$ and Y_{cr} as follows:

– $v_{*cr} = \sqrt{\frac{\tau_{cr}}{\rho}}$ where $\tau_{cr} = 0.0125 + 0.019D$.

$\tau_{cr} = 0.0125 + 0.019 \times 31 = 0.6 \frac{lb}{ft^2} \frac{ft}{s^2}$ and $\rho = 62.4 \frac{lb}{ft^3}$.

Therefore, $v_{*cr} = \sqrt{\frac{0.6}{62.4} \frac{ft}{s}} \times 0.3 \frac{m}{ft} = 0.03 \frac{m}{s}$

– $v_* = \sqrt{gSd} = \sqrt{9.81 \times 0.00048 \times 7.55} = 0.19 \frac{m}{s}$

– $X = \frac{v_* D}{v} = \frac{0.19 \times 31 \times 10^{-3}}{10^{-6}} = 5845.1$

- $Y = \frac{\rho v_*^2}{\gamma_s D} = \frac{1{,}000 \times 0.19^2}{16186.5 \times 31 \times 10^{-3}} = 0.07$

- $\zeta = \left(\frac{X^2}{Y}\right)^{\frac{1}{3}} = \left(\frac{5845.1^2}{0.07}\right)^{\frac{1}{3}} = 784.2$

- $d_* = \frac{d}{D} = \frac{7.55}{31 \times 10^{-3}} = 243.5$

- $Y_{cr} = 0.13\zeta^{-0.392} e^{-0.015\zeta^2} + 0.045\left[1 - e^{-0.068\zeta}\right]$

  $= 0.13 \times 784.2^{-0.392} e^{-0.015 \times 784.2^2} + 0.045\left[1 - e^{-0.068 \times 784.2}\right] = 0.045.$

Compute $(\eta)_{i+n}$ as

$$(\eta)_{i+n} = \frac{Y}{Y_{cr}} = \frac{0.07}{0.045} = 1.57.$$

Step 3.

Compute $(C)_{i+n}$.

- $\frac{\Lambda_{dunes}}{d} = 6\left[1 + 0.01\frac{(d_* - 40)(d_* - 400)}{d_*} \exp\left(-0.055\sqrt{d_*} - 0.04X\right)\right]$

  $= 6\left[1 + 0.01\frac{(243.5 - 40)(243.5 - 400)}{243.5} \exp\left(-0.055\sqrt{243.5} - 0.04 \times 5845.1\right)\right]$

  $= 6.0$

- $\frac{\Lambda_{ripples}}{d} = \frac{1{,}000}{d_*} = \frac{1{,}000}{243.5} = 4.1$

- $\left(\frac{\Delta}{\Lambda}\right)_{dunes} = 0.013\psi_{Xdune}(X)(\eta - 1)\exp\left[-\frac{(\eta - 1)}{13(1 - \exp(-0.0082))}\right],$

  where $\psi_{Xdunes}(X) = 1 - \exp\left[-\left(\frac{X}{10}\right)^2\right] = 1 - \exp\left[-\left(\frac{5845.1}{10}\right)^2\right] = 1.0.$
  Therefore,

  $$\left(\frac{\Delta}{\Lambda}\right)_{dunes} = 0.013 \times 1.0 \times (1.57 - 1)\exp\left[-\frac{(1.57 - 1)}{13(1 - \exp(-0.0082))}\right] = \sim 0$$

- $\left(\frac{\Delta}{\Lambda}\right)_{ripples} = 0.035\psi_{Xripples}(X)(\eta - 1)\exp\left[-0.1(\eta - 1)\right],$

  where, since $X > 2.5$, $\psi_{Xripples}(X) = \exp\left[-((X - 2.5)/14)^2\right]$
  $= \exp\left[-((5845.1 - 2.5)/14)^2\right] = \sim 0.$
  Therefore,

  $$\left(\frac{\Delta}{\Lambda}\right)_{ripples} = 0.035 \times 0.0 \times (1.57 - 1)\exp\left[-0.1(1.57 - 1)\right] = \sim 0$$

- $C_s = \frac{1}{\kappa}\ln\left(\psi(X)\frac{d_*}{2}\right),$

  where κ is 0.4 and $\psi(X) = \exp\left(\kappa w_s^{-1}\right),$

where w_s is calculated following Equation (10.53b) as

$$w_s = 8.5 + [2.5 \ln(2X) - 3] \exp\left(-0.217[(\ln(2X))]^2\right)$$
$$= 8.5 + [2.5 \ln(2 \times 5845.1) - 3] \exp\left(-0.217[(\ln(2 \times 5845.1))]^2\right) = 8.5.$$

Therefore, $\psi(X) = \exp\left(0.4 \times 8.5^{-1}\right) = 1.048$.

Finally, $C_s = \frac{1}{\kappa} \ln\left(\psi(X)\frac{d_*}{2}\right) = \frac{1}{0.4} \times \ln\left(1.048 \times \frac{243.5}{2}\right) = 12.12$

– Compute $(C)_{i+n}$ as

$$\frac{1}{C^2} = \frac{1}{C_s^2} + \sum_{j=d_u,\,r_i} a_j \left(\frac{\Delta}{\Lambda}\right)_j^{b_j} \frac{\Lambda_j}{d},$$

where $a_j = \frac{1}{2}$ and $b_j = 2$.

Therefore,

$$\frac{1}{C^2} = \frac{1}{12.12^2} + \frac{1}{2}\left(0.0^2 \times 6.0 + 0.0^2 \times 4.1\right).$$

Finally, $(C)_{i+n} = 12.12$.

Step 4.

Compute $(Q)_{i+n}$.

$$(Q)_{i+n} = (w)_{i+n}(d)_{i+n}(C)_{i+n}\sqrt{g(S)_{i+n}(d)_{i+n}}$$
$$= 241.5 \times 7.55 \times 12.12\sqrt{9.81 \times 0.00048 \times 7.55} = 4166.7\,\frac{\text{m}^3}{\text{s}}.$$

To obtain $(Q)_{i+R} \approx Q$, additional iterations were undertaken. When $(Q)_{i+R} \approx Q$ of $4{,}386\,\frac{\text{m}^3}{\text{s}}$, the regime characteristics were estimated as

$w_R = 243.7$ m, $d_R = 7.72$ m, $S_R = 0.00048$, $N_R = 10^{4.3}$, $(F_r)_R = 0.014$, $\eta_R = 1.6$

and $c_R = 12.2$.

where $N_R = \frac{Q}{w_R D v_* c_r} = \frac{4{,}386}{243.7 \times 31^{-3} \times 0.03} = 19709.4 = 10^{4.3}$ and $(F_r)_R = \frac{\frac{7}{4}Y_{cr}}{N_R}\left(C_R^2 \eta_R\right)^{\frac{3}{2}} = \frac{1.65 \times 0.045}{19709.4}\left(12.2^2 \times 1.6\right)^{\frac{3}{2}} = 0.014$.

References

Chang, H. H. (1979). Minimum stream power and river channel patterns. *Journal of Hydrology*, Vol. 41, No. 3–4, pp. 303–327.

Chang, H. H. (1980). Stable alluvial canal design. *Journal of Hydraulics Division, ASCE*, Vol. 106, No. HY5, pp. 873–891.

Davies, T. H. R. and Southerland, A. J. (1983). Extremal hypothesis for river behavior. *Water Resources research*, Vol. 19, No. 1, pp. 141–148.

Da Silva, A. M. F., and Yalin, M. S. (2017). *Fluvial Processes*. CRC Press, Boca Raton, FL.

AWhite, W. R., Bettes, R., and Paris, E. (1982). Analytical approach to river regime. *Journal of Hydraulics Division, ASCE*, Vol. 108, No. HY10, pp. 1179–1193.

Yalin, M. S. (1982). *River Mechanics*. Pergamon Press, Oxford.

Yalin, M. S. (1992). *River Mechanics*. Pergamon Press, Oxford.

Yalin, M. S. and Ferreira da Silva, A. M. (1999). Regime channels in cohesionless alluvium. *Journal of Hydraulic Research*, Vol. 37, No. 6, pp. 725–742.

Yang, C. T. (1984). Unit stream power equation for gravel. *Journal of Hydraulic Engineering, ASCE*, Vol. 110, No. 12, pp. 1783–1797.

Yang, C. T. (1987). Energy dissipation rate in river mechanics. In: *Sediment Transport in Gravel Bed Rivers*, edited by C. R. Thorne, J. C. Bathurst, and R. D. Hays, John Wiley & Sons, New York.

11

Similarity Principle

The similarity principle is based on the acknowledgment that a river, if left alone for a sufficiently long time with fixed values of water and sediment discharge loads, will adjust its width, depth, slope, and meandering pattern in a certain manner. If the values of water and sediment discharge loads imposed on the river change are different, the adjustment will be made similarly. This chapter derives the hydraulic geometry using this principle of similarity.

Notation

c = constant

c_d = drag coefficient

d = mean flow depth

d_b = thickness of the boundary layer

d_i = flow depth for system i (i=1, 2)

D = characteristic grain size or diameter

D_{35} = grain diameter corresponding to 35% of the grains being smaller than other grains

f = friction factor

f_b = friction in the boundary layer

f_d = friction factor due to dune

F_r = Froude number

g = acceleration due to gravity

h = dune height

k = sand roughness

L = dune length

q = flow discharge per unit channel width

q_B = rate of bed load transport (volume of material per unit width per unit time)

q_T = total sediment discharge (bed load and suspended load) per unit width

---

Q = water discharge
R_n = Reynolds number
S_e = energy slope
S_f = gradient due to friction
v = mean velocity
v_1 = velocity above the crest before expansion
v_2 = mean velocity downstream of the crest
v_b = mean velocity of the boundary layer
v_f = friction velocity
v_s = setting velocity or fall velocity
w = width
α = dimensionless coefficient depending on the flow geometry
δ = average thickness of the viscous sublayer
ΔH^e = loss of energy due to expansion
λ = scale ratio
λ_H = vertical length scale ratio
λ_L = horizontal length scale ratio
γ_s = specific weight of sediment
γ_{s*} $(= \gamma_s/\gamma)$ = specific gravity or relative density of grains
γ = specific weight of water
υ = kinematic viscosity
ϕ = dimensionless sediment transport rate
ψ = resistance coefficient depending on the relative dimensions of the dunes
ρ = mass density of water
τ_0 = total shear stress
τ_* = dimensionless form of bed shear τ_0
τ_{0b} = shear stress in the boundary layer
τ_{0d} = friction due to the surface of a dune
τ_{0f} = form drag on the dunes
τ_{*e} = dimensionless effective shear stress
τ_{*d} = dimensionless dune-covered bed shear
τ_{*c} = critical dimensionless shear stress

11.1 Introduction

The discussion here focuses on alluvial rivers, where the term "alluvial" refers to streams in which the bed sediment is the same as the moving sediment. However, in most natural rivers, a certain portion of sediment being carried is very fine sediment, called wash load, often of secondary importance, which is not represented in the bed. For alluvial stream hydraulics, the most significant sediment particles are in the range of sand (0.06 to 2 mm) and gravel (2 to 20 mm) and their size, shape, and specific gravity.

The river regime is shaped by the water and sediment load supplied by the watershed to the river. Sediment discharge of a river is primarily determined by the nature of its drainage area. This is also true of water discharge. In other words, both sediment and water discharges are factors that are essentially independent of the river itself and are determined exclusively by the hydrology, geology, land use/land cover, and topography of the drainage area. It is assumed that streams are in equilibrium if their characteristics have not changed notably in a number of years, with seasonal changes being discarded.

For a given alluvial river, water and sediment discharges increase in the downstream direction. The same is usually the case with width and depth. The slope and grain size usually decrease gradually from source to estuary. The grain size decreases approximately exponentially in the downstream direction. This chapter employs the similarity principle to derive at-a-station hydraulic geometry.

11.2 Dominant Discharge

For given water and sediment discharge loads, the river itself will tend to create the geometry of its stream channel in a specific manner. This means that if the river is left alone for a sufficiently long time with fixed values of water and sediment discharge loads, it will adjust its width, depth, slope, and meandering pattern in a definitive manner. Since the slope and the meandering pattern do not change rapidly enough to follow seasonal variations of discharge, it is natural to invoke some kind of "dominant" or "formative" values of discharge for these variables. The dominant discharge should then be defined as the steady discharge that gives rise to the same slope and meandering pattern as the annual sequence of discharges. Another more practical definition of dominant discharge would be the discharge that is only exceeded, for example, 25% of the time, because larger discharges are those responsible for the creation of slope and meandering pattern.

11.3 Basic Parameters

The regime of an alluvial river can be characterized by three dimensionless parameters, including (1) energy slope S_e; (2) d/D (where d is the mean flow depth and D is the characteristic grain size or diameter); and (3) dimensionless effective shear stress τ_{*e}. Another important parameter is the specific gravity or relative density of grains γ_{s*} ($= \gamma_s/\gamma$), which is, however, almost constant at 2.68, where γ is the weight density of water, and γ_s is the weight density of sediment.

If these parameters are common between two streams, then it can be regarded that the streams satisfy geometric and dynamic similarity. This implies that the streams exhibit the same hydraulic resistance, width w of the water surface, and development of macro-formations, such as meandering. However, in reality, this will seldom be true because topographic features that have a dominant influence will be different. It may also be noted that Froude number F_r is also a basic parameter. However, Hansen (1967) has shown that two rivers that have the three parameters in common must correspond to the same F_r. Therefore, one can also consider S_e, τ_{*e}, and F_r as three basic parameters.

11.4 Similarity Principle

A system of non-dimensional parameters is obtained to completely characterize the flow system. This is achieved by the use of similarity considerations, assuming that the effect of viscous shear is negligible, i.e., for dunes and the upper flow regime. In the case of ripples, a more or less pronounced scale effect can be expected. The approach based on the similarity principle was proposed by Engelund and Hansen (1967), and the discussion here closely follows their work.

11.4.1 Flow Velocity, Friction Velocity, and Settling Velocity

The mean velocity v along the vertical in open channel flow can be expressed as

$$\frac{v}{v_f} = 6 + 2.5 \ \ln \frac{d}{k}, \tag{11.1}$$

in which v_f is the friction or shear velocity, d is the flow depth, and k is the sand roughness approximated as equal to $2.5D$ (the mean fall diameter). The friction velocity can be expressed as

$$v_f = \sqrt{\tau_0/\rho}, \tag{11.2}$$

where ρ is the mass density of water, and τ_0 is the average bed shear stress.

The setting velocity or fall velocity v_s, which depends on the size and specific gravity of sediment grains, slope, and drag, can be defined as

$$v_s = \sqrt{\frac{4(\gamma_{s*} - 1)gD}{3c_d}}, \tag{11.3a}$$

where D is the sediment grain size or diameter; γ_{s*} is relative sediment weight density $= \gamma_s/\gamma$; γ is the specific weight of water; γ_s is the specific weight of sediment; g is the acceleration due to gravity; and c_d is the drag coefficient that for a single grain depends on the Reynolds number $R_n = (v_sD)/v$, where D is the mean fall diameter. For very small values of R_n, $c_d = 24/R_n$, where v is the kinematic viscosity. The fall or settling velocity then becomes

$$v_s = \frac{(\gamma_{s*} - 1)gD^2}{18v}.$$ (11.3b)

11.4.2 Bed Shear Stress

The average bed shear stress, τ_0, in uniform open channel flow (or tractive force) can be defined as

$$\tau_0 = \gamma dS_e,$$ (11.4)

in which S_e is the energy slope. The dimensionless bed shear stress, τ_*, can be defined using the balance of forces (skin friction and drag force, shear force, and immersed weight) as

$$\tau_* = \frac{\tau_0 D^2}{[(\gamma_s - \gamma)D^3]} = \frac{\tau_0}{[\gamma(\gamma_{s*} - 1)D]} = \frac{dS_e}{(\gamma_{s*} - 1)D} = \frac{v_f^2}{g(\gamma_{s*} - 1)D}.$$ (11.5)

The sediment particles begin to move when the dimensionless shear stress exceeds a critical value τ_{*c}.

In channel flow, the total shear stress in the fluid particle mixture can be divided into the shear stress transmitted through the intergranular fluid and the shear stress transmitted due to the interchange of momentum caused by the encountering of sediment particles. On the channel bed with dunes, the total bed shear stress, τ_0, given by Equation (11.4), can be divided into two different parts:

$$\tau_0 = \tau_{0d} + \tau_{0f},$$ (11.6)

where τ_{0d} is the friction due to the surface of a dune, and τ_{0f} is the form drag on the dunes. It may be recalled that dunes are a bed configuration that is not known beforehand. The hydraulic resistance therefore arises due to sediment grains on the surface of the dunes and due to the dunes themselves. Sediment grains constitute small-scale roughness elements, while dunes are macroscale roughness elements. Hence, τ_{0d} is the part due to the friction on the dune surface, and τ_{0f} is the part due to the form drag on the dunes where the water pressure is higher on the rear side than on the leeward side (Engelund and Hansen, 1967). The transport of the bed load is related to τ_{0d}, hence the reason for designating it as effective bed shear. Figure 11.1 shows a plot of bed shear stress against velocity.

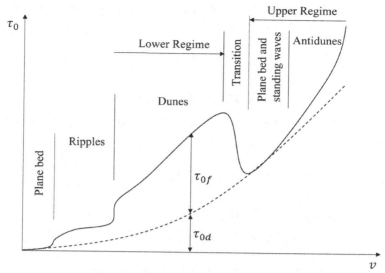

Figure 11.1 Bed shear stress against velocity

11.4.3 Energy Slope

The energy dissipation per unit length of the channel can be written as

$$\tau_0 v = \gamma(dv)S_e = \gamma q S_e, \tag{11.7}$$

where q is flow discharge per unit channel width. From the definition of friction factor, f, the total bed stress can be expressed as

$$\tau_0 = f\frac{1}{2}\rho v^2. \tag{11.8}$$

One can also write from the Darcy–Weisbach equation:

$$S_e = f\frac{1}{d}\frac{v^2}{2g}. \tag{11.9}$$

Using Equation (11.9), one can write velocity in terms of friction velocity as

$$v = \sqrt{\frac{2}{f}gdS_e} = v_f\sqrt{\frac{2}{f}}. \tag{11.10}$$

If Froude number F_r is defined as

$$F_r = \frac{v}{\sqrt{gd}}, \tag{11.11}$$

then Equation (11.9), with the use of Equation (11.11), can be written as

$$S_e = \frac{1}{2} f F_r^2. \tag{11.12}$$

11.4.4 Hydraulic Roughness

To determine if the channel bed is hydraulically smooth, the average thickness of the viscous sublayer, δ, is computed as

$$\delta = \frac{11.6 v}{v_f}. \tag{11.13}$$

If the thickness of the viscous sublayer is not distinguishable from the roughness elements of the bed, then the bed is said to be hydraulically rough. Otherwise, the bed will be hydraulically smooth.

Now the shear stress due to dunes can be expressed as

$$\tau_{0d} = f_d \rho \frac{v^2}{2}, \tag{11.14}$$

in which f_d is the friction factor due to the dune. Similarly, one can also express τ_{0f}.

11.4.5 Loss of Energy

The objective in the similarity approach is to characterize the flow system using the basic dimensionless parameters, which is plausible if the effect of viscous shear is neglected. To that end, the first step is to derive hydraulic resistance for a dune-covered bed. Over such a bed, the loss of mechanical energy is due to (1) friction and (2) flow expansion after each crest. In the opposite case of antidunes, although flow does not separate from the bed, breaking occurs at the water surface, which causes the loss of energy similar to that of expansion in the case of dunes. The discussion to follow focuses on the dune case but may apply to the upper flow regime as well.

The loss of energy due to expansion, ΔH^e, can be expressed using Carnot's formula (Rouse, 1950):

$$\Delta H^e = \alpha \frac{(v_1 - v_2)^2}{2g}, \tag{11.15}$$

where v_1 is the velocity above the crest before expansion, v_2 is the mean velocity downstream of the crest, which will be smaller than v_1, g is the acceleration due to gravity, and α is a dimensionless coefficient depending on the flow geometry.

Let the flow discharge per unit channel width be denoted by q, the dune height by h, and mean flow depth by d. Then, velocities v_1 and v_2 can be expressed as

$$v_1 = \frac{q}{d - \left(\frac{h}{2}\right)}; \quad v_2 = \frac{q}{d + \left(\frac{h}{2}\right)}, \tag{11.16}$$

where q is the water discharge per unit width ($Q/w = vd$). Substituting Equation (11.16) into Equation (11.15), one gets

$$\Delta H^e = \frac{\alpha q^2}{2g} \left[\frac{1}{d - \left(\frac{h}{2}\right)} - \frac{1}{d + \left(\frac{h}{2}\right)} \right]^2 \simeq \frac{\alpha v^2}{2g} \left(\frac{h}{d}\right)^2. \tag{11.17}$$

11.4.6 Energy Gradient

The energy gradient, denoted as S_e, defined as the loss of energy per unit length of the channel, comprising the sum of gradient due to friction (S_f) and that due to expansion, can be expressed as

$$S_e = S_f + \frac{\Delta H^e}{L}, \tag{11.18}$$

where L is the length of the dunes. Recalling that $\tau_0 = \gamma d S_e = \tau_{0d} + \tau_{0f}$ or

$$S_e = \frac{\tau_{0d}}{\gamma d} + \frac{\tau_{0f}}{\gamma d}. \tag{11.19}$$

Equation (11.19) can be expressed with the use of Equation (11.17) as

$$S_e = \frac{\tau_{0d}}{\gamma d} + \frac{1}{2} \frac{v^2}{gd} \frac{\alpha h^2}{dS_e}. \tag{11.20}$$

Hansen (1967) expressed Equation (11.20) as

$$S_e = \tau_{*d}(\gamma_{s*} - 1)\frac{D_{35}}{d} + \frac{v^2}{2gD} \left(\frac{d}{L}\right)\left(\frac{h}{d}\right)^2 \tag{11.21}$$

or

$$S_e = \tau_{*d}(\gamma_{s*} - 1)\frac{D_{35}}{d} + \frac{1}{2}\psi d^2, \tag{11.22}$$

where ψ is a resistance coefficient, depending on the relative dimensions of the dunes; τ_{*d} is the dimensionless dune-covered bed shear; v is the average flow velocity; D_{35} is the grain diameter corresponding to 35%; and γ_{s*} is the relative grain density.

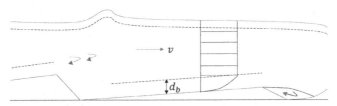

Figure 11.2 Flow over a dune-covered bed

Referring to Figure 11.2, it can be assumed that no exchange of energy occurs between the upper flow and the boundary layer flow. Therefore, the energy gradient of boundary layer flow, defined as the dissipated energy divided by unit weight and discharge, should equal that of the upper flow and that of the total flow, and can be written (Engelund and Hansen, 1967) as

$$S_e = \frac{\tau_{0b}}{\gamma v_b} \frac{v_b}{d_b} = \frac{f_b \rho v^2}{2 \gamma d_b} = \frac{f_b v^2}{2 g d_b}, \tag{11.23}$$

where v_b is the mean velocity of the boundary layer; d_b is the thickness of the boundary layer; f_b is the friction in the boundary layer; and τ_{0b} is the shear stress in the boundary layer. Since one can write the energy gradient as

$$S_e = \frac{f}{d} \frac{v^2}{2g}, \tag{11.24}$$

equating Equations (11.23) to Equation (11.24), one can obtain

$$\frac{f_b}{d_b} = \frac{f}{d}. \tag{11.25}$$

Recalling

$$\sqrt{\frac{2}{f}} = \frac{v}{\sqrt{g d S_e}}, \tag{11.26}$$

can obtain the expression

$$\sqrt{\frac{2}{f_b}} = \frac{v}{\sqrt{g d_b S_e}}. \tag{11.27}$$

The friction factor f_b for the boundary layer can be determined as

$$\sqrt{\frac{2}{f_b}} = c + 2.5 \ln \frac{d_b}{k}, \tag{11.28}$$

where k is the equivalent sand roughness as defined by Nikuradse (1933), and c is a constant.

Considering $f = f_b$ for $d = d_b$ and using $c = 6$ (Engelund and Hansen, 1967), one obtains

$$\frac{v}{\sqrt{gd_bS_e}} = 6.0 + 2.5\ln\frac{d_b}{k}. \tag{11.29}$$

Equation (11.29) was derived by Einstein (1950). Similar to Equation (11.8), one can write

$$\tau_{0b} = \frac{1}{2}f_b\rho v^2. \tag{11.30}$$

Using Equations (11.8), (11.23), (11.25), and (11.30), one obtains

$$\tau_{0b} = \gamma d_bS_e. \tag{11.31}$$

Equation (11.31) can also be written as

$$\tau_{0b} = \gamma dS_f = \gamma d_bS_e. \tag{11.32}$$

Dividing Equation (11.20) by $F_r^2/2$, $F_r =$ Froude number, friction factor f can be expressed as

$$f = f_b + \frac{\alpha h^2}{dL}, \tag{11.33}$$

where L is the dune length.

An important relationship between dimensionless parameters is expressed for the total stress $\tau = \tau_{0d} + \tau_{0f}$ by Equation (11.6), where

$$\tau_{0d} = \frac{d_bS_e}{(\gamma_s - 1)D} \tag{11.34}$$

$$\tau_{0f} = \frac{1}{2}F_r^2\frac{\alpha h^2}{(\gamma_s - 1)DL}, \tag{11.35}$$

where $\gamma_{s_*} = \left(\frac{\gamma_s}{\gamma}\right)$ is the relative density of sediment grains, and D is the mean fall diameter. τ_* is a dimensionless form of bed shear τ_0.

11.4.7 Geometric and Dynamic Similarity

Let there be two flow systems, 1 and 2, that are supposed to be geometrically similar, with the scale ratio $\lambda = d_1/d_2$, where d_i is the flow depth for system i ($i = 1, 2$), as shown in Figure 11.3. Assuming the bed configuration to be dunes, the conditions for complete similarity can be defined (Engelund and Hansen, 1967).

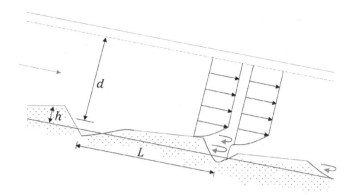

Figure 11.3 Two flow systems

For geometric and dynamic similarity, one can write

(1)
$$\frac{d_1}{D_1} = \frac{d_2}{D_2} \quad \text{or} \quad \lambda = \frac{d_1}{d_2} = \frac{D_1}{D_2}. \tag{11.36}$$

(2) Equal slopes

$$S_{e1} = S_{e2}; \quad \frac{S_{e1}}{S_{e2}} = \frac{f_1}{f_2}. \tag{11.37}$$

(3) The dimensionless effective bed shear stress must be the same:

$$\tau_{0e1} = \tau_{0e2}. \tag{11.38}$$

(4) The Froude number should be the same:

$$F_{r1} = \frac{v_1}{\sqrt{g_1 d_1}} = F_{r2} = \frac{v_2}{\sqrt{g_2 d_2}}. \tag{11.39}$$

The dune patterns will be similar when these conditions are similar. That is, $\tau_1 = \tau_2$ and $\tau_{0d1} = \tau_{0d2}$. This says that the expansion loss is the same fraction of the total loss of mechanical energy in both streams. The relative density is considered constant.

For two similar streams, the friction factor is the same. Because slopes are common, Froude numbers are identical. Thus, for a given sediment size D, the flow system is completely defined if d, S_e, and τ_{0d} are specified. Then, the scale ratios of sediment transport rates can be calculated as

$$\frac{q_{T1}}{q_{T2}} = \frac{v_{f1}}{v_{f2}} \frac{D_1}{D_2}, \tag{11.40}$$

where q_T is the total sediment discharge (bed load and suspended load) per unit width, v_f is the friction or shear velocity, and subscripts 1 and 2 refer to systems 1 and 2. The friction velocity can be written as

$$v_f = \sqrt{\tau_*}\sqrt{(\gamma_{s^*} - 1)gD}. \tag{11.41}$$

Then, Equation (11.40) can be written, with the use of Equation (11.41), as

$$\frac{q_{T1}}{q_{T2}} = \sqrt{\frac{(\gamma_{s*1} - 1)g_1 D_1^3}{(\gamma_{s*2} - 1)g_2 D_2^3}}. \tag{11.42}$$

From Equation (11.42), one can write the dimensionless sediment transport rate as

$$\phi_1 = \phi_2; \quad \phi = \frac{q_T}{\sqrt{(\gamma_{s^*} - 1)gD^3}}. \tag{11.43}$$

If q_B is the rate of bed load transport (volume of material per unit width per unit time), then the dimensionless form of bed load discharge can be expressed as

$$\phi_B = \frac{q_B}{\sqrt{(\gamma_{s^*} - 1)gD^3}}. \tag{11.44}$$

For small sediment discharge $\phi_B < 10$, Engelund and Hansen (1967) expressed

$$\phi_B = 8(\tau_* - 0.047)^{\frac{3}{2}}, \tag{11.45}$$

where τ_* is the dimensionless form of bed shear. Equation (11.45) shows that the dimensionless sediment transport rate ϕ is a function of dimensionless parameters d/D, S_e, and τ_* and can therefore be written as

$$\phi = \phi\left(\frac{d}{D}, S_e, \tau_{0d}\right) \quad \text{or} \quad \phi = \phi\left(\tau_*, \frac{d}{D}, \tau_{0d}\right). \tag{11.46}$$

For comparing streams of different slopes, it is assumed that effective bed shear stresses are equal, that is, $\tau_{0d1} = \tau_{0d2}$, which corresponds to dynamic similarity. Further, the loss of energy due to expansion must be the same fraction of the total energy loss, which means that the total bed shear is divided by the same proportion in both cases. This can be further explained as follows.

From Equation (11.33), one can write

$$f_1 = f_{b1} + \frac{a_1 h_1^2}{d_1 L_1}; \quad f_2 = f_{b2} + \frac{a_2 h_2^2}{d_2 L_2}, \tag{11.47}$$

where h is the dune height. From equal slope similarity, one can express

$$\frac{f_1}{f_2} = \frac{f_{b1}}{f_{b2}} = \frac{\lambda_H}{\lambda_L}, \tag{11.48}$$

where λ_L is the horizontal length scale ratio and λ_H is the vertical length scale ratio ($= d_1/d_2$). Equation (11.48) can also be written as

$$\frac{L_1 f_1}{h_1} = \frac{L_2 f_2}{h_2}. \tag{11.49}$$

One can also express for f_b in a similar fashion.

It is now hypothesized that alluvial streams adjust their bed roughness in accordance with the principle of similarity and distorted vertical scale. Since

$$\frac{f_b}{f} = \frac{\tau_{0b}}{\tau_*}, \tag{11.50}$$

it can be inferred that parameter τ_* must be common for the two streams. Experiments by Guy et al. (1966) show that

$$\tau_{0d} = 0.06 + 0.4\tau_*^2 \tag{11.51}$$

and for $\tau_* > 0.15$, Equation (11.51) can be approximated as

$$\tau_{0d} = 0.4\tau_*^2. \tag{11.52}$$

Now the sediment discharge (the total volume of moving sand particles per unit time) needs to be determined. As before, the focus will be on the dune case. It is assumed that the sediment is eroded from the stream side of the dune and is deposited on the lee side. This elevates the moving sediment load q_T to a height comparable to the dune height h. Using energy considerations, Engelund and Hansen (1967) derived

$$f q_T \left(\frac{h}{fL}\right) = \alpha \frac{(\tau_{0d} - \tau_{0c})}{(\gamma_s - \gamma)D} v_f D. \tag{11.53}$$

Using the expression for dimensionless shear τ_*,

$$\tau_* = \frac{\tau_{0d} D^2}{(\gamma_s - \gamma)D^3} = \frac{\tau_{0d}}{\gamma(\gamma_{s*} - 1)D} = \frac{dS_e}{(\gamma_{s*} - 1)D} = \frac{v_f^2}{(\gamma_{s*} - 1)gD}, \tag{11.54}$$

Equation (11.53) can be written as

$$f q_T \left(\frac{h}{fL}\right) = \alpha(\tau_{0d} - 0.06)\sqrt{\tau_*}\sqrt{(\gamma_{s*} - 1)gD^3}. \tag{11.55}$$

The dimensionless sediment discharge can be defined as

$$\phi = \frac{q_T}{\sqrt{(\gamma_{s*} - 1)gD^3}}. \tag{11.56}$$

Considering Equations (11.55) and (11.56), the transport equation can be written as

$$f\phi \sim \tau_*^{\frac{5}{2}}. \tag{11.57}$$

If the mean particle velocity is assumed to be proportional to the frictional velocity,

$$v_f = \sqrt{\frac{\tau_{0d}}{\rho}}. \tag{11.58}$$

Then

$$f\phi \sim (\tau_{0d} - 0.06)\sqrt{\tau_{0d}}. \tag{11.59}$$

This yields the sediment transport relation as

$$f\phi = 0.077\tau_*^2\sqrt{\tau_*^2 + 0.15}. \tag{11.60}$$

For small values of τ_*, Equation (11.60) can be approximated as

$$f\phi \sim \tau_*^2. \tag{11.61}$$

Equation (11.61) is applicable for bed load discharge only. For large values of τ_*, Equation (11.61) can be expressed as

$$f\phi \sim \tau_*^3. \tag{11.62}$$

Equation (11.62) corresponds to the case when suspended load is the dominant sediment discharge.

In order to derive the relations of depth, width, slope, and meander length with water and sediment discharge as well as grain size, hydraulic resistance, and transport capacity of alluvial streams are expressed as

$$v = 10.9D^{-\frac{3}{4}}d^{\frac{5}{4}}S_e^{\frac{9}{8}}, (\tau_* > 0.15 \text{ and for dunes only}) \tag{11.63}$$

$$f\phi = 0.1\,\tau_*^{\frac{5}{2}}, (\tau_* > 0.15). \tag{11.64}$$

Taking L for the wave or meander length, and w for width,

$$f\frac{L}{d} = c_1 \quad (c_1 = 14) \tag{11.65}$$

$$f\frac{w}{d} = c_2 \quad (c_2 = 1.4). \tag{11.66}$$

The principle of similarity predicts

$$\frac{L}{w} = 10 \tag{11.67}$$

$$f\frac{L}{d} = 14. \tag{11.68}$$

Utilizing the definitions of f given by Equation (11.37), τ_* given by Equation (11.5), ϕ given by Equation (11.57), and Equations (11.63), (11.64), (11.67), and (11.68), one gets

$$w = 0.78 D^{-0.316} Q^{0.525} \tag{11.69}$$

$$d = 0.108 \left(\frac{Q_T}{Q}\right)^{-\frac{2}{7}} D^{0.21} Q^{0.317} \tag{11.70}$$

$$S_e = 12.8 \left(\frac{Q_T}{Q}\right)^{\frac{4}{7}} D^{0.527} Q^{-0.212} \tag{11.71}$$

$$L = 7.8 D^{-0.316} Q^{0.525}. \tag{11.72}$$

11.5 Comparison with Regime Relations

Equations (11.69)–(11.72) are compared with regime equations proposed by Blench (1966), as shown in Table 11.1. In terms of the exponents of Q, the two sets of expressions are almost identical. However, in terms of the exponents of d, the deviations are noted.

Example 11.1: Compute the stage–discharge relation for Pigeon Roost Creek near Byhalia, whose slope (S_e) is given as 9×10^{-4}, the value of γ_{s*} is 2.68, grain size from sieve analysis $D_{35} = 0.35$ mm and $D_{65} = 0.45$ mm. The mean grain diameter $D = 0.38$ mm. The mean water depth d can be taken as 0.30 m. What will be the corresponding discharge?

Table 11.1. *Comparison of similarity-based hydraulic relations with regime relations*

Hydraulic variable	Variation with water discharge Q and grain size D	
	Similarity principle-based relations	Regime relations
Width, w	$d^{-0.316} Q^{0.525}$	$d^{-\frac{1}{4}} Q^{\frac{1}{2}}$
Depth, d	$d^{0.21} Q^{0.312}$	$Q^{\frac{1}{3}}$
Slope, S_e	$d^{0.527} Q^{-0.212}$	$d^{\frac{1}{2}} Q^{-\frac{1}{6}}$

Solution: 1. Compute the value of dimensionless shear stress

$$\tau_* = \frac{dS_e}{(\gamma_{s*} - 1)D} = \frac{0.3 \times 9 \times 10^{-4}}{(2.68 - 1) \times 0.38 \times 10^{-3}} = 0.422.$$

2. Compute τ_{0d} as

$$\tau_{0d} = 0.06 + 0.4\tau_*^2 = 0.06 + 0.4 \times 0.422^2 = 0.132.$$

3. Obtain the value of d_b as

$$d_b = \frac{\tau_{0d}}{\tau_*} d = \frac{0.132}{0.422} 0.30 = 0.094 \text{ m.}$$

4. Compute mean velocity v with surface roughness $k = 2.5D = 0.95$ mm as

$$\frac{v}{\sqrt{gd_bS_e}} = 6 + 2.5 \ln \frac{d_b}{k}.$$

so $v = \sqrt{gd_bS_e} \times \left(6 + 2.5 \ln \frac{d_b}{k}\right) = \sqrt{9.81 \times 0.094 \times 9 \times 10^{-4}} \times$
$\left(6 + 2.5 \ln \frac{94}{0.95}\right) = 0.5$ m/s.

5. Compute discharge per unit width,

$$q = vd = 0.50 \times 0.30 = 0.15 \text{ m}^2/\text{s.}$$

6. Compute the friction factor,

$$f = \frac{2S_e}{F_r^2} = \frac{2gdS_e}{v^2} = \frac{2 \times 9.81 \times 0.30 \times 9 \times 10^{-4}}{0.5 \times 0.5} = 0.021.$$

7. Compute ϕ using

$$\phi = \frac{1}{f}0.1 \tau_*^{\frac{5}{2}} = \frac{0.1}{0.021} \times (0.422)^{\frac{5}{2}} = 0.55.$$

8. Compute total sediment discharge q_T as

$$q_T = \phi\sqrt{(\gamma_{s*} - 1)gD^3} = 0.55\sqrt{(2.68 - 1) \times 9.81 \times (0.38 \times 10^{-3})^3}$$
$$= 1.65 \times 10^{-5} \text{ m}^2/\text{s.}$$

9. Now check whether the flow is in the dune regime. Compute v_f

$$v_f = v\sqrt{\frac{f}{2}} = 0.5\sqrt{\frac{0.021}{2}} = 0.051 \text{ m/s.}$$

10. Compute Reynolds number,

$$R = \frac{v_f D}{v} = \frac{0.051 \times 0.38 \times 10^{-3}}{1.3 \times 10^{-6}} = 15.$$

Since $R > 12$, the bed form is actually dune, not ripple.

Example 11.2 Compute the natural depth and slope of an alluvial channel 10 m wide carrying a water discharge of 6.0 m^3/s and sediment discharge of 2.6×10^{-4} m/s. The sediment particle size is 0.4 mm and $d/D = 2.4 \times 10^3$.

Solution: 1. Compute

$$\frac{vd}{\sqrt{g(\gamma_{s*} - 1)D^3}} = \frac{0.6}{\sqrt{9.81 \times (2.68 - 1) \times (0.4 \times 10^{-3})^3}} = 1.81 \times 10^4$$

2. Compute ϕ,

$$\phi = \frac{Q_T}{10 \times \sqrt{(\gamma_{s*} - 1)gD^3}} = \frac{2.6 \times 10^{-4}}{10 \times \sqrt{1.68 \times 9.81 \times (0.4 \times 10^{-3})^3}} = 0.8$$

3. Then, using $D = 0.4$ mm and the design chart $S_e = 3.0 \times 10^{-4}$ and $d = 0.96$ m.

References

Blench, T. (1966). *Mobile-bed fluviology*. Dept. Tech. Serv., Tech III Div., University of Alberta, Edmonton.

Einstein, H. A. (1950). *The bed-load function for sediment transportation in open channel flows*. Technical Bulletin No. 1026, U.S. Department of Agriculture, Washington, DC.

Engelund, F. and Hansen, E. (1967). *A monograph on sediment transport in alluvial streams*. Technical University of Denmark, Copenhagen.

Guy, H. P., Simons, D. B., and Richardson, E. V. (1966). *Summary of alluvial channel data from flume experiments, 1956–1961*. U.S. Geological Survey Professional Paper-462-I, Washington, DC.

Hansen, E. (1967). *On the formation of meanders as a stability criterion*. Basic Research Progress Report No. 13, Hydraulic Laboratory, Technical University of Denmark, Copenhagen.

Nikuradse, J. (1933). Stromunggsgesetze in rauhen Rohren. VDI-Forschungsheft, No. 361.

Rouse, H., editor (1950). *Engineering Hydraulics*. John Wiley and Sons, New York, pp. 413.

12

Channel Mobility Theory

Alluvial channels are continuously modified by sediment movement and exist in comparative equilibrium. The longitudinal profiles and cross-sections of these channels depend on hydraulic and sediment factors and boundary conditions that govern channel morphology. This chapter discusses the theory of channel mobility leading to stable hydraulic geometry.

Notation

A = cross-sectional area of the mean rate of flow at the mean level of tidal flow

a_0 = coefficient depending on the amplitude of fluctuation of low rate

b_0 = proportionality coefficient

C = Chezy's coefficient

d = mean channel depth

D = sediment size

f = friction factor

F = force

F_d = force characterizing the stability of bed sediment load

F_p = force representing the stability of channel sediment load

g = acceleration due to gravity

K = parameter

K_p = generalized index of channel mobility

m = correction coefficient

M_{maxn} = parameter of the fluctuation of stream velocity

n = ratio of the bottom velocity to the mean velocity

n_m = roughness coefficient

P_i = probabilities

Q = flow rate during T

Continued

Q_c = sediment transport capacity (competence) as turbidity

$\overline{Q_c}$ = mean sediment transport capacity

Q_n = flow discharge, equal to the discharge with an annual probability of 2%

$\widetilde{Q_s}$ = competence for the mean flow rate

R_k = spatial derivative of K_p

S = channel slope

T = time

T_0 = total duration of low tide during period T

v = mean stream velocity

V = volume of water in the estuary

V_Q = channel carrying capacity

V_s = stream transport capacity

V_{sL} = volume of sediment transported at low tide during period T_{OL}

V_T = volume of tidal water

V_{TL} = volume of water flow at low tide during period T_{OL}

V_{Ts} = volume of sediments

v_{cs} = critical flow velocity at which the sediment no longer remains in suspension

v_{cb} = critical flow velocity at which the displacement of the bottom sediment particles ceases

w = channel width at the water surface

α = relative stability coefficient

α_s = stability of bank slope

α_Δ = stability of bottom sediments

β = ratio of the mean turbidity at the bottom, determined by the turbidity distribution formula

γ = specific weight of water

γ_s = volume of the sediment particles by weight or the specific sediment weight

r_A = ratio of width to depth (w/d)

Δ = fluctuation

12.1 Introduction

When a reservoir is built on a river, the channel below the dam or spillway will be subject to deformation. This same phenomenon is also observed in tidal estuaries when containing embankments or sluices are constructed to block the tidal water. Channel deformation and hydro-construction on rivers and their estuaries are intimately connected. The alluvial channels of lowland rivers and their tidal estuaries are continuously modified by the action of sediment-laden currents. For a majority of rivers, the deformation is reversible and has a quasi-periodic character. All alluvial channels, except individual depositional or erosional ones, exist in comparative equilibrium. The shapes of longitudinal profiles and cross-sections of such channels depend on hydraulic and sediment factors and boundary conditions. These factors therefore govern the channel morphology.

Since channel deformation is closely connected with sediment movement, the channel morphology must depend on the dynamics of streamflow. There is reason to believe that all of the sediment load (i.e., sediment of all sizes) can take part in channel formation, and consequently the overall quantity of sediment is one of the factors determining channel morphology. Although the volume of bottom deposits influences channel morphology, except for a small number of streams, in the alluvial channels of lowland streams, suspended sediments comprise the major portion of sediment transport. The classification of sediments into bed and non-bed, based on usual methods, is quite arbitrary. The stability of hydraulic geometry depends on the movement of sediment on the bed and banks and subsequent transport. This chapter discusses the theory of channel mobility following Tou (1965).

12.2 Sediment Transport

If channel erosion and deposition can mutually compensate for each other and are equal during time T, then the channel-carrying capacity (V_Q) and the stream transport capacity (V_s) must correspond to the liquid and solid runoff of the basin:

$$V_Q = \int_0^T wdv\, dt = \overline{w}\overline{d}\overline{v}T \tag{12.1}$$

$$V_s = \int_0^T wdvQ_s\, dt = \overline{w}\overline{d}\overline{v}\overline{Q_c}T \tag{12.2}$$

where w is the channel width at the water surface, d is the mean channel depth, v is the mean stream velocity, and Q_c is the sediment transport capacity (competence) expressed as turbidity. An over bar on hydraulic variables denotes that their values correspond to mean flow rate; $\overline{Q_c}$ is the mean sediment transport capacity during the period T, which can be expressed as

$$\overline{Q_c} = m\widetilde{Q_s}, \tag{12.3}$$

where $\widetilde{Q_s}$ is the competence (sediment transport) for the mean flow rate, and m is the correction coefficient whose magnitude is determined by the hydrograph shape and other hydrologic factors.

This chapter follows the work of Tou (1965), who developed the following equation for sediment discharge:

$$Q_c = K \frac{v^3}{g d v_{cs}}, \tag{12.4}$$

where v_{cs} is the critical flow velocity at which the sediment no longer remains in suspension, and K is a parameter determined as

$$K = 0.055 \gamma_s n \beta \sqrt{g/C}, \tag{12.5}$$

where γ_s is the volume of the sediment particles by weight or the specific sediment weight; n is the ratio of the bottom velocity to the mean velocity, determined from the velocity distribution; β is the ratio of the mean turbidity at the bottom, determined by the turbidity distribution formula; g is the acceleration due to gravity; and C is Chezy's coefficient. The critical velocity corresponding to the cessation of sediment particle suspension is determined as

$$v_{cs} = \frac{2.24}{M_{maxn}} \sqrt{\frac{\gamma_s - \gamma}{\gamma}} gD, \tag{12.6}$$

where M_{maxn} is the parameter of the fluctuation of stream velocity, γ is the specific weight of water, and D is the sediment size. For $d/\Delta > 1,000$, one can assume that $M_{maxn} \simeq 1$ (Tou, 1965), where Δ is the fluctuation.

Let Q be the average flow rate during T, Q_c the average turbidity, that is, $Q = V_Q/T$ and $Q_c = V_S/QT$, then substituting Equation (12.4) in Equation (12.2) and solving simultaneously with Equation (12.1), one obtains

$$d = K \frac{v^3}{g \, Q_c v_{cs}} \tag{12.7}$$

$$w = \frac{g V_{cs} Q_c Q}{K v^4} \tag{12.8}$$

$$\frac{w}{d} = \frac{g^2 v_{cs}^2 Q_c^2 Q}{K^2 v^7}. \tag{12.9}$$

The mean flood flow rate is denoted by Q and the mean turbidity during the flood period by Q_c. Because of compensating influences of erosion and deposition over several years, Q is the mean multi-year water flow rate, Q_c is the multiannual turbidity, w is the mean width, and d is the mean depth during the average multiannual rate of water flow (discharge). It may be noted that for simplicity the overbar sign denoting mean value has been omitted. If the total change of the river channel approaches zero during the flood period, then T is determined by the duration of the flood. For most lowland rivers and tidal estuaries, $K = 3 \sim 5$.

12.3 Hypothesis of Channel Mobility

A channel can have different widths and depths for carrying a specific volume of water and sediment. This implies that the equations governing the movement of water and sediment lead only to a necessary but insufficient condition to uniquely determine channel morphology. For a given width, the governing equations produce the corresponding values of depth and velocity. Likewise, the width and velocity for a given depth can be determined from these equations. Under natural conditions without any outside interference, the continuing erosion and deposition of sediment, however, influence the width and depth of eroding channels, and neither depths nor velocities can be determined a priori. Analysis of the morphology of river channels shows that despite the varied forms of cross-sections and varied widths and depths that exist, the range of variation is small. For stable channels, cross-sections have closely similar forms.

A channel can have variable widths and depths for transporting the specific volumes of water and sediment. The question arises: Why does morphology of alluvial channels exhibit periodicity? The reason is that channels with different cross-sections possess different degrees of mobility for given water and sediment loads and boundary conditions. In the process of deformation the channel tends to take on a form so that its mobility would be minimal. Such a tendency of the channel can be called the principle of minimal channel mobility. It should, however, be noted that "minimum" means "relative minimum," that is, the minimal mobility under given conditions. This means that channels existing under different conditions exhibit different levels of minimal mobility.

12.3.1 Index of Mobility

The next question is how to quantify or index the channel mobility. The channel mobility can be indexed by the amplitude of the longitudinal and cross-sectional deformation of the channel. The amplitude of deformation depends on the amplitude of fluctuation of water and sediment loads, as well as on the rate at which the channel is eroded by the stream, and certain factors. For a given section, the channel mobility depends on the relative velocity on one hand and on the ratio of channel width to channel depth on the other. Under the condition of constant active cross-sectional area, the larger the ratio of width to depth, the greater the competence (turbidity) of the channel, and consequently the greater will be the volume of sediment transported by the stream, if the stream is not overloaded. It is observed that the wider and shallower the channel is the greater is its mobility. Therefore, the ratio of channel width to its depth is often considered as one of the indices of channel mobility.

At the same time, the greater the stream velocity and consequently the greater the current force acting on the channel, the less the channel mobility. Therefore, the ratio of active force F to the force F_p, which represents the stability of the channel sediment load, can be regarded as another index of channel mobility. The force F_p can be expressed as the force F_d, which characterizes the stability of the bed sediment load, and coefficient α, which takes into account the relative stability of the soil of bank slopes and bottom, that is, it takes on the equality:

$$F_{p,y} = \alpha^2 F_{d,y}. \tag{12.10}$$

Since F and F_p are, respectively, proportional to the square of the current velocity and the square of the critical velocity (v_{cb}) at which the displacement of the bottom sediment particles ceases, the following holds:

$$\frac{F}{F_d} = \frac{v^2}{\alpha^2 v_{cb}^2}, \tag{12.11}$$

where v_{cb} is computed from Equation (12.6) when the mean diameter of the bottom sediment particles D is known. Since the ratio $\left(\frac{v}{\alpha v_{cb}}\right)^2$ and the ratio w/d are two partial indices of channel mobility, it is natural to assume the generalized index of channel mobility by taking a version of the weighted average as:

$$K_p = a_0 \left[\left(\frac{v}{\alpha v_{cb}}\right)^2 + b_0 \frac{w}{d} \right], \tag{12.12}$$

where a_0 is a coefficient depending on the amplitude of the fluctuation of the flow rate and is constant for a given cross-section, and b_0 is a proportionality coefficient.

In the process of erosion and deposition, the channel tends to take on a form corresponding to least mobility. Therefore, the morphology of river channels must satisfy the following conditions:

$$\frac{\partial K_p}{\partial v} = 0; \quad \frac{\partial K_p}{\partial d} = 0; \quad \frac{\partial K_p}{\partial w} = 0. \tag{12.13}$$

These conditions lead to the same results, so anyone of them can be employed to complete the system of equations. If the first of these conditions is used, then w/d must be expressed as a function of velocity alone. Then, substituting Equation (12.9) in Equation (12.12) and then in Equation (12.13), the velocity for minimal channel mobility is obtained as:

$$v = \left(\frac{7b_0}{2}\right)^{\frac{1}{9}} \left[\frac{g^2 v_{cs}^2 Q_c^2 \alpha^2 v_{cb}^2 Q}{K^2} \right]^{\frac{1}{9}}. \tag{12.14}$$

Equation (12.14) needs to be validated using range of field data. It may be noted that when taking $\frac{\partial^2 K_p}{\partial v^2}$, K_p has a minimal value during $\frac{\partial K_p}{\partial v} = 0$.

12.4 Hydromorphometric Relationships

During deformation the channel corresponds to minimal mobility; therefore, the actual shape of the cross-section corresponds to the most stable form for given conditions. Therefore, to determine the stable channel form, one can utilize relationships based on the hypothesis of minimal channel mobility. Substitution of Equation (12.14) in Equations (12.7) and (12.8) produces the following relations:

$$d = \left(\frac{7b_0}{2}\right)^{\frac{1}{3}}\left(\frac{K\alpha^2 v_{cb}^2 Q}{gv_{cs}Q_c}\right)^{\frac{1}{3}} = fQ^{\frac{1}{3}}, \quad f = \left(\frac{7b_0}{2}\right)^{\frac{1}{3}}\left(\frac{K\alpha^2 v_{cb}^2}{gv_{cs}Q_c}\right)^{\frac{1}{3}}; d \propto Q^{\frac{1}{3}} \quad (12.15)$$

$$w = \left(\frac{2}{7b_0}\right)^{\frac{4}{9}}\left(\frac{gv_{cs}Q_c Q^5}{K\alpha^8 v_{cb}^8}\right)^{\frac{1}{9}} = kQ^{\frac{5}{9}}, \quad k = \left(\frac{2}{7b_0}\right)^{\frac{4}{9}}\left(\frac{gv_{cs}Q_c}{K\alpha^8 v_{cb}^8}\right)^{\frac{1}{9}}; w \propto Q^{\frac{5}{9}} \quad (12.16)$$

$$\frac{w}{d} = \left(\frac{2}{7b_0}\right)^{\frac{7}{9}}\left(\frac{g^4 v_{cs}^4 Q_c^4 Q^2}{K^4 \alpha^{14} v_{cb}^{14}}\right)^{\frac{1}{9}} = f_d Q^{\frac{2}{9}}, \quad f_d = \left(\frac{2}{7b_0}\right)^{\frac{7}{9}}\left(\frac{g^4 v_{cs}^4 Q_c^4}{K^4 \alpha^{14} v_{cb}^{14}}\right)^{\frac{1}{9}}; \frac{w}{d} \propto Q^{\frac{2}{9}} \quad (12.17)$$

$$A = \left(\frac{2}{7b_0}\right)^{\frac{1}{9}}\left(\frac{K^2 Q^8}{g^2 v_{cb}^2 Q_c^2 \alpha^2 v_{cs}^2}\right)^{\frac{1}{9}} = a_1 Q^{\frac{8}{9}}, \quad a_1 = \left(\frac{2}{7b_0}\right)^{\frac{1}{9}}\left(\frac{K^2}{g^2 v_{cb}^2 Q_c^2 \alpha^2 v_{cs}^2}\right)^{\frac{1}{9}}; A \propto Q^{\frac{8}{9}},$$

$$(12.18)$$

where A is the cross-sectional area of the mean rate of flow at the mean level of tidal flow. Based on analysis of field data on rivers, $b_0 = 0.15$.

The relative stability coefficient α can be defined as

$$\alpha = \frac{\alpha_s}{\alpha_\Delta}, \quad (12.19)$$

where α_s and α_Δ represent, respectively, the stability of the bank slope and bottom sediments. Table 12.1 shows the approximate values of these parameters. For most rivers, the channel bottom is composed of fine sand and bank slope of silt, so the values of α_s and α_Δ are of the same order of magnitude. Therefore, $\alpha \approx 1.0$.

In equations of Lacey (1930), Blench (1952, 1953), Leopold and Maddock (1953), Rybkin (1947), and Velikanov (1958), the channel depth is proportional to the rate of flow in the ratio 0.22~0.4, as shown in Table 12.2. The channel width is proportional to the rate of flow in the ratio 0.5~0.57. The cross-sectional area is of 0.83~0.9. In the derived Equations (12.15)–(12.19), the depth is proportional to the rate of water flow in the ratio of 0.33; the width is proportional to the rate of water flow in the ratio 0.56; and the cross-sectional area is proportional to the rate of

Table 12.1. *Approximate values of parameters*

Classification of bottom and bank sediments	Diameter of sediment particles (mm)	Stability of index	Remarks
1. Coarse sand	2.0–1.0	2.5–2.0	1. In the presence of vegetation cover, the stability must increase slightly.
2. Medium sand	1.0–0.25	2.0–1.2	
3. Fine sand	0.25–0.05	1.2–0.8	
4. Silt	0.05–0.01	1.2–0.8	
5. Silty clay	0.01–0.005	1.2–1.7	2. For channel with bank protection structures 3.0.
6. Clay	0.005	1.8–2.2	
7. Heavy clay	0.005	2.3–2.5	

Table 12.2. *Hydromorphological relations*

Author	d	V	A
Lacey	$Q^{0.33}$	$Q^{0.5}$	$Q^{0.83}$
Blench	$Q^{0.33}$	$Q^{0.5}$	$Q^{0.83}$
Leopold	$Q^{0.4}$	$Q^{0.5}$	$Q^{0.9}$
Rybkin	$Q^{0.22}$	$Q^{0.57}$	$Q^{0.79}$
Velikanov	$Q^{0.3}$ or $Q^{0.35}$	$Q^{0.54}$ or $Q^{0.40}$	$Q^{0.84}$ or $Q^{0.75}$
O'Brien	–	–	$Q^{0.85}$
Brunn	–	–	$Q^{0.1}$
Tou	$Q^{0.33}$	$Q^{0.56}$	$Q^{0.89}$

water flow in the ratio of 0.89. Figures 12.1–12.3 compare the results of Equations (12.15)–(12.19) with measured values of channel characteristics (after Tou, 1965). Equations (12.15)–(12.19), to an extent, generalize the results of many investigators.

The data used included a wide range of variation of the principal values, namely: the ratio of water flow varied from 0.11 to 200,000 m³/s, turbidity from 0.18 to 25 kg/m³, bottom particle diameter from 0.22 to 0.79 mm, channel width from 1.78 to 19,200 m, channel flow depth from 0.26 to 19.0 m, and the cross-sectional area from 0.46 to 148,000 m².

12.5 Tidal Estuaries Morphology

The channels of tidal estuaries exhibit cyclical deformation and differ from those of nontidal rivers by even greater complexity. The periodicity is not only determined by the transport of water and sediment loads but also by the tidal water flow and deposition. Because of the high and low tide currents and related factors changing

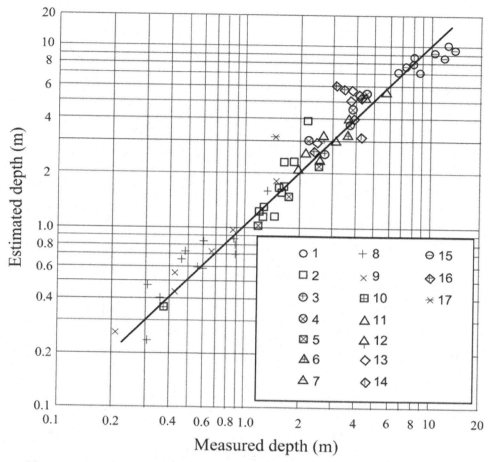

Figure 12.1 Comparison of computed and measured values of flow depth (after Tou, 1965)

constantly, the channel processes in tidal estuaries vary quite rapidly and intensively. Nevertheless, long-term observations suggest that the channels of tidal estuaries can be considered to be in a state of equilibrium, and channel morphology is governed principally by the flow of water and sediment from the basin and the sea.

If the total channel deformation of tidal estuaries reduces to zero during T_0, the channel should be able to convey the total volume of water and sediment during low tide. Let V_T and V_{Ts} denote the volume of tidal water and sediments. Then, one can write the following:

$$V + V_T = wdvT_0 \tag{12.20}$$

$$V_s + V_{Ts} = wdvQ_cT_0, \tag{12.21}$$

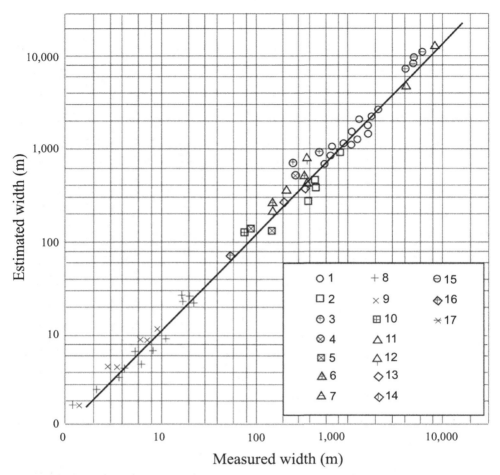

Figure 12.2 Comparison of computed and measured values of channel width (after Tou, 1965)

where V is the volume of water in the estuary, V_s is the volume of sediment in the estuary, w is the channel width at the mean water level, d is the mean depth, v is the mean velocity of the current at low tide, T_0 is the total duration of low tide during period T, and Q_c is the transport competence of the flow at low tide, which is determined by Equation (12.4). If we denote the volume of water flow at low tide during period T_{0L} by V_{TL}, and the volume of sediment transported at low tide by the current during the same period by V_{sL}, the mean turbidity by $\overline{Q_c}$,

$$V + V_{TL} = V_{TL} = QT_{0L} \tag{12.22}$$

$$V_s + V_{sL} = Q_{sL} = Q\overline{Q_c}\, T_{0L}. \tag{12.23}$$

During one year if the channel erosion and deposition offset each other, Q denotes the mean annual flow rate at low tide. Therefore, Q and Q_c should be considered as

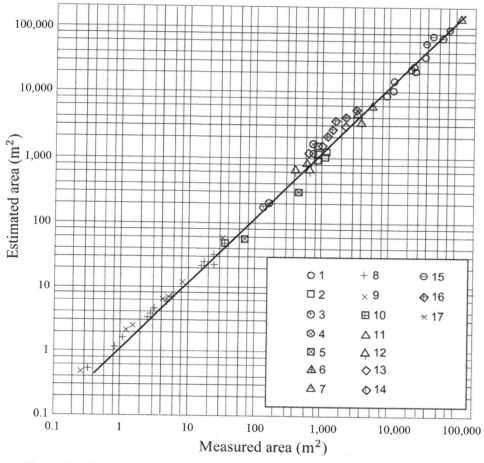

Figure 12.3 Comparison of computed and measured values of cross-sectional area (after Tou, 1965)

mean multiyear values. Equations (12.22) and (12.23) have the same form as Equations (12.1) and (12.2), and therefore hydromorphometric relations should be of the same form for estuarial sections.

12.6 Relation of Channel Width to Depth and Degree of Widening of the Channel of Tidal Estuaries

The ratio of channel width to flow depth depends principally on the water and sediment loads, that is, the value of w/d is proportional to $Q^{2/9}Q_c^{4/9}$, assuming that the bottom sediment particle sizes are constant, as shown by Equation (12.17). For rivers with similar suspended sediment load (turbidity), the greater the rate of water flow, the greater is *w/d*, and for rivers with equal rate of water flow, the greater the turbidity, the greater the *w/d* ratio. If the sediments of the bank slopes have high stability in comparison to the bottom (i.e., $\alpha > 1$), the channel will have a smaller value of *w/d*, other conditions being equal. On the contrary, if the bank slopes are easier to erode in comparison to the bottom (i.e., $\alpha < 1$), the channel has a greater value of *w/d*.

For tidal estuaries, as the rate of water flow increases toward the sea, the channel becomes wider and such channel widening along its course can be expressed as

$$\frac{\partial w}{\partial x} = 1.33 \ g^{\frac{1}{9}} \frac{\partial}{\partial x} \left(\frac{v_{cs} Q_c Q^5}{K \alpha^8 v_{cb}^8} \right)^{\frac{1}{9}}. \tag{12.24}$$

Equation (12.24) shows that the degree of channel widening along its course depends mainly on the amount of increase in the value $(Q_c Q^5)^{1/9}$. For a small increase in the rate of water flow along the course of nontidal rivers, their channel width increases slowly along the way, but a different picture is observed for tidal estuaries. The flow rate changes continuously along the way, depending on the variation of the "tidal prism," necessitating a considerable increase in width along the course toward the sea. The greater the amplitude of fluctuations of the tidal water level and estuary volume, the greater is the change in the rate of water flow along the course, and consequently the channel widens more rapidly.

12.7 Longitudinal Channel Profile

The slope of the longitudinal profile is determined by the hydromorphometric equations derived earlier. The flow equation for a stream with nonuniform flow can be expressed as

$$S = \frac{v^2}{C^2 d} + \frac{1}{2g} \frac{\partial v^2}{\partial x} + \frac{\partial d}{\partial x}, \tag{12.25}$$

where S is the channel slope, C is Chezy's friction factor, defined as $C = \left(\frac{1}{n_m}\right) d^{\frac{1}{6}}$, and n_m is the Manning roughness coefficient.

Substituting Equations (12.14) and (12.15) in Equation (12.25) yields

$$S = 1.15 \ n_m^2 \left(\frac{g^4 v_{cs}^4 Q_c^4}{K^4 \alpha^2 v_{cb}^2 Q} \right)^{\frac{2}{9}} + \frac{0.43}{g^{\frac{5}{9}}} \frac{\partial}{\partial x} \left(\frac{v_{cs}^2 Q_c^2 \alpha^2 v_{cb}^2 Q}{K^2} \right)^{\frac{2}{9}} + 0.807 \frac{\partial}{\partial x} \left(\frac{K \alpha^2 v_{cb}^2 Q}{g v_{cs} Q_c} \right)^{\frac{1}{3}}. \tag{12.26}$$

Owing to small changes in the rate of water flow and sediment along the course for nontidal streams, the first right term in Equation (12.26) becomes predominant such that the other two terms can be discarded. Therefore, the parameter in computing the gradient is the quantity $(Q_c^4/Q)^{2/9}$, showing that the greater the turbidity the greater the gradient.

In tidal estuaries, the rate of water flow is considerably larger than the rate of water flow in their upper reaches, and the turbidity is sometimes much smaller. However, the bottom in such estuaries is shallower than in upper reaches and sometimes even very shallow. Because of the rapid increase in turbidity along the reach in some tidal sectors, the value of Q/Q_c decreases. Therefore, the third term in Equation (12.26) becomes large and negative. Consequently, the channel gradient approaches zero, and sometimes even becomes negative, that is, a reverse gradient. But near the sea, with a small increase in turbidity or even a decrease in turbidity and a rapid increase of the rate of water flow, the third term of Equation (12.26) turns from negative to considerably positive. Therefore, the channel slope also changes from negative to positive.

If a section of a river has neither tributary nor distributary and the water flow rate and sediment do not vary significantly along its course, the channel gradient in this section must remain constant, following Equation (12.26), and the longitudinal profile must have a linear form. If the water flow rate and sediment content vary along the reach but the value of $Q_c^{8/9}/Q^{2/9}$ remains constant, then the longitudinal profile should have the same shape. If, along some section, the turbidity increases and the water flow rate decreases, or both factors vary along the course, but the value of $Q_c^{8/9}/Q^{2/9}$ increases along the way, the longitudinal profile has the form of a convex curve. When the channel bifurcates as a result of a sharp decrease in the water flow rate and the turbidity remains almost constant, the longitudinal channel profile also has the shape of a convex curve. The increase in the water flow rate and decrease in turbidity in the downstream direction is characteristic of lowland rivers, and as a result the value of $Q_c^{8/9}/Q^{2/9}$, and at the same time channel slope decreases along the way. Hence, the concave shape of the channel profile is most prevalent.

If the water flow rate increases rapidly along the way, even if the turbidity is constant or increases slightly, the value of $Q_c^{8/9}/Q^{2/9}$ will probably decrease in the downstream direction. In all cases the longitudinal channel is complex. If the relative stability of the banks and bottom (i.e., index α) increases along the way, the gradient will gradually decrease. If the diameter of the bottom sediment particles decreases along the way, the channel gradient often decreases because of the decrease of roughness coefficient.

12.7.1 Index of Mobility

The channel mobility depends principally on the relative size of fluctuations in the rate of water flow, relative effective force of the stream on the channel, and the relation of the channel width to its depth. It is observed that the greater the amplitude of fluctuations of flow rate, the greater the range of the seasonal deformation of the channel and consequently the less stable the channel. The coefficient "a_0" in Equation (12.12) is expressed as

$$a_0 = \frac{Q_n}{Q}, \tag{12.27}$$

where Q_n is the flow discharge, equal to the discharge with an annual probability of 2%, and Q is the multiannual water discharge. Therefore, the generalized mobility index equation can be expressed as:

$$K_n = \frac{Q_n}{Q}\left[\left(\frac{v}{\alpha v_{cb}}\right)^2 + 0.15\frac{w}{d}\right]. \tag{12.28}$$

If Equations (12.14) and (12.17) are inserted into Equation (12.28) we have

$$K_n = 1.11\frac{Q_n}{Q}\left(\frac{v_{cs}^2 Q_c^2 Q^1}{K^2 \alpha^7 v_{cb}^7}\right)^{\frac{2}{9}}. \tag{12.29}$$

Equation (12.29) shows that channel mobility depends not only on the amplitude of fluctuations of water discharge but also on the quantity of transported sediment and the ability of soil to resist erosion. The more extensive the displacement of sediment, the greater is the channel mobility index and the less stable is the channel. The more stable the soil, the smaller is the channel mobility index and the greater is the channel stability. When the requirement of minimal mobility is met, the relative velocity $(v/v_{cb})^2$ and the ratio w/d have identical function structures.

12.8 Another Analysis of Generalized Mobility Index

Channel deformation is closely connected with sediment movement, and during deformation a channel tends to take on a form that its mobility would be minimal under given conditions. A channel can have different widths and depths for carrying a specific volume of water and sediment. Channels with different cross-sections possess different degrees of mobility for given water and sediment loads and boundary conditions. Equation (12.12) defining a generalized index of channel mobility (K_p) can be expressed as:

$$K_p = a_0\left[\left(\frac{v}{\alpha v_{cb}}\right)^2 + b_0 r_A\right]. \tag{12.30}$$

It is hypothesized that in order for the channel to be stable the mobility index should be minimum or its spatial derivative should tend to zero. The spatial derivative of K_p can be expressed from Equation (12.30) as

$$\frac{\partial K_p}{\partial x} = a_0 \left[\frac{2v}{(av_{cb})^2} \frac{\partial v}{\partial x} + b_0 \frac{\partial r_A}{\partial x} \right]. \tag{12.31}$$

Thus, the spatial derivative of K_p (denoted as R_k) is made up of the spatial derivative of aspect (denoted as R_1) and the spatial derivative of velocity (denoted as R_2), that is,

$$R_k = R_1 + R_2, \tag{12.32}$$

where

$$R_k = \frac{\partial K_p}{\partial x} = a_0 \frac{\partial}{\partial x} \left[\left(\frac{v}{av_{cb}} \right)^2 + b_0 \frac{w}{d} \right]; R_1 = a_0 b_0 \frac{\partial r_A}{\partial x}; R_2 = \frac{2a_0 v}{(av_{cb})^2} \frac{\partial v}{\partial x}. \tag{12.33}$$

Dividing Equation (12.32) throughout by the left side, one obtains

$$1 = P_1 + P_2 = \frac{b_0 \frac{\partial r_A}{\partial x}}{\frac{\partial}{\partial x} \left[\left(\frac{v}{av_{cb}} \right)^2 + b_0 r_A \right]} + \frac{\frac{2v}{(av_{cb})^2} \frac{\partial v}{\partial x}}{\frac{\partial}{\partial x} \left[\left(\frac{v}{av_{cb}} \right)^2 + b_0 r_A \right]}. \tag{12.34}$$

Equation (12.34) suggests that for the minimization of mobility index: $P_1 = P_2$, and the channel may adjust its geometry in accordance with this possibility. Then, from Equation (12.34) one can write

$$\frac{b_0 \frac{\partial r_A}{\partial x}}{\frac{\partial}{\partial x} \left[\left(\frac{v}{av_{cb}} \right)^2 + b_0 r_A \right]} = \frac{\frac{2v}{(av_{cb})^2} \frac{\partial v}{\partial x}}{\frac{\partial}{\partial x} \left[\left(\frac{v}{av_{cb}} \right)^2 + b_0 r_A \right]} \quad \text{or} \quad b_0 \frac{\partial r_A}{\partial x} = \frac{2v}{(av_{cb})^2} \frac{\partial v}{\partial x}. \tag{12.35}$$

Depending on the conditions imposed on the channel, aspect and velocity may not equally contribute to the minimization of mobility index. Integration of Equation (12.35) with the condition that $v = 0$ when $r_A = 0$ yields

$$b_0 r_A = \frac{v^2}{(av_{cb})^2}. \tag{12.36}$$

Equation (12.36) states that the minimization of mobility index is accomplished by the adjustment in aspect (width/depth) r_A and velocity v. Inserting Equation

(12.36) in Equation (12.9), we obtain an expression for velocity as a function of discharge as

$$v = \left[\frac{g^2 v_{cb}^2 v_{cs}^2 \alpha^2 b_0 Q_c^2}{K^2} \right]^{\frac{1}{9}} Q^{\frac{1}{9}} \quad \text{or} \quad v \propto Q^{\frac{1}{9}}. \tag{12.37}$$

Introducing Equation (12.37) in Equation (12.7) leads to an expression for depth as a function of discharge as

$$d = K^{\frac{1}{3}} \frac{v_{cb}^{\frac{2}{3}}}{g^{\frac{1}{3}} v_{cs}^{\frac{1}{3}} Q_c^{\frac{1}{3}}} b_0^{\frac{1}{3}} \alpha^{\frac{2}{3}} Q^{\frac{1}{3}} \quad \text{or} \quad d \propto Q^{\frac{1}{3}}. \tag{12.38}$$

Similarly, an expression for w is obtained by inserting Equation (13.37) in Equation (12.8) as

$$w = \frac{g^{\frac{1}{9}} v_{cs}^{\frac{1}{9}} Q_c^{\frac{1}{9}}}{b_0^{\frac{4}{9}} \alpha^{\frac{8}{9}} v_{cb}^{\frac{8}{9}} K^{\frac{1}{9}}} Q^{\frac{5}{9}} \quad \text{or} \quad w \propto Q^{\frac{5}{9}}. \tag{12.39}$$

Now the ratio of width to depth (w/d), r_A, can be expressed as a function discharge as

$$\frac{w}{d} = \frac{g^{\frac{4}{9}} v_{cs}^{\frac{4}{9}} Q_c^{\frac{4}{9}}}{b_0^{\frac{7}{9}} \alpha^{\frac{14}{9}} v_{cb}^{\frac{14}{9}} K^{\frac{4}{9}}} Q^{\frac{2}{9}} \quad \text{or} \quad r_A \propto Q^{\frac{2}{9}}. \tag{12.40}$$

The exponents for w, d, and v, respectively, are 0.56, 0.33, and 0.11, which compare favorably with 0.5, 0.4, and 0.1 for Leopold and Maddock. Tou (1965) derived the exponent values of w, d, and r_A as 0.56, 0.33, and 0.22, which exactly match the ones derived using entropy.

Alternatively, one can also consider the mobility index as a function of width and depth in place of aspect, besides velocity. Then, its spatial derivative can be written as

$$\frac{\partial K_p}{\partial x} = a_0 \left[\frac{2v}{(\alpha v_{cb})^2} \frac{\partial v}{\partial x} - \frac{b_0 w}{d_2} \frac{\partial d}{\partial x} + \frac{b_0}{d} \frac{\partial w}{\partial x} \right]. \tag{12.41}$$

Thus, the spatial derivative of K_p (denoted as R_k) is made up of the spatial derivative of width (denoted as R_1), the spatial derivative of depth (denoted as R_2), and the spatial derivative of velocity (denoted as R_3), that is,

$$R_k = R_1 + R_2 + R_3, \tag{12.42}$$

where

$$R_k = \frac{\partial K_p}{\partial x} = a_0 \frac{\partial}{\partial x}\left[\left(\frac{v}{av_{cb}}\right)^2 + b_0\frac{w}{d}\right]; R_1 = \frac{a_0 b_0}{d}\frac{\partial w}{\partial x};$$

$$R_2 = -\frac{a_0 b_0}{d^2}w\frac{\partial d}{\partial x}; R_3 = \frac{2a_0 v}{(av_{cb})^2}\frac{\partial v}{\partial x}.$$

(12.43)

Dividing Equation (12.42) throughout by the left side, one obtains

$$1 = P_1 + P_2 + P_3 = \frac{\frac{b_0}{d}\frac{\partial w}{\partial x}}{\frac{\partial}{\partial x}\left[\left(\frac{v}{av_{cb}}\right)^2 + b_0\frac{w}{d}\right]} - \frac{\frac{b_0}{d^2}w\frac{\partial d}{\partial x}}{\frac{\partial}{\partial x}\left[\left(\frac{v}{av_{cb}}\right)^2 + b_0\frac{w}{d}\right]} + \frac{\frac{2v}{(av_{cb})^2}\frac{\partial v}{\partial x}}{\frac{\partial}{\partial x}\left[\left(\frac{v}{av_{cb}}\right)^2 + b_0\frac{w}{d}\right]}.$$

(12.44)

Following the same procedure as before, Equation (12.44) suggests that there can be four possibilities for the minimization of mobility index: Possibility 1: $P_w = P_d$; Possibility 2: $P_d = P_v$; Possibility 3: $P_w = P_v$; and Possibility 4: $P_w = P_d = P_v$. The channel may adjust its geometry in accordance with any of these four possibilities, depending on the conditions imposed on the channel. However, the contribution due to the change in depth is negative, which is not realistic because the probability cannot be negative. This means that possibilities 1, 2, and 4 are not practical. The only possibility that is feasible is possibility 3, which entails adjustment in width and velocity for the minimization of mobility. Unless the depth is kept constant, which is not practical, this case will lead to the same results as derived considering the aspect. Therefore, this approach will not be pursued further.

References

Blench, T. (1952). Regime theory for self-formed sediment bearing channels. *Transactions, American Society of Civil Engineers*, Vol. 117, No. 1, pp. 383–400.

Blench, T. (1953). Regime theory equations applied to a tidal river estuary. Proceedings, Minnesota International Hydraulics Convention, pp. 77–83.

Brunn, P. and Gerritssen, P. (1958). Stability of coastal inlets. *Proceedings, Division of Waterways, Harbors and Ports, ASCE*, Vol. 84, No. WW3, pp. 1–49.

Lacey, G. (1930). Stable channels in alluvium. *Minutes of Proceedings, Institution of Civil Engineers, London*, Vol. 229, No. 1930, pp. 259–292.

Leopold, L. B. and Maddock, T. (1953). *The hydraulic geometry of stream channels and some physiographic implications*. Geological Survey Professional Paper 252, U.S. Geological Survey, Washington, DC.

O'Brien, M. P. (1931). Estuary tidal prisms related to entrance areas. Civil Engineering.

Rybkin, S. I. (1947). *Morphometric classification of streams*. Meteorologiya and Gidrologiya.

Tou, K.-J. (1965). Hydromorphology of alluvial channels of lowland river and tidal estuaries. *Scientia Sinica*, Vol. XIV, No. 8, pp. 121–127.

Velikanov, G. (1958). Fluvial processes. Str., pp. 61–70.

13

Maximum Sediment Discharge and Froude Number Hypothesis

It is hypothesized that river morphology is governed by the dominant discharge, saturation of sediment discharge, and maximization of Froude number leading to the minimum amount of energy dissipation. The minimum energy dissipation rate may be achieved by the adjustment of sediment transport rate, friction factor, or Froude number of the flow under some special conditions. This chapter discusses the derivation of river geometry based on the minimization of energy dissipation rate or the aforementioned factors.

Notation

A = cross-sectional area of flow

b = exponent

C = a constant

C^{ste} = a factor related to head loss

C_s = total sediment concentration

d = mean depth of flow cross-section

D = median size of the channel boundary sediment

D_m = mean grain diameter

D_{m0} = mean grain diameter at origin

D_{ml} = mean grain diameter at x_l

D_{90} = maximum diameter

D_* = dimensionless sediment diameter

f = exponent

f_s = friction factor

g = acceleration due to gravity

G_0, G_1 = center of gravity
H = amplitude of meanders
K = a constant
KE = kinetic energy
K_s = head loss induced by the friction on the bottom
L = wave length of meander
m = exponent
n = Manning roughness coefficient
n_b = the number of elementary beds
P = wetted perimeter
P_* = dimensionless hydraulic radius
PE = potential energy
Q = bankfull discharge
Q_* = dimensionless dominant discharge
Q_l = water discharge at x_l
Q_s = total bed load discharge
Q_{s*} = dimensionless sediment transport
R = hydraulic radius
R_* = dimensionless hydraulic radius
S_0 = valley slope
S_w = water surface slope
S_* = dimensionless slope
v = final velocity
v_0 = initial velocity
w = width of flow cross-section
x = distance from the upstream origin
Z_A = altitude of A
Z_B = altitude of B
α = a coefficient
β = aspect ratio
β_o = aspect ratio at origin
β_l = aspect ratio at downstream
α_0 = angle between the axis of the river bed and the main axis of the valley
γ = specific weight of water
γ_s = sediment weight density
ρ = specific mass of water
ρ_s = sediment density
τ = tractive force
τ_c = critical tractive force
$\tau_* = \frac{\tau}{\tau_c}$ = dimensionless parameter for tractive force
υ = kinematic viscosity

13.1 Introduction

Ramette (1980) hypothesized that the ideal morphological formation of a water course is governed by (a) flow at full volume of dominant discharge, (b) no sediment discharge or saturation of sediment discharge to ensure stability, and (c) maximization of Froude number so that the minimum amount of energy dissipation is used for channel excavation. The first hypothesis is equivalent to the old regime concept that a channel is formed by a dominant water discharge. The second hypothesis is equivalent to the hypothesis of maximum sediment discharge. The third hypothesis is equivalent to achieving the minimization of energy dissipation rate by maximization of Froude number under some special conditions. The discussion in this chapter follows the work of Ramette (1980). Millar and Quick (1998) formulated a model assuming that the stable width of a gravel-bed river with cohesive banks corresponds to the maximum bed load transporting capacity. Pickup (1976) showed that bed-load channels adjusted their cross-sections such that given slope, roughness, and sediment load, the channel shape approached the optimum for bed-load-transport.

13.2 Hypotheses

An alluvial channel system can adjust its channel geometry, slope, roughness, bed form, velocity distribution, channel pattern, longitudinal bed profile, and sediment transport rate, either individually or collectively, to minimize its rate of energy dissipation under certain constraints imposed on the system. The three hypotheses indicate that the state of minimum rate of energy dissipation may be achieved by adjusting sediment transport rate, friction factor, or Froude number of the flow under some special conditions. The objective of these adjustments is to minimize the rate of energy dissipation. The application of these hypotheses is limited to fluvial systems where these factors are allowed to adjust.

A riverbed can be locally defined by the mean geometric parameters, such as width w and mean depth d of flow cross-section, wave length of meanders $2L$, amplitude of meanders $2H$, and river slope S_0. These are five geometric parameters, depending on the distance from the upstream origin x. The "external" parameters influencing the fluvial processes are bankfull discharge Q, valley slope S_0, and grain size distribution of sediment characterized by mean diameter D_m, and maximum diameter D_{90} or $\delta = D_{90}/D_m$. Thus, there are four external parameters, also depending on the distance, which can be used to express the five geometric parameters.

13.2.1 Flow Resistance

There is the well-known Manning–Strickler resistance equation for fully rough flow:

$$Q = K_s A R^{\frac{2}{3}} S_0^{\frac{1}{2}}, \tag{13.1}$$

where $K_s = 1/n$ is the Manning–Strickler coefficient, n is the Manning roughness coefficient, R is the hydraulic radius, and A is the cross-sectional area of flow. K_s depends on D_m, d, and S_0, when dunes and ripples appear on the bottom of the river. For a rectangular section,

$$Q = K_s \frac{(wd)^{\frac{5}{3}}}{(w+2d)^{\frac{2}{3}}} S_0^{\frac{1}{2}} \tag{13.2}$$

or

$$Q = \varphi(w, d, D_m, S_0). \tag{13.3}$$

The number of unknown parameters is six: K_s, w, d, S_0, $2L$, $2H$. And thus, five other relations are needed.

The present state of a riverbed depends on its past history. Initially, the bed was flat and the flow had excavated it progressively allowing a full bank discharge. The flow had spent some potential energy to finally recover a higher kinetic energy. It can be surmised that the flow follows the line of least resistance, meaning thereby that potential energy spent is minimum and kinetic energy recovered is maximum, and if possible the river slope is equal to the valley slope but not more.

A river conveys water as well as sediment. In the upper part of the river, the bed load is supplied by numerous tributaries with higher slopes, which bring extensive sediment discharge. It is then supposed that for a given water discharge, the sediment discharge is maximum all along the bed. Thus, the river bed must organize itself according to the following rules: (1) Manning–Strickler relation for a full-bank discharge, (2) maximum water slope equal to the valley slope, (3) maximum bed load discharge, and (4) maximum energy yield ratio of potential energy (for excavating the bed to the final features) and kinetic energy the channel recovers.

13.2.2 Sediment Transport

It is assumed that the sediments are sufficiently rough so an appropriate bed load formula is applicable. For illustration, the Meyer-Peter bed load formula is employed here, which can be expressed as:

$$Q_s = Kw(\tau - \tau_c)^{\frac{3}{2}}, \tag{13.4}$$

where Q_s is total bed load discharge, w is the width, K is a constant, τ is the tractive force, τ_c is the critical tractive force, $\tau = \gamma R S_w \alpha$, where γ is the specific weight of water, R is the hydraulic radius, S_w is the water surface slope, $\tau_c \approx 0.5(\gamma_s - 1)D_m$, where γ_s is the sediment weight density, and $\alpha < 1$ is a coefficient.

Combining Equations (13.3) and (13.4), the result is

$$Q_s = \varphi(w), \tag{13.5}$$

if Q, S_w, and D_m are given. The maximum bed load transport is implied by the differential equation:

$$\varphi'(w) = 0. \tag{13.6}$$

Thus, w and d can be calculated. Using the dimensionless parameters

$$\beta = \frac{w}{d} \tag{13.7}$$

$$\tau_* = \frac{\tau}{\tau_c}, \tag{13.8}$$

and the assumption that the roughness coefficient K_s varies only slightly with β, the aspect ratio β can be written as

$$\beta = \frac{30\tau_* - 12}{10 - \tau_*}. \tag{13.9}$$

The Manning–Strickler equation can be expressed in terms of these dimensionless parameters:

$$\frac{\tau_*(\beta + 2)^{\frac{3}{4}}}{\beta^{\frac{3}{8}}} = \frac{C}{K_s^{\frac{3}{8}}} \delta^{\frac{1}{4}} Q^{\frac{3}{8}} D_m^{-\frac{3}{4}} S_w^{\frac{13}{16}}, \tag{13.10}$$

where C is a constant, and $\delta = D_{90}/D_m$.

13.2.3 Energy Equation

Now we consider the energy yield. During excavation from the ground of the valley to the final stage of the river system, the center of gravity is lowered from G_0 to G_1, as shown in Figure 13.1. Thus, the potential energy (*PE*) spent on excavating the bed can be written as

$$PE = \rho g Q \frac{d}{2}, \tag{13.11}$$

where ρ is the specific mass of water or mass density.

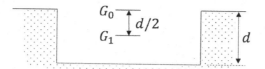

Figure 13.1 Lowering of center of gravity

During the same phase, if v_0 was the initial velocity and v is the final velocity, the kinetic energy (*KE*) recovered can be expressed as

$$KE = \rho Q \frac{v^2 - v_0^2}{2}.$$ (13.12)

Thus, the maximum energy yield is obtained when $\frac{v^2 - v_0^2}{gd}$ is maximum. Since v_0 is an initial constant and $v = Q/(wd)$, it is inferred that $wd^{3/2}$ has to be minimum. Therefore,

$$\frac{dw}{dd} = -\frac{3}{2}\frac{w}{d} = -\frac{3\beta}{2}.$$ (13.13)

From Equation (13.2) it is deduced that

$$d\left[\frac{(wd)^{\frac{5}{3}}}{(w+2d)^{\frac{2}{3}}}\right] = 0$$ (13.14)

or

$$\frac{dw}{dd} = -\beta\frac{5}{3}\frac{\beta+6}{\beta+10}.$$ (13.15)

From Equations (13.13) and (13.15), one obtains

$$\beta = 18$$ (13.16)

and from Equation (13.9)

$$\tau_* = 4.$$ (13.17)

Thus, the maximum energy is yielded for the theoretical value of $\beta = 18$.

13.2.4 Coefficient K_s

Now consider K_s, which expresses the head loss induced by the friction on the bottom, flat or possibly covered with dunes and ripples. Chollet and Cunge (1979) expressed

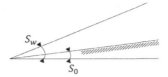

Figure 13.2 Water surface slope and bed

$$K_s = 11.3\sqrt{gD_m}\left(\frac{\beta+2}{\beta d}\right)^{\frac{2}{3}}\left[0.06 + 0.16\left(\frac{dS_w}{D_m}\right)^2\right]^{\frac{5}{8}} S_w^{-\frac{5}{8}}, \tag{13.18}$$

where S_w is the slope of the water surface. Equation (13.18) is equivalent to

$$K_s = C^{ste}\left(\frac{10\beta+12}{30+\beta}\right)^{0.907}\frac{(\beta+2)^{0.667}}{\beta^{0.306}}\frac{D_m^{0.737}}{Q^{0.361}S_w^{0.727}}, \tag{13.19}$$

where C^{ste} is a factor that can be determined by Equation (13.18) to Equation (13.19). Equation (13.19) shows that K_s depends only slightly on β.

13.2.5 Water Surface Slope

Consider, now, the water surface slope and bed feature, as shown in Figure 13.2. The maximum value of S_w is S_0. Suppose that $S_w > S_0$ locally. Then, flow excavates uselessly the downstream part of its bed, which is not compatible with the line of least resistance. If $S_w = S_0$, the bed is straight and follows the line of higher slope of the valley if it is taken as an inclined plane. If $S_w < S_0$, the bed cannot be straight. This will correspond to an increase of the length of the bed coming out of meanders.

To conclude, the riverbed will remain straight as long as it will be able to carry the full bank discharge and maximum bed load discharge. This will generally take place in its upstream part. As the sediment diameter decreases by attrition, the bed cross-section will progressively tend toward the ideal equilibrium $\beta = 18$. Then, meanders will appear.

13.3 Case 1: Straight and Meandering Single Beds

For a single bed, Figure 13.3 shows β as a function of τ_* [using Equation (13.9)] and also the Manning–Strickler relation [using Equation (13.10)] with the slope of the valley (S_0), mean sediment diameter in the upstream part, and K_s given by Equation (13.19). Let I_0 be the first curve of Equation (13.10) agreeing with Equation (13.9) at $\beta_0 < 18$. This point characterizes the bed at its upstream limit.

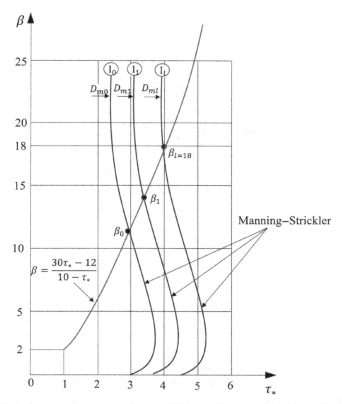

Figure 13.3 β as a function of τ_* and the Manning–Strickler relation for a single bed

As the sediment diameter decreases from D_{m0} to D_{ml} downstream by attrition, the new characteristic point is given by the point $\beta_1 > \beta_0$. Further downstream $D_m = D_{ml}$, Equations (13.10) and (13.9) intersect at $\beta_l = 18$, the equilibrium point. The value of D_{ml} is obtained by using $\beta = 18$, $\tau_* = 4$, $K_s = f(18)$, $\delta = \delta_0$, and $S_w = S_0$, in Equation (13.10).

The corresponding value of the abscissa (x_l) in terms of D_{ml} can be constructed from the relation of attrition such as

$$D_{ml} = D_{m0} \exp(-Kx). \tag{13.20}$$

From the origin to the abscissa x_l, the bed is straight because $S_w = S_0$ and $D_m < D_{ml}$, therefore, if $x > x_l$.

From Equation (13.10), since $\beta = 18$ and $\tau_* = 4$, one gets

$$Q^{\frac{3}{8}}D_m^{-\frac{3}{4}}S_w^{\frac{13}{16}} = C^{ste} = Q^{\frac{3}{8}}D_{ml}^{-\frac{3}{4}}S_0^{\frac{13}{16}} \tag{13.21}$$

or

$$S_w = S_0 \left(\frac{D_m}{D_{m_l}} \right)^{\frac{12}{13}}.$$

(13.22)

Thus, since $D_m < D_{m_l} \rightarrow S_w < S_0$ and meanders form.

13.4 Case 2: Multiple Meandering Bed

In Figure 13.4 curve II is a second aspect of Equation (13.10), which agrees with Equation (13.9) at $\beta_0 \gg 18$. This point characterizes the bed at its upstream limit as previously. Consider the initial phase of excavation for $Q_l < Q$ to which curve 2 corresponds in agreement with curve 3 at the equilibrium point $\beta_l = 18$. Therefore, an equilibrium partial bed occurs to convey Q_l. To convey the complementary water discharge $(Q–Q_l)$ the flow has to excavate a second, third, or nth bed (see Figure 13.5). The value of Q_l can be obtained from Equation (13.10), in which $\beta = 18$, $\tau_* = 4$, $\delta = \delta_0$, $D_m = D_{m0}$, and $S_w = S_0$.

The number of elementary beds (n_b) is therefore

$$n_b = \frac{Q}{Q_l}.$$

(13.23)

In each of these elementary beds the diameter of the sediment will decrease from the upstream to the downstream, and meanders will form. For $n_b \geq 10$ there will be a composite bed presenting islands, braids, etc.

13.5 Case 3: Meanders

The river width w is obtained from Equation (13.2), in which $d = \frac{w}{\beta}$, and $K_s = f(\beta)$ from Equation (13.19). Referring to Figure 13.6,

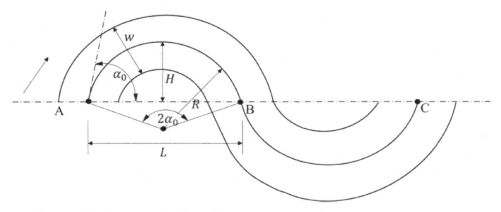

Figure 13.4 Constant local curvature

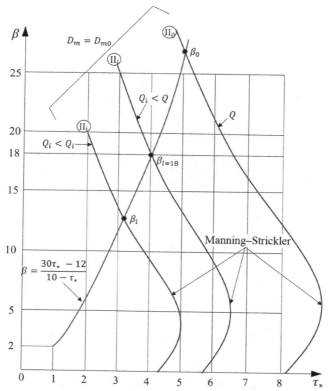

Figure 13.5 β as a function of τ_* and the Manning–Strickler relation for multiple beds

$$S_w = \frac{Z_A - Z_B}{\widehat{AB}} \quad \text{and} \quad S_0 = \frac{Z_A - Z_B}{\overline{AB}}, \tag{13.24}$$

where Z_A is the altitude of A, and Z_B is the altitude of B. If α_0 is the angle between the axis of the riverbed and the main axis of the valley, then

$$\frac{S_w}{S_0} = \frac{\overline{AB}}{\widehat{AB}} = \frac{2R\sin\alpha_0}{2R\alpha_0}, \tag{13.25}$$

where α_0 is in radians, but

$$\frac{S_w}{S_0} = \left(\frac{D_m}{D_{m_l}}\right)^{\frac{12}{13}}. \tag{13.26}$$

Therefore,

$$\frac{\sin\alpha_0}{\alpha_0} = \left(\frac{D_m}{D_{m_l}}\right)^{\frac{12}{13}}, \tag{13.27}$$

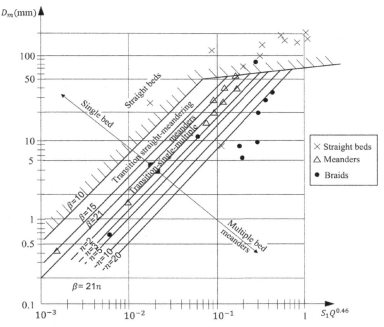

Figure 13.6 Relation between sediment diameter and water surface slope or channel classification

where α_0 can be calculated from the above relation. In terms of R and α_0,

$$L = 2R\sin\alpha_0 \quad \text{Half wave length} \tag{13.28}$$

$$H = R(1 - \cos\alpha_0) \quad \text{Half meander width.} \tag{13.29}$$

If R can be assumed to be as $R \approx 2.5w$,

$$L \approx 5w\sin\alpha_0 \tag{13.30}$$

$$H \approx 2.5w(1 - \cos\alpha_0). \tag{13.31}$$

To summarize, Figure 13.6 displays a scheme that classifies single, straight, meandering, and multiple beds. With $K_s = 32$ and $15 \leq \beta \leq 21$, one obtains

$$D_{90} = \frac{0.35}{0.20}\delta^{1.33}S_w^{1.1}Q^{0.5} \tag{13.32}$$

$$w = \frac{1.9}{1.5}S_w^{-0.19}Q^{0.375} \quad \text{for single bed} \tag{13.33}$$

$$w = \frac{1.9}{1.5}n^{0.625}S_w^{-0.19}Q^{0.375} \quad \text{for multiple beds; } n \text{ single channels} \tag{13.34}$$

$$w = \frac{1.3}{1.05} Q^{0.46} D_m^{-0.17} \delta^{0.06}, \quad \delta = \frac{D_{50}}{D_m} \qquad (13.35)$$

$$7.1 < \frac{2L}{w} < 10, \frac{\pi}{4} < \alpha_0 < \frac{\pi}{2} \qquad (13.36)$$

$$2L \approx 12.8 Q^{0.46} D_m^{-0.17} \delta^{0.06} \qquad (13.37)$$

$$2L = 52 Q^{0.46} \delta^{0.06} \qquad (13.38)$$

$$\frac{2L}{2w} = \frac{2 \sin \alpha_0}{1 - \cos \alpha_0} = 2.5, \alpha_0 = 77^\circ \qquad (13.39)$$

$$S_1 \sim Q^{-\frac{1}{7.5}} \qquad (13.40)$$

$$A = wd \approx 2.2 D_m^2 \left(\frac{Q}{D_m^{\frac{5}{2}} \sqrt{(\rho_s - 1)g}} \right)^{0.855} \quad \text{for stable channels,} \qquad (13.41)$$

where ρ_s is sediment density.

The theoretical relations derived are compared with experimental derivations reported in the literature as given in Table 13.1.

For straight sandy channels having uniform boundary material and constant dominant discharge, Farias (2000) employed maximum efficiency in sediment transport and obtained a menu of theoretical regime equations for design of stable alluvial channels. In so doing he employed appropriate equations for alluvial friction and sediment transport and analyzed 355 data sets that covered ranges of: discharge from 0.07 to 407.76 m³/s, median sediment size from 0.029 to 0.805 mm, wetted perimeter from 1.7 to 111.8 m, hydraulic radius from 0.180 to 3.718, slope from 0.00058 to 0.000473, channel width from 1.5 to 112.8 m, depth from 0.244 to 4.084 m, and sediment concentration from 16 to 9.0 ppm using regression analysis. He presented these regime equations as power type relations for dimensionless wetted perimeter (P_*), dimensionless hydraulic radius (R_*), and dimensionless slope (S_*) as functions of dimensionless dominant discharge (Q_*), dimensionless sediment transport (Q_{s*}), total sediment concentration $(C_{s*} = C_s)$, and dimensionless sediment diameter (D_*) defined as follows:

$$P_* = \frac{P}{D}; R_* = \frac{R}{D}; S_* = \frac{S}{\Delta}; \Delta = \frac{(\rho_s - \rho)}{\rho}; D_* = \left[g \frac{\Delta}{v^2} \right]^{\frac{1}{3}} D;$$

$$Q_* = \frac{Q}{\left[g \Delta D^5 \right]^{\frac{1}{2}}}; Q_{s*} = \frac{Q_s}{\left[g \Delta D^5 \right]^{\frac{1}{2}}}; C_{s*} = C_s = \frac{Q_s}{Q}, \qquad (13.42)$$

Table 13.1. *Comparison of theoretically derived relations with experimental derivations reported in the literature*

Theoretical relations	Remarks	Experimental relations	References
$D_{90} = \dfrac{0.35}{0.20} \delta^{1.33} S_w^{1.1} Q^{0.5}$	$\delta = \dfrac{D_{90}}{D_m}$	$d_m = 2.7 S_w Q^{0.44}$	Larras (1963)
$w = \dfrac{1.9}{1.5} S_w^{-0.19} Q^{0.375}$	Single bed	$w = \dfrac{1.7}{0.75} S_w^{-0.2} Q^{0.5}$	Prus-Chacinski (1958)
$w = \dfrac{1.9}{1.5} n^{0.625} S_w^{-0.19} Q^{0.375}$	Multiple beds; n single channels	$w = \dfrac{1.7}{0.75} S_w^{-0.20} Q^{0.50}$	Prus-Chacinski (1958)
$w = \dfrac{1.3}{1.05} Q^{0.46} D_m^{-0.17} \delta^{0.06}$	$\delta = \dfrac{D_{50}}{D_m}$		
$7.1 < \dfrac{2L}{w} < 10$	$\dfrac{\pi}{4} < \alpha_0 < \dfrac{\pi}{2}$	$7.0 < \dfrac{2L}{w} < 11$	Wolman and Leopold (1957)
$2L \approx 12.8 Q^{0.46} D_m^{-0.17} \delta^{0.06}$			
$2L \approx 52 Q^{0.46} \delta^{0.06}$	with $D_m = 0.3$ mm	$2L = 61 Q^{0.47}$ $D_m = 0.3$ mm	Ackers and Charlton (1970)
$\dfrac{2L}{2w} = \dfrac{2\sin\alpha_0}{1 - \cos\alpha_0} = 2.5$	with $\alpha_0 = 77°$	$\dfrac{2L}{2w} \approx 2.5$	Leopold and Wolman (1957)
$S_1 \sim Q^{-\frac{1}{75}}$	β, D_m, K_s constant	$Q^{-\frac{1}{9}} \sim S_1 \sim Q^{-\frac{1}{6}}$	Bogardi (1978)
$A = wd \approx 2.2 D_m^2 \left(\dfrac{Q}{D_m^{\frac{5}{2}} \sqrt{(s-1)g}} \right)^{0.855}$	Stable alluvial channels	$A = 2.2 D_m^2 \left(\dfrac{Q}{D_m^{\frac{5}{2}} \sqrt{(s-1)g}} \right)^{0.855}$	Kondap and Garde (1979)

in which Q is the dominant discharge, Q_s is the sediment transport rate, C_s is the total sediment concentration, D is the median size of the channel boundary sediment, P is the wetted perimeter, R is the hydraulic radius, S is the slope, g is the acceleration due to gravity, ρ is the water density, ρ_s is the sediment density, and v is the kinematic viscosity. Using the maximum sediment transport efficiency, Farias (1993) found that the aspect ratio [width (w)/depth (d)] as a power function of dimensionless discharge where the exponent and the coefficient of proportionality was nearly constant. Then, he determined power relations for P_*, R_*, and S_* using appropriate formulas for alluvial friction and sediment transport, as given in Table 13.2. Plots of these relations were quite close to each other.

Table 13.2. *Theoretical regimes equations [Code:* F *for alluvial friction;* T *for sediment transport]*

Code	Theoretical regime equations
E-H: Engelund and Hansen (1967) [F & T]	$P_* = 0.950 D_*^{0.068} C_{s*}^{-0.009} Q_*^{0.497}$
	$R_* = 0.954 D_*^{-0.216} C_{s*}^{-0.022} Q_*^{0.399}$
	$S_* = 9.766 D_*^{-0.040} C_{s*}^{0.666} Q_*^{-0.200}$
A-W: F from White et al. (1982) and T from Ackers and White (1973), Ackers (1972)	$P_* = 0.076 D_*^{0.666} C_{s*}^{-0.023} Q_*^{0.555}$
	$R_* = 1.213 D_*^{-0.432} C_{s*}^{-0.120} Q_*^{0.357}$
	$S_* = 0.918 D_*^{0.419} C_{s*}^{0.521} Q_*^{-0.193}$
B-B: F and T from Brownlie (1983)	$P_* = 0.202 D_*^{0.253} C_{s*}^{-0.067} Q_*^{0.524}$
	$R_* = 0.624 D_*^{-0.141} C_{s*}^{-0.110} Q_*^{0.368}$
	$S_* = 11.097 D_*^{-0.098} C_{s*}^{0.605} Q_*^{-0.225}$
V-R: F and T from van Rijn (1993)	$P_* = 0.584 D_*^{0.210} C_{s*}^{-0.050} Q_*^{0.492}$
	$R_* = 0.203 D_*^{-0.085} C_{s*}^{-0.139} Q_*^{0.398}$
	$S_* = 4.680 D_*^{-0.210} C_{s*}^{0.504} Q_*^{-0.233}$
P-P: F and T from Peterson (1970)	$P_* = 0.210 D_*^{0.199} C_{s*}^{-0.104} Q_*^{0.520}$
	$R_* = 0.709 D_*^{-0.131} C_{s*}^{-0.077} Q_*^{0.375}$
	$S_* = 23.342 D_*^{0.130} C_{s*}^{0.779} Q_*^{-0.213}$
K-K: F and T from Karim and Kennedy (1982)	$P_* = 0.488 D_*^{0.213} C_{s*}^{-0.043} Q_*^{0.500}$
	$R_* = 0.239 D_*^{-0.134} C_{s*}^{-0.172} Q_*^{0.381}$
	$S_* = 8.467 D_*^{0.010} C_{s*}^{0.642} Q_*^{-0.221}$

13.6 Maximum Sediment Efficiency

Kirkby (1977) hypothesized that an alluvial channel, in the medium term (say several decades), adjusts over a period of time to transport as efficiently as possible the sediment load supplied to it. One point needs further explanation, which relates to selecting an appropriate time scale for the application of this hypothesis. This is because for alluvial channels sediment transportation is a limiting process. In the short term the channel selectively transports its load and in the long term it reaches equilibrium. Therefore, in the medium term it carries the sediment load supplied to it from upstream and from side slopes such that the sediment does not accumulate indefinitely.

For application of this hypothesis, it is assumed that that for a proportion of time that may vary downstream, the flow occurs at a constant dominant rate and the dominant size of sediment transported downstream is more or less constant, despite the appreciable change in roughness element downstream. Further, the time span corresponds to the values of water and sediment discharge, dominant sediment grain size, and valley slope (equal to maximum channel slope). The discussion here closely follows Kirkby (1977).

13.6.1 Governing Equations

The maximum sediment efficiency hypothesis for the derivation of downstream hydraulic geometry entails equations for water discharge, Darcy–Weisbach equation for roughness, bedload transport equation, most efficient channel cross-section, and roughness diameter. Thus, the governing equations can be expressed as follows:

$$\text{Flow discharge: } Q = wRv, \tag{13.43}$$

in which Q is the total dominant water discharge, R is the hydraulic radius, w is the channel width, and v is the cross-sectionally averaged velocity.

$$\text{Darcy–Weisbach equation: } S = \frac{f_s}{R}\frac{v^2}{2g}, \tag{13.44}$$

in which f_s is the friction factor, S is the channel slope, and g is the acceleration due to gravity.

Equations (13.43) and (13.44) can be combined to replace v as

$$Q = \left(\frac{2g}{f_s}\right)^{\frac{1}{2}} wR^{\frac{3}{2}}S^{\frac{1}{2}}. \tag{13.45}$$

Friction factor and sediment diameter relation: $\dfrac{1}{\sqrt{f_s}} = 1.77 \ln \left(\dfrac{R}{D_{84}} \right) + 2.0,$

$$(13.46)$$

in which D_{84} is the bed material size or diameter.

Bedload transport equation: $Q_s = 8 \left([g(\gamma_s - \gamma)D^3] \right) w \left(\dfrac{RS}{D(\gamma_s - \gamma)} - 0.047 \right)^{\frac{3}{2}}.$

$$(13.47)$$

Equation (13.47) is the Meyer-Peter and Muller (1948) equation.

Kirkby (1977) expressed volumetric sediment concentration as

$$r_2 C_s = 3.43 - \frac{(r_1 r_2 S - 0.077)^{\frac{3}{2}}}{r_1^{\frac{3}{2}} (r_2 S)^{\frac{1}{2}} (1.77 \ln r_1 + 2.0)}, \qquad (13.48)$$

where $C_s = Q_s/Q$; $r_1 = R/D_{84}$; and $r_2 = D_{84}/D$. Equation (13.48) gives a family of curves plotting $r_2 S$ against r_1 for different values of $r_2 C_s$ and displays three regions: (1) no sediment transport, (2) stable channels, and (3) unstable channels as well as the locus of maximum efficiency. It also shows that there is a minimum slope for sediment transport and the minima lie on the maximum efficiency curve:

$$r_2 Q_s = \frac{0.115 \ln r_1 + 0.208}{r_1}. \qquad (13.49)$$

Equation (13.45) shows that if discharge is given, the channel cross-section decreases steadily with the increase in depth along a line of constant slope. On the right side of the locus of maximum efficiency, a small increase in depth results in the decrease of both sediment concentration (for constant slope) and channel cross-section. These two decreases together make the channel unstable, because there will be loss of sediment from the flow filling the channel, which will then become deeper and narrower leading to the collapse of banks. Likewise, if there is a local decrease of channel depth then the channel will become shallower and wider in an unstable manner. Thus, the entire region to the right of the locus is seen to be unstable. Braiding is an example of this phenomenon. To the left of the locus, local increases of depth and concentration and a decrease in cross-section counterbalance one another, leading to channel stability, as seen for meandering channels.

To further explore, a plot of slope S against $r_1 r_2 = R/D$ with the assumption of a dominant grain size $D = 10$ mm shows that bed material size exercises only a slight influence on the channel stable/unstable transition and displays the effect of varying r_2 on concentration curves, pointing out that the coarsest bed material is the most efficient way of carrying the whole of its load, even if the dominant size is much smaller.

13.6.2 At-a-Station Hydraulic Geometry

For this case, the channel slope and sediment diameter may be regarded as constant. From Equation (13.46), one can express

$$\frac{1}{\sqrt{f_s}} \propto R\beta_0, \tag{13.50}$$

where β_0 may have an average value of 1/6 for large rivers $\left(10 < R/D_{84} < 10^4\right)$ but may be greater for small rivers $(\beta_0 = 0.48$ for $1 < R/D_{84} < 10)$. Inserting Equation (13.50) into Equation (13.45), one gets

$$b + (1.5 + \beta_0)f = 1 \tag{13.51}$$

$$m = (0.5 + \beta_0)f. \tag{13.52}$$

Figure 13.7 shows the relation expressed by Equation (13.51), in which the conventional exponents have been employed: $w \propto Q^b; R \propto Q^f; v \propto Q^m$. Points above $b = 0.4-0.6$ and $f = 0.2$ correspond to the channel within channel cross-sections, which are primarily governed by the frequency distribution of flows.

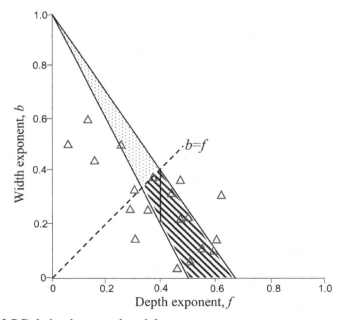

Figure 13.7 Relation between b and f

Typical channel cross-sections are governed by bank stability. For trapezoidal sections, the average value of β_0 is about 1/6 and may approach zero for very large R. The width becomes independent of R for small depths and increases proportionally to R for large depths. This suggests that $w/f_s^{0.5}$ from Equation (13.45) behaves as $R^{1.5}$ for a large range of depths, where the increase in the width exponent balances the decrease in the exponent of $1/f_s^{0.5}$. Therefore, from equation (13.45), one can write

$$Q \propto \frac{w}{f_s^{0.5}} R^{\frac{3}{2}} \propto R^{2.5}, \tag{13.53}$$

which leads to the depth exponent of 0.4 on average. Therefore, from Equation (13.50), $m = 0.2-0.4$ with an average of 0.27 corresponding to $\beta_0 = 1/6$. Then, by subtraction $b = 0.2-0.4$ with the corresponding average of 0.33. Thus, the average values of exponents are: $b = 0.33$, $f = 0.40$, and $m = 0.27$.

13.6.3 Downstream Hydraulic Geometry

The dominant grain size D may be assumed constant and the maximum efficiency can be approximated from Equation (13.49) as

$$r_2 Q_s \propto r_1^{-(1-\beta_0)}. \tag{13.54}$$

Supposing

$$R \propto Q^f, \tag{13.55}$$

in which Q is the dominant discharge. Equation (13.54) can be expressed as

$$\left(\frac{R}{D_{84}}\right)^{-(1-\beta_0)} \propto \left(\frac{D_{84}}{D}\right) S \tag{13.56}$$

or

$$S \propto Q^{-(1-\beta_0)} D_{84}^{-\beta_0} D. \tag{13.57}$$

From Equation (13.45),

$$w \propto Q f_s^{\frac{1}{2}} R^{-\frac{3}{2}} S^{-\frac{1}{2}} \propto Q \left(\frac{R}{D_{84}}\right)^{-\beta_0} R^{-\frac{3}{2}} S^{-\frac{1}{2}} \propto Q^{1-(1.5+\beta_0)f+\frac{1}{2}(1-\beta_0)f} D_{84}^{1.58\beta_0}$$

$$\propto Q^{1-f(1+1.5\beta_0)} D_{84}^{1.5\beta_0}. \tag{13.58}$$

One can express D_{84} as

$$D_{84} \propto Q^{-b_0}, \tag{13.59}$$

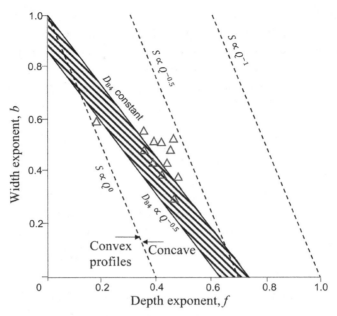

Figure 13.8 Downstream variation of hydraulic geometry exponents

where b_0 is usually a positive value. Then, exponent b can be expressed as

$$b = 1 - (1 + 1.5)f - 1.5\beta_0 b_0. \tag{13.60}$$

Equation (13.60) gives the locus of probable values of b downstream for various values of β_0 and b_0, as shown in Figure 13.8. For large sand-bed rivers, the value of D_{84} corresponds to dune rather than grain size, which may increase downstream, resulting in negative values of b_0. The small velocity exponent corresponds to high sediment efficiency.

For small rivers, with larger values of β_0 up to 0.5 or greater, the exponent values spread more widely and become more sensitive to the downstream variation in D_{84}. If the profile is convex, then from Equation (13.57),

$$(1 - \beta_0)f - \beta_0 b_0 > 0. \tag{13.61}$$

Inserting Equation (13.61) in Equation (13.60),

$$b > 1 - 2.5f. \tag{13.62}$$

In large rivers, the slope exponent is small and

$$Q_s \propto Q^{-0.3}. \tag{13.63}$$

The efficiency criterion can be expressed as

$$r_2 C_s = \frac{1}{r_1} \left(\frac{1.77 \ln r_1 + 2.0}{1.77 \ln r_1 + 3.18} \right)^{\frac{1}{2}}, \tag{13.64}$$

which is a measure of sediment yield per unit area as bed load. Equation (13.64) expresses the proportionality with r_1 raised to a power of -0.93 to -0.997, which is approximately equal to 1 for $1 < r_1 < 10^4$. Inserting for r_1, one gets

$$C_s \propto Q^{-f}. \tag{13.65}$$

The negative exponent has a larger value than the exponent for the total sediment yield. Following Brune (1948),

$$C_s \propto A^{-0.15} \propto Q^{-0.2}. \tag{13.66}$$

Kirkby (1977) reasoned that the values of bedload and slope are externally determined and their downstream variation determines the other hydraulic geometry exponents. If the channel slope and bedload concentration are constant then $b = 1$ and $f = 0$, which correspond to a widening channel with constant depth and constant D_{84}. Likewise, if the channel slope is constant but its bedload concentration has a normal variation $\left(C_s \propto Q^{-0.4} \right)$, then $b = 0$ and $f = 0.4$, which correspond to a deepening channel of constant width and a marked decrease in sediment grain size downstream.

References

Ackers, P. (1972). River regime: research and application. *Journal of the Institution of Water Engineers*, Vol. 26, pp. 257–274.

Ackers, P. and Charlton, F. G. (1970). Meander geometry arising from varying flows. *Journal of Hydrology*, Vol. 11, No. 3, pp. 230–252.

Ackers, P. and White, W. R. (1973). Sediment transport: New approach and analysis. *Journal of Hydraulics Division, ASCE*, Vol. 99, pp. 2041–2060.

Bogardi, J. (1978). *Sediment Transport in Alluvial Streams*. pp. 635–636, Akademiai Kiado, Budapest.

Brownlie, W/R. (1983). Flow depth in sand bed channels. *Journal of Hydraulics Division, ASCE*, Vol. 109, No. 7, pp. 959–990.

Brune, G. (1948). *Rates of sediment production in midwestern United States*. Soil Conservation Service, Technical Publication 65, U.S. Department of Agriculture, Washington, DC.

Chollet, J. P., and Cunge, J. A. (1979). New interpretation of some head loss-flow velocity relationships for deformable movable beds. *Journal of Hydraulic Research*, Vol.17, No. 1, pp. 1–13.

Engelund, F. and Hansen, E. (1967). *A monograph on sediment transport in alluvial streams*. pp. 63, Teknisk Forlag-Copenhagen, Denmark.

Farias, H. D. (1993). *Hydraulic geometry of sand-silt channels in regime.* Water: The Lifeblood of Africa, International Symposium, Victoria Falls, Zimbabwe.

Farias, H. D. (2000). Physical and mathematical modeling of alluvial channels in regime. *HYDRA*, Vol. 1, pp. 348–353.

Karim, M. F. and Kennedy, J. F. (1982). *IALLUVIAL: A computer-based flow and sediment routing model for alluvial streams and its application to the Missouri River.* Iowa Institute of Hydraulic Research, Report No. 250, The University of Iowa, Iowa City.

Kirkby, M. J. (1977). Maximum sediment efficiency as a criterion of alluvial channels. Chapter 27 in: *River Channel Changes*, edited by K. J. Gregory, pp. 427–442, John Wiley & Sons, New York.

Kondap, D. M. and Garde, R. J. (1979). *Design of stable channels.* Irrigation and Power, No. 4, New Delhi.

Larras, J. (1963). Profondeurs maximales d'erosion des fonds mobiles autour des piles en riviere. Ann. *ponts et chaussées*, Vol. 133, No. 4, pp. 411–424.

Millar, R. G. and Quick, M. C. (1998). Stable width and depth of gravel-bed rivers with cohesive banks. *Journal of Hydraulic Engineering*, Vol. 124, No. 10, pp. 1005–1013.

Peterson, D. F. (1970). *Progress report on studies of hydraulic geometries of large bed element streams.* PRWG 92-1, pp. 34, Utah Water research Laboratory, Utah State University, Logan.

Pickup, G. (1976). Adjustment of stream-channel shape to hydrologic regime. *Journal of Hydrology*, Vol. 30, pp. 365–373.

Prus-Chacinski, T. M. (1958). Proceedings of Institution of Civil Engineers, Vol. 11.

Ramette, M. (1980). *A theoretical approach on fluvial processes.* Proceedings, International Symposium on River Sedimentation, pp. C16–1 to C-16-18, Beijing, China.

van Rijn, L. C. (1993). *Principles of Sediment Transport in Rivers, Estuaries and Coastal Seas.* Aqua Publications, Amsterdam.

White, W. T., Bettes, R. and Paris, E. (1982). Analytical approach to river regime. *Journal of Hydraulics Division, ASCE*, Vol. 108, pp. 1179–1193.

Wolman, M. G. and Leopold, L. B. (1957). *River flood plains: Some observations on their formation*, U.S. Geological Survey Professional Paper, 282C, 107 pp., Washington, DC.

14

Principle of Minimum Froude Number

The stability of river channel adjustment toward equilibrium is controlled by a set of factors governed by the Froude number. When the channel is subjected to water discharge, sediment load, and sediment particle size, it tends to attain stability that corresponds to the minimum Froude number and minimum bed material motion. Field observations and computer simulations for sand-bed rivers show that a river in equilibrium tends to acquire a stable geometry with minimum Froude number. This leads to the hypothesis of minimum Froude number, which is presented in this chapter.

Notation

A = cross-sectional area

C = sediment concentration

C_t = total sediment concentration (suspended and bed load)

C_1 = morphological constant

C_2 = morphological constant

d = flow depth

d_m = measured depth

d_p = predicted depth

D_s = sediment size

F_r = Froude number

F_{rmin} = minimum Froude number

F_{rm} = Froude number computed using measured data

F_{rp} = predicted Froude number

g = gravitational acceleration

H = Shannon entropy

Continued

J = weighting factor
K = parameter related to sediment diameter
n = Manning roughness coefficient
P_i = probabilities
q = discharge per unit width
Q = water discharge
R_1 = contribution of change in width to the change in Froude number
R_2 = contribution of change in flow depth to the change in Froude number
S = slope
v = average flow cross-sectional velocity
v_c = average flow velocity corresponding to incipient sediment motion
v_f = settling velocity in water
vS = unit stream power
w = width
x = distance
α = exponent
υ = kinematic viscosity

14.1 Introduction

The river channel behavior is characterized by five dependent variables, including mean flow velocity, channel gradient (slope), channel width, water flow depth, and resistance factor. These variables adjust in response to two independent variables, including discharge and mean sediment size. These variables are related by equations of conservation of mass, resistance, and sediment transport. However, these three equations are not sufficient to describe river behavior. Therefore, another hydraulic relationship that describes a mechanism that controls the river adjustment behavior is needed. This is suggested to be the minimum Froude number. This chapter discusses the principle of minimum Froude number. Jia (1990) conducted computer simulations to analyze the stability of rivers in equilibrium and found that the minimum Froude number was also a mechanism for river channel adjustment. He also tested the hypothesis of minimum Froude number on a number of natural sand-bed rivers and canals. This chapter draws on the work of Jia (1990).

14.2 River Stability and Froude Number

It is hypothesized that river stability is connected with Froude number, as will be discussed in what follows.

14.2.1 Channel Stability

In general, stability and equilibrium correspond to each other, that is, a stable channel must attain equilibrium. The implication here is that the channel must not experience significant erosion or deposition of sediment but can experience short-term hydrologic fluctuations. In reality, rivers attain a stable or equilibrium condition for a certain length of time, and again adjust their geometry in response to dynamic sediment load and water discharge. Thus, they attain dynamic stability.

14.2.2 Role of Potential Energy

In general, the stability of a river channel represents the equilibrium, with no significant erosion and deposition or no net sediment transport. A natural channel profile is from a higher elevation (potential energy) to a lower one (potential energy), and flow occurs under the influence of gravity. As potential energy decreases, the channel tends to attain stability and the lower energy corresponds to the higher stability. For a unit river segment, the potential energy relative to the adjustment downstream of the river segment is represented by the water surface slope. Under fluctuating natural conditions, the potential energy decreases and the river channel tends toward minimum water slope and higher stability. This is also implied by minimum stream power for constant channel-forming discharge. Maximum stability corresponds to equilibrium, which can aid the identification of channel form. Despite varying patterns and stabilities, rivers are at or near maximum stability if they are in equilibrium (Schumm, 1981).

14.2.3 Role of Sediment Movement

Channel stability is governed by bed sediment movement. If the movement is greater, then the channel stability is lower, even if the sediment transport is balanced by the sediment input from the upstream. In sand-bed channels, flow governs the sediment movement and consequent bed form development: Higher flow causes higher sediment movement and more developed bed forms.

14.2.4 Role of Froude Number

If the flow regime is proportional to the Froude number, slope, and stream power per unit area (Simon and Senturk, 1977), then the channel stability or equilibrium can be denoted by the minimum values of these variables. The flow regime will increase with stream power per unit length and per unit weight of water. For the minimum values of these hydraulic variables the bed material movement will be the lowest and the stability the highest. However, from field measurements, it is not possible to get the minimum values of these variables and computer simulation seems to be the only resort.

14.3 Modeling and Simulation

For the minimization of Froude number, it is pertinent to first discuss the assumptions and governing equations.

14.3.1 Assumptions

It is assumed that the channel is alluvial with sand bed, and the adjustments of slope and cross-section form are in accord with water and sediment transport equilibrium conditions. The flow resistance is influenced by bed forms, which in turn are influenced by sediment movement. At channel-forming discharge, both bed load and sediment load are important and interactive. The bankfull discharge is the dominant discharge and does not change much over the long term, if climate does not change much. This means that over the time scale of channel adjustment the sediment discharge and water discharge remain more or less constant.

14.3.2 Mass Conservation

The conservation of mass can be expressed as

$$Q = Av = wdv, \tag{14.1}$$

where Q is the water discharge, A is the cross-sectional area, w is the width, d is the flow depth, and v is the average flow cross-sectional velocity. Discharge per unit width Q/w defines specific discharge q.

14.3.3 Flow Resistance

The flow resistance depends on the bed form, sediment movement, and sediment particle size. It can be expressed by Manning's equation (in SI system of units) as:

$$v = \frac{1}{n} d^{\frac{2}{3}} S^{\frac{1}{2}}, \tag{14.2}$$

where n is the Manning roughness coefficient related to sediment size D_s:

$$n = \frac{D_s^{\frac{1}{6}}}{A}. \tag{14.3}$$

The flow cross-sectional area A is constant where there is no sediment movement and the bed is flat.

14.3.4 Sediment Transport

Sediment moves and ripples form with increasing velocity and hence the flow resistance increases and cross-sectional area A decreases. This suggests that A is also a function of sediment transport. If v_c is the average flow velocity corresponding to incipient sediment motion and C_t is the total sediment concentration (suspended and bed load), then A is related to v/v_c (Li and Liu, 1961), as shown in Figure 14.1. Up to the value of v/v_c equal to one, the value of A remains constant; for v/v_c exceeding 1.0, A decreases and bed form resistance increases. Beyond the critical value the bed form resistance decreases with v/v_c and A gets large again. With the change of A, the bed form changes from plane to ripple, then to dune and to plane again, as shown in Figure 14.2.

The sediment concentration C, flow velocity v, slope S, flow depth d, sediment size D_s, and flow discharge Q are expressed by relationships that are formulated differently (Simons and Senturk, 1977). The differences in the relationships are the reason that the resulting hydraulic equations are different. Therefore, empirically verified relationships must be utilized.

A sediment transport equation, developed by Yang (1988), can be chosen, for it accounts for the total sediment load:

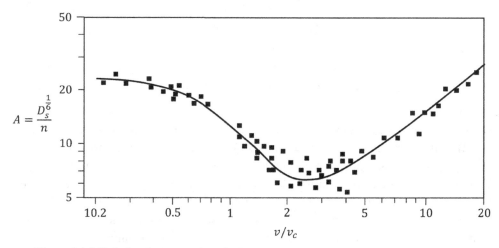

Figure 14.1 Relation between A and v/v_c

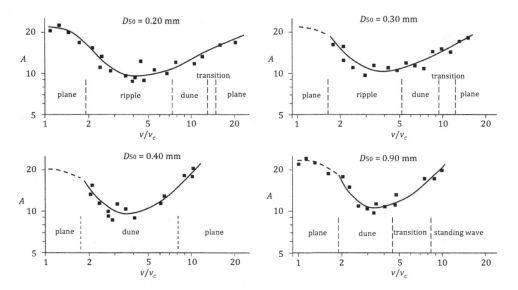

Figure 14.2 Relation between A and v/v_c and bed forms

$$\log C_t = 5.435 - 0.236 \log\left(\frac{v_f D_s}{v}\right) - 0.457 \log\left(\frac{U_*}{v_f}\right)$$

$$+ \left[1.799 - 0.409 \log\left(\frac{v_f D_s}{v}\right) - 0.314 \log\left(\frac{U_*}{v_f}\right)\right] \log\left(\frac{vS}{v_f} - \frac{v_c S}{v_f}\right),$$

$$(14.4)$$

in which

$$U_* = (gdS)^{\frac{1}{2}} \tag{14.5}$$

$$\frac{v_c}{v_f} = \frac{2.5}{\log\left(\frac{U_* D_s}{v}\right) - 0.06} + 0.66, \quad 1.2 < \frac{U_* D_s}{v} < 70 \tag{14.6}$$

$$\frac{v_c}{v_f} = 2.05, 70 < \frac{U_* D_s}{v}, \tag{14.7}$$

where v_f is the settling velocity in water, and v is the kinematic viscosity. Equations (14.2) and (14.4) can be utilized to generate families of curves for unit stream power (vS) – stream power per unit length per unit weight – Manning's n and Froude number against discharge. Here water discharge per unit width is employed to express the adjustment in channel cross-section for constant sediment size and concentration for each family of curves. The stream power is smaller when the channel cross-section is narrower and deeper. Interestingly, Manning's n curves are convex having maximum values, whereas Froude number curves are concave having minimum values. Curves point out that the adjustment of channel cross-section influences unit stream power, flow resistance, and Froude number and that slope and unit stream power have no minimum over a range of specific discharge, meaning these cannot specify the hydraulic condition at which the channel will reach equilibrium. Toward the end of curves, the values of slope and unit stream power correspond to very low values of resistance, which may not be realistic. However, it can be shown that the Froude number does have a minimum value, which can be employed to decipher the channel equilibrium condition.

14.4 Minimization of Froude Number

The relationship between F_{rmin} and unit stream power (vS) for different values of sediment concentration and sediment size shows that for fixed sediment size the unit stream power and Froude number increase with the increase in sediment load. If the sediment concentration is fixed the unit stream power and Froude number do not increase all the time with the increase in sediment size. Equation (14.4) shows that sediment concentration and unit stream power are closely connected. Therefore, sediment concentration may not be used along with the unit stream power. Because hydraulic geometry relations correspond to minimum Froude number, Jia (1990) showed, using regression analysis, the relation between minimum Froude number and unit stream power as

$$F_{rmin} = K(vS)^{\alpha}, \tag{14.8}$$

in which K varies with sediment size, and α is exponent.

Parameter K is related to sediment diameter D_s as shown in Figure 14.3, as

$$K = 4.49 D_s^{-0.186}. \tag{14.9}$$

Thus, Froude number can be expressed as a function of both sediment size and unit stream power as

$$F_{rmin} = 4.49 D_s^{-0.186} (vS)^{0.377}. \tag{14.10}$$

Equation (14.10) is appropriate for stable alluvial channels. Figure 14.4 shows the variation of Froude number with stream power for different sediment diameters.

14.5 Testing

Jia (1990) tested Equation (14.10) using data from 42 alluvial rivers and stable water canals collected from the literature (Einstein and Barbarosa, 1952; Colby and

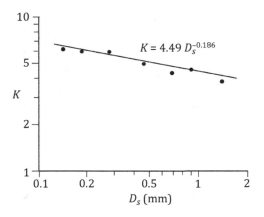

Figure 14.3 Relation between K and D_S.

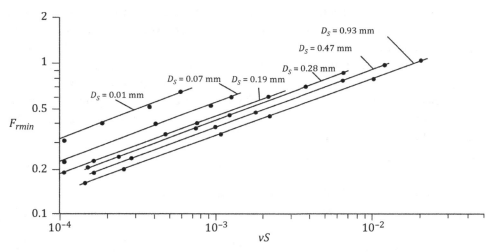

Figure 14.4 Variation of Froude number with unit stream power for various values of sediment diameter

Hembree, 1955; Leopold and Wolman, 1957; Hubbell and Matejka, 1959; Simons and Albertson, 1963; Brownlie, 1983). These data were with sediment finer than 2 mm and Froude number less than one. Figure 14.5 compares the values of Froude number computed from the data, F_{rm}, with the values of Froude number obtained from Equation (14.10), F_{rp}, with mean error of less than 15%.

Equation (14.10) can be recast as

$$d = 0.00507 \, D_s^{0.371} S^{-0.755} v^{1.25}. \tag{14.11}$$

Figure 14.6 compares the values of depth obtained from Equation (14.11), d_p, with the values of measured depth d_m. There is 27.7% error in the predicted values.

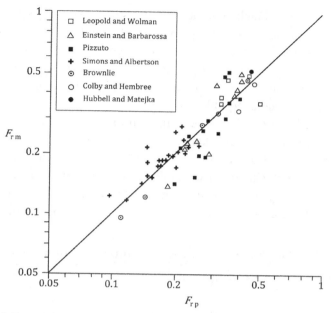

Figure 14.5 Comparison between the values of Froude number computed from the data, F_{rm}, with the values of Froude number obtained from Equation (14.10) F_{rp}

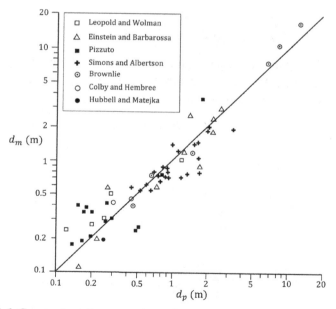

Figure 14.6 Comparison between the values of depth obtained from Equation (14.11), d_p, with the values of measured depth d_m

Natural sand-bed channels and canals have stable geometry that tends toward minimum Froude number. The deviations exhibited by Figures 14.5 and 14.6 from the real equilibrium conditions may be due to errors in measurement, approximations of the equations used, and natural processes. Natural processes slow down or diminish as the Froude number tends to minimum and the channel becomes stable and attains equilibrium.

It may be noted that the theory of minimum Froude number is the same as the theory of stream power per unit length and per unit weight of water, for Froude number identifies the state of flow (i.e., subcritical, critical, or supercritical), whereas other factors cannot. The minimum Froude number mechanism is a macroscopic governing mechanism of channel adjustment. The minimum Froude number does not correspond to the maximum resistance condition, however.

14.6 Another Look at Froude Number Minimization

Froude number, F_r, can be expressed as

$$F_r = \frac{v}{\sqrt{gd}} = \frac{Q}{A\sqrt{gd}} = \frac{Q}{\sqrt{g}wd^{\frac{3}{2}}}. \tag{14.12}$$

Differentiating Equation (14.2) with respect to distance x for a discharge of given frequency,

$$\frac{\partial F_r}{\partial x} = -\frac{Q}{\sqrt{g}}\left[\frac{1}{w^2 d^{\frac{3}{2}}}\frac{\partial w}{\partial x} + \frac{1}{w}\frac{3}{2}\frac{1}{d^{\frac{5}{2}}}\frac{\partial d}{\partial x}\right]. \tag{14.13}$$

Equation (14.13) explains the change in Froude number along with the change due to change in channel width and flow depth. Let the change in F_r be represented as R then it is composed of two components, R_1 and R_2, that is,

$$R = R_1 + R_2, \tag{14.14}$$

where

$$R = \frac{\partial F_r}{\partial x} \tag{14.15}$$

$$R_1 = -\frac{Q}{\sqrt{g}}\frac{1}{w^2 d^{\frac{3}{2}}}\frac{\partial w}{\partial x} \tag{14.16}$$

$$R_2 = -\frac{Q}{\sqrt{g}}\frac{1}{w}\frac{3}{2}\frac{1}{d^{\frac{5}{2}}}\frac{\partial d}{\partial x}, \tag{14.17}$$

where R_1 expresses the contribution of change in width to the change in Froude number, and R_2 expresses the contribution of change in flow depth to the change in

Froude number. Since there are only two variables involved in the definition of Froude number, the change in Froude number is due to these two parts. Since each part of Equation (14.15) represents a part of the total change in F_r it can be expressed as a proportion as

$$P_1 = \frac{R_1}{R} = \frac{-\frac{Q}{\sqrt{g}} \frac{1}{w^2} \frac{1}{d^{\frac{3}{2}}} \frac{\partial w}{\partial x}}{\frac{\partial F_r}{\partial x}} \tag{14.18}$$

$$P_2 = \frac{R_2}{R} = \frac{-\frac{Q}{\sqrt{g}} \frac{1}{w} \frac{3}{2} \frac{1}{d^{\frac{5}{2}}} \frac{\partial d}{\partial x}}{\frac{\partial F_r}{\partial x}}. \tag{14.19}$$

Quantities P_i, $i = 1$ and 2, can be interpreted as probabilities. Therefore, the Shannon entropy of the flow system can be written as

$$H = -\sum_{i=1}^{2} P_i \log P_i. \tag{14.20}$$

If there are no constraints imposed on the flow system, then the principle of maximum entropy would yield for minimum Froude number:

$$P_1 = P_2. \tag{14.21}$$

Equation (14.21) suggests that this is only one possibility for the minimization of Froude number:

The channel may adjust its geometry in accordance with this possibility. The minimization of Froude number is due to the changes in channel flow depth and width, which seems viable. Equating Equations (14.18) and (14.19), one gets

$$\frac{-\frac{Q}{\sqrt{g}} \frac{1}{w^2} \frac{1}{d^{\frac{3}{2}}} \frac{\partial w}{\partial x}}{\frac{\partial F_r}{\partial x}} = \frac{-\frac{Q}{\sqrt{g}} \left(\frac{1}{w}\right) \left(\frac{3}{2}\right) \frac{1}{d^{\frac{5}{2}}} \frac{\partial d}{\partial x}}{\frac{\partial F_r}{\partial x}}. \tag{14.22}$$

Equation (14.22) gives

$$\frac{3}{2} \frac{\partial d}{d} = \frac{\partial w}{w}. \tag{14.23}$$

Equation (14.23) states the manner in which changes in depth and width will occur in order to accomplish the minimization of Froude number. However, in reality, the minimization of F_r may not occur in accordance with Equation (14.23). Therefore, from a practical standpoint, it may be desirable to introduce a weighting factor $0 < J < 1$, which will define the proportion in which the minimization of F_r will be accomplished by the adjustment of d and w. Then, Equation (14.23) will become

$$\frac{3}{2}J\frac{\partial d}{d} = \frac{\partial w}{w}.$$

(14.24)

On integration, the solution of Equation (14.24) can be expressed as

$$w = C_1 d^{\frac{3J}{2}}$$

(14.25)

or

$$d = C_2 w^{\frac{2}{3J}},$$

(14.26)

where C_1 and C_2 are regarded as morphological constants.

Now we derive relations for d, w, S, and v. Substituting Equation (14.26) in Equation (14.12),

$$w^{1+\frac{1}{J}} = \frac{Q}{F_r\sqrt{g}(C_2)^{\frac{2}{3}}}.$$

(14.27)

Equation (14.27) can be expressed as

$$w = \frac{Q^{\frac{J}{1+J}}}{\left[F_r\sqrt{g}(C_2)^{\frac{2}{3}}\right]^{\frac{J}{1+J}}} = aQ^{\frac{J}{1+J}} \propto Q^{\frac{J}{1+J}}; \quad a = \left[F_r\sqrt{g}(C_2)^{\frac{2}{3}}\right]^{-\frac{J}{1+J}}.$$

(14.28)

It is interesting that if $J = 1$, then

$$w \propto Q^{0.5}.$$

(14.29)

The equation is close to what is observed in field, but not exactly.

In a similar manner, substituting Equation (14.25) in Equation (14.12),

$$d^{\frac{3}{2}(1+J)} = \frac{Q}{F_r\sqrt{g}C_1}.$$

(14.30)

Equation (14.30) can be expressed as

$$d = \frac{Q^{\frac{2}{3(1+J)}}}{\left[F_r\sqrt{g}C_1\right]^{\frac{2}{3(1+J)}}} = cQ^{\frac{2}{3(1+J)}} \propto Q^{\frac{2}{3(1+J)}}; \quad c = \left[F_r\sqrt{g}C_1\right]^{-\frac{2}{3(1+J)}}.$$

(14.31)

If $J = 1$, then Equation (14.31) leads to

$$d \propto Q^{\frac{2}{6}},$$

(14.32)

which is close to what is observed in field, but not exactly.

From the continuity equation expressed by Equation (14.1), with the use of Equations (14.28) and (14.31), one can write

$$v = \frac{1}{ac} Q^{1 - \frac{J}{1+J} - \frac{2}{3(1+J)}}; \quad v = f \, Q^{1 - \frac{J}{1+J} - \frac{2}{3(1+J)}}; v \propto Q^{1 - \left(\frac{J}{1+J}\right) - \frac{2}{3(1+J)}}; v \propto Q^{\frac{1}{3(1+J)}}. \quad (14.33)$$

If $J = 1$, then Equation (14.33) yields

$$v \propto Q^{\frac{1}{6}}, \quad (14.34)$$

which is close is what is observed in field.

An expression for slope can be obtained by using Manning's equation:

$$v = \frac{1}{n} d^{\frac{2}{3}} S^{\frac{1}{2}}. \quad (14.35)$$

Substituting Equations (14.31) and (14.33) in Equation (14.35), one gets

$$f Q^{1 - \left(\frac{J}{1+J}\right) - \left(\frac{2}{3(1+J)}\right)} = \frac{1}{n} c^{\frac{2}{3}} Q^{\frac{4}{9(1+J)}} S^{\frac{1}{2}}. \quad (14.36)$$

Equation (14.36) yields

$$S = \left(\frac{nf}{c^{\frac{2}{3}}}\right)^{2} Q^{2 \frac{-1}{9(1+J)}} \propto Q^{2 \frac{-1}{9(1+J)}}. \quad (14.37)$$

If $J = 1$, Equation (14.37) yields

$$S \propto Q^{\left(-\frac{1}{9}\right)}, \quad (14.38)$$

which is too deviant from what has been reported in the literature.

References

Brownlie, W. R. (1983). Flow depth in sand-bed channels. *Journal of Hydraulic Engineering*, Vol. 108, No. 7, pp. 959–990.

Colby, B. R. and Hembree, C. H. (1955). Computation of total sediment discharge, Nebraska River near Cody, Nebraska. U.S. geological Survey Water-Supply Paper 1387, pp. 187.

Einstein, H. A. and Barbarosa, N. I. (1952). River channel roughness. *Transaction of ASCE*, Vol. 117, pp. 1121–1132.

Hubbell, D. W. and Matejka, D. Q. (1959). Investigations of sediment transportation, Middle Loup River, at Dunning, Nebraska. U.S. Geological Survey Water-Supply Paper1476, pp. 123.

Jia, Y. (1990). Minimum Froude number and the equilibrium of alluvial sand rivers. *Earth Surface Processes and Landforms*, Vol. 15, pp. 199–208.

Leopold, L. B. and Wolman, M. G. (1957). River channel patterns braided, meandering and straight. U.S. Geological Survey Professional Paper 282-B.

Li, C. and Liu, J. (1961). The Resistance of Alluvial Rivers (Chinese).

Schumm, S. A. (1981). Evolution and response of fluvial system, sedimentologic implications. In: F. G. Ethridge and R. M. Flores (eds.), *Society of Economic Paleontologists and Mineralogists*, Special Publications, Tulsa, OK, 31, pp. 19–29.

Simons, D. B. and Alberton, M. L. (1963). Uniform water conveyance channels in alluvial material. *Transactions of ASCE*, Vol.128, Part I, pp. 65–107.

Simons, D. B. and Senturk, F. (1977). *Sediment Transport Technology*. Water Resources Publications, Highland Ranch, CO.

Yang, C. T. (1988). Sediment transport and unit stream power. *Civil Engineering Practice*, Vol. 5, No. 8, pp. 265–289, edited by P. N. Cheremisinoff, N. P. Cheremisinoff, and S. L. Cheng, Technomie Publishing Co., Lancaster, PA.

15

Hypothesis of Maximum Friction Factor

The hypothesis of maximum friction factor states that the channel geometry evolves to a stable nonplanar shape when the friction factor reaches a local maximum. It is supported by published data on bed forms, channels with artificial roughness elements, meandering channels, and bed armoring. This hypothesis can be regarded as an extremal hypothesis. However, this hypothesis may not be invariably true.

Notation

C_1 = morphological constant
C_2 = morphological constant
C_* = morphological constant
C_{**} = morphological constant
C^* = morphological constant
C^{**} = morphological constant
d = channel flow depth
D = medium size of sediments
f = friction factor or coefficient of resistance to flow
F_1 = subsidiary condition or constraint
g = acceleration due to gravity
H = Shannon entropy
J = weighting factor
L = Lagrangian function
P_i, $i = 1, 2$, and 3, = probabilities
q_s = sediment discharge per unit width
Q = discharge

Continued

Q_s = sediment discharge
r = weighting factor
R_1 = contribution of change in channel flow depth to the change in friction
R_2 = contribution of change in channel width depth to the change in friction
R_3 = contribution of change in channel slope depth to the change in friction
S = mean channel bed slope
v = mean flow velocity
w = mean flow width
W = weighting factor
x = distance
γ = weight density of water
γ_s = weight density of sediment
a_0 = Lagrange multiplier

15.1 Introduction

Davies and Sutherland (1980) proposed the hypothesis of maximum friction factor. They stated,

If the flow of a fluid past an original plane boundary is able to deform the boundary to a non-linear shape, it will do so in such a way that the friction factor increases. The deformation will cease when the shape of the boundary is that which gives rise to a local maximum friction factor. Thus, the equilibrium shape of a non-planar self-deformed flow boundary or channel corresponds to a local maximum factor.

Bridge (1981) found that the friction factors were maximum at bankfull stage in meander bends. In laboratory channels, Whittaker and Jaeggi (1982) observed friction factors to achieve maximum values during the development of step-pool structures. Thus, the hypothesis of maximum friction factor was inferred from the examination of published data on bed forms, channels with artificial roughness elements, meandering channels, and bed armoring. It can be shown that under certain conditions or combinations of independent variables, this hypothesis is equivalent to the extremal hypotheses. However, this hypothesis may not invariably be true.

15.2 Formulation of Maximum Friction Factor Hypothesis

The friction factor, f, or coefficient of resistance to flow, can be defined as

$$f = \frac{8gdS}{v^2},$$ (15.1)

where d is the channel flow depth, S is the mean channel bed slope, v is the mean flow velocity, and g is the acceleration due to gravity. The continuity equation for flow can be expressed as

$$Q = wdv, \tag{15.2}$$

where Q is discharge, and w is the mean flow width. Inserting Equation (15.2) into Equation (15.1) for eliminating v, the friction factor can be expressed as

$$f = \frac{8gw^2d^3S}{Q^2}. \tag{15.3}$$

Now the objective is to maximize friction in accordance with the hypothesis of maximum friction. Equation (15.3) shows that maximization of f implies the maximization of (w^2d^3S/Q^2). The maximization can be accomplished by employing the method of Lagrange multipliers, where the Lagrangian function L can be expressed as

$$L = f + a_0F_1, \tag{15.4}$$

where $F_I = 0$ is a subsidiary condition or constraint that is supplied by sediment transport equation, and a_0 is the Lagrange multiplier. For a given discharge, a necessary condition for an extreme value is provided by

$$\frac{\partial L}{\partial w} = 0; \quad \frac{\partial L}{\partial d} = 0; \quad \frac{\partial L}{\partial S} = 0, \tag{15.5}$$

together with the subsidiary condition.

A suitable sediment transport formula is the modified Parker (1978) bed load formula (Chang, 1980), which can be expressed as

$$\Phi = 6.62 \left[\frac{1}{\psi} - 0.03 \right]^5 \psi^{3.9}, \tag{15.6}$$

in which

$$\frac{1}{\psi} = \frac{dS}{(S_s - 1)D} \tag{15.7}$$

is the Shields entrainment function and

$$\Phi = \frac{q_s}{\left[g(S_s - 1)D^3 \right]^{0.5}} \tag{15.8}$$

is the Einstein sediment discharge, where q_s is the sediment discharge per unit width, $S_s = \frac{\gamma_s}{\gamma}$, γ_s is the weight density (specific weight) of sediment, γ is the weight density of water, and D is the medium size of sediments. With

$$Q_s = q_s w \tag{15.9}$$

one can write from Equations (15.8) and (15.9),

$$\Phi_b = \Phi w = \frac{Q_s}{\left[g(S_s - 1)D^3\right]^{0.5}}. \tag{15.10}$$

Therefore, from Equations (15.6) and (15.10), one may write the subsidiary condition F_1 as

$$F_1 = 6.62w\left[\frac{1}{\psi} - 0.03\right]^5 \psi^{3.9} - \Phi_b = 0. \tag{15.11}$$

Thus, the Lagrangian function $L\,(w,\,d,\,S)$ can be expressed as

$$L(w, d, S) = \frac{8gw^2d^3S}{Q^2} + a_0F_1. \tag{15.12}$$

The solution of Equation (15.5) yields $(1/\psi) > 0$ and $(1/\psi) < 0$. Thus, the required extreme value does not exist. When the friction factor is plotted against the constrained dependent variables, it may not be possible to identify the extrema as turning points on the plots. Thus, the maximum friction factor (MFF) hypothesis is open to question.

15.3 Maximization

For maximizing f given by Equation (15.3) with respect to distance x for the same frequency of discharge, one can write

$$\frac{\partial f}{\partial x} = \frac{8g}{Q^2}\left[3d^2\frac{\partial d}{\partial x}Sw^2 + d^3S2w\frac{\partial w}{\partial x} + d^3w^2\frac{\partial S}{\partial x}\right]. \tag{15.13}$$

Equation (15.13) states that the change in friction with distance in the flow direction is constituted by the sum of three parts, designated as R_1, R_2, and R_3, where

$$R = \frac{\partial f}{\partial x} = \frac{8g}{Q^2}\frac{\partial}{\partial x}\left[d^3Sw^2\right] = R_1 + R_2 + R_3 \tag{15.14}$$

$$R_1 = \frac{8g}{Q^2} 3d^2 Sw^2 \frac{\partial d}{\partial x} \tag{15.15}$$

$$R_2 = \frac{8g}{Q^2} d^3 S2w \frac{\partial w}{\partial x} \tag{15.16}$$

$$R_3 = \frac{8g}{Q^2} d^3 w^2 \frac{\partial S}{\partial x}. \tag{15.17}$$

Each part of Equation (15.14) represents a part of the total change in f and can be expressed as a proportion as

$$P_1 = \frac{R_1}{R} = \frac{\left(\frac{8g}{Q^2}\right) \frac{\partial d}{\partial x} 3d^2 Sw^2}{\left(\frac{8g}{Q^2}\right) \partial (d^3 Sw^2)/\partial x} = \frac{3d^2 Sw^2 (\partial d/\partial x)}{\partial (d^3 Sw^2)/\partial x} \tag{15.18}$$

$$P_2 = \frac{R_2}{R} = \frac{\left(\frac{8g}{Q^2}\right) \frac{\partial w}{\partial x} d^3 2wS}{\left(\frac{8g}{Q^2}\right) \partial (d^3 Sw^2)/\partial x} = \frac{2Sd^3 w (\partial w/\partial x)}{\partial (d^3 Sw^2)/\partial x} \tag{15.19}$$

$$P_3 = \frac{R_3}{R} = \frac{\left(\frac{8g}{Q^2}\right) \frac{\partial S}{\partial x} d^3 w^2}{\left(\frac{8g}{Q^2}\right) \partial (d^3 Sw^2)/\partial x} = \frac{d^3 w^2 (\partial S/\partial x)}{\partial (d^3 Sw^2)/\partial x}. \tag{15.20}$$

Quantities P_i, $i = 1$, 2, and 3, can be interpreted as probabilities. Therefore, the Shannon entropy of the flow system can be written as

$$H = -\sum_{i=1}^{3} P_i \log P_i. \tag{15.21}$$

If there are no constraints imposed on the flow system, then the principle of maximum entropy would yield for maximum friction:

$$P_1 = P_2 = P_3. \tag{15.22}$$

Equation (15.22) suggests that there can be four possibilities for the maximization of friction:

Possibility 1: $P_1 = P_2$
Possibility 2: $P_2 = P_3$
Possibility 3: $P_1 = P_3$
Possibility 4: $P_1 = P_2 = P_3$.

The channel may adjust its geometry in accordance with any of these four possibilities depending on the conditions imposed on the channel.

Possibility 1: $P_1 = P_2$. The maximization of friction is due to the changes in channel flow depth and width. This case seems most viable. Equating Equations (15.18) and (15.19), one gets

$$\frac{3d^2Sw^2(\partial d/\partial x)}{\partial(d^3Sw^2)/\partial x} = \frac{2d^3wS(\partial w/\partial x)}{\partial(d^3Sw^2)/\partial x}. \tag{15.23}$$

Equation (15.23) gives

$$\frac{3}{2}\frac{\partial d}{d} = \frac{\partial w}{w}. \tag{15.24}$$

Equation (15.24) states the manner in which changes in depth and width will occur in order to accomplish the maximization of friction. However, in reality, the maximization of f may not occur in accordance with Equation (15.24). Therefore, from a practical standpoint, it may be desirable to introduce a weighting factor, $0 < W < 1$, which will define the proportion in which the maximization of f will be accomplished by the adjustment of d and w. Then, Equation (15.24) will become

$$\frac{3}{2}W\frac{\partial d}{d} = \frac{\partial w}{w}. \tag{15.25}$$

On integration, the solution of Equation (15.25) can be expressed as

$$w = C_1 d^{\frac{3W}{2}} \tag{15.26a}$$

or

$$d = C_2 w^{\frac{2}{3W}}, \tag{15.26b}$$

where C_1 and C_2 are regarded as morphological constants.

Now we derive relations for d, w, S, and v. Substituting Equation (15.26a) in Equation (15.3),

$$C_1^2 d^{3+3W} = \frac{Q^2 f}{8gS}. \tag{15.27}$$

Equation (15.27) leads to

$$d = \left(\frac{f}{C_1^2 8gS}\right)^{\frac{1}{3(1+W)}} Q^{\frac{2}{3(1+W)}} \quad \text{or} \quad d = C_3 Q^{\frac{2}{3(1+W)}}. \tag{15.28}$$

It is interesting to note that if $W = 1$, Equation (15.28) becomes

$$d = C_3 Q^{\frac{1}{3}}, \quad C_3 = \left(\frac{f}{C_1^2 8 g S} \right)^{\frac{1}{6}}. \tag{15.29}$$

Now inserting Equation (15.26b) in Equation (15.3),

$$w^2 \left(C_2 w^{2/3W} \right)^3 = \frac{Q^2 f}{8 g S}. \tag{15.30}$$

Equation (15.30) results in

$$w^{\left(2+\frac{2}{w}\right)} = \frac{f}{8 g s (C_2)^3} Q^2 \quad \text{or} \quad w = C_4 Q^{\frac{W}{1+W}}, \quad C_4 = \left(\frac{f}{8 g S (C_2)^3} \right)^{\frac{W}{2(1+W)}}. \tag{15.31}$$

If $W = 1$, Equation (15.31) becomes

$$w = C_4 Q^{\frac{1}{2}}. \tag{15.32}$$

Recalling the definition of velocity,

$$v = \frac{Q}{w d}. \tag{15.33}$$

The friction factor f can be written as

$$f = \frac{8 g S d}{v^2}. \tag{15.34}$$

The velocity v can be written from Equation (15.34) as

$$v = \left(\frac{8 g S}{f} \right)^{\frac{1}{2}} d^{\frac{1}{2}}. \tag{15.35}$$

Inserting Equation (15.28) in Equation (15.35),

$$v = \left(\frac{8 g S}{f} \right)^{\frac{1}{2}} (C_3)^{\frac{1}{2}} Q^{\frac{1}{3(1+W)}} \quad \text{or} \quad v = C_5 Q^{\frac{1}{3(1+W)}}, \quad C_5 = (C_3)^{\frac{1}{2}} \left(\frac{8 g s}{f} \right)^{\frac{1}{2}}. \tag{15.36}$$

If $W = 1$, Equation (15.36) reduces to

$$v = C_5 Q^{\frac{1}{6}}. \tag{15.37}$$

Checking the exponents of d, w, and v, it is noted that

$$\frac{1}{3} + \frac{1}{2} + \frac{1}{6} = 1. \tag{15.38}$$

Equation (15.38) shows that Equations (15.29), (15.32), and (15.37) satisfy the flow continuity Equation (15.2).

Now the channel slope can be written as

$$S = \frac{fQ^2}{8gd^3w^2}.$$

(15.39)

Inserting Equations (15.28) and (15.31) in Equation (15.39),

$$S = \left(\frac{f}{8g}\right)Q^2\left(\frac{1}{c_3}\right)^3 Q^{-\frac{6}{3(1+W)}}\left(\frac{1}{C_4}\right)^2 Q^{-\frac{2W}{1+W}} \quad \text{or} \quad S = C_6 Q^0 = C_6.$$

(15.40)

Equation (15.40) states that the bed slope is independent of discharge.

Expressing friction in terms of discharge by inserting Equations (15.28) and (15.31),

$$f = 8gSw^2d^3Q^{-2} = 8gS(C_3)^3 Q^{\frac{2}{1+W}}(C_4)^2 Q^{\frac{2W}{1+W}}Q^{-2} = C_7 Q^0 = C_7.$$

(15.41)

Equation (15.41) shows that friction factor is independent of discharge. The exponents of w, d, v, and S are, respectively, 1/2, 1/3, 1/6, and 0, which compare well with the values obtained by Leopold and Maddock (1953) as 0.5, 0.4, and 0.1.

Possibility II: $P_2 = P_3$. This possibility states that the maximization of friction factor f is accomplished by the adjustment in width w and slope S. Equating Equation (15.19) to Equation (15.20), the result is

$$\frac{2Sd^3w(\partial w/\partial x)}{\partial(d^3Sw^2)/\partial x} = \frac{d^3w^2(\partial S/\partial x)}{\partial(d^3Sw^2)/\partial x}.$$

(15.42)

Equation (15.42) leads to

$$\frac{\partial w}{w} = \frac{\partial S}{2S}.$$

(15.43)

Equation (15.43) reflects the change in width and slope for the maximization of friction. To generalize the nature of change, one can introduce a weighting factor $0 < J < 1$ for apportioning the adjustment of width and slope for maximizing friction. Thus, Equation (15.43) can be written as

$$\frac{\partial w}{w} = J\frac{\partial S}{2S}.$$

(15.44)

Solution of Equation (15.44) can be given as

$$w = C_* S^{J/2}$$

(15.45a)

or

$$S = C_{**}w^{\frac{2}{J}},$$ (15.45b)

where C_* and C_{**} are morphological constants.

Inserting Equation (15.45a) in Equation (15.3), one obtains

$$S\left(C_* S^{J}\right)^2 = \frac{f}{8gd^3}Q^2 \quad \text{or} \quad S = C_{***}Q^{\frac{2}{(1+J)}}, \quad C_{***} = \left(\frac{f}{8gd^3 C_*^2}\right)^{\frac{1}{1+J}}.$$ (15.46)

If $J = 1$, then Equation (15.46) reduces to

$$S = C_{***}Q.$$ (15.47)

Equation (15.47) states that slope increases or decreases linearly with the increase or decrease of discharge. Inserting Equation (15.45b) in Equation (15.3),

$$w^2 C_{**}w^{\frac{2}{J}} = \frac{f}{8gd^3}Q^2.$$ (15.48)

Equation (15.48) can be written as

$$w = C_{IV}Q^{\frac{2J}{2+2J}}, \quad C_{IV} = \left(\frac{f}{8gd^3 C_{**}}\right)^{\frac{J}{2+2J}}.$$ (15.49)

If $J = 1$, Equation (15.49) reduces to

$$w = C_{IV}Q^{\frac{1}{2}}.$$ (15.50)

Now for the friction factor,

$$f = 8gd^3 C_*^2 S^{1+\frac{J2}{2}}Q^{-2} = C_V Q^{-2}.$$ (15.51a)

Equation (15.51a) decreases inversely with the square of discharge.

Substituting Equations (15.50) and (15.47) in Equation (15.3), an expression for d is derived as

$$d = KQ^{\frac{1}{6}}, \quad K = \left(\frac{f}{8gC_{***}C_{IV}^2}\right)^{\frac{1}{3}}.$$ (15.51b)

Glancing at the exponents of w, d, and S and the conditions of maximization, it seems clear that possibility is not a practical one.

Possibility III: $P_1 = P_3$. This possibility states that the maximization of friction factor is achieved by the adjustment of flow depth and channel slope. Equating Equation (15.18) to Equation (15.20),

$$\frac{3d^2Sw^2(\partial d/\partial x)}{\partial(d^3Sw^2)/\partial x} = \frac{d^3w^2(\partial S/\partial x)}{\partial(d^3Sw^2)/\partial x}. \tag{15.52}$$

Equation (15.52) leads to

$$\frac{\partial d}{d} = \frac{1}{3}\frac{\partial S}{S}. \tag{15.53}$$

Equation (15.53) provides the changes in depth and slope for maximizing friction. In order to generalize the apportionment of their contribution to the maximization of friction, a weighting factor $0 < r < 1$ can be introduced. Then, Equation (15.53) can be written as

$$\frac{\partial d}{d} = \frac{r}{3}\frac{\partial S}{S}. \tag{15.54}$$

Solution of Equation (15.54) can be given as

$$d = C^*S^{\frac{r}{3}} \tag{15.55a}$$

or

$$S = C^{**}d^{\frac{3}{r}}. \tag{15.55b}$$

Substituting Equation (15.55b) in Equation (15.3),

$$d^3C^{**}d^{\frac{3}{r}} = \frac{f}{8gw^2}Q^2. \tag{15.56}$$

Equation (15.56) yields

$$d = \left(\frac{f}{8gw^2C^{**}}\right)^{\frac{r}{3(1+r)}}Q^{\frac{2r}{3(1+r)}} = C^{***}Q^{\frac{2r}{3(1+r)}}, \quad C^{***} = \left(\frac{f}{8gw^2C^{**}}\right)^{\frac{r}{3(1+r)}}. \tag{15.57}$$

If $r = 1$, then Equation (15.57) reduces to

$$d = C^{***}Q^{\frac{1}{3}}. \tag{15.58}$$

Now, from Equation (15.3),

$$w^2 = \left(\frac{f}{8gSd^3}\right)Q^2. \tag{15.59}$$

Inserting Equation (15.57) in Equation (15.59),

$$w = \left(\frac{f}{8gS}\right)^{\frac{1}{2}}C^{***-3/2}Q^{1-\frac{r}{1+r}} = C^{IV}Q^{\frac{1}{1+r}}. \tag{15.60}$$

If $r = 1$, Equation (15.60) reduces to

$$w = C^{IV} Q^{\frac{1}{2}}.$$ (15.61)

From Equation (15.3), the flow velocity v can be expressed as

$$v = \left(\frac{8gS}{f}\right)^{\frac{1}{2}} d^{\frac{1}{2}}.$$ (15.62)

Inserting Equation (15.57) in Equation (15.62),

$$v = \left(\frac{8gS}{f}\right)^{\frac{1}{2}} C^{***1/2} Q^{\frac{r}{3(1+r)}} = C^{IV} Q^{\frac{r}{3(1+r)}}, \quad C^{IV} = \left(C^{***} \frac{8gS}{f}\right)^{\frac{1}{2}}.$$ (15.63)

If $r = 1$, Equation (15.63) reduces to

$$v = C^{IV} Q^{\frac{1}{6}}.$$ (15.64)

This possibility yields the exponents of w, d, and v as 1/2, 1/3, and 1/6, which compare favorably with those of Leopold and Maddock (1953) as 1/2, 2/5, and 1/10.

Possibility IV: $P_1 = P_2 = P_3$. This possibility states that the maximization of friction factor f is achieved by the adjustment of width, depth, and slope. For deriving an expression for d, Equation (15.3) can be written as

$$d^3 w^2 S = \frac{f}{8g} Q^2.$$ (15.65)

Eliminating w and S using Equations (15.26a) and (15.55b), Equation (15.65) can be written as

$$d^3 \left(c_1 d^{\frac{3W}{2}}\right)^2 C^{**} d^{\frac{3}{r}} = \frac{f}{8g} Q^2.$$ (15.66)

Equation (15.66) yields

$$d = \left(\frac{f}{8gC_1^2 C^{**}}\right)^{\frac{r}{3(1+Wr+r)}} Q^{\frac{2r}{3(1+Wr+r)}} = \lambda_0 Q^{\frac{2r}{3(1+Wr+r)}}, \quad \lambda_0 = \left(\frac{f}{8gC_1^2 C^{**}}\right)^{\frac{r}{3(1+Wr+r)}}.$$ (15.67)

If $r = W = 1$, Equation (15.67) reduces to

$$d = \lambda_0 Q^{\frac{2}{9}}.$$ (15.68)

Now, eliminating S and d from Equation (15.65) with the use of Equations (15.26b) and (15.45b), Equation (15.65) becomes

$$C^{**}d^{\frac{3}{r}}w^2\left(C_2w^{\frac{2}{3W}}\right)^3 = C^{**}\left(C_2w^{\frac{2}{3W}}\right)^{\left(\frac{3}{r}\right)}w^2\left(C_2w^{\frac{2}{3W}}\right)^3 = C^{**}(C_2)^{\frac{3}{r}+3}w^{\frac{2}{Wr}+2+\frac{2}{W}} = \frac{f}{8g}Q^2.$$

(15.69)

Equation (15.69) can be written as

$$w = \lambda_1 Q^{\frac{2Wr}{2+2Wr+2r}}, \quad \lambda_1 = \left(\frac{f}{8g}\frac{1}{C^{**}C_2^{\left(\frac{3}{r}+3\right)}}\right)^{\frac{Wr}{2+2Wr+2r}}.$$

(15.70)

If $r = W = 1$,

$$w = \lambda_1 Q^{\frac{1}{3}}.$$

(15.71)

If Equations (15.71) and (15.68) are inserted in the continuity equation, then an expression for v is obtained as

$$v = \left(\frac{1}{\lambda_0\lambda_1}\right)Q^{\frac{4}{9}}.$$

(15.72)

The exponents of w, d, and v are, respectively, 1/3, 2/9, and 4/9, which compare well with those of Leopold and Maddock (1953) as 1/2, 2/5, and 1/10.

References

Bridge, J. S. (1981). *A summary of flow and sedimentary process studies on the River South Esk, Glen Clova, Scotland, Abstract. Modern Ancient Fluvial Systems*, pp. 19, University of Keele.

Chang, H. (1980). Geometry of gravel streams. *Journal of Hydraulics Division, ASCE*, Vol. 106, No. 9, pp. 1443–1456.

Davies, T. R. and Sutherland, A. J. (1980). Resistance to flow past deformable boundaries. *Earth Surface Processes*, Vol. 5, pp. 175–179.

Leopold, L. B. and Maddock, J. T. (1953). The hydraulic geometry of stream channels and some physiographic implications. U.S. Geological Survey Professional Paper 252, pp. 1–57.

Parker, G. (1978). Self-formed straight rivers with equilibrium banks and mobile bed, Part 1.The sand-silt river. *Journal of Fluid Mechanics*, Vol. 89, No. 1, pp. 109–125.

Whittaker, J. G. and Jaeggi, M. N. R. (1982). On the origin of step-pool systems in mountain streams. *Journal of Hydraulics Division, ASCE*, Vol. 97, No. HY4, pp. 505–522.

16

Maximum Flow Efficiency Hypothesis

Design of a stable alluvial channel is based on the hypothesis that the equilibrium state of a channel corresponds to maximum flow. The channel design can then be accomplished by employing the continuity equation, resistance law, sediment transport equation, and the channel cross-section shape. This chapter derives the channel hydraulic geometry for, primarily, three cross-sections, namely trapezoidal, rectangular, and triangular.

Notation

A = flow cross-sectional area

C_d = coefficient

d = flow depth

D = sediment particle size

n = Manning roughness coefficient

Q = discharge

Q_s = sediment (bed load) discharge

Q_{smax} = maximum sediment transport capacity

P = wetted perimeter

r = a dimensionless shape parameter

r_{max} = a dimensionless shape parameter where Q_s is maximum

R = hydraulic radius

S = energy slope

v = average flow velocity

w = channel width

γ = specific weight of water

τ_c = critical shear stress

τ_0 = flow shear stress

16.1 Introduction

A stable or relatively stable alluvial channel is a channel that transports water and sediment stemming from its drainage area or otherwise externally imposed on it without causing net erosion or deposition. The channel adjusts its form in response to the water and sediment load, and the adjustment is preceded by bank and bed erosion and deposition, and with no net erosion or deposition, the channel is considered to be in equilibrium. This is consistent with the proposition of Gilbert (1914), which states that rivers with a large supply of bed load adjust their cross-sections to transport it as efficiently as possible. It was then employed by Mackin (1948) and Schumm (1960). Although it has long been recognized that the adjustment is driven mainly by flow efficiency (Mackin, 1948), there is no consensus as to what constitutes flow efficiency. Griffith (1927) and Kirkby (1977) proposed the maximum sediment transporting capacity (MSTC) to be an appropriate form of flow efficiency, whereas Rubey (1952) stated the maximum hydraulic radius (MHR) to be the maximum flow efficiency (MFE). Chang (1979a, b, 1988) reasoned that MHR and minimum stream power (MSP) would coincide at the same point, where he defined MSP from the principle of virtual work or energy for a mechanical system. Huang and Nanson (2000) described the optimum condition for flow efficiency as the maximum sediment transport capacity per unit available stream power and stated that the conditions of MFE and the general form of MHR would coincide.

The geometry of a stable alluvial channel can be derived from three basic relationships, which are continuity equation, resistance equation, and sediment transport equation. The objective of this chapter is to show that maximum flow efficiency corresponds to a stable channel geometry, as shown by Huang and Nanson (2000), and the discussion here follows their work.

16.2 Basic Relations

Three basic relations are employed here: (1) temporally and spatially integrated continuity equation, (2) resistance equation or Manning's equation, and (3) sediment transport equation, such as DuBoys equation. Each of these relations is now described.

16.2.1 Continuity Equation

The lumped continuity equation states that discharge Q is equal to the product of average flow velocity v and flow cross-sectional area A and can be expressed as

$$Q = Av. \tag{16.1}$$

If the channel is assumed rectangular then the cross-sectional area A can be expressed as

$$A = wd, \tag{16.2}$$

where w is channel width and d is the flow depth.

16.2.2 Resistance Equation

There are several resistance equations that have been used in deriving hydraulic geometry relations, such as Chezy's, Manning's, or Lacey's (Lacey, 1958). In this discussion, Manning's equation is employed, for it is perhaps the most widely used equation in open channel hydraulics.

In SI units, it can be expressed as

$$v = \frac{1}{n}R^{\frac{2}{3}}S^{\frac{1}{2}}, \tag{16.3}$$

where v is the average cross-sectional flow velocity, R is the hydraulic radius ($= A/P$), P is the wetted perimeter, S is the energy slope, often expressed as equal to bed slope, and n is the Manning roughness coefficient. For a rectangular channel, the hydraulic radius can be expressed as

$$R = \frac{A}{P} = \frac{wd}{w + 2d}. \tag{16.4}$$

16.2.3 Sediment Transport

Several sediment transport equations, such as Peter-Meyer–Muller formula and DuBoys formula, have been used in hydraulic geometry relations. In this discussion the Duboys sediment transport formula is employed, which can be expressed as

$$Q_s = C_d \tau_0 (\tau_0 - \tau_c) w, \tag{16.5}$$

in which Q_s is the sediment (bed load) discharge, τ_0 is the flow shear stress, τ_c is the critical shear stress, w is the channel width, and C_d is the coefficient.

The critical shear stress is often determined from sediment particle size (D) in mm (ASCE, 1975; Chang, 1980a, b, 1988) as

$$\tau_c = 0.061 + 0.093D \tag{16.6}$$

and coefficient C_d as a function of sediment grain size as

$$C_d = 0.17D^{-\frac{3}{4}}. \tag{16.7}$$

The flow shear stress is expressed as

$$\tau_0 = \gamma RS, \tag{16.8}$$

where γ is the specific weight of water.

16.2.4 Shape Parameter

Hydraulic geometry entails relations for four variables, including depth, width, velocity, and slope, to discharge but there are only three basic governing Equations (16.1), (16.3), and (16.5). In order to reduce the number of independent variables, a dimensionless shape parameter r (ratio of width to depth–aspect ratio) can be introduced as:

$$r = \frac{w}{d}. \tag{16.9}$$

Equation (16.9) has been used to classify rivers into narrow and wide rivers.

16.3 Derivation of Hydraulic Geometry Relations

The channel geometry can be expressed in terms of r as

$$w = rd; \; A = wd = rd^2; \; R = \frac{A}{P} = \frac{wd}{w + 2d} = \frac{wd}{rd + 2d} = \frac{rd^2}{rd + 2d} = \frac{r}{r + 2}d. \tag{16.10}$$

Now Equation (16.3) can be expressed in terms of r as

$$\frac{Q}{rd^2} = \frac{1}{n}\left(\frac{r}{r + 2}d\right)^{\frac{2}{3}} S^{\frac{1}{2}}. \tag{16.11}$$

The relation for d can be obtained from Equation (16.11) as

$$d = \frac{n^{\frac{3}{8}}}{S^{\frac{3}{16}}}(r + 2)^{\frac{1}{4}}\left(\frac{1}{r}\right)^{\frac{5}{8}} Q^{\frac{3}{8}}. \tag{16.12}$$

Likewise, the relation for w can be expressed in terms of r as

$$w = \frac{n^{\frac{3}{8}}}{S^{\frac{3}{16}}}(r + 2)^{\frac{1}{4}}(r)^{\frac{3}{8}} Q^{\frac{3}{8}}. \tag{16.13}$$

Similarly, the bed shear stress can be expressed as a function of r as

$$\tau_0 = \gamma n^{\frac{3}{8}}\frac{S^{\frac{13}{16}}}{(r + 2)^{\frac{3}{4}}}(r)^{\frac{3}{8}} Q^{\frac{3}{8}}. \tag{16.14}$$

Now the sediment transport equation can be expressed in terms of r as

$$Q_s = C_d \gamma n^{\frac{3}{8}} S^{\frac{13}{16}} \frac{r^{\frac{3}{8}}}{(r+2)^{\frac{3}{4}}} Q^{\frac{3}{8}} \left(\gamma n^{\frac{3}{8}} S^{\frac{13}{16}} \frac{r^{\frac{3}{8}}}{(r+2)^{\frac{3}{4}}} Q^{\frac{3}{8}} - \tau_c \right) \frac{n^{\frac{3}{8}}}{S^{\frac{3}{16}}} r^{\frac{3}{8}} (r+2)^{\frac{1}{4}} Q^{\frac{3}{8}}. \quad (16.15)$$

Equation (16.15) can be written as

$$Q_s = K_1 \frac{r^{\frac{3}{4}}}{(r+2)^{\frac{1}{2}}} \left(K_2 \frac{r^{\frac{3}{8}}}{(r+2)^{\frac{3}{4}}} - \tau_c \right), \quad (16.16)$$

where

$$K_1 = C_d \gamma S^{\frac{5}{8}} n^{\left(\frac{3}{4}\right)} Q^{\frac{3}{4}} \quad (16.17)$$

$$K_2 = \gamma S^{\frac{13}{16}} n^{\left(\frac{3}{8}\right)} Q^{\frac{3}{8}}. \quad (16.18)$$

In order to determine the maximum Q_s, it will be instructive to plot the right side of Equation (16.16) against r, as shown in Figure 16.1, which shows the value of r (say r_{max}) where Q_s is maximum. Further, differentiating Equation (16.16) with respect to r and equating the derivative to zero yield the condition that must be satisfied for achieving Q_{smax}.

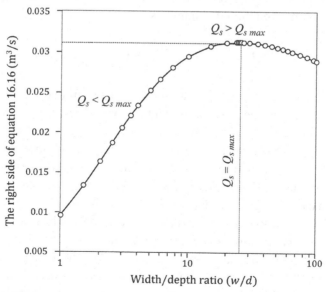

Figure 16.1. Variation of sediment discharge with variation in width-to-depth ratio [sediment size = 0.3 mm, slope = 0.0005, sediment density = 2.65×10^3 kg/m^3, and $Q = 500$ m^2/s]

The right side (*RS*) of Equation (16.16) can be written as

$$RS = K_1 K_2 \frac{r^{\frac{9}{8}}}{(r+2)^{\frac{5}{4}}} - K_1 \tau_c \frac{r^{\frac{3}{4}}}{(r+2)^{\frac{1}{2}}}. \tag{16.19}$$

Differentiating Equation (16.19) with respect to *r* and equating the derivative to 0, one obtains

$$\frac{dRS}{dr} = 0 = K_2 \frac{9}{8} \frac{r^{\frac{1}{8}}}{(r+2)^{\frac{5}{4}}} - K_2 \frac{5}{4} \frac{r^{\frac{9}{8}}}{(r+2)^{\frac{9}{4}}} - \tau_c \left(\frac{3}{4} \frac{1}{r^{\frac{1}{4}}(r+2)^{\frac{1}{2}}} - \frac{1}{2} \frac{r^{\frac{3}{4}}}{(r+2)^{\frac{3}{2}}} \right). \tag{16.20}$$

Equation (16.20) can be simplified as

$$\frac{r^{\frac{3}{8}}}{(r+2)^{\frac{3}{4}}} - \frac{\tau_c}{K_2} \frac{2r+12}{18-r} = 0. \tag{16.21}$$

Equation (16.21) yields the condition

$$\left| \frac{r^{\frac{3}{8}}}{(r+2)^{\frac{3}{4}}} = \frac{\tau_c}{K_2} \frac{2r+12}{18-r} \right|_{r=r_{max}}. \tag{16.22}$$

Substituting Equation (16.22) in Equation (16.16), one gets the maximum sediment transport capacity of the alluvial channel as

$$Q_{smax} = K_1 \tau_c \frac{r^{\frac{3}{4}}}{(r+2)^{\frac{1}{2}}} \left(\frac{3r-6}{18-r} \right). \tag{16.23}$$

Equation (16.23) shows that Q_{smax} varies with *r* between two limiting conditions: (1) $r = 2$, which defines the lower limit, and $r = 18$, which defines the upper limit. Thus, one can conclude that three situations arise: (1) $Q_s < Q_{smax}$, (2) $Q_s = Q_{smax}$, and (3) $Q_s > Q_{smax}$, as shown in Figure 16.1. These conditions have a physical meaning. If $Q_s < Q_{smax}$ then the alluvial channel can maintain its equilibrium by adjusting its form alone. However, Figure 16.1 shows that two channels of different forms can transport the given sediment discharge, that is, the channel on the left side has a lower width to depth ratio (narrower channel), while the channel on the right side has a higher ratio (wider channel). The narrower deeper channel will become unstable and may have erodible banks. On the other hand, the wider shallower channel will become unstable and may have bed erosion, and may be narrowing because of localized flow concentration. In other words, the channel may not be stable under this situation. If $Q_s > Q_{smax}$ then the channel will be capable of transporting the imposed sediment load by adjusting the form alone and will have to adjust slope, roughness or flow discharge, or all three. It has been shown that at bankfull conditions, the channel is capable of transporting the imposed sediment load while maintaining its geometric stability (Wolman and Miller, 1960).

16.3.1 Optimum Hydraulic Geometry Relations

Equation (16.23) can be rewritten with the use of Equation (16.17) as

$$Q_s = C_d \gamma \tau_c S^{\frac{5}{8}} n^{\frac{3}{4}} Q^{\frac{3}{4}} \frac{r^{\frac{3}{4}}}{(r+2)^{\frac{1}{2}}} \left(\frac{3r-6}{18-r} \right). \tag{16.24}$$

From Equation (16.24), the slope can be expressed as

$$S = \left(\frac{Q_s}{C_d \tau_c \gamma n^{\frac{3}{4}}} \right)^{\frac{8}{5}} \frac{(r+2)^{\frac{4}{5}}}{(r)^{\frac{6}{5}}} \left(\frac{18-r}{3r-6} \right)^{\frac{8}{5}} Q^{-\frac{6}{5}}. \tag{16.25}$$

With the substitution of Equation (16.25) in Equation (16.12), the expression for flow depth can be expressed as

$$d = n^{\frac{3}{5}} (\gamma \tau_c C_d)^{\frac{3}{10}} \left(\frac{1}{r} \right)^{\frac{2}{5}} (r+2)^{\frac{1}{10}} \left(\frac{3r-6}{18-r} \right)^{\frac{3}{10}} Q_s^{-\frac{3}{10}} Q^{\frac{3}{5}}. \tag{16.26}$$

Inserting Equation (16.25) in Equation (16.13), one gets an expression for width as

$$w = n^{\frac{3}{5}} (\gamma \tau_c C_d)^{\frac{3}{10}} (r)^{\frac{3}{5}} (r+2)^{\frac{1}{10}} \left(\frac{3r-6}{18-r} \right)^{\frac{3}{10}} Q_s^{-\frac{3}{10}} Q^{\frac{3}{5}}. \tag{16.27}$$

Equations (16.15)–(16.27) are restricted by two limits or thresholds, so hydraulic geometric relations are now derived corresponding to each threshold.

16.3.2 Lower Threshold Geometry Relations

The lower threshold is defined by $r = 2$. At this value, $Q_{smax} = 0$ from Equation (16.24), which means $Q_s = 0$ from Equation (16.16). Thus, from Equation (16.16) one can write

$$\left| K_2 \frac{r^{\frac{3}{8}}}{(r+2)^{\frac{3}{4}}} - \tau_c \right|_{r=r_m=2} = 0. \tag{16.28}$$

Inserting Equation (16.18) in Equation (16.28) and solving, one gets

$$\gamma S^{\frac{13}{16}} n^{\frac{3}{8}} \frac{2^{\frac{3}{8}}}{(4)^{\frac{3}{4}}} Q^{\frac{3}{8}} - \tau_c = 0. \tag{16.29}$$

Equation (16.29) leads to

$$S = \frac{4^{\frac{12}{13}}}{2^{\frac{6}{13}}} \left(\frac{\tau_c}{\gamma n^{\frac{3}{8}}}\right)^{\frac{16}{13}} Q^{-\frac{6}{13}}. \tag{16.30}$$

Equation (16.30) states that

$$S \propto Q^{-\frac{6}{13}} (= Q^{-0.462}). \tag{16.31}$$

Substituting Equation (16.30) in Equation (16.12) for obtaining the relation of d, one obtains

$$d = \frac{4^{\frac{1}{13}}}{2^{\frac{7}{13}}} n^{\frac{6}{13}} \left(\frac{\tau_c}{\gamma}\right)^{-\frac{3}{13}} Q^{\frac{6}{13}}. \tag{16.32}$$

Equation (16.32) states that

$$d \propto Q^{\frac{6}{13}} (= Q^{0.462}). \tag{16.33}$$

Inserting Equation (16.30) in Equation (16.13), the result is

$$w = 4^{\frac{1}{13}} 2^{\frac{6}{13}} n^{\frac{6}{13}} \left(\frac{\tau_c}{\gamma}\right)^{-\frac{3}{13}} Q^{\frac{6}{13}}. \tag{16.34}$$

Equation (16.34) states that

$$w \propto Q^{\frac{6}{13}} (= Q^{0.462}). \tag{16.35}$$

It may be interesting to compare Equations (16.31), (16.33), and (16.35) with the relations obtained by Lane (1952) using the idealized threshold theory, in which it is assumed that sediment at each point on the wetted perimeter is in a state of impending motion. For this theory the exponents for the width and depth relations are 0.46 and the exponent for the slope relation is −0.46. These are almost in perfect agreement with those derived using the maximum flow efficiency for the width–depth ratio of 2, as shown in Table 16.1.

16.3.3 Upper Threshold Geometry Relations

Equation (16.23) says that if $r = 18$, the maximum sediment discharge becomes infinite for given discharge and slope. This condition, almost nonexistent, can theoretically occur if the critical shear stress is much much smaller than the bed stream stress or tractive force, that is, $\tau_c \lll \tau_0$. This means that the critical shear stress plays virtually no role in sediment transport. Under such a condition, the DuBoys sediment transport formula is expressed as

Table 16.1. *Comparison of lower threshold channel geometry relations with threshold theory*

Channel geometry parameter	Maximum flow efficiency	Threshold theory (Lane, 1952)
Width (w)	$w \propto Q^{0.462}$	$w \propto Q^{0.46}$
Depth (d)	$d \propto Q^{0.462}$	$d \propto Q^{0.46}$
Slope (S)	$S \propto Q^{-0.462}$	$S \propto Q^{-0.46}$
Width–depth ratio (w/d)	2.5	7.05–8.61 (with angle of repose between 30° and 35°)

$$Q_s = C_d \tau_0^2 w. \tag{16.36}$$

Now taking τ_c as 0 in Equation (16.16) and inserting Equations (16.17) and (16.18) in Equation (16.16), and taking $r = 18$, one obtains

$$Q_s = C_d \gamma^2 n^{\frac{9}{8}} \frac{18^{\frac{9}{8}}}{(20)^{\frac{5}{4}}} S^{\frac{23}{16}} Q^{\frac{9}{8}}. \tag{16.37}$$

Equation (16.37) yields the relation for S as

$$S = \left(\frac{Q_s 20^{\frac{5}{4}}}{C_d \gamma^2 n^{\frac{9}{8}} 18^{\frac{9}{8}}} \right)^{\frac{16}{23}} Q^{-\frac{18}{23}}. \tag{16.38}$$

Inserting Equation (16.38) in Equation (16.12), the relation for d is obtained as

$$d = n^{\frac{12}{23}} 20^{\frac{2}{23}} 18^{-\frac{11}{23}} \gamma^{\frac{6}{23}} (C_d)^{\frac{3}{23}} Q^{\frac{12}{23}} Q_s^{-\frac{3}{23}}. \tag{16.39}$$

Substituting Equation (16.38) in Equation (16.13) yields

$$w = n^{\frac{12}{23}} 20^{\frac{2}{23}} 18^{\frac{12}{23}} \gamma^{\frac{6}{23}} (C_d)^{\frac{3}{23}} Q^{\frac{12}{23}} Q_s^{-\frac{3}{23}}. \tag{16.40}$$

Equations (16.38) to (16.40) provide a set of geometry relations corresponding to the condition that bed shear stress is very high by comparison with critical shear stress.

16.3.4 Average Channel Geometry Relations

The channel geometry adjusts within the range defined by the two thresholds, that is, $r = 2$ and $r = 18$. One can take any value between these thresholds and compute the hydraulic geometry relations. For simplicity one can assume integral values of r from 3 to 17. Equation (16.24) can be written with all the terms involving r on one side as

$$\frac{r^{\frac{3}{4}}}{(r+2)^{\frac{1}{2}}}\left(\frac{3r-6}{18-r}\right) = \left(\frac{Q_s}{C_d\gamma\tau_c S^{\frac{5}{8}}n^{\frac{3}{4}}Q^{\frac{3}{4}}}\right). \tag{16.41}$$

The left side of Equation (16.41) can be expressed as a power function of r as:

$$\frac{r^{\frac{3}{4}}}{(r+2)^{\frac{1}{2}}}\left(\frac{3r-6}{18-r}\right) \approx (7.5e-3)r^{2.90}, \tag{16.42}$$

as shown in Figure 16.2.

Likewise, for Equation (16.25) for S one can write

$$\frac{(r+2)^{\frac{4}{5}}}{(r)^{\frac{6}{5}}}\left(\frac{18-r}{3r-6}\right)^{\frac{8}{5}} \approx 144r^{-3.09}, \tag{16.43}$$

as shown in Figure 16.3.

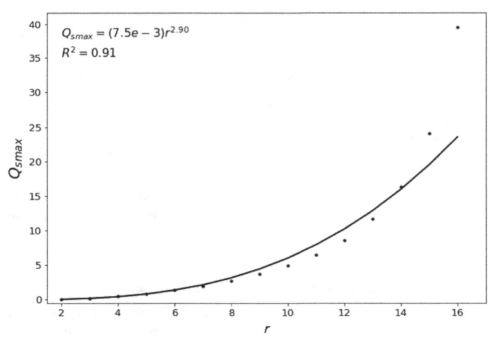

Figure 16.2 Sediment discharge as function of aspect ratio r

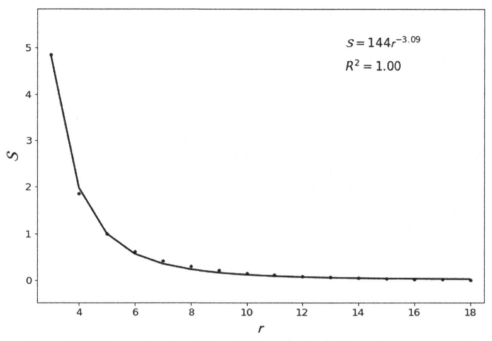

Figure 16.3 Slope as function of aspect ratio r

Similarly, for Equation (16.26) for flow depth d, one can express

$$\left(\frac{1}{r}\right)^{\frac{2}{5}}(r+2)^{\frac{1}{10}}\left(\frac{3r-6}{18-r}\right)^{\frac{3}{10}} \approx 0.15 r^{0.71}, \tag{16.44}$$

as shown in Figure 16.4.

In a similar vein, for Equation (16.27) for width w, one can write

$$(r)^{\frac{3}{5}}(r+2)^{\frac{1}{10}}\left(\frac{3r-6}{18-r}\right)^{\frac{3}{10}} \approx 0.07 r^{1.99}, \tag{16.45}$$

as shown in Figure 16.5. Then, the average optimum hydraulic geometry relations can be derived as follows.

From Equations (16.41) and (16.42), one can write

$$r = \left(\frac{1,000}{7.5}\right)^{\frac{1}{2.9}}\left(\frac{Q_s}{C_d\gamma\tau_cS^{\frac{5}{8}}n^{\frac{3}{4}}Q^{\frac{3}{4}}}\right)^{\frac{1}{2.9}} = 5.4\left(\frac{Q_s}{C_d\gamma\tau_cS^{\frac{5}{8}}n^{\frac{3}{4}}Q^{\frac{3}{4}}}\right)^{\frac{1}{2.9}}. \tag{16.46}$$

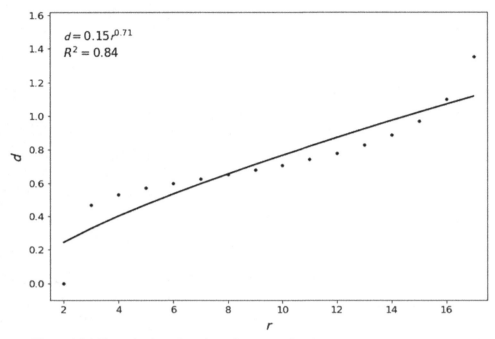

Figure 16.4 Flow depth as function of aspect ratio r

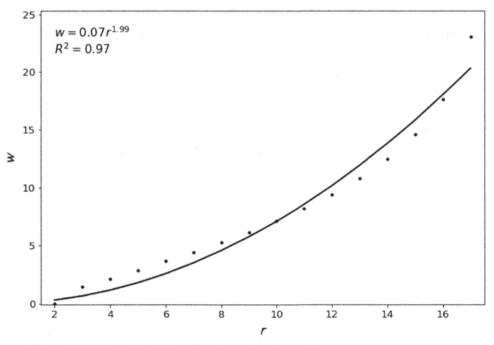

Figure 16.5 Channel width as function of aspect ratio r

From Equations (16.25) and (16.43) the slope S can be expressed as

$$S = 144r^{-3.09}\left(\frac{Q_s}{C_d\tau_c\gamma n^{\frac{3}{4}}}\right)^{\frac{8}{5}}Q^{-\frac{6}{5}}. \tag{16.47}$$

With the use of Equation (16.46), S from Equation (16.47) can be written as

$$S = 0.87\frac{Q_s^{0.31}}{(C_d\gamma\tau_c)^{0.31}n^{0.24}}Q^{-0.24}. \tag{16.48}$$

From Equations (16.44) and (16.26), the expression for flow depth can be expressed as

$$d = 0.15r^{0.71}n^{\frac{2}{5}}(\gamma\tau_c C_d)^{\frac{3}{10}}Q_s^{-\frac{3}{10}}Q^{\frac{3}{5}}. \tag{16.49}$$

With the substitution of Equation (16.46), Equation (16.49) for flow depth d can be written as

$$d = 0.5(C_d\gamma\tau_c)^{0.06}S^{-0.15}n^{0.42}Q_s^{-0.06}Q^{0.42}. \tag{16.50}$$

From Equations (16.27) and (16.45), one gets an expression for width as

$$w = 0.07r^{1.99}n^{\frac{2}{5}}(\gamma\tau_c C_d)^{\frac{3}{10}}Q_s^{-\frac{3}{10}}Q^{\frac{3}{5}}. \tag{16.51}$$

Inserting Equation (16.46), Equation (16.52) for width w can be written as

$$w = 2.0(C_d\gamma\tau_c)^{-0.39}S^{-0.43}n^{0.09}Q_s^{0.39}Q^{0.09}. \tag{16.52}$$

In Equations (16.48), (16.50), and (16.52), the quantities n, C_d, and τ_c can be expressed in terms of sediment grain size D and can therefore be treated as constant, even though their combined effect on channel geometry is complicated with channel geometry and bank strength (Millar and Quick, 1993; Huang and Nanson, 1998). Merging the constant quantities with constant of proportionality, Equations (16.48), (16.50), and (16.52) can be written as

$$S \propto Q_s^{0.31}Q^{-0.24} \propto \left(\frac{Q_s}{Q}\right)^{0.31}Q^{0.07} \tag{16.53}$$

$$d \propto Q_s^{-0.06}Q^{0.42} \propto \left(\frac{Q_s}{Q}\right)^{-0.06}Q^{0.48} \tag{16.54}$$

$$w \propto Q_s^{0.39}Q^{0.09} \propto \left(\frac{Q_s}{Q}\right)^{0.39}Q^{0.48}. \tag{16.55}$$

In Equations (16.53)–(16.55), it can be assumed that the sediment concentration (Q_s/Q) can be assumed either constant or varying within a limited range (less than

Table 16.2. *Downstream hydraulic geometry relations*

Huang and coworkers (Huang and Warner, 1995; Huang and Nanson, 1995, 1998; Huang, 1996)	Julien and Wargadalam (1995)	Maximum low efficiency
$w \propto Q^{0.501} S^{-0.156}$	$w \propto Q^{0.4-0.5} S^{-(0.2-0.25)}$	$w \propto Q^{0.39} S^{1.25}$
$d \propto Q^{0.299} S^{-0.206}$	$d \propto Q^{0.4-0.25} S^{-(0.2-0.125)}$	$d \propto Q^{0.37} S^{-0.19}$
$v \propto Q^{0.201} S^{0.3626}$	$v \propto Q^{0.2-0.25} S^{0.4-0.375}$	$v \propto Q^{0.24} S^{-1.06}$

500 ppm) (Simons and Albertson, 1960). Since the sediment discharge Q_s is not known in many cases, channel slope S can be used as an alternative in Equations (16.53)–(16.55). Therefore, these equations can be expressed as

$$d \propto Q^{0.37} S^{-0.19} \tag{16.56}$$

$$w \propto Q^{0.39} S^{1.25} \tag{16.57}$$

$$v \propto Q^{0.24} S^{-1.06}. \tag{16.58}$$

Equation (16.58) follows from the continuity equation with Equations (16.56) and (16.57).

Comparison of Equations (16.56) to (16.58) with downstream hydraulic geometry relations (Huang and Warner, 1995; Huang and Nanson, 1995, 1998; Julien and Wargadalam, 1995), given in Table 16.2, shows that the three sets of relations obtained from entirely different approaches are quite close to each other, and there are only minor differences between the exponents of discharge and channel slope. This suggests that natural alluvial rivers adjust their channel forms in order to reach an optimal or equilibrium state regardless of the environment in which they occur.

16.4 Physical Evidence for Maximum Flow Efficiency

Flow continuity, flow resistance, and sediment transport govern channel adjustment and hence determine the condition of maximum sediment transport capacity, Q_{smax}, defined by Equation (16.23), subject to the constraints of flow discharge, channel slope, and sediment size. Since there are four variables involved: flow discharge, sediment discharge, channel slope, and sediment size, Equation (16.23) can also mathematically define three situations:

(1) It can define minimum slope subject to the constraints of flow discharge, sediment discharge, and sediment size.

(2) It can define minimum flow discharge subject to the constraints of sediment discharge, channel slope, and sediment size.

(3) It can define an optimum sediment size subject to the constraints of flow discharge, sediment discharge, and channel slope.

Equation (16.23) can be expressed as

$$\left. \frac{Q_s}{K_1 \tau_c} \right|_{r=r_m} = \text{a maximum} \left[= \frac{r^{\frac{3}{4}}}{(r+2)^{\frac{1}{2}}} \left(\frac{3r-6}{18-r} \right) \right]. \tag{16.59}$$

If it is assumed that the sediment size does not change, then inserting Equation (16.17) in Equation (16.59) yields

$$\left. \frac{Q_s}{C_d \gamma S^{\frac{5}{8}} n^{\left(\frac{3}{4}\right)} Q^{\frac{3}{4}} \tau_c} \right|_{r=r_m} = \text{a maximum} \left[= \frac{r^{\frac{3}{4}}}{(r+2)^{\frac{1}{2}}} \left(\frac{3r-6}{18-r} \right) \right]. \tag{16.60}$$

The major part of Equation (16.60) can be written as

$$\frac{Q_s}{S^{\frac{5}{8}} Q^{\frac{3}{4}}} = \text{a maximum}. \tag{16.61}$$

Equation (16.61) expresses the maximum sediment transport capacity per unit of approximate total stream power (actual stream power is γQS, $\gamma =$ specific weight of water). This can be construed as the MFE regime for the following reasons. First, MFE embodies the following optimum conditions:

(1) For fixed S and Q, $Q_s = $ a maximum, as hypothesized by Pickup (1976), Kirkby (1977), and White et al. (1982).

(2) For fixed Q and Q_s, $S = $ a minimum, as hypothesized by Chang (1979a, b, 1988).

(3) For fixed Q_s, $S^{\frac{5}{8}} Q^{\frac{3}{4}} = $ a minimum or $SQ^{1.2} = $ a minimum, close to $\gamma QS = $ a minimum, as hypothesized by Chang (1980a, b, 1986, 1988).

This discussion shows that the maximum sediment transporting capacity (MSTC) hypothesis proposed by Pickup (1976), Kirkby (1977), and White et al. (1982) and minimum stream power (MSP) hypothesis proposed by Chang (1980a, b, 1986, 1988) should yield stable channel relations consistent with the MFE hypothesis.

Second, the MFE and MSTC and MSP hypotheses coincide with the general form of the maximum hydraulic radius (MHR), where the maximum sediment discharge Q_{smax} is in a critical state of zero for given flow discharge, channel slope, and sediment size. From Equation (16.23) it is seen that this situation occurs when channel shape $r = 2$. Assuming $Q_s = Q_{smax} = 0$ at that point, Equation (16.5) yields

$$\tau_0|_{r=r_m=2} = \tau_c. \tag{16.62}$$

However, the $\tau_0 - r$ relationship from Equation (16.14) shows that at $r = r_m = 2$, the following is satisfied:

$$\tau_0|_{r=r_m=2} = \text{a maximum.} \tag{16.63}$$

Noting the definition of $\tau_0 = \gamma RS$, Equation (16.41) yields for a given channel slope

$$R|_{r=r_m=2} = \text{a maximum.} \tag{16.64}$$

Equation (16.14) shows that for constant channel slope and sediment size a maximum hydraulic radius given by Equation (16.44) yields a minimum flow discharge, whereas the condition of maximum hydraulic radius defined in open flow hydraulics results in a maximum flow discharge. This difference can be explained by noting that the maximum hydraulic radius is subject to the condition of constant channel cross-sectional area, but Equation (16.64) is free of this condition.

Third, the combination of the relation between channel width and channel shape factor given by Equation (16.13) and the quasi-parabolic (rising and falling) relation between average flow shear stress and shape factor given by Equation (16.14) provides the physical evidence for MFE. Equation (16.16) shows that for given flow discharge and channel slope an increase in r leads to an increase in Q_s if r is small but to a decrease in Q_s if r is large (see Figure 16.1). When the full range of variation of r is considered, a maximum sediment transport rate occurs at a medium value of r.

The bed load sediment transport depends on the interaction between the tractive force and the force resisting sediment movement. When τ_0 is very large such that τ_c is negligible in Equation (16.16) then Q_s approaches a maximum value when $r = 18$. On the other hand, if τ_0 is small and τ_c is very large to a threshold value of $\tau_0 = \tau_c$ then Q_s approaches a maximum value at $r = 2$. This illustrates that the interaction between the tractive force and resistance to the sediment movement causes the variation of optimum channel shape in the range between $r = 2$ and $r = 18$.

Fourth, although the relations derived here are based on Manning's flow resistance and DuBoys sediment transport relation, the results should hopefully be consistent with other relations.

References

ASCE (1975). *Sedimentation Engineering*. ASCE Press, New York.

Chang, H. H. (1979a). Geometry of rivers in regime. *Journal of the Hydraulics Division, ASCE*, Vol. 105, pp. 691–706.

Chang, H. H. (1979b). Minimum stream power and river channel; patterns. *Journal of Hydrology*, Vol. 41, pp. 303–327.

Chang, H. H. (1980a). Stable alluvial canal design. *Journal of the Hydraulics Division, ASCE*, Vol. 106, pp. 873–891.

Chang, H. H. (1980b). Geometry of gravel streams. *Journal of the Hydraulics Division, ASCE*, Vol. 106, pp. 1443–1456.

Chang, H. H. (1986). River channel changes: Adjustment of equilibrium. *Journal of Hydraulic Engineering*, Vol. 112, pp. 43–55.

Chang, H. H. (1988). *Fluvial Processes in River Engineering*. Wiley, New York.

Gilbert, G. K. (1914). *The transport of debris by running water*. U.S. Geological Survey Professional Paper 86, Washington, DC.

Griffith, W. M. (1927). A theory of silt and scour. *Proceedings of the Institution of Civil Engineers, London*, Vol. 223, pp. 243–263.

Huang, H. Q. (1996). Discussion of "Alluvial channel geometry: theory and application by Julien and Wargadalam." *Journal of Hydraulic Engineering*, Vol. 122, pp. 750–751.

Huang, H. Q. and Nanson, G. C. (1995). On a multivariate model of channel geometry. *Proceedings of the XXVIth Congress of the International Association of Hydraulic Research*, Vol. 1, pp. 510–515, Telford.

Huang, H. Q. and Nanson, G. C. (1998). The influence of bank strength on channel geometry. *Earth Surface Processes and Landforms*, Vol. 23, pp. 865–876.

Huang, H. Q. and Nanson, G. C. (2000). Hydraulic geometry and maximum flow efficiency as products of the principle of least action. *Earth Surface Processes and Landforms*, Vol. 25, pp. 1–16.

Huang, H. H. and Warner, R. F. (1995). The multivariate controls of hydraulic geometry: A causal investigation in terms of boundary shear distribution. *Earth Surface Processes and Landforms*, Vol. 20, pp. 115–130.

Julien, P. Y. and Wargadalam, J. (1995). Alluvial channel geometry: Theory and applications. *Journal of Hydraulic Engineering*, Vol. 121, pp. 312–325.

Kirkby, M. J. (1977). Maximum sediment efficiency as a criterion for alluvial channels. In: *River Channel Changes*, edited by K. J. Gregory, Wiley, Chichester, pp. 429–442.

Lacey, G. (1958). Flow in alluvial channels with sandy mobile beds. *Proceedings of the Institution of Civil Engineers, London*, Vol. 11, pp. 145–164.

Lane, E. W. (1952). *Progress report on results of studies on design of stable channels*. Hydraulic Laboratory Report HYD-352, U.S. Bureau of Reclamation, Denver, CO.

Mackin, J. H. (1948). Concept of the graded river. *Geological Society of America Bulletin*, Vol. 59, pp. 39–64.

Millar, R. G. and Quick, M. C. (1993). The effect of bank stability on geometry of gravel rivers. *Journal of Hydraulic Engineering*, Vol. 119, pp. 1343–1363.

Pickup, G. (1976). Adjustment of stream channel shape to hydrologic regime. *Journal of Hydrology*, Vol. 30, pp. 365–373.

Rubey, W. W. (1952). *Geology and mineral resources of the Hardin and Brussels Quadrangles (in Illinois)*. U.S. Geological Survey Professional Paper 218, Washington, DC.

Schumm, S. A. (1960). *The shape of alluvial channels in relation to sediment type*. Professional Paper vol. 352, US Geological Survey doi:10.3133/pp352b, Washington, DC.

Simons, D. B. and Albertson, M. L. (1960). Uniform water conveyance channels in alluvial materials. *Journal of the Hydraulic Division, ASCE*, Vol. 86, pp. 33–71.

White, W. R., Bettes, R. and Paris, E. (1982). Analytical approach to river regime. *Journal of the Hydraulics Division, ASCE*, Vol. 108, pp. 1179–1193.

Wolman, M. G. and Miller, J. P. (1960). Magnitude and frequency of forces in geomorphic processes. *Journal of Geology*, Vol. 68, No. 1, pp. 54–74.

17

Principle of Least Action

Rivers tend to follow the path of least action for transporting the sediment and water loads imposed on them. Because regime hydraulic geometry relations entail more unknowns than the equations of continuity, resistance, and sediment transport, optimization is utilized to determine the preferred cross-section from among many possible cross-sections, and this cross-section satisfies the path of least action. This chapter discusses this principle and derives the hydraulic geometry based on this principle.

Notation

a = constant

A = mean cross-section area

a_0 = distance from center of the flow cross section to the end point of wetting perimeter

$A_g(x)$ = general bed shape as an arbitrary function

c = constant

C_d = coefficient

d = mean channel depth

D = sediment size

E_k = kinetic energy

E_p = potential energy

F = functional

F_e = efficiency of flow in transporting sediment load

F_{er} = erosive force due to flow

F_g = gravitational force

F_r = resistance force

F_{rtotal} = total resistance force

---

g = acceleration due to gravity

l = quantity of action

L_f = Lagrangian function

L = reach length over which the variation occurs

m = fluid mass

N_a = constant associated with sediment size

P = wetted perimeter

P_s = total stream power per unit length

Q = flow discharge

Q_s = sediment transport

r = cross-sectional shape factor

R = hydraulic radius

r_m = corresponding to stable equilibrium or regime condition

S = channel slope or energy gradient

t_1 = starting point of motion

t_2 = ending points of motion

v_r = path position variable

v_{rm} = position satisfying minimum values of both E_p and l

v_t = velocity of movement following the path

w = channel width

ΔH = elevation difference

Δt = fixed time scale

τ_0 = flow shear stress

τ_c = critical shear stress for the incipient motion of sediment

α = parameter

β = constant

γ = weight density of water

λ = constant

ϕ_r = constraints

θ = angle of erosive force with the vertical

17.1 Introduction

Rivers are characterized by relatively stable hydraulic geometry. This geometry is not the same all along the river; nevertheless, river cross-sections tend to maintain characteristic shapes. To derive the regime hydraulic geometry relations, continuity, resistance, and sediment transport equations are employed, but these equations are not sufficient because hydraulic geometry entails more unknowns than the number of equations that are stated here. This leads to the concept of optimization, by which one preferred cross-section from among many possible cross-sections can be selected. Since nature always follows the path of least action or least expenditure of energy, rivers also tend to follow this path for transporting maximum sediment and water. This defines the least action principle (LAP) discussed by Huang et al. (2002), which seems inherent in the self-adjusting behavior of alluvial channels. The least action principle is also called Maupertuis principle, which states that nature is thrifty in all its actions. One interpretation of LAP is that flow of water occurs with least resistance and erosion. This means that stable hydraulic geometry can be determined using LAP, implying that the stable hydraulic cross-section should correspond to the cross-section that minimizes the loss of friction energy and channel erosion. This chapter discusses LAP following closely the work of Huang et al. (2002) and Ohara and Yamatani (2019).

17.2 Adjustment of Alluvial Cross-Sections

The least action principle allows only scalar quantities (energy and work) to be used, without resorting to forces, for describing the possible trajectories or paths. Invoking Hamilton's principle of least action, one can write

$$l = \int_{t_1}^{t_2} (E_k - E_p)\,dt = \int_{t_1}^{t_2} L_f\,dt = \text{a minimum,} \qquad (17.1)$$

which is required to satisfy

$$\frac{d}{dt}\left(\frac{\partial L_f}{\partial v_t}\right) - \frac{\partial L_f}{\partial v_r} = 0, \qquad (17.2)$$

where l is the quantity of action, E_k is the kinetic energy, E_p is the potential energy, $L_f\,(=E_k - E_p)$ is the Lagrangian function, v_r is the path position variable, v_t is the velocity of movement ($v_t = dv_r/dt = \dot{v}_r$) following the path, and t_1 and t_2 define the path interval or the starting and ending points of motion.

If the river behavior is independent of time and space, then Equation (17.2) reduces to

$$\frac{\partial E_p}{\partial v_r} = 0. \qquad (17.3)$$

In the case of a river, the potential energy decreases in the direction of flow. The position v_{rm} satisfying Equation (17.3) shows minimum values of both E_p and l, or

$$l = \text{a minimum} \propto E_p = \text{a minimum}$$

$$v_r = v_{rm} \qquad\qquad v_r = v_{rm}. \tag{17.4}$$

For a river not subject to any external constraints, Equation (17.4) is satisfied by the stationary potential energy. If there are constraints in the form of $\phi_r(v_r) = 0$, then the minimum value of potential energy depends on the constraints and may be greater than zero. The principle of minimum potential energy (MPE) is applied as a necessary and sufficient condition for the river to maintain a dynamic equilibrium (Riley and Sturges, 1993).

The flow in alluvial channels satisfies the continuity equation, resistance relation, and sediment transport equation, which satisfy the constraints expressed as $\phi_r(v_r) = 0$. The energy of steady uniform flow is derived from gravitational potential energy. These relations include two position variables of width and depth for given flow discharge, sediment load, channel slope, and channel boundary composition, and will apply to cross-sections of any shape. If the form factor, width (w)–depth (d) ratio ($r = w/d$), is introduced then the two position variables can be coupled into one for a rectangular section. The velocity of movement v_t in the direction of v_r, that is, r, will be 0 because the flow moves uniformly only in a direction perpendicular to the variation of cross-sectional shape factor r. These conditions satisfy the minimum energy principle (MEP). Thus, steady uniform flow can be written as

$$E_p(r)\rfloor_{(\phi_r(v_r)=0, r=r_m)} = \text{a minimum}, \tag{17.5}$$

where r_m corresponds to stable equilibrium or regime condition (non-erosional and non-depositional).

Equation (17.5) states that the number of channel sections can be large if all relevant cross-sections are considered. However, Equation (17.5) will be satisfied by only one cross-section corresponding to the dynamic equilibrium cross-section. In an alluvial channel the flow is maintained by an elevation difference ΔH. The potential energy then equals

$$E_p = mg\Delta H = \gamma Q \Delta t \Delta H = \gamma Q \Delta t L S, \tag{17.6}$$

where L is the reach length over which the variation occurs, and S is the energy slope, m is the fluid mass, γ is the weight density of water, Δt is the fixed time scale, g is the acceleration due to gravity, and Q is the flow discharge. Equation (17.5) can be derived by solving

$$P_s(r)\rfloor_{\phi_r(r)=0, r=r_m} = \text{a minimum}, \tag{17.7}$$

where P_s is the total stream power per unit length or $P_s = \gamma Q S$. Since Q is imposed, Equation (17.7) becomes

$$S(r)\rfloor_{\phi_r(r)=0,dQ=0,r=r_m} = \text{a minimum.} \tag{17.8}$$

The rate of sediment transport Q_s in channels reflects the rate of work done by the available stream power P_s. This suggests that Q_s can serve as a surrogate of energy available for sediment transport (Bagnold, 1966). Hence, the flow efficiency for sediment transport can be expressed as

$$\frac{Q_s}{\gamma QS} = F_e, \tag{17.9}$$

in which F_e is the efficiency of flow in transporting sediment load and is a function of r, that is, $F_e = F_e(r)$. Equation (17.9) states that the only way to make P_s minimum is by changing r, that is,

$$F_e(r)\rfloor_{\phi_r(r)=0,dQ_s=0,r=r_m} = \text{a maximum.} \tag{17.10}$$

The only way to keep P_s minimum and constant over the whole range of variation of r is to maximize Q_s, while balancing the maximum flow efficiency:

$$Q_s(r)\rfloor_{\phi_r(r),dQ=0,r=r_m} = \text{a maximum.} \tag{17.11}$$

It is seen from this discussion of LAP that the principle of minimum stream power (PMSP) and the principle of maximum sediment transporting capacity (MSTC) are equivalent and are special cases of the principle of maximum flow efficiency (MFE).

17.3 Validity

Using a case study, Huang et al. (2002) provided a justification of the theoretical inferences of LAP. They employed DuBoys' (1879) sediment transport formula, and Lacey's (1958) flow resistance equation. Recalling Lacey's flow resistance relation,

$$\frac{Q}{A} = \frac{1}{N_a} d^{\frac{1}{4}}\sqrt{RS} \quad \text{(SI units)}, \tag{17.12}$$

where N_a is related to sediment size D as

$$N_a = 0.0253D^{\frac{1}{8}}, \tag{17.13}$$

in which D is in mm, d is the mean channel depth (m), R is the hydraulic radius (m), A is the mean cross-section (m^2), and S is the channel slope or energy gradient.

DuBoys' sediment transport equation can be expressed as

$$Q_s = C_d\tau_0(\tau_0 - \tau_c)w, \tag{17.14}$$

where

$$\tau_0 = \gamma RS \tag{17.15}$$

is the flow shear stress, τ_c is the critical shear stress for the incipient motion of sediment, and C_d and τ_c are evaluated with sediment size D (mm) as (Chang, 1980)

$$C_d = 0.17D^{\frac{3}{4}} \tag{17.16}$$

$$\tau_c = 0.061 + 0.093D. \tag{17.17}$$

Assuming the channel to be rectangular on a constant slope and straight and flow to be uniform, the transport of sediment load Q_s and water discharge Q, with stream power γQS, Equations (17.12) and (17.14) are functions of two geometric variables, w and d. These equations can be written as

$$w = \left(\frac{QN_a}{\sqrt{S}}\right)^{\frac{4}{11}} r^{\frac{5}{11}}(r+2)^{\frac{2}{11}}; \quad \frac{dw}{dr} = \frac{w}{11} \frac{7r+10}{11r(r+2)} \tag{17.18}$$

$$\tau_0 = \gamma N_a^{\frac{4}{11}} S^{\frac{9}{11}} Q^{\frac{4}{11}} \frac{r^{\frac{5}{11}}}{(r+2)^{\frac{9}{11}}}; \quad \frac{d\tau_0}{dr} = \frac{4}{11} \frac{\tau_0(2.5-r)}{r(r+2)}. \tag{17.19}$$

Incorporating Equations (17.18) and (17.19) into the derivative of Equation (17.14),

$$\frac{dQ_s}{dr} = Q_s \left[\frac{1}{w}\frac{dw}{dr} + \frac{2\tau_0 - \tau_c}{\tau_0 - \tau_c}\frac{d\tau_c}{dr}\right] = \frac{Q_s}{11r(r+2)} \frac{(30-r)\tau_0 - (20+3r)\tau_c}{\tau_0 - \tau_c}. \tag{17.20}$$

Since $dQ_s/dr = 0$ at $r = r_m$ for given Q_s, Equations (17.18), (17.19), and (17.20) lead to

$$\frac{\tau_0}{\tau_c} = \frac{20+3r_m}{30-r_m} \quad \text{or} \quad \frac{\tau_0 - \tau_c}{\tau_c} = \frac{4(r_m - 2.5)}{30-r_m} \tag{17.21}$$

$$Q_s = 4K_1\tau_c \frac{r_m^{\frac{10}{11}}}{(r_m+2)^{\frac{17}{11}}} \frac{r_m - 2.5}{30-r_m}, \tag{17.22}$$

where

$$K_1 = \gamma C_d N_a^{\frac{8}{11}} S^{\frac{7}{11}} Q^{\frac{9}{11}}. \tag{17.23}$$

From Equation (17.20), $dQ_s/dr > 0$ for $r < r_m$, but for $r > r_m$. $dQ_s/dr > 0$. At this point r_m, Q_s, although constant, is actually the maximum that alluvial channels are capable of transporting:

$$Q_s|_{r=r_m} = \text{a maximim.} \tag{17.24}$$

For DuBoys' formula with Lacey's flow resistance equation it is found that

$$F_e = \frac{Q_s}{P_s^{\frac{4}{5}}}. \tag{17.25}$$

Now Equation (17.14) reduces to

$$Q_s = C_d \tau_0 (\tau_0 - \tau_c) w = C_d \frac{\tau_0^{1+x}(\tau_0 - \tau_c)}{\tau_0^x} w = P_s^{\lambda} \frac{(\tau_0 - \tau_c)}{\tau_0^r} w(r). \tag{17.26}$$

Using Equations (17.18) and (17.19),

$$\lambda = \frac{4}{5}, \quad x = \frac{1}{5}, \quad w(r) = C_d \gamma^{\frac{2}{5}} N_a^{\frac{4}{5}} \frac{r}{(r+2)^{\frac{4}{5}}}. \tag{17.27}$$

The sediment transport equation can now be written as

$$\frac{Q_s}{P_s^{\frac{4}{5}}} = F_e = w(r) \frac{\tau_0 - \tau_c}{\tau_0^{\frac{1}{5}}}. \tag{17.28}$$

Hence, the change in F_e with respect to r can be expressed as

$$\frac{dF_e}{dr} = F_e \left[\frac{1}{w(r)} \frac{dw(r)}{dr} + \left(\frac{1}{\tau_0 - \tau_c} - \frac{1}{5\tau_0} \right) \frac{d\tau_c}{dr} \right] = F_e \left[\frac{(30 - r)\tau_0 - (20 + 3r)\tau_c}{11r(r+2)} \right]. \tag{17.29}$$

It can be noted from Equations (17.20) and (17.29) that

$$\frac{dQ_s/dr}{Q_s} = \frac{dF_e/dr}{F_e}. \tag{17.30}$$

The generalized optimal flow condition at the point $r = r_m$ can be expressed as Equation (17.25). It is noted that F_e in Equation (17.9) is different from F_e in Equation (17.25). This difference permits definition of F_e in a more general form as

$$F_e = \frac{Q_s}{P_s^\alpha}, \quad P_s = \gamma Q S, \tag{17.31}$$

in which α varies with the combination of the resistance and sediment transport equations: $\alpha = 1.0$ for Bagnold's (1966) sediment transport equation and $\alpha = 4/5$ for DuBoys' (1879) sediment transport equation under Lacey's resistance equation.

Stream power is the power of available energy for maintaining steady uniform flow. The variation of P_s with channel cross-section shape reflects the degree of resistance of the section to the transport of Q_s. Thus, the section at which P_s is minimum is the one for which energy expenditure in overcoming channel boundary resistance is the least, which renders the channel maximally efficient, validating the applicability of LAP to the alluvial channel for adjustment.

In order to derive hydraulic geometry relations, Equations (17.18), (17.19), (17.21), and (17.22) can be combined to yield

$$\frac{(3r_m + 20)(r_m + 2)^{\frac{9}{11}}}{(30 - r_m)r^{\frac{5}{11}}} \propto Q^{\frac{4}{11}}S^{\frac{9}{11}} \tag{17.32}$$

$$w = Q^{\frac{4}{11}}S^{\frac{9}{11}}r^{\frac{5}{11}}_m(r_m + 2)^{\frac{2}{11}}, \tag{17.33}$$

in which r varies in the range of 2.5 and 30. The traditional hydraulic geometry relations are simplified forms of true relations.

To simplify Equations (17.32) and (17.33), Hung et al. (2002) derived, based on field data,

$$r^{1.915} \propto Q^{\frac{4}{11}}S^{\frac{9}{13}} \qquad (r = 0.911) \tag{17.34}$$

$$w \propto Q^{\frac{4}{11}}S^{-\frac{2}{11}}r^{0.605} \qquad (r = 1.00). \tag{17.35}$$

Equations (17.34) and (17.35) determine the averaged hydraulic geometry relations, as shown in Table 17.1.

17.4 Comparison

Hydraulic geometry relations derived using LAP are compared with those derived by Julien and Wargadalam (1995), Huang and Warner (1995), Huang (1996), and Hunag and Nanson (1995, 1998) as shown in Table 17.1. Although these workers used different principles, the differences are limited, suggesting that an alluvial channel form adjusts in accord with LAP, MPE, and MFE.

Equations (17.21), (17.22), and (17.23) yield a relation among channel slope, Q, Q_s, and r_m that can be averaged as (Huang et al., 2002)

$$S \propto Q^{\frac{7}{11}}_s Q^{\frac{-8}{7}}r^{-4.6500}_m. \tag{17.36}$$

Table 17.1. *Hydraulic geometry relations*

Huang and co-workers[*]	Julien and Wargadalam (1995)	LAP
$w \propto Q^{0.501}S^{-0.156}$	$w \propto Q^{0.4-0.5}S^{-0.2-0.25}$	$w \propto Q^{0.478}S^{0.076}$
$d \propto Q^{0.299}S^{-0.206}$	$d \propto Q^{0.4-0.25}S^{-0.20-0.125}$	$d \propto Q^{0.289}S^{0.350}$
$v \propto Q^{0.200}S^{0.362}$	$v \propto Q^{0.2-0.25}S^{0.40-0.375}$	$\propto Q^{0.232}S^{0.274}$

[*] Huang (1996), Huang and Warner (1995); Huang and Nanson (1995, 1998).

Combining Equations (17.34), (17.35), and (17.36), the result is the channel geometry relations:

$$w \propto Q_s^{0.04} Q^{0.426} \propto \left(\frac{Q_s}{Q}\right)^{0.04} Q^{0.466} \tag{17.37}$$

$$d \propto Q_s^{-0.184} Q^{0.526} \propto \left(\frac{Q_s}{Q}\right)^{-0.184} Q^{0.342} \tag{17.38}$$

$$S \propto Q_s^{0.526} Q^{-0.678} \propto \left(\frac{Q_s}{Q}\right)^{0.526} Q^{-0.152}. \tag{17.39}$$

If the effect of (Q_s/Q) is neglected then Equations (17.37)–(17.39) are close to the three regime relations $w \propto Q^{\frac{1}{2}}$, $d \propto Q^{\frac{1}{3}}$, and $S \propto Q^{\frac{1}{7}}$ (Lacey, 1929, 1933, 1946, 1958). Equations (17.34) and (17.35) show that the excess shear stress $\tau_0 - \tau_c$ causes the variation in the optimal channel shape r_m. If $\tau_0 = \tau_c$ at the incipient motion of sediment, then $r_m = 2.5$. If τ_0 is very large, τ_c is negligible, then $r_m = 30$. This suggests that bed load is critical in the development of stable channel geometry, and in bed load transport $\tau_0 - \tau_c$ is the determining factor.

Although these relations are based on Lacey's resistance equation and DuBoys' sediment transport equation, they contain equations that encompass a wide range of resistance and sediment transport equations. The maximal flow efficiency of straight alluvial channels is based on the principle of least action, which is general for flow in rivers and canals over wide-ranging environments and is physically based.

17.5 Stable Hydraulic Section Using Calculus of Variation

In the stable section, the channel does not erode at a minimum water area for a given discharge, because the tractive force exerted by the flow on the slope is balanced by the resistance force (F_r). Let F_{er} be the erosive force due to flow, as shown in Figure 17.1. Erosion occurs on the channel banks and bottom. The force of resistance to erosion on the bed must overcome the gravitational force besides the tractive force (vertical), which acts on banks as well (horizontal force). Ohara and Yamatami (2019) derived the stable hydraulic section using LAP with the aid of calculus of variation, and their work is followed here.

Let a general channel shape be expressed as $A_g(x)$, as shown in Figure 17.1. Let the vertical erosion force on the bed require α times more force than the riverbed resistance force due to gravity. Assuming the erosive force has an angle θ with the vertical, then the resistance forces in the horizontal (x) and vertical (y) directions can be expressed as

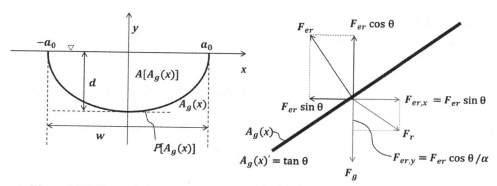

Figure 17.1 General channel cross-section and depiction of forces

$$F_{r,x} = F_{er} \sin \theta; \quad F_{r,y} = \frac{F_{er}}{\alpha} \cos \theta. \tag{17.40}$$

The α parameter can be expressed by noting that the erosive force balances the vertical resistance force as well as the gravitation force as

$$\alpha = 1 + \frac{F_g}{F_{r,y}}, \tag{17.41}$$

where F_g is the gravitational force.

The resultant resistance force acting on the particles on the channel bed denoted by A_g can be expressed with the use of Equation (17.40) as

$$F_r = \sqrt{F_{er,x}^2 + F_{r,y}^2} = \sqrt{(F_{er} \sin \theta)^2 + \left(\frac{F_{er} \cos \theta}{\alpha}\right)^2}$$

$$= \frac{F_{er}}{\alpha \sqrt{1 + \left(\frac{dA_g}{dx}\right)^2}} \sqrt{1 + \left(\alpha \frac{dA_g}{dx}\right)^2}. \tag{17.42}$$

Since erosion occurs on the bed as well as sides (or banks), the total resistance force, F_{rtotal}, can be obtained by integrating Equation (17.42) along the channel perimeter as

$$F_{rtotal} = \int_{-a_0}^{a_0} F_r \sqrt{1 + \left(\frac{dA_g}{dx}\right)^2} \, dx = \frac{F_e}{\alpha} \int_{-a_0}^{a_0} \sqrt{1 + \left(\alpha \frac{dA_g}{dx}\right)^2} \, dx = \frac{F_r}{\alpha} P[A_g(x)], \tag{17.43}$$

in which P is the wetted perimeter expressed as

$$P[A_g(x)] = \int_{-a_0}^{a_0} \sqrt{1 + \left(\alpha \frac{dA_g}{dx}\right)^2} \, dx, \tag{17.44}$$

where a_0 is the distance from the center of flow cross-section to the end point of wetting perimeter. In Equation (17.44), parameter α is the weighting factor on the vertical line element (dA_g/dx), which accounts for the anisotropy in channel erodibility.

The cross-sectional area A for a general channel bed shape A_g can be expressed as

$$A[A_g(x)] = \int_{-a_0}^{a_0} A_g(x)dx. \tag{17.45}$$

The objective is to derive the stable hydraulic section $A_g(x)$ that minimizes the wetted perimeter $P[A_g(x)]$, because the maximum conveyance for a given section occurs when the wetted perimeter is minimum, assuming no water surface friction. For erodible channels, Equation (17.41) shows that the stable hydraulic section corresponds to the best hydraulic section. The minimization of the wetted perimeter can be achieved by employing the calculation of variation as

$$F[A_g(x)] = \lambda A[A_g(x)] + P[(A_g(x)] = \int_{-a_0}^{a_0} \left[\lambda A_g(x) + \sqrt{1 + \left(\alpha \frac{dA_g(x)}{dx} \right)^2} \right] dx, \tag{17.46}$$

where F is a functional, and λ is a constant. If

$$f\left[A_g(x), \frac{dA_g(x)}{dx}\right] = \lambda A_g(x) + \sqrt{1 + \left(\alpha \frac{dA_g(x)}{dx} \right)^2} \tag{17.47}$$

then the Euler–Lagrange equation becomes

$$\frac{\delta F[A_g(x)]}{\delta A_g(x)} = \frac{\partial}{\partial A_g} f\left[(x), \frac{dA_g(x)}{dx}\right] - \frac{d}{dx} \left[\frac{\partial f\left[A_g(x), \frac{dA_g(x)}{dx}\right]}{\partial \left(\frac{dA_g(x)}{dx}\right)} \right] = 0 \tag{17.48}$$

$$= \lambda - \frac{d\left[\frac{\alpha^2 \frac{dA_g(x)}{dx}}{\sqrt{1 + \left(\alpha \frac{dA_g(x)}{dx}\right)^2}} \right]}{dx} = 0. \tag{17.49}$$

If the channel is assumed to be uniform, straight, and horizontally symmetric, then Equation (17.49) has the solution:

$$A_g(x) = \beta - \sqrt{\left(\frac{1}{\alpha}\right)^2 - \left(\frac{x}{\alpha}\right)^2}, \tag{17.50}$$

Table 17.2. *Parameters of regime equations*

Aspect ratio (r)	With exponent b	Depth exponent f	Authors	Remarks
8.2–8.4	0.26	0.40	Leopold and Maddock (1953)	20 river gaging stations from the US
–	0.5	0.40	Leopold and Maddock (1953)	20 river gaging stations from the US
22.86	0.5	0.34	Rhoads (1991)	252 gaging stations from Missouri River basin
–	0.38	0.46	Nanson and Huang (2017)	Gwydir, Namoi, and Barwon rivers
32–61.5	0.40	0.38	Fukuoka (2010)	Sacramento, Alberta, and Joganji rivers
	0.50	0.333	Lacey (1933), Blench (1951)	Regime theory, Canals, India
2.5	0.44	0.44	Huang and Nanson (2000)	Trapezoidal section
2.5–30	0.478	0.28	Huang and Nanson (2000)	Trapezoidal section
Flexible	0.375	0.375	Ohara and Yanatami (2019)	Ellipse section

where β is a constant. Equation (17.50) shows that the stable hydraulic section should be an ellipse. When the channel is a half ellipse, the hydraulic radius (A/P) is maximum, that is, $\alpha = a_0/d$. For $\lambda = 1/d$, and $\beta = 0$, Equation (17.50) becomes

$$A_g(x) = -d\sqrt{1 - \left(\frac{x}{a_0}\right)^2} \quad \text{or} \quad \left(\frac{x}{a_0}\right)^2 + \left[\frac{A_g(x)}{d}\right]^2 = 1. \tag{17.51}$$

The anisotropy parameter α is about half of the aspect ratio:

$$\alpha = \frac{1}{2}\frac{w}{d}. \tag{17.52}$$

With the use of Manning's equation, the regime equations can be expressed as

$$d = cQ^{\frac{3}{8}} \tag{17.53}$$

$$w = aQ^{\frac{3}{8}}, \tag{17.54}$$

in which

$$c = \left[\frac{2n}{\pi\alpha}\left(\frac{4K}{\pi}\right)^{\frac{2}{3}}S^{-\frac{1}{2}}\right]^{\frac{3}{8}} \tag{17.55}$$

$$K = 1 + \frac{1}{2}\left(\ln(4\alpha) - \frac{1}{1\cdot 2} \right)\frac{1}{\alpha^2} + \cdots \qquad (17.56)$$

$$a = 2\alpha c. \qquad (17.57)$$

Equation (17.56) shows that K is not a function of depth. Ohara and Yantami (2019) computed the wetted perimeter using the Cayley formula, which has a large aspect ratio that holds true for large rivers. Then, the hydraulic radius becomes proportional to depth at the center. They compared the exponents in equations with those reported in the literature for field observations, as shown in Table 17.2. They suggested that the channel cross-section may approach a semi-circle with the increase of erosive forces during high flows that dominate the gravity force. Consequently, the anisotropy parameter tends to unity. However, cross-sections are seldom elliptical but tend toward trapezoidal or rectangular shapes.

References

Bagnold, R. A. (1966). *An approach to the sediment transport problem from general physics.* U.S. Geological Survey Professional Paper 422-I, Washington, DC.

Blench, T. (1951). *Hydraulics of Sediment Bearing Canals and Rivers*, Vol. 1. Evans Industries, Original from the University of Wisconsin, Madison; Digitized, August 10, 2007; Length, 260 pages.

Chang, H. H. (1980). Geometry of gravel streams. *Journal of the Hydraulics Division*, Vol. 106, No. 9, pp. 1443–1456.

DuBoys, P. (1879). Le Rhone et les Rivieres a Lit Affouilable. *Annales des Ponts et Chousses*, Vol. 18, Series 5, pp. 141–195.

Fukuoka, S. (2010). Determination method of river width and cross-section for harmonization between flood control and river environment (Japanese). Advances in River Engineering, pp. 5–10.

Huang, H. Q. (1996). Discussion of alluvial channel geometry theory and applications by Julien and Wargadalam. *Journal of Hydraulic Engineering*, Vol. 122, pp. 750–751.

Huang, H. Q. and Nanson, G. C. (1995). On a multivariate model of channel geometry. *Proceedings of the XXVI Congress of the International Association for Hydraulic Research*, Vol. 1, pp. 510–515, Thames Telford.

Huang, H. Q. and Nanson, G. C. (1998). The influence of bank strength on channel geometry: an integrated analysis of some observations. *Earth Surface Processes and Landforms*, Vol. 24, pp. 865–876.

Huang, H. Q. and Nanson, G. C. (2000). Hydraulic geometry and maximum flow efficiency as products of the principle of least action. *Earth Surface Processes and Landforms*, Vol. 25, pp. 1–16.

Huang, H. Q. and Warner, R. F. (1995). The multivariate controls of hydraulic geometry: A causal investigation in term of boundary shear distribution. *Earth Surface Processes and Landforms*, Vol. 15, pp. 115–130.

Huang, H. Q., Nanson, G. C., and Fagar, S. D. (2002). Hydraulic geometry of straight alluvial channels and principle of least action. *Journal of Hydraulic Research*, Vol. 40, No. 2, pp. 153–160.

Julien, P. Y. and Wargadalam, J. (1995). Alluvial channel geometry theory and applications. *Journal of Hydraulic Engineering*, Vol. 121, pp. 312–325.

Lacey, G. (1929). Stable channels in alluvium. *Proceedings of the Institution of Civil Engineers*, Vol. 229, pp. 16–47.

Lacey, G. (1933). Uniform flow in alluvial rivers and canals. *Proceedings of the Institution of Civil Engineers*, Vol. 237, pp. 421–423.

Lacey, G. (1946). A general theory of flow in alluvium. *Proceedings of the Institution of Civil Engineers*, Vol. 27, pp. 16–47.

Lacey, G. (1958). Flow in alluvial channels with sandy stable beds. *Proceedings of the Institution of Civil Engineers*, Vol. 9, Discussion 11, pp. 145–164.

Leopold, L. B. and Maddock, T. (1953). *The hydraulic geometry of stream channels and some physiographic implications*. Geological Professional Paper, 252, U.S. Government Printing Office, Washington, DC.

Nanson, G. C. and Huang, H. Q. (2017). Self-adjustment in rivers: Evidence for least action as the primary control of alluvial channel form and process. *Earth Surface Processes and Landforms*, Vol. 42, No. 4, pp. 575–594.

Ohara, N. and Yamatani, K. (2019). Theoretical stable hydraulic section based on the principle of least action. *Scientific Reports*, Vol. 9, 7957, https://doi.org/10.1038/s41598–019-44347-4.

Rhoads, B. L. (1991). A continuously varying parameter model of downstream hydraulic geometry. *Water Resources Research*, Vol. 27, No. 8, pp. 1865—1872.

Riley, W. F. and Sturges, L. D. (1993). *Engineering Mechanics-Statics*. John Wiley & Sons, New York.

18

Theory of Minimum Energy Dissipation Rate

A river constantly adjusts its geometry and morphology in response to the water and sediment load it receives from its watershed and to human activities, such as straightening, dredging, cutoff, levee construction, restoration, and diversion. The adjustment requires dissipation of energy. When the energy dissipation reaches a minimum rate the river tends to reach equilibrium. This chapter discusses the theory of minimum energy dissipation rate for deriving the hydraulic geometry when the river is in equilibrium state.

Notation

a = constant

A = flow cross-sectional area

b = exponent

B = base width

c = constant

C = Chezy's roughness coefficient

d = flow depth

d_a = average flow depth

f = exponent

k = constant

I = a coefficient

i = constant

J = a coefficient

K = a constant

L = the channel reach length

L_f = Lagrangian function

m = exponent

---

m_0 = exponent
n = Manning's roughness coefficient
P = dimensionless unit stream power
P_c = critical dimensionless unit stream power required at the incipient motion
P_w = wetted perimeter
Q = discharge
Q_s = sediment discharge
r = ratio of base width to flow depth
R = hydraulic radius
r_a = average width–depth ratio
r_w = ratio of top width to flow depth
s = subscript corresponding to sediment
S = energy slope
t = a constant
v = velocity
v_f = average fall velocity of sediment particles
w = water surface width
x = an exponent
Z = horizontal
α = coefficient
γ = specific weight of water
γ_s = specific weight of sediment
λ = Lagrange multiplier
Φ = total rate of energy dissipation
Φ_w = rate of energy dissipation for the flow of water
Φ_s = rate of energy dissipation for sediment transport

18.1 Introduction

A system is in an equilibrium condition when its rate of energy dissipation is at its minimum value, which depends on the constraints the system has to satisfy. When the system is not in equilibrium, its rate of energy dissipation is not minimum. However, the system will attempt to adjust in such a manner that its rate of energy dissipation reduces until it reaches the minimum and regains equilibrium.

A natural river constantly adjusts itself in response to varying discharge, sediment load, and other factors, including anthropogenic influences. Although a true minimum may never occur, the river will adjust in such a manner that the rate of energy dissipation approaches a minimum value and thus tends toward an equilibrium condition. Yang et al. (1981) developed the theory of minimum rate of energy dissipation for hydraulic geometry and showed that it can be reduced, under special and simple conditions, to the theory of minimum stream power and the theory of minimum unit stream power. This chapter draws on their work and discusses the derivation of hydraulic geometry using this theory.

Hydraulic geometry relations are often expressed as

$$w = aQ^b; \quad d = cQ^f; \quad v = kQ^m, \tag{18.1}$$

where Q is the water discharge; w is the channel width; d is the average water depth; v is the cross-sectionally averaged velocity; a, c, and k are proportionality constants; and b, f, and m are exponents.

There are many ways by which a river system minimizes its rate of energy dissipation. It is assumed that the minimization process is accomplished through the adjustment of channel geometry at a given river section. The water and sediment discharges are the known constraints that are determined by the watershed characteristics. The bed and bank materials are equally erodible. The total fall or the slope-length product of a reach is given. Thus, a river will adjust its width, depth, velocity, roughness, and slope to minimize its rate of energy dissipation. Because the time required to change channel geometry is much longer than the time required to change bed forms, and the observed channel geometry represents the long-term cumulative results of past transport phenomena, the total channel roughness over different bed forms can be treated as a constant.

18.2 Theory of Minimum Energy Dissipation Rate

The theory of minimum energy dissipation rate consists of five basic governing equations: (1) Flow resistance equation, (2) rate of energy dissipation for water transport, (3) rate of energy dissipation for sediment transport, (4) total energy dissipation rate, and (5) sediment concentration as a function of stream power. Each of these equations is now presented.

18.2.1 Flow Resistance Equation

The flow resistance equation can be expressed as

$$v = \alpha R^{m_0 - 1}, \tag{18.2}$$

where v is the velocity of flow, R is the hydraulic radius equal to flow cross-sectional area (A) divided by wetted perimeter (P_w), α is a coefficient as a function of energy slope (S) and friction coefficient, and m_0 is an exponent, equal to 5/3 for

Manning's equation and 3/2 for Chezy's equation. The α coefficient can be expressed as

$$\alpha = \frac{S^{0.5}}{n} \text{ for Manning's equation and } \alpha = C\sqrt{S} \text{ for Chezy's equation,} \quad (18.3)$$

in which n is Manning's roughness coefficient and C is Chezy's roughness coefficient.

Using the continuity equation, the flow discharge Q can be expressed as

$$Q = Av. \quad (18.4)$$

Inserting Equation (18.2) in Equation (18.4), one can write:

$$Q = \alpha A R^{m_0 - 1}. \quad (18.5)$$

18.2.2 Energy Dissipation Rate for Water Transport

The rate of energy dissipation (Φ_w) for the flow of water can be expressed as

$$\Phi_w = Q\gamma LS, \quad (18.6)$$

where γ is the specific weight of water, L is the channel reach length, and S is the energy slope. Note that the product of L and S yields the energy head.

18.2.3 Energy Dissipation Rate for Sediment Transport

The rate of energy dissipation (Φ_s) for sediment transport can be expressed as

$$\Phi_s = Q_s\gamma_s LS, \quad (18.7)$$

in which Q_s is the sediment discharge, and γ_s is the specific weight of sediment. Here subscript s corresponds to sediment.

18.2.4 Total Rate of Energy Dissipation

The total rate of energy dissipation (Φ) required to transport water Q and sediment Q_s through a reach of length L can be written as the sum of rate of energy dissipation for flow of water given by Equation (18.6) and the rate of energy dissipation for sediment transport given by Equation (18.7):

$$\Phi = \Phi_w + \Phi_s = (Q\gamma + \gamma_s Q_s)LS. \quad (18.8)$$

18.3.5 Sediment Concentration

The sediment concentration can be expressed by the dimensionless unit stream power (P) equation in its power form (Yang, 1973; Yang and Molinas, 1982) as

$$\frac{Q_s}{Q} = I(P - P_c)^J, \qquad P > P_c \qquad (18.9)$$
$$= 0, \qquad P \leq P_c,$$

where

$$P = \frac{vS}{v_f}, \qquad (18.10)$$

in which P is the dimensionless unit stream power, which is the rate of potential energy dissipation per unit weight of water, v_f is the average fall velocity of sediment particles, P_c is the critical dimensionless unit stream power required at the incipient motion, and I and J are coefficients. It may be noted that the numerator of Equation (18.10), vS, expresses the unit stream power. Equation (18.10) is applicable only when the sediment is at the state of incipient motion or the sediment concentration is greater than zero.

18.3 Derivation of Hydraulic Geometry

For deriving hydraulic geometry relations, the shape of channel geometry and flow resistance equation need to be specified. Normally, trapezoidal, rectangular, and triangular shapes are assumed. For flow resistance, the Manning and Chezy equations are often employed. Since the trapezoidal shape is a more general shape that specializes into both rectangular and triangular shapes, it is discussed first.

18.3.1 Trapezoidal Section

For a trapezoidal section, let the base width be B, flow depth be d, and water surface width be w, and let the side slope be defined by one vertical to Z horizontal (i.e., 1: Z). The flow cross-sectional area (A) for the trapezoidal section, as shown in Figure 18.1, can then be expressed as

$$A = (B + Zd)d \qquad (18.11)$$

and the wetted perimeter (P_w) as

$$P_w = B + 2\sqrt{1 + Z^2}d. \qquad (18.12a)$$

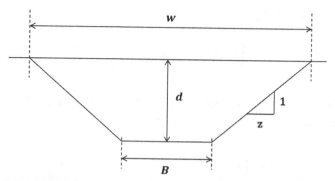

Figure 18.1 Trapezoidal section

The hydraulic radius R can now be expressed as

$$R = \frac{A}{P_w} = \frac{(B+Zd)d}{B+2d\sqrt{1+Z^2}}.$$ (18.12b)

Using Manning's equation, the flow discharge Q can be written as

$$Q = \frac{A}{n}R^{\frac{2}{3}}S^{\frac{1}{2}} = \frac{1}{n}(B+Zd)d\left[\frac{(B+Zd)d}{B+2d\sqrt{1+Z^2}}\right]^{\frac{2}{3}}S^{\frac{1}{2}},$$ (18.13)

in which S is the slope of the energy line, often used as the slope of the channel bed in uniform flow, and n is the Manning roughness coefficient.

From Equation (18.13), the energy slope S can be expressed as

$$S = \left[\frac{Qn}{(B+Zd)d}\right]^2\left[\frac{B+2d\sqrt{1+Z^2}}{(B+Zd)d}\right]^{4/3}.$$ (18.14)

Inserting Equation (18.14) in Equation (18.8), the total energy dissipation rate can be expressed as

$$\Phi = (\gamma Q + \gamma_s Q_s)L\left[\frac{Qn}{(B+Zd)d}\right]^2\left[\frac{B+2d\sqrt{1+z^2}}{(B+Zd)d}\right]^{4/3}.$$ (18.15)

Equation (18.9), with v replaced by Q/A, can be written as

$$\left(\frac{Q_s}{IQ}\right)^{\frac{1}{J}} + P_c = \frac{Q}{v_f A}S = \frac{Q}{(B+Zd)dv_f}S.$$ (18.16)

From equation (18.16), the energy slope S can be written as

$$S = \left[\frac{(B + Zd)dv_f}{Q}\right]\left[P_c + \left(\frac{Q_s}{IQ}\right)^{\frac{1}{j}}\right]. \tag{18.17}$$

Equating Equation (18.14) to Equation (18.17), one obtains the constraint for minimizing the rate of energy dissipation as:

$$\left[\frac{Q^3 n^2}{v_f(B + Zd)^{\frac{13}{3}}d^{\frac{13}{3}}}\right]\left(B + 2d\sqrt{1 + Z^2}\right)^{\frac{4}{3}} = \left[P_c + \left(\frac{Q_s}{IQ}\right)^{\frac{1}{j}}\right]. \tag{18.18}$$

Now, in order to determine B, d, and Z, the rate of energy dissipation given by Equation (18.15) is to be minimized subject to Equation (18.18). It should be noted that Equation (18.15) is only one equation, whereas there are three unknowns. One way to do the minimization is to use the method of Lagrange multipliers or eliminate Q_s in Equation (18.15) using Equation (18.18) and then differentiate the resulting equation partially with respect to B, d, and Z separately, which will lead to three equations that can be solved for B, d, and Z.

To simplify the algebra, let the ratio r of base width to flow depth be defined as

$$r = \frac{B}{d} \tag{18.19}$$

and the ratio of top width to flow depth be defined as

$$r_w = \frac{w}{d} = \frac{B + 2Zd}{d} = r + 2Z. \tag{18.20}$$

It may be noted that in general the side slope component denoted by Z is not zero, and the base width B and the top width w are not the same.

Inserting Equation (18.19) in Equation (18.15), the energy dissipation rate can be written as

$$\Phi = \frac{n^2 Q^2 L(\gamma Q + \gamma_s Q_s)\left(r + 2\sqrt{1 + Z^2}\right)^{\frac{4}{3}}}{d^{\frac{16}{3}}(r + Z)^{\frac{10}{3}}}. \tag{18.21}$$

Likewise, inserting Equation (18.19) in Equation (18.18), the constraint equation can be written as

$$\frac{n^2 Q^3 \left(r + 2\sqrt{1 + Z^2}\right)^{\frac{4}{3}}}{v_f(r + Z)^{\frac{13}{3}}d^{\frac{22}{3}}} = P_c + \left(\frac{Q_s}{IQ}\right)^{\frac{1}{j}}. \tag{18.22}$$

From Equations (18.21) and (18.22), d can be eliminated as follows. From Equation (18.22), d can be expressed as

$$\frac{n^{3/11}Q^{9/22}\left(r+2\sqrt{1+Z^2}\right)^{\frac{2}{11}}}{(v_f)^{\left(\frac{3}{22}\right)}(r+Z)^{\frac{13}{22}}\left[P_c+\left(\frac{Q_s}{IQ}\right)^{\frac{1}{j}}\right]^{\frac{3}{22}}}=d. \tag{18.23}$$

Inserting Equation (18.23) into Equation (18.21), the result is the energy dissipation rate equation without d:

$$\Phi=\frac{n^2Q^2L(\gamma Q+\gamma_s Q_s)\left(r+2\sqrt{1+Z^2}\right)^{\frac{4}{3}}(v_f)^{\frac{8}{11}}(r+Z)^{\frac{104}{33}}\left[P_c+\left(\frac{Q_s}{IQ}\right)^{\frac{1}{j}}\right]^{\frac{8}{11}}}{(r+Z)^{\frac{10}{3}}\left[n^{\frac{6}{11}}Q^{9/22}\left(r+2\sqrt{1+Z^2}\right)^{\frac{2}{11}}\right]^{\left(\frac{16}{3}\right)}}. \tag{18.24}$$

Equation (18.24) can be simplified as

$$\Phi=\frac{n^{6/11}L(\gamma Q+\gamma_s Q_s)\left(r+2\sqrt{1+Z^2}\right)^{\frac{4}{11}}(v_f)^{\frac{8}{11}}\left[P_c+\left(\frac{Q_s}{IQ}\right)^{\frac{1}{j}}\right]^{\frac{8}{11}}}{Q^{\frac{2}{11}}(r+Z)^{\frac{2}{11}}}. \tag{18.25}$$

Equation (18.25) can be written in simple form as

$$\Phi=\frac{K\left(r+2\sqrt{1+Z^2}\right)^{\frac{4}{11}}}{(r+Z)^{\frac{2}{11}}}, \tag{18.26}$$

where K is a constant defined as

$$K=\frac{n^{\frac{6}{11}}v_f^{8/11}}{Q^{\frac{2}{11}}}L(\gamma Q+\gamma_s Q_s), \tag{18.27}$$

which shows that K is a function of v_f, L, Q, n, and Q_s.

For trapezoidal sections, the surface water width is larger than the base width. Hence, Equation (18.26) can be expressed with the use of Equation (18.20) as

$$\Phi=\frac{K\left(r_w-2Z+2\sqrt{1+Z^2}\right)^{\frac{4}{11}}}{(r_w-Z)^{\frac{2}{11}}}. \tag{18.28}$$

The minimum value of Φ can be obtained by the optimum width–depth ratio. Now three cases can be identified: (1) the trapezoid is unconstrained, (2) the width-depth ratio is unrestricted, and (3) the side slope is unrestricted. Each of these cases is now discussed in what follows.

18.3.1.1 Case 1: Unrestricted Trapezoid

In this case, Φ is minimized by adjusting both the width–depth ratio r_w and the side slope Z. Differentiating Equation (18.28) partially with respect to r_w and equating the derivative to 0,

$$\frac{\partial \Phi}{\partial r_w} = 0 = \frac{4K\left(r_w - 2Z + 2\sqrt{1+Z^2}\right)^{-\frac{7}{11}}}{11(r_w - Z)^{\frac{2}{11}}} - \frac{2K\left(r_w - 2Z + 2\sqrt{1+Z^2}\right)^{\frac{4}{11}}}{11(r_w - Z)^{\frac{13}{11}}}.$$

(18.29)

Equation (18.29) yields

$$\frac{2(r_w - Z) - \left(r_w - 2Z + 2\sqrt{1+Z^2}\right)}{(r_w - Z)\left(r_w - 2Z + 2\sqrt{1+Z^2}\right)} = 0.$$

(18.30)

Equation (18.30) produces

$$r_w = 2\sqrt{1 + Z^2}.$$

(18.31)

Equation (18.31) is plotted in Figure 18.2.

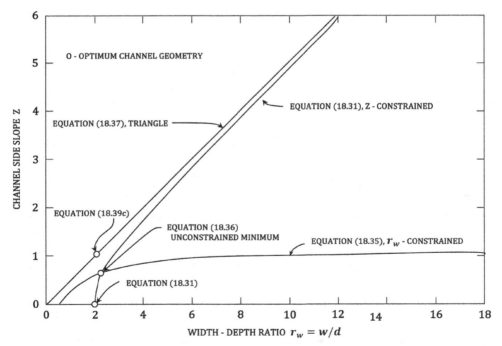

Figure 18.2 Relation between channel side slope and width-depth ratio

Now differentiating Equation (18.28) partially with respect to Z and equating the derivative to 0, the result is

$$\frac{\partial \Phi}{\partial Z} = 0$$

$$= \frac{4K\left(r_w - 2Z + 2\sqrt{1+Z^2}\right)^{-\frac{7}{11}}\left(-2 + \frac{2Z}{\sqrt{1+Z^2}}\right)}{11(r_w - Z)^{\frac{2}{11}}}$$

$$+ \frac{2K\left(r_w - 2Z + 2\sqrt{1+Z^2}\right)^{\frac{4}{11}}}{11(r_w - Z)^{\frac{13}{11}}}.$$
(18.32)

Equation (18.32) can be written as

$$\frac{4K\left(r_w - 2Z + 2\sqrt{1+Z^2}\right)^{\frac{4}{11}-1}\left(-2\sqrt{1+Z^2}+2Z\right)}{11(r_w - Z)^{\frac{2}{11}}\sqrt{1+Z^2}}$$

$$+ \frac{2K\left(r_w - 2Z + 2\sqrt{1+Z^2}\right)^{\frac{4}{11}}}{11(r_w - Z)^{\frac{13}{11}}} = 0.$$
(18.33)

Equation (18.33) can be simplified as

$$r_w\left(4Z - 3\sqrt{1+Z^2}\right) = 2\left(Z^2 - 1 - Z\sqrt{1+Z^2}\right).$$
(18.34)

Hence, Equation (18.34) yields

$$r_w = \frac{2\left(Z^2 - 1 - Z\sqrt{1+Z^2}\right)}{4Z - 3\sqrt{1+Z^2}}.$$
(18.35)

Equation (18.35) is graphed in Figure 18.2.

Equations (18.31) and (18.35) are solved for r_w and Z and the solution yields

$$Z = \frac{1}{\sqrt{3}}; \quad r_w = \frac{4}{\sqrt{3}}.$$
(18.36)

Equation (18.36) plots as a point in Figure 18.2.

18.3.1.2 Case 2: Width-Constrained Trapezoid

In this case the minimization of Φ is achieved by adjusting only the side slope Z. Differentiating Equation (18.28) partially with respect to Z and equating the derivative to zero, Equation (18.35) is the result that specifies the relation between r_w and Z.

18.3.1.3 Case 3: Side-Slope Constrained Trapezoid

In this case the minimization of Φ is achieved by adjusting only the width–depth ratio r_w. Differentiating Equation (18.28) partially with respect to r_w and equating the derivative to zero, Equation (18.31) is the result that specifies the relation between r_w and Z.

18.3.2 Triangular Geometry

For a triangular section, $B = 0$. Therefore, the r_w and Z quantities are related as

$$r_w = 2Z. \tag{18.37}$$

Equation (18.37) is graphed in Figure 18.2. Inserting Equation (18.37) in Equation (18.28), the result is

$$\Phi = \frac{K\left(2\sqrt{1+Z^2}\right)^{\frac{4}{11}}}{Z^{\frac{2}{11}}} = K(2)^{\frac{4}{11}}(1+Z^2)^{\frac{4}{22}}Z^{-\frac{2}{11}}. \tag{18.38}$$

Taking the partial derivative of Φ in Equation (18.38) with respect to Z and setting the derivative to zero, one obtains

$$\frac{d\Phi}{dZ} = 0 = K(2)^{\frac{4}{11}}\left[\frac{2}{11}(1+Z^2)^{\frac{4}{22}-1}(2Z) - (1+Z^2)^{\frac{4}{22}}\left(\frac{2}{11}\right)Z^{-\frac{2}{11}-1}\right]. \tag{18.39a}$$

This equation leads to

$$-1 + Z^2 = 0, \tag{18.39b}$$

whose solution is

$$Z = 1, \quad r_w = 2. \tag{18.39c}$$

Equation (18.39c) shows that the most efficient triangular section is half of a square and plots as a point in Figure 18.2.

18.3.3 Rectangular Geometry

For rectangular geometry, the flow cross-sectional area is

$$A = wd \tag{18.40}$$

and the wetted perimeter is

$$P_w = w + 2d. \tag{18.41}$$

The water discharge Q using Manning's equation for a rectangular channel can be expressed as

$$Q = \frac{1}{n}\left[\frac{wd}{w+2d}\right]^{\frac{2}{3}} S^{\frac{1}{2}} wd . \tag{18.42}$$

Slope S can be expressed from Equation (18.42) as

$$S = \left(\frac{Qn}{wd}\right)^2 \left(\frac{w+2d}{wd}\right)^{\frac{4}{3}}. \tag{18.43}$$

Using Equation (18.43), the energy slope S can be eliminated between Equations (18.6) and (18.7) and consequently Equation (18.8). The total rate of energy dissipation can therefore be written as

$$\Phi = Q^2(Q\gamma + Q_s\gamma_s)\frac{Ln^2}{(wd)^2}\left[\frac{w+2d}{wd}\right]^{\frac{4}{3}} = Q^2(Q\gamma + Q_s\gamma_s)Ln^2\frac{(w+2d)^{\frac{4}{3}}}{(wd)^{\frac{10}{3}}}. \tag{18.44}$$

Substituting Equation (18.10) in Equation (18.9) and then using Equations (18.44) and (18.42), one can write

$$\frac{n^2Q^3(w+2d)^{\frac{4}{3}}}{v_f(wd)^{\frac{13}{3}}} = P_c + \left(\frac{Q_s}{IQ}\right)^{\frac{1}{j}}. \tag{18.45}$$

In order to determine w and d, the Φ value given by (18.44) should be minimized, subject to the constraint given by Equation (18.45). To that end, the corresponding Lagrangian function can be written as

$$L_f = \Phi + \lambda\left[P_c + \left(\frac{Q_s}{IQ}\right)^{\frac{1}{j}} - \frac{n^2Q^3(w+2d)^{\frac{4}{3}}}{v_f(wd)^{\frac{13}{3}}}\right], \tag{18.46}$$

where λ is the Lagrange multiplier. With the use of Equation (18.44), Equation (18.46) can be written as

$$L_f = Q^2(Q\gamma + Q_s\gamma_s)Ln^2\frac{(w+2d)^{\frac{4}{3}}}{(wd)^{\frac{10}{3}}} + \lambda\left[P_c + \left(\frac{Q_s}{IQ}\right)^{\frac{1}{j}} - \frac{n^2Q^3(w+2d)^{\frac{4}{3}}}{v_f(wd)^{\frac{13}{3}}}\right]. \tag{18.47}$$

Taking the partial derivative of L_f with respect to d and with respect to w separately and setting the resulting derivative equations to zero, the following is obtained:

$$Lv_f(\gamma Q + \gamma_s Q_s)(12d + 10w)wd = \lambda Q(18d + 13w) \qquad (18.48)$$

$$Lv_f(\gamma Q + \gamma_s Q_s)(20d + 6w)wd = \lambda Q(26d + 9w). \qquad (18.49)$$

Dividing Equation (18.48) by Equation (18.49), the result is

$$\frac{12d + 10w}{20d + 6w} = \frac{18d + 13w}{26d + 9w}, \qquad (18.50)$$

which has a solution as

$$w = 2d \qquad (18.51)$$

for a rectangular open channel. Equation (18.51) is the best hydraulic or most efficient section. Thus, the theoretical self-adjusted channel cross-section transports a given amount of water and sediment most effectively with a minimum rate of energy dissipation.

Inserting Equation (18.51) into Equations (18.48) and (18.49) yields the Lagrangian multiplier,

$$\lambda = \frac{16}{11}Lv_f\left(\gamma + \gamma_s\frac{Q_s}{Q}\right)d^2. \qquad (18.52)$$

In order to ensure that $w = 2d$ is the value that can make Φ a minimum, let it be assumed that

$$w = rd, \qquad (18.53)$$

in which r is the width–depth ratio. Substituting Equation (18.53) into Equation (18.44), the total energy dissipation rate becomes

$$\Phi = Q^2(Q\gamma + Q_s\gamma_s)Ln^2\frac{(r+2)^{\frac{4}{3}}}{(r)^{\frac{10}{3}}d^{\frac{16}{3}}}. \qquad (18.54)$$

Similarly, substituting Equation (18.53) in Equation (18.45), one can write

$$\frac{n^2Q^3(r+2)^{\frac{4}{3}}}{v_f(r)^{\frac{13}{3}}d^{\frac{22}{3}}} = P_c + \left(\frac{Q_s}{IQ}\right)^{\frac{1}{j}}. \qquad (18.55)$$

Eliminating d from Equations (18.54) and (18.55), one obtains

$$\Phi = K\frac{(2+r)^{\frac{4}{11}}}{r^{\frac{2}{11}}}, \qquad (18.56)$$

where K is the parameter for terms other than w and d in Equations (18.54) and (18.55). Equation (18.56) has a minimum value at $r = 2$. Thus, $w = 2d$ corresponds to the minimum rate of energy dissipation, as shown in Figure 18.3.

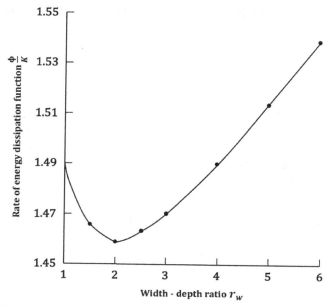

Figure 18.3 Relationship between energy dissipation rate and width-depth ratio

Let it be assumed that the average particle fall velocity of sediment being transported is expressed as

$$v_f = tQ^x,$$ (18.57)

where t is a constant and x is an exponent. Substituting Equation (18.57) into Equation (18.45) yields

$$\frac{n^2 Q^3}{t(2)^{\frac{5}{3}} Q^x d^{\frac{22}{3}}} = P_c + \left(\frac{Q_s}{IQ}\right)^{\frac{1}{7}}.$$ (18.58)

The right side of (18.58) shows that it should be a dimensionless given value. This equation shows that d can be expressed as

$$d \propto Q^{\frac{9-3x}{22}}.$$ (18.59a)

Comparing with Equation (18.1), it is seen that

$$f = \frac{9 - 3x}{22}.$$ (18.59b)

In order to satisfy Equation (18.51), exponents b and f must be the same, i.e.,

$$b = \frac{9 - 3x}{22}.$$ (18.60)

Substituting Equations (18.59b) and (18.60) into the continuity equation ($Q = vdw$), one obtains

$$b + f + m = 1, \tag{18.61}$$

which yields

$$m = \frac{2 + 3x}{11}. \tag{18.62}$$

Now let it be assumed that

$$S = iQ^j. \tag{18.63}$$

Substituting Equations (18.1), (18.2), (18.59b), (18.62), and (18.63) into the Manning equation, one gets

$$j = \frac{-2 + 8x}{11}. \tag{18.64}$$

Thus,

$$\frac{vS}{v_f} = \frac{ki}{t}. \tag{18.65}$$

Equation (18.65) states that a fixed amount of dimensionless unit stream power is required to transport a given amount of water and sediment under the equilibrium condition.

For uniform sediment particles the average particle size or the average fall velocity remains a constant with varying discharge. That is, x should be zero for uniform materials and Equations (18.59b), (18.60), (18.62), and (18.64) become

$$b = f = \frac{9}{22} \tag{18.66}$$

$$m = \frac{2}{11} \tag{18.67}$$

$$j = -\frac{2}{11}. \tag{18.68}$$

The value given in Equation (18.68) is close to the value of $-1/6$ given by Blench (1957) in his regime formula for channel slope. It may be noted that the discussion here considers that either both the channel side slope and the width–depth ratio are free to adjust or only one of them is free to adjust and the other is specified as a constraint.

18.4 Analysis and Application

In the derivations shown in this chapter, the width–depth ratio was based on the maximum depth. However, in natural streams, flow may be nonuniform, for which the width–depth ratio can be defined using the average flow depth, d_a. In that case, for the trapezoid,

$$d_a = \frac{(w - Zd)d}{w} = \frac{(r_w - Z)d}{r_w} \tag{18.69}$$

$$r_a = \frac{w}{d_a} = \frac{r_w^2}{(r_w - Z)}. \tag{18.70}$$

Substituting Equation (18.70) in Equations (18.31) and (18.35), the relationships for r_w- and Z-constrained cases can be derived, as shown in Figure 18.2. Inserting Equation (18.70) in Equations (18.36), (18.39), and (18.51), the optimal relationships for r_a for trapezoid, triangular, and rectangular sections are obtained as

$$r_a = 3.078, \quad Z = \frac{1}{\sqrt{3}} \qquad \text{for trapezoid} \tag{18.71}$$

$$r_a = 4, \quad Z = 1 \qquad \text{for triangular} \tag{18.72}$$

$$r_a = 2, \quad Z = 0 \qquad \text{for rectangular.} \tag{18.73}$$

Equations (18.28), (18.35), and (18.70) show that the depth ratios r_w and r_a are sensitive to the variation in Z. In the case of alluvial channels with noncohesive material, the channel side slope depends on the bank material particle size and the angle of repose. In general, the bank material decreases in the downstream direction and both width–depth ratios increase in the downstream direction. It is seen that hydraulic exponents b, m, and j are not dependent on the change in side slope Z but are dependent on the degree of uniformity of bed material being transported. This may explain a relatively narrow range of variations of b and f for rivers with different values of Z. The theoretical values of b and f are in close agreement with measured values from laboratory flumes (Yang et al., 1981).

For natural rivers the side slope depends on the angle of repose and bank stabilization due to vegetation. The channel width–depth ratio depends on the valley width and the percentage of silt-clay of bank and bed materials (Schumm, 1968), the past history of channel movement, and other natural or manmade constraints. These factors explain that the trapezoidal cross-section of a stable natural alluvial channel is not half of a hexagon, which corresponds to the absolute minimum rate of energy dissipation. In order for this unconstrained condition to be attained, the entire boundary must be composed of noncohesive material with an angle of repose of 60 degrees under the channel-forming condition. In natural conditions this is not the

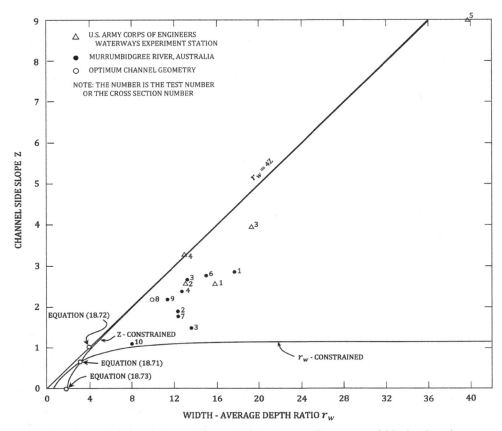

Figure 18.4 Relation between channel side slope and average width-depth ratio

case, and the optimum natural channel geometry would reflect certain combinations of Z and r_w. This suggests that the optimum value of width–depth ratio should be between Z- and r_w-constraint curves, as shown in Figure 18.4.

Friedkin (1945) conducted a series of tests on alluvial meandering channels in the laboratory at the US Waterways Experiment Station, Vicksburg, Mississippi. Some of the cross-sections were trapezoidal and some looked parabolic, which were located halfway between gages and are summarized in Table 18.1. Figures 18.4 shows that test 4 agrees with the Z-constrained case. In the case of test 4, this was because the material was completely sand, for which the optimal geometry is completely determined by the angle of repose or the Z-value under the given flow condition. For test 5 with 100% granular coal, Figure 18.3 shows that its geometry is dominated by the Z-constraint. For tests 1, 2, and 3, the Z-constraint is not sufficient and the observed Z and r_w should be between the Z-constraint and r-constraint curves, as shown in Figure 18.3.

Table 18.1. *Laboratory meandering channel cross-sections at crossings measured at US Army Engineer Experiment Station*

Test	Bed and bank materials	Slope (S)	Discharge Q (cfs)	Bottom width (B) (ft)	Top width (w) (ft)	Depth (d) (ft)	Side slope (Z)	Width–average depth ratio (r_w)
1	75% crushed coal 25% loess	0.003	0.24	1.7	2.8	0.22	2.5	15.84
2	75% crushed coal 25% loess	0.003	0.30	1.3	2.7	0.28	2.5	13.02
3	65% sand 35% silt	0.007	0.24	1.1	2.5	0.18	3.9	19.31
4	100% sand	0.006	0.24	0.2	1.6	0.22	3.2	12.99
5	100% granular coal	0.006	0.24	1.2	3.9	0.15	9.0	39.80

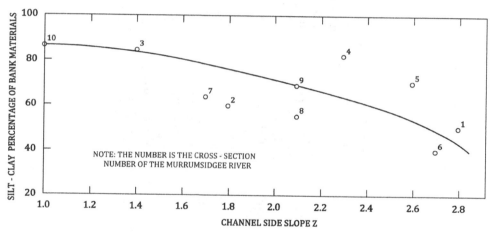

Figure 18.5 Relation between silt-clay percentage of bank materials and channel side slope

To better understand the relative influence of Z and r_w on the river geometry, the relationship between silt-clay percentage of bank material and channel side slope can be plotted, as shown for Murrumbidge River in Figure 18.5 using the data given in Table 18.2 (Schumm, 1968). The figure shows that Z decreases with the increase in the silt-clay percentage in the bank material. The implication here is that as the silt-clay percentage increases the material becomes more cohesive and the back becomes steeper, as shown in Figure 18.5. Under this condition, the bank becomes less conducive to adjustment. This means that the width–depth ratio must

Table 18.2. *Cross-sections along the Murrumbidgee River, Australia*

Station	Silt-clay (%) Bank	Channel	Particle size	Slope (S)	Discharge Q (cfs)	Bottom width (B) (ft)	Top width (w) (ft)	Depth (d) (ft)	Side slope (Z)	Width–average depth ratio (r_w)
1. Wantabadgery	50	0.07	0.88	0.00028	13,000	147	241	17	2.8	17.67
2. Wagga Wagga	60	15	0.40	0.00017	25,000	174	273	27	1.8	12.30
3. Currawarna	85	17	0.63	0.00011	14,000	195	252	21	1.4	13.58
4. Narrandera	82	27	0.90	0.00021	20,000	132	246	25	2.3	12.84
5. Darlington Point	70	23	0.71	0.00013	11,000	100	220	23	2.6	13.14
6. Yarrada Lagoon	40	8	0.52	0.00013	10,000	122	230	20	2.7	15.03
7. Bringagee	64	0.12	0.47	0.00013	9,000	130	195	19	1.7	12.30
8. Carrathool	55	18	1.30	0.00013	16,000	82	213	31	2.1	9.90
9. Hay	69	23	0.95	0.00011	20,000	125	245	29	2.1	11.24
10. Maude	87	40	0.57	0.00008	9,000	120	165	24	1.0	8.04

adjust, and therefore the channel width or the ratio becomes more important. Figure 18.3 shows that stations 10 and 3 have a larger percentage of silt-clay and smaller Z values, where the channel geometry is dominated by width, as Figure 18.2 shows.

References

Blench, T. (1957). *Regime Behavior of Canals and Rivers*. Butterworths, London.

Friedkin, J. F. (1945). *A laboratory study of the meandering of alluvial rivers*. U.S. Army Corps of Engineers Waterways Experiment Station, Vicksburg, MS.

Schumm, S. A. (1968). *River adjustment to altered hydrologic regime-Murrumbidgee River and paleochannels, Australia*. U.S. Geological Survey Professional Paper 598, Washington, DC.

Yang, C. T. (1973). Incipient motion and sediment transport. *Journal of the Hydraulics Division, ASCE*, Vol. 99, No. HY10, pp. 1697–1704.

Yang, C. T. and Molinas, A. (1982). Sediment transport and unit stream power function. *Journal of the Hydraulics Division, ASCE*, Vol. 108, No. 6, pp. 774–793.

Yang, C. T., Song, C. C. S., and Woldenberg, M. J. (1981). Hydraulic geometry and minimum rate of energy dissipation. *Water Resources Research*, Vol. 17, No. 4, pp. 1014–1018.

19

Entropy Theory

Hydraulic geometry, described by depth, width, velocity, slope, and friction, is determined using three equations of continuity, resistance, and sediment transport and by satisfying the condition of minimum production of entropy in the channel system. This chapter discusses the methodology based on this condition.

Notation

A = cross-sectional area

b = exponent

C = sediment concentration

C_1 = constraint

C_2 = constraint

d = flow depth

D = sediment size (or diameter)

E_i = energy of each reach

f = exponent

F_i = individual probabilities of functions

H = Shannon entropy

$h(x)$ = flow depth as a function of lateral distance

h_c = flow depth at the centerline

J = exponent

K = constant of proportionality

L = length

m = exponent

n = friction factor at a cross-section along a river channel

N = number of reaches

P_w = wetted perimeter

p_i = probability of E_i

Q = mean discharge
Q_s = sediment transport
r_i = ratio of the units of energy
s = transverse slope
S = slope of energy line
s_0 = submerged coefficient of friction
SP = stream power per unit weight of water
v = mean velocity
w = surface width
W = half channel width from the centerline
x_* = dimensionless distance
y_* = dimensionless elevation
y = exponent
Y = total fall
z = exponent
σ_m = standard deviation of depth
σ_f = standard deviation of velocity
σ_y = standard deviation of roughness
σ_z = standard deviation of slope
λ_i = Lagrange multipliers
r = power function exponent

19.1 Introduction

Leopold and Maddock (1953) provided a set of hydraulic geometry relations as:

$$w \propto Q^b \tag{19.1}$$

$$d \propto Q^f \tag{19.2}$$

$$v \propto Q^m \tag{19.3}$$

$$S \propto Q^z \tag{19.4}$$

$$n \propto Q^y, \tag{19.5}$$

where v is the mean velocity; d is the flow depth; w is the surface width; S is the slope of energy line; n is the friction factor at a cross-section along a river channel where the mean discharge is Q; and b, f, m, z, and y are the exponents that, respectively, describe the variability in velocity, depth, width, slope, and friction along a channel but do not uniquely determine the magnitudes of these variables. The objective is to determine the exponents at a station with bankfull discharge as well as in the downstream direction with increasing discharge of uniform frequency. To that end we need a set of equations.

19.2 Formulation

Equations (19.1) to (19.5) have five unknowns, velocity, depth, width, slope, and friction, whose evaluation is sought. However, only three equations connecting these variables are available, and the remaining equations are obtained by satisfying the condition of the maximum probable distribution of energy in the channel system: the basic concept underlying the minimum production of entropy in the open channel system.

The first equation is given by the continuity of water flow, which for a rectangular cross-section can be expressed as:

$$Q = vdw. \tag{19.6a}$$

Substituting the geometry Equation (19.1) to (19.3) in Equation (19.6), one obtains

$$Q^1 = KQ^{b+f+m}, \tag{19.6b}$$

where K is the constant of proportionality. This yields the relations:

$$b + f + m = 1. \tag{19.7}$$

A second equation is obtained from an expression for hydraulic friction, which can be expressed using Manning's equation as

$$v \propto \frac{d^{\frac{2}{3}} S^{\frac{1}{2}}}{n}. \tag{19.8}$$

Since

$$Q^m \propto \frac{Q^{\frac{2f}{3}} Q^{\frac{z}{2}}}{Q^y}, \tag{19.9}$$

one gets by equating the exponents on both sides of Q:

$$m = \frac{2}{3}f + \frac{1}{2}z - y. \tag{19.10}$$

The third equation is one of the transport of sediment per unit mass of water, which is assumed to be uniform. The assumption of uniformity is a close approximation of the conclusion reached by Leopold and Maddock (1953) from an analysis of data on suspended sediment transport. Following Bagnold (1960), the stream power at flows sufficiently high to be effective in shaping the river channel is directly related to the transport of sediment, whose movement governs the channel morphology. Denoting the stream power per unit weight of water, SP, as

$$SP = vdS, \tag{19.11}$$

sediment transport in excess of the point of incipient motion can be written as

$$Q_s \propto (vdS)^{1.5}. \tag{19.12}$$

The sediment concentration, C, defined as sediment discharge per unit water discharge, can be expressed as

$$C \propto (vd)^{0.5} S^{1.5} n, \tag{19.13}$$

which is consistent with Bagnold (1960). The effect of sediment size (or diameter), D, needs to be considered. Sediment transport varies inversely as about the 0.8 power of the particle size D. Then,

$$C \propto (vd)^{0.5} S^{1.5} n D^{-0.8}. \tag{19.14}$$

According to the Strickler equation, the friction factor, n, varies as the 1/6 power of the sediment particle size:

$$n \propto D^{\frac{1}{6}} \quad \text{or} \quad D \propto n^6. \tag{19.15}$$

Thus,

$$C \propto (vd)^{0.5} S^{1.5} n \frac{1}{(n^6)^{0.8}} \quad \text{or} \quad C \propto (vd)^{0.5} S^{1.5} \frac{1}{n^{3.8}} \approx C \propto (vd)^{0.5} S^{1.5} \frac{1}{n^4}. \tag{19.16}$$

Note that sediment transport and friction factor are also influenced by bed form; and cohesiveness, sorting, and texture of the material. However, to keep the analysis simple, only particle size is considered here.

Considering sediment transport per unit discharge as independent of the hydraulic geometry of a river in dynamic equilibrium, the sediment concentration may be considered constant. Therefore, from Equation (19.16),

$$0.5m + 0.5f + 1.5z - 4y = 0. \tag{19.17}$$

The other equations are derived from a consideration of the most probable distribution of energy and total energy in the river system.

Consider a channel divided into N reaches, $1, 2, \ldots, N$. Each reach is associated with its own energy E_i, $I = 1, 2, \ldots, N$, and the total energy E is associated with the whole channel. The ratios of the units of energy, $r_1 = \frac{E_1}{E}, r_2 = \frac{E_2}{E}, \ldots, r_N = \frac{E_N}{E}$, are presumed to be statistically independent. The most probable distribution of energy exists when the entropy ϕ of the channel is a maximum. That is,

$$\phi = -\sum_{i=1}^{N} \ln p_i + \text{constant}, \tag{19.18}$$

where p_i is the probability of E_i. Thus, the joint probability that a particular combination of energies exists among reaches can be written as

$$P \propto \exp(-r_1)\exp(-r_2)\cdots\exp(-r_N). \tag{19.19}$$

The joint probability would be maximum when all probabilities are equal, that is,

$$r_1 = r_2 = \cdots = r_N \quad \text{or} \quad E_1 = E_2 = \cdots = E_N. \tag{19.20}$$

Thus, energy tends to be equal in each unit length of the channel.

The equable distribution of energy corresponds to a tendency toward the uniformity of hydraulic properties along a river system. Considering the distribution of internal energy, for example, uniform distribution of internal energy per unit mass is achieved when the depth and velocity tend toward uniformity in the river system. This means that the exponents f and m are as near zero as possible. Then, following the Manning equation

$$v = \frac{1}{n}d^{\frac{2}{3}}S^{\frac{1}{2}} \tag{19.21}$$

one obtains

$$\frac{1}{2}z - y = 0. \tag{19.22}$$

Since the energy is spent principally at the bed, the equable distribution of energy requires that the stream power per unit of bed area tends toward uniformity. That is, QS/w tends toward uniformity. This, then, yields

$$f + m + z = 0. \tag{19.23}$$

Prigogine's principle states that for an open stable system the production of energy is minimized. This leads to the condition that the total rate of work in the system tends to be minimum:

$$\sum QS\Delta Q \rightarrow \text{minimum} \tag{19.24}$$

or

$$\sum Q^{1+z}\Delta Q \rightarrow \text{minimum}. \tag{19.25}$$

For a drainage basin, this condition will be fulfilled by z becoming an increasingly large negative value. For the basin, the average slope $\sum S\Delta Q/\sum \Delta Q$ must remain finite. Therefore, the value of z must be greater than -1 or z must approach -1 or $1 + z$ must approach zero. Or

$$1 + z = 0. \tag{19.26}$$

These three conditions are based on the energy distribution. The condition of minimum total work tends to make the profile concave, whereas the condition of uniform distribution of internal energy tends to straighten the profile. Hence, the most probable state is sought.

The most probable condition is achieved when the product of probabilities of deviations from expected values is a maximum. Assuming normality, it can be stated that the product of individual probabilities of functions is a maximum when their variances are equal. Let the functions be $F_1, F_2, \ldots$, etc. as stated previously. Then,

$$\left(\frac{F_1}{\sigma_{F_1}}\right)^2 = \left(\frac{F_2}{\sigma_{F_2}}\right)^2 = \left(\frac{F_3}{\sigma_{F_3}}\right)^2 = \cdots = \text{etc.} \tag{19.27}$$

Therefore,

$$\left(\frac{F_1}{\sigma_{F_1}} + \frac{F_2}{\sigma_{F_2}}\right)\left(\frac{F_1}{\sigma_{F_1}} - \frac{F_2}{\sigma_{F_2}}\right) = 0, \tag{19.28}$$

for which two possibilities exist:

$$\frac{F_1}{\sigma_{F_1}} + \frac{F_2}{\sigma_{F_2}} = 0 \tag{19.29}$$

$$\frac{F_1}{\sigma_{F_1}} - \frac{F_2}{\sigma_{F_2}} = 0. \tag{19.30}$$

Of these, only one solution would be a valid one. Recalling

$$F_1 = \frac{1}{2}z - y \tag{19.31}$$

$$F_2 = m + f + z \tag{19.32}$$

$$F_3 = 1 + z, \tag{19.33}$$

their standard deviations are

$$\sigma_{F_1} = \sqrt{\left(\frac{\sigma_z}{2}\right)^2 + \sigma_y^2} \tag{19.34}$$

$$\sigma_{F_2} = \sqrt{\sigma_m^2 + \sigma_y^2 + \sigma_z^2} \tag{19.35}$$

$$\sigma_{F_3} = \sqrt{\sigma_z^2} = \sigma_z. \tag{19.36}$$

The standard deviations $\sigma_m, \sigma_f, \sigma_z,$ and σ_y represent the variability of the several factors occurring along a river, and are taken to be proportional to their respective values. Thus,

$$\left(\frac{m+f+z}{\sqrt{3}}\right)^2 = \left(\frac{\frac{z}{2}-y}{\sqrt{1.5}}\right)^2 \tag{19.37}$$

$$\left(\frac{\frac{z}{2}-y}{\sqrt{1.5}}\right)^2 = \left(\frac{1+z}{1}\right)^2 \tag{19.38}$$

$$\left(\frac{m+f+z}{\sqrt{3}}\right)^2 = \left(\frac{1+z}{1}\right)^2. \tag{19.39}$$

The solution is found to be

$$y = -\frac{1}{2}(m+f) \tag{19.40}$$

$$z = -0.53 + 0.93y, \tag{19.41}$$

which, together with three hydraulic conditions derived earlier, lead to a solution for $m, f, b, z,$ and y. The values turn out to be

$$b = 0.55 \tag{19.42}$$

$$f = 0.36 \tag{19.43}$$

$$m = 0.09 \tag{19.44}$$

$$z = -0.74 \tag{19.45}$$

$$y = -0.22. \tag{19.46}$$

19.3 Derivation of At-a-Station Hydraulic Geometry Using Entropy

Employing the Shannon entropy it is possible to derive Equations (19.1)–(19.5). For illustration, the depth–discharge Equation (19.2) is derived.

Depth–discharge power function: It is assumed that the depth corresponding to mean annual flow at any cross-section is a random value and the values from different cross-sections from a random sample have, therefore, a probability distribution. The depth will range from a minimum value to a maximum value. If the sample is large, the minimum value will be much smaller than the maximum value, and for simplicity it can be assumed that the minimum value is negligible and can be taken as zero.

The first step is to determine this distribution. Let the probability density function of d be denoted as $f(d)$. This can be accomplished by maximizing the Shannon entropy of depth, $H(d)$:

$$H(d) = - \int_0^{d_{max}} f(d) \ln f(d) dd. \tag{19.47}$$

Equation (19.47) expresses a measure of uncertainty about $f(d)$ or the average information content of sampled d. Maximizing $H(d)$ is equivalent to minimizing $\{f(d)\ln[f(d)]\}$. In order to determine $f(d)$ that is least biased toward what is not known as regards flow depth, the principle of maximum entropy (POME) developed by Jaynes (1957a, b, 1982) is invoked, which requires the specification of certain information, called constraints, on the flow depth. According to POME, the most appropriate probability distribution is the one that has the maximum entropy or uncertainty, subject to these constraints.

For deriving the power law depth distribution, only the law of mass conservation is found to be sufficient. Of course, the total probability law must always be satisfied. Therefore, the first constraint, C_1, on the probability density function of flow depth can be written as:

$$C_1 = \int_0^{d_{max}} f(d) dd = 1. \tag{19.48}$$

For the power law depth distribution, following Singh (1998) the constraints to be specified are Equation (19.48) and

$$C_2 = \int_0^{d_{max}} \ln d f(d) dd = \overline{\ln d}. \tag{19.49}$$

Equation (19.49) is the mean of the logarithmic depth values and is the second constraint C_2.

In order to obtain the least-biased probability distribution of d, $f(d)$, the Shannon entropy, given by Equation (19.47), is maximized following POME, subject to Equations (19.48)–(19.49). To that end, the method of Lagrange multipliers can be employed. The Lagrangian function then becomes

$$L = - \int_0^{d_{max}} f(d) \ln f(d) dd - (\lambda_0 - 1) \left(\int_0^{d_{max}} f(d) dd - C_1 \right)$$

$$- \lambda_1 \left(\int_0^{d_{max}} [\ln d] f(d) dd - C_2 \right), \tag{19.50}$$

where λ_0 and λ_1 are the Lagrange multipliers. One can also write Equation (19.50) or the Lagrangian function L as

$$L = -f(d)\ln f(d) - (\lambda_0 - 1)f(d) - \lambda_1[\ln d]f(d). \tag{19.51}$$

Recalling the Euler–Lagrange equation of calculus of variation and differentiating Equation (19.50) or (19.51) with respect to $f(d)$ and equating the derivative to zero, one obtains

$$\frac{\partial L}{\partial f(d)} = 0 = -\ln f(d) - \lambda_0 - \lambda_1 \ln d. \tag{19.52}$$

Equation (19.52) leads to the entropy-based probability density function (PDF) of the flow depth as

$$f(d) = \exp[-\lambda_0 - \lambda_1 \ln d] \quad \text{or} \quad f(d) = \exp(-\lambda_0)d^{-\lambda_1}. \tag{19.53}$$

The PDF of d contains the Lagrange multipliers λ_0 and λ_1, which can be determined using Equations (19.48) and (19.49).

The Lagrange multipliers can be determined by substituting Equation (19.53) in Equations (19.48) and (19.49). Substitution of Equation (19.53) in Equation (19.48) yields

$$\exp(\lambda_0) = \frac{d_{max}^{-\lambda_1+1}}{-\lambda_1 + 1} \quad \text{or} \quad \lambda_0 = -\ln(-\lambda_1 + 1) + \ln\left(d_{max}^{-\lambda_1+1}\right), \tag{19.54}$$

where d_{max} is the depth at $d = d_{max}$. Let $r = -\lambda_1 + 1$. Then, substitution of Equation (19.54) in Equation (19.53) yields

$$f(d) = r\frac{d^{r-1}}{d_{max}^r}. \tag{19.55}$$

Equation (19.55) is the probability density function of the power law depth distribution. For $r > 1$, the probability density function (PDF) monotonically increases from zero to an upper value $[f/d_{max}]$. The cumulative probability distribution of velocity becomes

$$F(d) = \frac{d^r}{d_{max}^r}. \tag{19.56}$$

It is then assumed that all values of flow depth d measured at any cross-section between minimum flow depth of zero and maximum flow depth d_{max} are equally likely. In reality this is not unlikely because at different cross-sections different values of flow depth do occur. The cumulative probability distribution of depth can be expressed as the ratio of the flow discharge to the point where depth is to be considered and the maximum mean annual discharge. Then, the probability of depth being equal to or less than d is Q/Q_{max} at any depth (measured from the minimum depth) less than d; thus the cumulative distribution function of depth, $F(d) = P(depth \leq d), P = $ probability, can be expressed as

$$F(d) = \frac{Q}{Q_{max}}. \tag{19.57}$$

The probability density function is obtained by differentiating Equation (19.57) with respect to d:

$$f(d) = \frac{dF(d)}{dd} = \frac{1}{Q_{max}} \frac{dQ}{dd} \quad \text{or} \quad f(d) = \left[(Q_{max}) \frac{dd}{dQ}\right]^{-1}. \tag{19.58}$$

The term $f(d) \, dd = F(d + dd) - F(d)$ denotes the probability of depth being between d and $d + dd$. Since Equation (19.57) constitutes the fundamental hypothesis employed for deriving the depth distribution using entropy, it will be useful to evaluate its validity. This hypothesis (i.e., the relation between the cumulative probability $F(d)$ and the ratio $Q/(Q_{max})$) should be tested for a number of river cross-sections.

Substituting Equation (19.55) in Equation (19.58) and integrating, one obtains

$$\frac{d}{d_{max}} = \left(\frac{Q}{Q_{max}}\right)^{\frac{1}{r}} \quad \text{or} \quad d = \frac{d_{max}}{(Q_{max})^f} Q^f, \quad f = \frac{1}{r}. \tag{19.59}$$

Equation (19.59) gives the power function for depth–discharge relation used in hydraulic geometry. It has an exponent ($1/r = f$) that needs to be determined.

Differentiating Equation (19.54) with respect to λ_1, one obtains

$$\frac{\partial \lambda_0}{\partial \lambda_1} = -\ln d_{max} + \frac{1}{-\lambda_1 + 1}. \tag{19.60}$$

One can also write from Equations (19.48) and (19.53):

$$\lambda_0 = \ln \left[\int_0^{d_{max}} d^{-\lambda_1} dd\right]. \tag{19.61}$$

Differentiating Equation (19.61) with respect to λ_1 and using Equation (19.53), one obtains

$$\frac{\partial \lambda_0}{\partial \lambda_1} = -\frac{\int_0^{d_{max}} [\ln d] d^{-\lambda_1} dd}{\int_0^{d_{max}} d^{-\lambda_1} dd} = \frac{-\int_0^{d_{max}} [\ln d] \exp(-\lambda_0) d^{-\lambda_1} dd}{\int_0^{d_{max}} \exp(-\lambda_0) d^{-\lambda_1} dd} = -\overline{\ln d}. \tag{19.62}$$

Equating Equation (19.60) to Equation (19.62) leads to an estimate of λ_1:

$$\lambda_1 = 1 - \frac{1}{\ln d_{max} - \overline{\ln d}}. \tag{19.63}$$

Therefore, the power function exponent $r = -\lambda_1 + 1$ becomes

$$r = \frac{1}{\ln d_{max} - \overline{\ln d}}. \tag{19.64}$$

Equation (19.64) shows that the exponent, r, of the power function can be estimated from the values of the logarithm of depth and the average of the logarithmic values of flow depths. The higher the difference between these logarithm values the lower will be the exponent.

The entropy (in Napiers) of the depth distribution can be obtained by substituting Equation (19.55) in Equation (19.47):

$$H = f\left(\ln d_{max} - \overline{\ln d}\right) + \overline{\ln d} - \ln r. \tag{19.65}$$

Equation (19.65) shows that for a given value of r, the uncertainty increases as the difference between logarithm of the maximum depth and the average of the logarithmically transformed values of the depth increases. Taking advantage of Equation (19.64), Equation (19.65) becomes

$$H = 1 + \overline{\ln d} - \ln r. \tag{19.66}$$

Equation (19.66) shows that higher average logarithmic depth will lead to higher uncertainty. In such situations, it will be desirable that depth is sampled more frequently.

Width–discharge power function: The width–discharge power function can also be derived using the methodology as described for the depth–discharge function. Therefore, the steps will not be repeated and only the final results are given here.

$$f(w) = r\frac{w^{r-1}}{w^r_{max}} \qquad r = -\lambda_1 + 1 \tag{19.67}$$

$$\frac{w}{w_{max}} = \left(\frac{Q}{Q_{max}}\right)^{\frac{1}{r}} \quad \text{or} \quad w = \frac{w_{max}}{(Q_{max})^b}Q^b, \quad b = \frac{1}{r} \tag{19.68}$$

$$r = \frac{1}{\ln w_{max} - \overline{\ln w}}. \tag{19.69}$$

Velocity–discharge power function: The velocity–discharge power function can also be derived using the methodology as described for the depth–discharge function. Therefore, the steps will not be repeated and only the final results are given here.

$$f(v) = r\frac{v^{r-1}}{v_{max}^r} \qquad r = -\lambda_1 + 1 \tag{19.70}$$

$$\frac{v}{v_{max}} = \left(\frac{Q}{Q_{max}}\right)^{\frac{1}{r}} \quad \text{or} \quad v = \frac{v_{max}}{(Q_{max})^m} Q^m, \quad m = \frac{1}{r} \tag{19.71}$$

$$r = \frac{1}{\ln v_{max} - \overline{\ln v}}. \tag{19.72}$$

Slope–discharge power function: The slope–discharge power function can also be derived using the methodology as described for the depth–discharge function. Therefore, the steps will not be repeated and only the final results are given.

$$f(S) = r\frac{S^{r-1}}{S_{max}^r} \qquad r = -\lambda_1 + 1 \tag{19.73}$$

$$\frac{S}{S_{max}} = \left(\frac{Q}{Q_{max}}\right)^{\frac{1}{r}} \quad \text{or} \quad S = \frac{S_{max}}{(Q_{max})^z} Q^z, \quad z = \frac{1}{r} \tag{19.74}$$

$$r = \frac{1}{\ln S_{max} - \overline{\ln S}}. \tag{19.75}$$

Roughness–discharge power function: The roughness-discharge power function can also be derived using the methodology as described for the depth–discharge function. Therefore, the steps will not be repeated and only the final results are given here.

$$f(n) = r\frac{n^{r-1}}{n_{max}^r} \qquad r = -\lambda_1 + 1 \tag{19.76}$$

$$\frac{n}{n_{max}} = \left(\frac{Q}{Q_{max}}\right)^{\frac{1}{r}} \quad \text{or} \quad n = \frac{n_{max}}{(Q_{max})^y} Q^y, \quad y = \frac{1}{r} \tag{19.77}$$

$$r = \frac{1}{\ln n_{max} - \overline{\ln n}}. \tag{19.78}$$

Sediment–discharge power function: The sediment–discharge power function can also be derived using the methodology as described for the depth–discharge function. Therefore, the steps will not be repeated and only the final results are given here.

$$f(Q_s) = r \frac{Q_s^{r-1}}{Q_{smax}^r} \qquad r = -\lambda_1 + 1 \qquad (19.79)$$

$$\frac{Q_s}{Q_{smax}} = \left(\frac{Q}{Q_{max}}\right)^{\frac{1}{r}} \quad \text{or} \quad Q_s = \frac{Q_{smax}}{(Q_{max})^J} Q^J, \quad J = \frac{1}{r} \qquad (19.80)$$

$$r = \frac{1}{\ln Q_{smax} - \overline{\ln Q_s}}. \qquad (19.81)$$

19.4 Derivation of Downstream Geometry Using Entropy

The downstream hydraulic geometry can be derived by maximizing the Shannon entropy in the same manner as the at-a-station hydraulic geometry. The forms of equations will be identical. The only difference is that the data will be from different stations along the same river reach.

19.5 Cross-Sectional Shape

Stable channel cross-section shapes can be divided into two types, as shown in Figures 19.1a and 19.1b. Figure 19.1a represents a typical cross-sectional shape consisting of curved banks and a constant depth zone inserted between them. Irrigation canals and alluvial rivers are often of this shape. Figure 19.1b represents another cross-sectional shape with two curved banks joining at the middle or centerline and is a lower limit of a stable channel, which can be referred to as threshold channel. The shape on either side of the centerline can be assumed to be the same for purposes of simplicity. The distribution of transverse slopes, as shown in Figure 19.2, which will lead to the bank profile of the

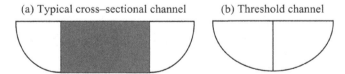

Figure 19.1 (a) Typical cross-sectional channel and (b) threshold channel

Figure 19.2 Distribution of transverse slopes

channel cross-section, will need to be determined. The dimensions and shape are assumed to depend on discharge and the boundary sediment size or the angle of repose of particles.

For determining the distribution of slope, let the elevation from the horizontal datum to the bank be denoted by y varying from 0 to d at the water surface $(0 \leq y \leq d)$. Let x be the lateral distance from the centerline, varying from 0 to W, where W is the half channel width from the centerline. Then, the total width of the channel would be $2W$ or w. Let the flow depth be denoted by h, varying from 0 to h_c, which equals d, where d is the bankfull flow depth at the centerline. Let the transverse slope be denoted by $s = \tan\theta = dy/dx$ varying from 0 to s_0, where s_0 is the submerged coefficient of friction, y is the elevation of the bank at x, s_0 is the maximum slope equal to the angle of repose ϕ, the angle of internal friction for sediment, as shown in Figure 19.2. Since the transverse slope increases monotonically from the centerline to the water surface and is found to vary from one cross-section to another, it is assumed to be a random variable denoted by S.

The Shannon entropy (Shannon, 1948) of the transverse slope can be expressed as

$$H(s) = -\int_0^{s_0} f(s) \ln f(s) ds, \tag{19.82}$$

where s is the value of random variable S, $f(s)$ is the probability density function (PDF) of S, and H is the entropy of S or $f(s)$. The PDF $f(s)$ can be derived by maximizing H, subject to specified constraints, in accordance with the principle of maximum entropy (POME) (Jaynes, 1957a, b). For purposes of simplicity, let there be no constraint that $f(s)$ must satisfy. Therefore,

$$\int_0^{s_0} f(s) ds = 1, \tag{19.83}$$

which is the total probability theorem. The least-biased probability distribution function $f(s)$ is obtained by maximizing entropy given by Equation (19.82), subject to Equation (19.83), using the method of Lagrange multipliers, where the Lagrangean function L can be expressed as

$$L = -\int_0^{s_0} f(s) \ln f(s) ds - (\lambda_0 - 1) \left(\int_0^{s_0} f(s) ds - 1 \right), \tag{19.84}$$

where λ_0 is the Lagrange multiplier. Differentiating Equation (19.84) with respect to $f(s)$, noting that $f(s)$ is variable and s is parameter, and equating the derivative to zero, one obtains

$$\frac{\partial L}{\partial f(s)} = 0 \Rightarrow -\ln f(s) - \lambda_0 = 0. \tag{19.85}$$

Equation (19.85) yields

$$f(s) = \exp(-\lambda_0). \tag{19.86}$$

Equation (19.86) is a uniform density function (PDF) of transverse slope S. The cumulative distribution function (CDF) of S is obtained by integrating Equation (19.86):

$$F(s) = \exp(-\lambda_0)s. \tag{19.87}$$

Equation (19.87) is linear, implying that all values of slope are equally likely.

The maximum entropy of S is obtained by inserting Equation (19.86) in Equation (19.82):

$$H(s) = \lambda_0 \exp(-\lambda_0)s_0, \tag{19.88}$$

which is expressed in terms of the Lagrange multiplier λ_0 and constant s_0. Equation (19.88) reflects the maximum uncertainty about the side slope.

Substitution of Equation (19.86) in Equation (19.83) yields

$$\exp(-\lambda_0) = \frac{1}{s_0}. \tag{19.89}$$

Therefore,

$$\lambda_0 = \ln s_0. \tag{19.90}$$

Substitution of Equation (19.89) in Equation (19.86) leads to the PDF of S as

$$f(s) = \frac{1}{s_0}. \tag{19.91}$$

Equation (19.91) states that the PDF of S is uniform. Likewise, substitution of Equation (19.89) in Equation (19.87) yields the CDF of S:

$$F(s) = \frac{s}{s_0}. \tag{19.92}$$

Equation (19.92) states that the CDF of S is linear bounded by the upper limit, s_0. Similarly, substitution of Equations (19.89) and (19.90) in Equation (19.88) yields

$$H(s) = \ln s_0. \tag{19.93}$$

It is interesting to note that the uncertainty about the slope depends only on the knowledge of the upper limit of S, s_0. Therefore, it is important to specify the value of s_0 as accurately as possible.

At any lateral distance from the centerline less than x, the transverse slope at that distance is less than s. It can then be reasoned that all values of x between 0 and W along the x-axis are equally likely to be sampled or have the same probability. Then, the probability of the transverse slope being equal to or less than s is x/W. The cumulative distribution function (CDF) of S can then be expressed in terms of the lateral or transverse distance as

$$F(s) = \frac{x}{W}.$$

(19.94)

Equation (19.94) is just a hypothesis, and its validity in field needs to be tested.

Differentiating Equation (19.94) yields the PDF of S as

$$f(s)ds = \frac{1}{W}dx \quad \text{or} \quad f(s) = \frac{1}{W}\frac{dx}{ds} = \left(W\frac{ds}{dx}\right)^{-1}.$$

(19.95)

The PDF given by Equation (19.95) must satisfy the constraint defined by Equation (19.83). Inserting Equation (19.91) in Equation (19.95), one gets

$$\frac{1}{s_0} = \frac{1}{W}\frac{dx}{ds}.$$

(19.96)

Integration of Equation (19.96) yields

$$s = \frac{s_0}{W}x.$$

(19.97)

Equation (19.97) expresses the distribution of transverse slope as a function of transverse distance, and satisfies the condition that $s = 0$ at $x = 0$ and $s = s_0$ at $x = W$. Equation (19.97) assumes that the side slope of the channel varies only with the transverse distance, given the values of the half width and the maximum slope. The values of s_0 and W can be determined beforehand for a given river.

The geometric parameters of a channel cross-section of interest are: shape, centerline flow depth, flow depth, aspect ratio, cross-section area, wetted perimeter, and hydraulic radius. Each parameter is now derived.

Shape function: Recalling the definition of slope, $s = dy/dx$, and equating it to Equation (19.97) one gets

$$\frac{dy}{dx} = \frac{s_0}{W}x.$$

(19.98)

Integrating Equation (19.98) with the condition that $s = 0$ at $x = 0$, the bank profile becomes

$$y = \frac{s_0}{2W}x^2.$$

(19.99a)

Equation (19.99a) gives the elevation of the right bank or water margin as a function of transverse distance x from the centerline up to $x = W$. For a given cross-section, there is a finite width at the bottom, and therefore $x = x_0$ at $s = 0$. Hence, for computing the flow depth, Equation (19.99a) should be modified as

$$y = \frac{s_0}{2W}(x - x_0)^2. \tag{19.99b}$$

Equation (19.99b) is the derived shape function $y = y(x)$.

Centerline flow depth: At $x = W$, $y = d$, which is also the flow depth at the centerline h_c and is given by Equation (19.99a) as

$$d = h_c = \frac{s_0 W}{2}. \tag{19.100}$$

Flow depth: Subtracting Equation (19.99a) from Equation (19.100), one gets the lateral distribution of local water flow depth as

$$h = d - y = \frac{s_0}{2}\left(W - \frac{x^2}{W}\right). \tag{19.101a}$$

Equation (19.100) specifies the depth at the centerline and Equation (19.101a) satisfies the boundary elevation at the bed. The difference of these two equations results in the lateral distribution of flow depth $h(x)$ as a function of lateral distance as

$$h(x) = \frac{s_0 W}{2}\left[1 - \left(\frac{x}{W}\right)^2\right]. \tag{19.101b}$$

For computing depth, Equation (19.101b) should be modified as

$$h(x) = \frac{s_0 W}{2}\left[1 - \left(\frac{x - x_0}{W}\right)^2\right]. \tag{19.101c}$$

Aspect ratio: The aspect ratio $w/d = 2W/d$ then becomes

$$\frac{w}{d} = \frac{4}{s_0}. \tag{19.102}$$

Cross-section area: Noting that $h\,dx = dA$, the cross-sectional area A can be obtained by integrating Equation (19.101b) as

$$A = \frac{1}{3}s_0 W^2. \tag{19.103}$$

Wetted perimeter: The wetted perimeter P_w is computed as follows: Consider an arc element of wetted perimeter $dP_w = (dx^2 + dy^2)^{1/2}$. From Equation (19.99a), $dy = x^2 s_0/(2W)$. Therefore, $dy = x s_0 dx/W$, and

$$dP_w = \left[dx^2 + \left(\frac{xS_0}{W} dx \right)^2 \right]^{1/2} = \left[1 + \frac{x^2 S_0^2}{W^2} \right]^{1/2} dx = \left(\frac{S_0}{W} \right) \left[\frac{W^2}{S_0^2} + x^2 \right]^{1/2} dx.$$

(19.104)

Integrating Equation (19.104) leads to

$$P_w = \frac{W}{S_0} \left\{ S_0 \sqrt{1 + S_0^2} + \ln \left[S_0 + \sqrt{1 + S_0^2} \right] \right\}$$

$$= \frac{2h_c}{S_0^2} \left\{ S_0 \sqrt{1 + S_0^2} + \ln \left[S_0 + \sqrt{1 + S_0^2} \right] \right\}.$$

(19.105)

Hydraulic radius: The hydraulic radius R can be expressed as

$$R = \frac{S_0 W^2}{3 P_w} = \frac{4 h_c^2}{3 S_0 P_w}.$$

(19.106)

19.6 Energy Gradient or Channel Slope

The entropy theory can also be applied to derive the longitudinal energy slope or channel slope (Cao and Chang, 1988). The energy gradient for a specified discharge Q represents the rate of power expenditure, which depends on sediment size, sediment load, river morphology, and human activities. Here the energy gradient in the streamwise direction and the longitudinal profile of an alluvial river are considered, assuming that the channel-bed slope is almost equal to the energy slope.

For deriving the energy gradient along the direction x measured from the upstream, let a river reach be of length L, with a total fall Y over L in the vertical plane. At any point x, let the elevation above the downstream point be denoted by y. It may be convenient to non-dimensionalize these variables and express slope as follows:

Dimensionless distance: $x_* = \frac{x}{L}$
Dimensionless elevation: $y_* = \frac{y}{Y}$
Energy gradient: $S = \frac{dx_*}{dy_*}$.

In natural rivers, as discharge increases in the downstream direction, the sediment size and concentration decrease and so does the energy gradient, meaning the maximum value of energy gradient (S_{max}) will be at the upstream and the minimum (S_{min}) at the downstream. If the energy gradient at any point x_* is S then at any point greater than this distance the energy gradient will be less than S. Since all values of x_* are equally likely to occur, the energy gradient values associated with the x_* values can be considered as random values. Thus, the probability of the energy gradient equal to or less than S can be expressed as $(1 - x_*)$. This leads to the expression for the cumulative distribution function (CDF) as

$$F(S) = 1 - x_*(S). \tag{19.107}$$

The derivative of Equation (19.107) gives the probability density function (PDF):

$$f(S) = \frac{dF(S)}{dS} = \frac{dF(S)}{dx_*} \frac{dx_*}{dS} = -\left(\frac{dS}{dx_*}\right)^{-1}. \tag{19.108}$$

For deriving the PDF using the Shannon entropy, one can write

$$H(S) = - \int_{S_{max}}^{S_{min}} f(S) \ln f(S) dS. \tag{19.109}$$

For maximizing $H(S)$, the following constraints can be expressed for simplicity:

$$\int_{S_{max}}^{S_{min}} f(S) dS = 1 \tag{19.110}$$

$$\int_{S_{max}}^{S_{min}} Sf(S) dS = \overline{S}. \tag{19.111}$$

Using the method of Lagrange multipliers, the Lagrangian function (L_f) can be expressed as

$$L_f = - \int_{S_{max}}^{S_{min}} f(S) \ln f(S) dS - (\lambda_0 - 1) \left[\int_{S_{max}}^{S_{min}} f(S) dS - 1 \right] - \lambda_1 \left[\int_{S_{max}}^{S_{min}} Sf(S) dS - \overline{S} \right], \tag{19.112}$$

where λ_0 and λ_1 are the unknown Lagrange multipliers. Differentiation of Equation (19.112) with respect to $f(S)$ and equating the derivative to 0 yield:

$$- \ln f(S) - 1 - (\lambda_0 - 1) - \lambda_1 S = 0. \tag{19.113}$$

Equation (19.113) leads to

$$f(S) = \exp(-\lambda_0 - \lambda_1 S). \tag{19.114}$$

The Lagrange multipliers can be determined using Equations (19.110) and (19.111). Inserting Equation (19.114) in Equation (19.110), one gets

$$\exp(\lambda_0) = \int_{S_{max}}^{S_{min}} \exp(-\lambda_1 S) dS, \tag{19.115}$$

which can be solved as

$$\lambda_0 = \ln\left\{\frac{1}{\lambda_1}[\exp(-\lambda_1 S_{max}) - \exp(-\lambda_1 S_{min})]\right\}, \tag{19.116}$$

which simplifies to

$$\lambda_0 = -\ln\lambda_1 + \ln[\exp(-\lambda_1 S_{max}) - \exp(\lambda_1 S_{min})]. \tag{19.117}$$

Differentiation of Equation (19.115) with respect to λ_1 leads to

$$\frac{\partial\lambda_0}{\partial\lambda_1} = -\frac{\int_{S_{max}}^{S_{min}} S\exp(-\lambda_0 - \lambda_1 S) dS}{\int_{S_{max}}^{S_{min}} \exp(-\lambda_0 - \lambda_1 S) dS} = -\overline{S}. \tag{19.118}$$

Differentiating Equation (19.117), one gets

$$\frac{\partial\lambda_0}{\partial\lambda_1} = -\frac{1}{\lambda_1} - \frac{S_{max}\exp(-\lambda_1 S_{max}) - S_{min}\exp(-\lambda_1 S_{min})}{\exp(-\lambda_1 S_{max}) - \exp(-\lambda_1 S_{min})}. \tag{19.119}$$

Equating Equations (19.118) and (19.119),

$$\frac{1}{\lambda_1} + \frac{S_{max}\exp(-\lambda_1 S_{max}) - S_{min}\exp(-\lambda_1 S_{min})}{\exp(-\lambda_1 S_{max}) - \exp(-\lambda_1 S_{min})} = \overline{S}. \tag{19.120}$$

For known S_{min} and S_{max}, Lagrange multipliers λ_0 and λ_1 can be determined using Equations (19.117) and (19.120).

Now equating Equations (19.108) and (19.114),

$$-\left(\frac{dS}{dx_*}\right)^{-1} = \exp(-\lambda_0 - \lambda_1 S). \tag{19.121}$$

Integration of Equation (19.121) with the condition that at $x_* = 0$, $S = S_{max}$ yields

$$x_* = \frac{\exp(-\lambda_0)}{\lambda_1} \left[\exp(-\lambda_1 S) - \exp(-\lambda_1 S_{max}) \right]. \tag{19.122}$$

Inserting Equation (19.116) in Equation (19.122),

$$x_* = \frac{\exp(-\lambda_1 S) - \exp(-\lambda_1 S_{max})}{\left[\exp(-\lambda_1 S_{max}) - \exp(-\lambda_1 S_{min}) \right]}. \tag{19.123}$$

Let $C_0 = \left[\exp(-\lambda_1 S_{max}) - \exp(-\lambda_1 S_{min}) \right]$. Then, Equation (19.123) can be expressed for S in terms of x_* as

$$S = -\frac{1}{\lambda_1} \ln \left[\exp(-\lambda_1 S_{max}) + C_0 x_* \right]. \tag{19.124}$$

References

Bagnold, R. A. (1960). Sediment discharge and stream power: A preliminary announcement. U.S. Geological Survey, Professional Paper 421.

Cao, S. and Chang, H. (1988). Entropy as a probability concept in energy-gradient distribution. *Proceedings. National Congress on Hydraulic Engineering, Colorado Springs, CO, August 8–12*, ASCE, New York, pp. 1013–1018.

Jaynes, E. T. (1957a). Information theory and statistical mechanics. *Physical Review*, Vol. 106, No. 4, pp. 620.

Jaynes, E. T. (1957b). Information theory and statistical mechanics II. *Physical Review*, Vol. 108, No. 2, pp. 171.

Jaynes, E. T. (1982). On the rationale of maximum-entropy methods. *Proceedings of the IEEE*, Vol. 70, No. 9, pp. 939–952.

Leopold, L. B. and Maddock, T. (1953). *The hydraulic geometry of stream channels and some physiographic implications*. Geological Survey Professional Paper 252, U.S. Geological Survey, Washington, DC.

Shannon, C. E. (1948). A mathematical theory of communication. *The Bell System Technical Journal*, Vol. 27, No. 3, pp. 379–423.

Singh, V. P. (1998). *Entropy-Based Parameter Estimation in Hydrology*. Kluwer Academic Publishers (now Springer), Dordrecht.

20

Minimum Energy Dissipation and Maximum Entropy Theory

This chapter employs the theory comprising the principle of maximum entropy (POME) and the principle of minimum energy dissipation or its simplified minimum stream power for deriving hydraulic geometry relations. The theory leads to four families of downstream hydraulic geometry relations and eleven families of at-a-station hydraulic geometry relations. The principle of minimum energy dissipation rate states that the spatial variation of the stream power of a channel for a given discharge is accomplished by the spatial variation in channel form (flow depth and channel width) and hydraulic variables, including energy slope, flow velocity, and friction. The principle of maximum entropy states that the change in stream power is distributed among the changes in flow depth, channel width, flow velocity, slope, and friction, depending on the constraints (boundary conditions) the channel has to satisfy. The conditions under which the families of hydraulic geometry relations can occur in field are discussed.

Notation

A = cross-sectional area

a, c, k, N, s = numerical constants

B = weighting factor

b, f, m, p, y = exponents of flow discharge Q in empirical hydraulic geometry relationships

C = Chezy's roughness coefficient

$C_1, C_{1*}, C_2, C_{2*}, C_3, C_{3*}$ = primary morphological coefficients

$C_s, C_{wS}, C_{vS}, C_{nS}, C_{dS}, C_w, C_d, C_{wd}, C_{wd*}, C_v$ = coefficients of morphological equations

d = mean flow depth

f = Darcy–Weisbach friction factor

Continued

g = acceleration due to gravity
J = weighting factor
m = exponent in Manning Formula
n = Manning's roughness coefficient
P = wetted perimeter
P_{us} = unit stream power
P_a, P_n = proportion of the adjustment of stream power by friction
P_w = proportion of the adjustment of stream power by channel width
P_d = proportion of the adjustment of stream power by flow depth
P_S = proportion of the adjustment of stream power by slope
Q = flow discharge
r = weighting factor
R = hydraulic radius
R_1 = spatial rate of adjustment of friction
R_2 = spatial rate of adjustment of channel width
R_3 = spatial rate of adjustment of flow depth
R_s = spatial rate of adjustment of SP along a river
S = channel slope
SP = stream power
v = average flow velocity
w = water surface width
x = distance along the flow direction
z = exponent of discharge Q in empirical relation between S and Q
α = roughness measure
β = exponent of flow depth d
ϕ, φ = parameters in basic form of regime equation
γ = weight density of water

20.1 Introduction

The hydraulic geometry relations expressed by Leopold and Maddock (1953) for a channel can be expressed in the form of power functions of discharge as

$$w = aQ^b, \quad d = cQ^f, \quad v = kQ^m, \tag{20.1a}$$

where w is the channel width; d is the flow depth; v is the flow velocity; Q is the flow discharge; and a, b, c, f, k, and m are parameters. Also added to Equation (20.1a) are:

$$n = NQ^p, S = sQ^y, \tag{20.1b}$$

where n is Manning's roughness factor, S is slope, and N, p, s, and y are parameters. Exponents b, f, m, p, and y represent, respectively, the rate of change of the hydraulic variables w, d, v, n, and S as Q changes; and coefficients a, c, k, N, and s are scale factors that define the values of w, d, v, n, and S when $Q = 1$.

The exponents and coefficients of hydraulic geometry relations of Equations (20.1a, b) vary from location to location on the same river and from river to river, as well as from high flow range to low flow range. This is because the influx of water and sediment and the constraints (boundary conditions) that the river channel is subjected to vary from location to location as well as from river to river. For a fixed influx of water and sediment a channel will exhibit a family of hydraulic geometry relations in response to the constraints imposed on the channel. It is these constraints that force the channel to adjust its allowable hydraulic variables. For example, if a river is leveed on both sides, then it cannot adjust its width and is, therefore, left to adjust other variables, such as depth, friction, slope, and velocity. Likewise, if a canal is lined, then it cannot adjust its friction.

This chapter applies the principles of minimum energy dissipation rate and maximum entropy to derive downstream and at-a-station hydraulic geometry relations. Inherent in the derivation is an explanation for the self-adjustment of channel morphology. It is shown that by combining the hypotheses based on the principles of maximum entropy and minimum energy dissipation rate a family of hydraulic geometry relations is obtained. This family may encompass many of the hydraulic geometry relations derived using other hypotheses.

20.2 Downstream Hydraulic Geometry Relations

Equations (20.1a, b) correspond to the case when the channel is in equilibrium state. Langbein (1964) hypothesized that when a channel adjusts its hydraulic variables corresponding to this state, the adjustment is shared equally among the hydraulic variables. Employing the principle of maximum entropy and minimum stream power, Deng and Zhang (1994) derived morphological equations, assuming that for a given discharge the flow depth and width were independent variables among five hydraulic variables. However, in practice the adjustment among hydraulic variables will seldom be equal. Depending on the adjustment of hydraulic variables, there can then be a family of hydraulic geometry relations, and the adjustment can explain the variability in the parameters (scale and exponents) of these relations.

Yang (1972) defined the unit stream power (USP) as the time rate of potential energy expenditure per unit cross-sectional area per unit weight of water in an alluvial channel and it can be expressed as the unit velocity-slope product having the dimensions of power per unit weight of water. Thus, USP, denoted as P_{us}, can be expressed as

$$P_{us} = vS, \qquad (20.2)$$

where v is the average flow velocity, and S is the energy slope. Stream power (SP) is the rate of energy dissipation due to the flow of water:

$$SP = Q\gamma S, \qquad (20.3a)$$

where γ is the weight density of water, and Q is discharge of water. SP can be obtained by integrating USP over a given cross-section. A channel responds to the influx of water and sediment coming from its watershed by the adjustment of SP. Yang (1972) found USP to be the dominating factor in the determination of total sediment concentration. Yang (1986, 1996) also related sediment load and channel geometry adjustment to SP.

The spatial rate of adjustment of SP along a river, R_s, can be expressed as

$$R_s = \frac{d(SP)}{dx} = \frac{d(Qrs)}{dx}, \qquad (20.3b)$$

where x is the space coordinate along the direction of flow. Cheema et al. (1997) determined stable width of an alluvial channel using the hypothesis that an alluvial channel attains a stable width when the rate of change of unit stream power with respect to its width is a minimum. Thus, an alluvial channel with stable cross-section has the ability to vary its width at a minimum expenditure of energy per unit width per unit time.

If a channel is assumed rectangular with d as the depth of flow and w as the width of flow, then the flow cross-sectional area $A = wd$, the wetted perimeter $P = w + 2d$, and the hydraulic radius $R = A/P = (wd)/(w + 2d)$. If the channel is wide rectangular, then $R \cong d = $ depth of flow. The flow discharge in Equation (20.3b) can be obtained from either Manning's or Chezy's or the Darcy–Weisbach equation. For wide rectangular channels, these equations can be written, respectively, as

$$Q = \frac{1}{n}AR^{2/3}S^{1/2} = \frac{1}{n}wd^{5/3}S^{1/2} \qquad (20.4a)$$

$$Q = CA\sqrt{RS} = Cwd^{3/2}\sqrt{S} \qquad (20.4b)$$

$$Q = 2\sqrt{\frac{2g}{f}}A\sqrt{RS} = 2\sqrt{\frac{2g}{f}}wd^{3/2}\sqrt{S}, \qquad (20.4c)$$

where n is Manning's roughness coefficient, C is Chezy's roughness coefficient, and f is the Darcy–Weisbach friction factor, and g is acceleration due to gravity. Clearly,

$$C = \frac{1}{n}wd^{1/6}; C = 2\sqrt{2g/f}. \qquad (20.4d)$$

Equations (20.4a) to (20.4c) can be expressed in a general from as:

$$Q = \alpha w d^{\beta} \sqrt{S},$$
(20.4e)

in which α is a roughness measure, and β is an exponent. For Manning's equation, $\alpha = 1/n$, and $\beta = 5/3$; for Chezy's equation, $\alpha = C$, and $\beta = 3/2$; and for Darcy–Weisbach equation, $\alpha = 2(2g/f)^{0.5}$, and $\beta = 3/2$.

The energy slope S can be expressed from Equation (20.4e) as

$$S = \frac{Q^2}{\alpha^2 w^2 d^{2\beta}}.$$
(20.5)

Thus, using Equations (20.3a) and (20.5), the stream power of a channel is expressed as

$$SP = \frac{\gamma Q^3}{\alpha^2 w^2 d^{2\beta}}.$$
(20.6)

In Equation (20.6), there are five variables: Q, S, w, α, and d; of these variables, Q, α, d, and w are on the right side of the equation, and Q and S through SP on the left side. Three of these variables, including α, w, and d, are controlling variables or constraints for a given discharge. The slope term S is not an independent variable here, because it is imbedded in the stream power, and hence it is not considered as a controlling variable. From a practical point of view, a natural river can easily adjust its width, depth, velocity, and roughness due to changing discharge. The longitudinal slope takes a very long time, say years if not centuries, to adjust (Yang, 1996). Therefore, one generally treats the longitudinal profile or slope as constant over a short period of time. Because of this time scale difference, S is not considered as a variable when compared with velocity, depth, width, and roughness. Thus, it is hypothesized that for a given influx of discharge from the watershed the channel will adjust or minimize its stream power by adjusting these three controlling variables. This hypothesis is similar to the theory of minimum variance (Langbein, 1964).

Substitution of Equation (20.6) in Equation (20.3b) yields

$$R_s = \frac{d(Q\gamma S)}{dx} = \gamma \frac{d}{dx}\left(\frac{Q^3}{\alpha^2 w^2 d^{2\beta}}\right) = \gamma Q^3 \frac{d}{dx}\left(\frac{1}{\alpha^2 w^2 d^{2\beta}}\right).$$
(20.7)

Equation (20.7) gives

$$R_s = \frac{-2\gamma Q^3}{\alpha^3 w^2 d^{2\beta}}\frac{d\alpha}{dx} - \frac{2\gamma Q^3}{\alpha^2 w^3 d^{2\beta}}\frac{dw}{dx} - \frac{2\beta\gamma Q^3}{\alpha^2 w^2 d^{2\beta+1}}\frac{dd}{dx}.$$
(20.8)

The right side of Equation (20.8) has three parts, designated as R_1, R_2, R_3:

$$R_1 = -\frac{2\gamma Q^3}{\alpha^3 w^2 d^{2\beta}} \frac{d\alpha}{dx} \tag{20.9}$$

$$R_2 = -\frac{2\gamma Q^3}{\alpha^2 w^3 d^{2\beta}} \frac{dw}{dx} \tag{20.10}$$

$$R_3 = -\frac{2\beta\gamma Q^3}{\alpha^2 w^2 d^{2\beta+1}} \frac{dd}{dx}. \tag{20.11}$$

Equation (20.9) can be interpreted as the spatial rate of adjustment of friction, Equation (20.10) as the spatial rate of adjustment of width, and Equation (20.11) as the spatial rate of adjustment of flow depth. Dividing Equations (20.9) to (20.11) by the total spatial rate of adjustment of *SP*, one gets

$$P_\alpha = \frac{R_1}{R_s} = -\frac{2\gamma Q^3}{\alpha^3 w^2 d^{2\beta}} \frac{[d\alpha/dx]}{[d(SP)/dx]} \tag{20.12}$$

$$P_w = \frac{R_2}{R_s} = -\frac{2\gamma Q^3}{\alpha^2 w^3 d^{2\beta}} \frac{[dw/dx]}{[d(SP)/dx]} \tag{20.13}$$

$$P_d = \frac{R_3}{R_s} = -\frac{2\beta\gamma Q^3}{\alpha^2 w^2 d^{2\beta+1}} \frac{[dd/dx]}{[d(SP)/dx]}. \tag{20.14}$$

Equation (20.12) can be interpreted as the proportion of the adjustment of stream power by friction, Equation (20.13) as the proportion of the adjustment of stream power by channel width, and Equation (20.14) as the proportion of the adjustment of stream power by flow depth.

Any system in equilibrium state under steady constraints tends to maximize its entropy in accordance with the principle of maximum entropy (POME) (Jaynes, 1957). When a river reaches a dynamic (or quasi-dynamic) equilibrium, the entropy should attain its maximum value. POME states that the entropy of a system is maximum when all probabilities are equal, that is, the probability distribution is uniform. Therefore,

$$P_\alpha = P_w = P_d. \tag{20.15}$$

Equation (20.15) holds under the stipulation that there are no constraints imposed on the channel and can be interpreted to mean that the self-adjustment of *SP* (γQS) is equally shared among α, w, and d. From an analysis of data from 165 gaging stations, Williams (1967, 1978) found that a channel adjusted all its hydraulic parameters (w, d, S, v) in response to changes in the influx of water and sediment and that self-adjustments were realized in an evenly distributed manner among factors. Equation (20.15) is similar to the concept embodied in the minimum variance theory (Langbein, 1964).

Equation (20.15) involves probabilities of three variables, meaning that any two of the three cases of adjustment in hydraulic variables may coexist as well as that of all three cases may coexist. These configurations of adjustment do indeed occur in nature (Wolman, 1955). Thus, the equality among three probabilities raises four possibilities and hence leads to four sets of equations: (1) $P_\alpha = P_w$, (2) $P_w = P_d$, (3) $P_\alpha = P_d$, and (4) $P_\alpha = P_w = P_d$. All four possibilities can occur in the same river in different reaches or in the same reach at different times, or in different rivers at the same time or at different times. In order to enumerate the consequences of these possibilities, one can either employ the general discharge Equation (20.4e) or employ either Manning's Equation (20.4a) or Chezy's Equation (20.4b) or Darcy–Weisbach Equation (20.4c). It is, however, more informative to use a specific discharge–resistance relation than the general discharge–resistance relation. Survey of literature shows that Manning's equation is more commonly employed than Chezy's C or Darcy–Weisbach's f in investigations on hydraulic geometry (Leopold and Wolman, 1957; Stall and Fok, 1968) and hence Manning's equation is used here. Consequently, replacing possibility P_α by P_n, Equations (20.12) to (20.14) are expressed for Manning's equation as

$$P_n = \frac{R_1}{R_s} = \frac{2n\gamma Q^3}{w^2 d^{10/3}} \frac{[dn/dx]}{[d(SP)/dx]} \tag{20.12a}$$

$$P_w = \frac{R_2}{R_s} = -\frac{2\gamma n^2 Q^3}{w^3 d^{10/3}} \frac{[dw/dx]}{[d(SP)/dx]} \tag{20.13a}$$

$$P_d = \frac{R_3}{R_s} = -\frac{10\gamma n^2 Q^3}{3w^2 d^{13/3}} \frac{[dd/dx]}{[d(SP)/dx]}. \tag{20.14a}$$

20.2.1 Primary Morphological Equations

The four possibilities for the spatial stream power adjustment lead to primary morphological equations, which are needed for deriving the downstream hydraulic geometry relations. The morphological equations are, therefore, derived first.

Possibility I: $P_w = P_n$: Here P_n is given by Equation (20.12a) and P_w by Equation (20.13a). Equating these two equations, one gets

$$\frac{dn}{dx} = -\frac{n}{w} \frac{dw}{dx}. \tag{20.16a}$$

Equation (20.16a) hypothesizes that the spatial change in stream power is accomplished by an equal spatial adjustment between flow width w and resistance expressed by Manning's n. The hypothesis can be considered as a limiting case and will presumably hold under the equilibrium condition. However, such a condition is not always achieved, and therefore the spatial change in stream power will be accomplished by an unequal adjustment between w and n. Hence, equation (20.16a) is modified as

$$\frac{dn}{dx} = -\frac{Bn}{w}\frac{dw}{dx},$$ (20.16b)

where B is a weighting factor, $0 \le B$, which accounts for the proportion in which the adjustment in stream power is shared between w and n. For the special case where the adjustment is shared equally between w and n, $B = 1$.

Integration of Equation (20.16b) yields

$$nw^b = C_1 \quad \text{or} \quad w = C_{1*}n^{-1/B},$$ (20.17a)

where C_1 and C_{1*} are constants of integration. For the limiting case ($B = 1$), Equation (20.17a) becomes

$$nw = C_1 \quad \text{or} \quad w = C_{1*}n^{-1}.$$ (20.17b)

Parameter C_1 or C_{1*} can be labeled as a primary morphological coefficient and Equation (20.17a) or (20.17b) as a primary morphological equation.

Possibility II: $P_w = P_d$: Here P_w is given by Equation (20.13a) and P_d by Equation (20.14a). Equating these two equations, one gets

$$\frac{dd}{dx} = \frac{3}{5}\frac{d}{w}\frac{dw}{dx}.$$ (20.18a)

Equation (20.18a) hypothesizes that the spatial variation in stream power is accomplished by an equal spatial adjustment between flow depth and flow width. This possibility occurs in channels where the roughness is fixed, say by lining, and the controlling variables are flow depth and width. The hypothesis can be considered as a limiting case and will hold under the equilibrium condition. Such a condition is, however, seldom achieved, and therefore the spatial change in stream power is accomplished by an unequal adjustment between d and w. Equation (20.18a) is now modified as

$$\frac{dd}{dx} = \frac{3r}{5}\frac{d}{w}\frac{dw}{dx},$$ (20.18b)

where r is a weighting factor, $0 \le r$, which accounts for the proportion in which the adjustment of stream power is shared between d and w. For the special case, where the adjustment is equally shared, $r = 1$.

Integration of Equation (20.18b) yields

$$\frac{w}{d^{\frac{5}{3r}}} = C_2 \quad \text{or} \quad d = C_{2*}w^{\frac{3r}{5}},$$ (20.19a)

where C_2 and C_{2*} are constants of integration. For the limiting case ($r = 1$), Equation (20.19a) reduces to

$$\frac{w}{d^{\frac{5}{3}}} = C_2 \quad \text{or} \quad d = C_{2*}w^{\frac{3}{5}}.$$ (20.19b)

Parameter C_2 or C_{2*} can be designated as a primary morphological coefficient and Equation (20.19a) or (20.19b) as a primary morphological equation. Equation (20.19a) resembles the basic form of regime equation expressed as

$$\frac{w\varphi}{d} = \phi,$$ (20.19c)

where $\phi = 3/5$ and $\varphi = 1/C_{2*}$.

Possibility III: $P_n = P_d$: Here P_n is given by Equation (20.12a) and P_d by Equation (20.14a). Equating these two equations, one gets

$$\frac{dn}{dx} = -\frac{5n}{3d}\frac{dd}{dx}.$$ (20.20a)

Equation (20.20a) hypothesizes that the spatial variation in stream power is accomplished by an equal spatial adjustment between flow depth and resistance. This hypothesis can be considered as a limiting case and will hold under the equilibrium condition. Such a condition is not always attained, and therefore the spatial change in stream power is accomplished by an unequal adjustment between d and n. Equation (20.20a) is now modified as

$$\frac{dn}{dx} = -\frac{5nJ}{3d}\frac{dd}{dx},$$ (20.20b)

where J is a weighting factor, $0 \le J$, which accounts for the proportion in which the adjustment of stream power is shared between n and d. For the special case where the adjustment is equally shared, $J = 1$.

Integration of Equation (20.20b) yields

$$n = C_3 d^{-\frac{5J}{3}} \quad \text{or} \quad d = C_{3*}n^{-\frac{3}{5J}},$$ (20.21a)

where C_3 and C_{3*} are constants of integration. For the limiting case ($J = 1$), Equation (20.21a) reduces to

$$nd^{5/3} = C_3 \quad \text{or} \quad d = C_{3*}n^{-3/5}.$$ (20.21b)

Parameter C_3 or C_{3*} can be considered as a primary morphological coefficient and Equation (20.21a) or (20.21b) as a primary morphological equation.

Possibility IV: $P_n = P_w = P_d$: Equation (20.17a) relates n and w, Equation (20.19a) relates w and d, and Equation (20.21a) relates n and d. The first two equations can be employed to eliminate n and d in Equation (20.4a) and express w as a function of Q. Similarly, Equations (20.19a) and (20.21a) can be used to eliminate w and n in Equation (20.4a) and express d as a function of Q. Likewise, all three equations can be used to express v as a function of Q.

Thus, three primary morphological Equations (20.17a or 20.17b), (20.19a or 20.19b), and (20.21a or 20.21b); and their three corresponding primary morphological coefficients, C_1, C_2, and C_3 (or C_{1*}, C_{2*}, and C_{3*}), are obtained. It should be noted that Equation (20.19b) can also be obtained by combining Equations (20.17b) and (20.21b) or Equation (20.17b) can be obtained by combining Equations (20.19b) and (20.21b).

20.2.2 Downstream Hydraulic Geometry Equations for a Given Discharge

If the discharge Q and slope S of a river are known, then substitution of primary morphological Equations (20.17a), (20.19a), and (20.21a) in Equation (20.4a) leads to the equations for hydraulic geometry of the river. Derivation of the equations under possibility II is discussed in what follows.

Possibility II: $P_w = P_d$: $w = C_2 d^{5/3r}$. This possibility leads to the hydraulic geometry relations for w, d, and v in terms of Q. These relations are derived from Equations (20.19a) and (20.4a) for gravel and alluvial rivers:

$$w = C_{wS} Q^{\frac{6}{5(1+r)}} \quad \text{for gravel rivers} \tag{20.22a}$$

$$w = C_{wS} Q^{\frac{13}{12(1+r)}} \quad \text{for sandy rivers} \tag{20.22b}$$

$$d = C_{dS} Q^{\frac{18r}{25(1+r)}} \quad \text{for gravel rivers} \tag{20.23a}$$

$$d = C_{dS} Q^{\frac{13r}{20(1+r)}} \quad \text{for sandy rivers} \tag{20.23b}$$

$$v = C_{vS} Q^{\frac{7r-5}{25(1+r)}} \quad \text{for gravel rivers} \tag{20.24a}$$

$$v = C_{vS} Q^{\frac{21r-5}{60(1+r)}} \quad \text{for sandy rivers,} \tag{20.24b}$$

where $C_{wS} = \frac{C_w}{C_S^{1/2(1+r)}}$, $C_w = [n(C_2)^r]^{1/(1+r)}$, $C_{dS} = C_d C_S^{-3r/[2(5+5r)]}$, $C_d = (nC_2)^{3r/(5+5r)}$, $C_{vS} = C_v C_S^{[3(1+r)+2]/[10(1+r)]}$, and $C_v = n^{-3/5}(C_w)^{-2/5}$.

This is the most investigated possibility. In this case, the exponent, b, of discharge is found to vary from 0 to 1, and the scale factor C_w varies with flow resistance. The precise value of b depends on the weighting factor r, which specifies the proportion for adjustment of stream power between w, d, and v. For the special case $r = 1$, where the adjustment is equally proportioned, $b = 0.5$. The width–discharge relation is found to depend on the slope of the channel, S. If S is expressed as a function of discharge with an exponent of $-1/6$, then the range of b becomes 0 to 13/12. These values of b encompass the entire range of values reported in the literature.

The value of exponent f varies from 0 to 3/5 (when r ranges from 0 to ∞), with the scale factor C_d being dependent on the flow resistance. The precise value depends on the value of r. For the special case $r = 1$, the value of f is 3/10. If the slope is expressed in terms of discharge with the power of $-1/6$, then the value of f ranges from 0 to 13/20 (when r ranges from 0 to ∞). These derived exponent values encompass the reported range.

The value of exponent, m, varies from 0 to 2/5, with the scale factor C_v being dependent on the flow depth. The exact value of m depends on the value of r. For the special case, $r = 1$, the value of m is 1/5. If the slope is expressed in terms of discharge with the power of $-1/6$, then the m exponent varies from $-1/12$ to 7/20 (when r ranges from 0 to ∞). Thus, the derived exponent values are seen to envelope the reported range.

For the downstream geometry of 72 streams from a variety of exponents, Park (1977) reported the range of b as 0.03 to 0.89 with modal class as 0.4 to 0.5; the range of f as 0.09 to 0.70 with modal class as 0.3 to 0.4; and the range of m as -0.51 to 0.75 with modal class as 0.1 to 0.2. Thus, the derived exponents are in the reported ranges. This discussion illustrates that the values of exponents, b, f, and m, do not possess fixed values; rather, they vary over certain ranges dictated by the way the adjustment of stream power is distributed among variables. Furthermore, the scale parameters are variant, depending on the channel hydraulics. This observation should be helpful with regionalization of scale factors.

20.3 At-a-Station Hydraulic Geometry Relations

The at-a-station stream behavior can change significantly over a short period of time as a result of irregular or more gradual adjustment of channel form to flow conditions. Boundary geology, geomorphology, and discharge have dominant influences among all independent variables. The exponents and coefficients of at-a-station hydraulic geometry relations of Equations (20.1a and b) vary with time as well as from high flow range to low flow range. The variations occur because the influx of water and sediment and the boundary conditions (called constraints) that the river is subjected to vary with time. This means that for varying influx of water and sediment, a channel at a given station may exhibit a range of exponents and scale factors in hydraulic geometry relations in response to the constraints imposed on the channel. It is these constraints that force the channel to adjust its allowable hydraulic variables. For example, if a river section has a pavement at the bottom, then it cannot adjust its depth and is, therefore, left to adjust other variables, such as width. Likewise, if the section is lined, then it cannot adjust its width and friction.

Using variations in at-a-station hydraulic geometry, Knighton (1974) explained interactions of measurable hydraulic variables with changes in discharge and modification in channel form at systematically selected stations in a single river system. The rate of width adjustment is influenced by the cohesiveness of the bank material, and through its effect on the energy loss it, in turn, influences the rate of change of the mean velocity. Resistance decreases rapidly with increasing discharge where the grain roughness is the dominant element but is less dominant where channel curvature and form roughness exert additional effects that persist for greater increases of depth. The rate of velocity increase varies accordingly. The degree of intersectional variation in hydraulic relations casts doubts on the validity of defining a mean value. The variation is systematically related to channel pattern, straight reaches being distinguishable from meander and braided reaches in terms of the rates of change of width, velocity, and resistance and slope. Pools and riffles may add a further distinction. Based on the principles of maximum entropy and minimum energy dissipation rate, a family of hydraulic geometry relations is obtained. This family may encompass many of the hydraulic geometry relations derived using different hypotheses.

Langbein (1964) hypothesized that when a channel adjusts its hydraulic variables corresponding to the equilibrium state, the adjustment is shared equally among the hydraulic variables. For varying discharge morphological relations involve five hydraulic variables: width, depth, velocity, friction, and slope. In practice, the adjustment among hydraulic variables will be unequal. Therefore, it can be hypothesized that hydraulic geometry relations will depend on the adjustment of hydraulic variables and the adjustment can explain the variability in the parameters (scale and exponents) of these relations or the variation in at-a-station hydraulic geometry.

Following Yang (1972), the temporal rate of adjustment of SP in a river cross-section, R_s, can be expressed as

$$R_S = \frac{d(SP)}{dt} = \frac{d(Q\gamma S)}{dt},$$
(20.25)

where t is time. Using Equations (20.4e) and (20.6), the stream power of a channel is expressed as

$$SP = \gamma \alpha w d^{\beta} S^{3/2}. \tag{20.26}$$

On the right side of Equation (20.26), there are four variables: w, d, α, and S. When discharge changes, a river cross-section can adjust its width, depth, velocity, roughness, and slope or a combination thereof. Thus, it is hypothesized that for time-varying influxes of discharge the channel cross-section will adjust or minimize its stream power by adjusting these four variables. Substitution of Equation (20.26) in Equation (20.25) yields

$$\frac{d(SP)}{dt} = R_S = \gamma \alpha d^{\beta} S^{3/2} \frac{dw}{dt} + \gamma \alpha \beta w d^{\beta-1} S^{3/2} \frac{dd}{dt} + \gamma w d^{\beta} S^{3/2} \frac{d\alpha}{dt} + \frac{3}{2} \gamma \alpha w d^{\beta} S^{1/2} \frac{dS}{dt}. \tag{20.27}$$

Equation (20.27) expresses the change in stream power in time. It is constituted by four parts, designated as R_1, R_2, R_3, and R_4, and can be recast as

$$\frac{d(SP)}{dt} = R_S = R_1 + R_2 + R_3 + R_4, \tag{20.28}$$

where R_i, $i = 1, 2, 3$, and 4, is defined as

$$R_1 = \gamma \alpha d^{\beta} S^{3/2} \frac{dw}{dt}; \quad R_2 = \gamma \alpha \beta w d^{\beta-1} S^{3/2} \frac{dd}{dt}; \quad R_3 = \gamma w d^{\beta} S^{3/2} \frac{d\alpha}{dt};$$

$$R_4 = \frac{3}{2} \gamma \alpha w d^{\beta} S^{1/2} \frac{dS}{dt}.$$

R_1 expresses the temporal adjustment of width, R_2 the temporal adjustment of depth, R_3 temporal adjustment of friction, and R_4 temporal adjustment of slope. Dividing by the total rate of adjustment SP, R_S, one gets

$$P_w = \frac{R_1}{R_S} = \frac{\gamma \alpha d^{\beta} S^{3/2} dw/dt}{d(SP)/dt} \tag{20.29}$$

$$P_d = \frac{R_2}{R_S} = \frac{\gamma \alpha \beta w d^{\beta-1} S^{3/2} dd/dt}{d(SP)/dt} \tag{20.30}$$

$$P_{\alpha} = \frac{R_3}{R_S} = \frac{\gamma w d^{\beta} S^{3/2} d\alpha/dt}{d(SP)/dt} \tag{20.31}$$

$$P_S = \frac{R_4}{R_S} = \frac{3}{2} \frac{\gamma \alpha w d^{\beta} S^{1/2} dS/dt}{d(SP)/dt}. \tag{20.32}$$

In Equations (20.29) to (20.32), respectively, P_w can be interpreted as the proportion of the temporal change of *SP* due to the temporal rate of adjustment of width, P_d the proportion of the temporal change of *SP* due to the temporal rate of adjustment of depth, P_α the proportion of the temporal change of *SP* due to the temporal rate of adjustment of friction, and P_S the proportion of the temporal change of *SP* due to the temporal rate of adjustment of slope.

According to the principle of maximum entropy (Jaynes, 1957), any system in equilibrium state under steady constraints tends to maximize its entropy. When a river cross-section reaches a dynamic (or quasi-dynamic) equilibrium, the entropy should attain its maximum value. The principle of maximum entropy (POME) states that the entropy of a system is maximum when all probabilities are equal, that is, the probability distribution is uniform. Application of this principle to a river cross-section in dynamic equilibrium, Equation (20.28), therefore, suggests:

$$P_w = P_d = P_\alpha = P_S. \tag{20.33}$$

Equation (20.33) holds, of course, under the stipulation that there are no constraints imposed on the channel section and can be interpreted to mean that the self-adjustment of *SP* is equally shared among *w*, *d*, *α*, and *S*. This interpretation is supported by Williams (1967, 1978), who found from an analysis of data from 165 gaging stations in the United States that a channel cross-section adjusted all its hydraulic parameters (*w*, *d*, *S*, and *v*) in response to changes in the influx of water and sediment and that self-adjustments were realized in an evenly distributed manner among factors.

Equation (20.33) involves probabilities of four variables, meaning that any adjustment in hydraulic variables in combinations of two, three, or four may occur. These give rise to different configurations of adjustment, which do indeed occur in nature (Wolman, 1955). Thus the equality among four probabilities yields eleven possibilities and hence leads to eleven sets of equations: (1) $P_w = P_d$, (2) $P_w = P_\alpha$, (3) $P_w = P_S$, (4) $P_d = P_\alpha$, (5) $P_d = P_S$, (6) $P_\alpha = P_S$, (7) $P_w = P_d = P_\alpha$, (8) $P_w = P_\alpha = P_S$, (9) $P_w = P_d = P_S$, (10) $P_d = P_\alpha = P_S$, and (11) $P_w = P_d = P_\alpha = P_S$.

It should be noted that all eleven possibilities can occur in the same river cross-section at different times, or in different river cross-sections at the same time or at different times. These different possibilities reflect the dynamic and morphological responses of channels to changing discharge and sediment influx. Considering velocity, depth, shear, resistance, and stream power as independent variables for investigating the minimum variance theory, Williams (1978) explored 11 cases that are similar in spirit to the 11 possibilities. For grouping channels based on such response, Rhodes (1977) partitioned the *b-m-f* diagram into 10 areas based on five lines representing constant values of (1) width/depth ratio ($b = f$), (2) competence ($m = f$), (3) Froude number ($m = f/2$), (4) velocity/cross-sectional area ratio ($m = b + f$), and (5) slope/roughness ratio ($m = 2f/3$). The channels represented by the points in the same area respond similarly to changes in discharge regardless of the specific values of the exponents. Discounting possibility (11) (the most general one), the 10 possibilities have interestingly some relationships to the 10 areas of Rhodes (1977).

Field observations by Leopold and Maddock (1953), Wolman (1955), among others show that the variation in slope S is small compared with velocity, depth, width, and roughness. Therefore, it may be assumed as constant. Under this assumption, Equation (20.33) reduces to

$$P_w = P_d = P_a. \tag{20.34}$$

Equation (20.34) suggests that adjustment in hydraulic variables will occur in four possible configurations: (1) $P_w = P_d$, (2) $P_w = P_a$, (3) $P_d = P_a$, and (4) $P_w = P_d = P_a$. This constitutes a special case of the general case presented earlier. In other words, configurations 1 and 2 are the same as possibilities 1 and 2, configuration 3 is the same as possibility 4, and configuration 4 is the same as possibility 7 of the general case. These possibilities can also be visualized from physical reasoning. Width and depth are related to the energy expenditure within a channel and are related to boundary sediment and sediment discharge (Maddock, 1969; Colby, 1961). The width–depth ratio is closely related to sediment transport (Schumm, 1968) and to boundary sediments (Schumm, 1960), especially for non-cohesive bed sediments (Riley, 1978). This ratio is also used as a measure of channel shape. Hydraulic radius or flow depth for wide rectangular channels is a measure of hydraulic efficiency. Channel slope is controlled by bed material strength and flow impinging force. Breaks in channel slope along a cross-section reflect channel boundary roughness, channel size, and geometric shape. Channel roughness is a function of bed form and grain size constituting the bed and determines the energy loss. Different possibilities reflect the variations in these channel characteristics in response to the boundary conditions and the influx of flow and sediment discharge. Riley (1978) divided 19 variables into four groups, which can be related to four possibilities in the special case enumerated earlier.

20.3.1 Morphological Equations

Morphological equations, reflecting the adjustment of hydraulic variables for accommodating the temporal adjustment of stream power, arise when only two hydraulic variables are considered at a time. This leads to six possibilities and the corresponding number of primary morphological equations, which are derived in this section. In general, it may be noted that the shape and longitudinal profile of a river in quasi-equilibrium are determined by the discharge and sediment concentration and the characteristics of the bank material. The initial size, shape, and resistance of the material provided by the river depends on the sediment concentration (Wolman, 1955). A river exhibits such local variations as being alternately wide and narrow, straight and meandering, deep and shallow. River cross-sections tend to be approximately semi-elliptical, trapezoidal, or triangular, and increasing discharge results in increased velocity, depth, and width. It is interesting to note that Williams (1978) grouped 165 at-a-station cross-sections into five classes, depending on the bank firmness. In all five cases the slope was considered constant. It may emphasize that the form of equations will be similar if the adjustment of stream power were considered in the spatial domain.

Possibility 1: $P_w = P_d$: Here P_w is given by Equation (20.29) and P_d by Equation (20.30). Equating these two equations, one obtains

$$\frac{dd}{dt} = \frac{1}{\beta} \frac{d}{w} \frac{dw}{dt}. \tag{20.35a}$$

Equation (20.35a) hypothesizes that the temporal change in stream power is accomplished by an equal temporal adjustment between flow depth d and flow width w. This possibility occurs when the cross-section has a fixed roughness and flow is controlled by width and depth. This hypothesis can be considered as a limiting case and will hold under the equilibrium condition. Such a condition is, however, seldom achieved, and therefore temporal change in stream power is accomplished by an unequal adjustment between w and d. To account for the proportion in which the adjustment of stream power is shared between d and w, Equation (20.35a) is modified by introducing a weighting factor, r, $r \geq 0$:

$$\frac{dd}{dt} = \frac{r}{\beta} \frac{d}{w} \frac{dw}{dt}. \tag{20.35b}$$

Integration of Equation (20.35b) yields

$$\frac{w}{d^{\beta/r}} = C_{wd} \quad \text{or} \quad d = C_{wd*} w^{r/\beta}, \tag{20.36a}$$

where C_{wd} or C_{wd*} is a constant of integration. For the limiting case, $r = 1$, Equation (20.36a) reduces to

$$w = C_{wd} d^{\beta} \quad \text{or} \quad d = C_{wd*} w^{1/\beta}. \tag{20.36b}$$

Parameter C_{wd} or C_{wd*} can be designated as a primary morphological coefficient and Equation (20.36a) or (20.36b) as a primary morphological equation.

20.3.2 Derivation of At-a-Station Hydraulic Geometry Relations

Substitution of morphological Equations (20.36a) in Equation (20.5) leads to equations for at-a-station hydraulic geometry. These equations are shown for one possibility and they can be derived under the remaining ten possibilities in a similar manner.

Possibility I: Hydraulic Geometry Relations for Width, Depth and Velocity:
$P_w = P_d$, $w = C_{wd}d^{\beta/r}$ or $d = C_{wd^*}w^{r/\beta}$.

The general morphological equation for this possibility is given by Equation (20.36a) or the special form by Equation (20.36b). If Manning's Equation (20.4a) is employed, then with $\alpha = 1/n$ and $\beta = 5/3$:

$$w = C_w Q^{\frac{1}{1+r}}, \quad C_w = \left(\frac{n(C_{wd})^r}{S^{1/2}}\right)^{\frac{1}{1+r}} \tag{20.37}$$

$$d = C_d Q^{\frac{3r}{(1+r)5}}, \quad C_d = \left(\frac{n}{C_{wd}S^{1/2}}\right)^{\frac{3r}{(1+r)5}} \tag{20.38}$$

$$v = C_v Q^{\frac{2r}{(1+r)5}}, \quad C_v = \left(\frac{1}{n}\right)^{\frac{5+3r}{5(1+r)}}\left(\frac{1}{C_{wd}}\right)^{\frac{2r}{(1+r)5}} S^{\frac{5+3r}{10(1+r)}}. \tag{20.39}$$

For the special case, $r = 1$, Equations (20.37) to (20.39) become

$$w = C_w Q^{\frac{1}{2}}, \quad C_w = \left(\frac{nC_{wd}}{S^{1/2}}\right)^{\frac{1}{2}} \tag{20.40}$$

$$d = C_d Q^{\frac{3}{10}}, \quad C_d = \left(\frac{n}{C_{wd}S^{1/2}}\right)^{\frac{3}{10}} \tag{20.41}$$

$$v = C_v Q^{\frac{2}{10}}, \quad C_v = \left(\frac{1}{n}\right)^{\frac{4}{5}}\left(\frac{1}{C_{wd}}\right)^{\frac{1}{5}} S^{\frac{2}{5}}. \tag{20.42}$$

This is perhaps the most investigated and frequently occurring possibility. Equations (20.37) to (20.39) show that the channel width varies with discharge raised to the power $b = 1/(1+r)$, the depth varies with discharge raised to the power $f = 3r/[5(1+r)]$, and velocity varies with discharge raised to the power $m = 2r/[5(1+r)]$. The values of exponents b, f, and m depend on the weighting factor r, which specifies the proportion for adjustment of stream power between w, d, and v. When the weighting factor r tends to 0, b, f, and m, respectively, take on 1, 0, and 0, and when r tends to ∞, b, f, and m take on, respectively, 0, 3/5, and 2/5 [see Equations (20.37)–(20.39)]. For the special case $r = 1$, where the adjustment is equally proportioned, $b = 0.5, f = 3/10$ and $m = 1/5$. Scale factors, C_w, C_d, and C_v vary with flow resistance and slope. Thus, the derived equations show that the values of exponents, b, f, and m, do not possess fixed values; rather, they vary over certain ranges dictated by the way the adjustment of stream power is distributed among variables.

For the at-a-station hydraulic geometry of 139 streams from a variety of environments, Park (1977) reported the range of b as 0.0 to 0.59 with modal class as 0.0 to 0.1; the range of f as 0.06 to 0.73 with modal class as 0.3 to 0.4; and the range of m as 0.07 to 0.71 with modal class as 0.4 to 0.5. Thus, the derived exponents are in the reported ranges.

To facilitate discussion, the values of exponents, b, f, m, p, and y for three cases (one special and two limiting cases) when the weighting factors are zero, unity, and infinity. It should be pointed out that the limiting case of infinity is only a theoretically generalized case for the factors r, B, and J, respectively, for the lack of knowledge of the values of their upper limits, which should be far less than infinity. In the discussion that follows, only Manning's equation is considered for economy of space.

References

Cheema, M. N., Marino, M. A., and DeVries, J. J. (1997). Stable width of an alluvial channel. *Journal of Irrigation & Drainage Engineering*, Vol. 123, No. 1, pp. 55–61.

Colby, B. R. (1961). *The effect of depth of flow on discharge of bed material*. U.S. Geological Survey Water Supply Paper 1498-D, Washington, DC.

Deng, Z. and Zhang, K. (1994). Morphologic equations based on the principle of maximum entropy. *International Journal of Sedimentaion Research*, Vol. 9, No. 1, pp. 31–46.

Jaynes, E. T. (1957). Information theory and statistical mechanics, I. *Physical Review*, Vol. 106, pp. 620–630.

Knighton, A. D. (1974). Variation in width–discharge relation and some implications for hydraulic geometry. *Geological Society of America Bulletin*, Vol. 85, pp. 1069–1076.

Langbein, W. B. (1964). Geometry of river channels. *Journal of Hydraulics Division, ASCE*, Vol. 90, No. HY2, pp. 301–311.

Leopold, L. B. and Maddock, T. J. (1953). *Hydraulic geometry of stream channels and some physiographic implications*. U.S. Geol. Survey Prof. Paper 252, pp. 55, Washington, DC.

Leopold, L. B. and Wolman, L. B. (1957). *River channel patterns: braided, meandering and straight, U.S. Geol.* Survey Professional Paper 282-B, U.S. Government Printing Office, Washington, DC.

Maddock, T. (1969). *The behavior of straight open channels with movable beds*. U.S. Geological Survey Professional Paper 622-A, Washington, DC.

Park, C. C. (1977). World-wide variations in hydraulic geometry exponents of stream channels: An analysis and some observations. *Journal of Hydrology*, Vol. 33, pp. 133–146.

Rhodes, D. D. (1977). The b-f-m diagram: Graphical representation and interpretation of at-a-station hydraulic geometry. *American Journal of Science*, Vol. 277, pp. 73–96.

Riley, S. J. (1978). The role of minimum variance theory in defining the regime characteristics of the lower Namoi-Gwydir basin. *Water Resources Bulletin*, Vol. 14, pp. 1–11.

Schumm, S. A. (1960). *The shape of alluvial channels in relation to sediment type.* U.S. Geological Survey Professional Paper 352B, pp. 17–30, Washington, DC.

Schumm, S. A. (1968). *River adjustment to altered hydrologic regime-Murrumbridgee River and Paleo channels, Australia.* U.S. Geological Survey Professional Paper 598, Washington, DC.

Stall, J. B. and Fok, Y.-S. (1968). *Hydraulic geometry of Illinois streams. University of Illinois Water Resources Center*, Research Report No. 15, pp. 47, Urbana.

Williams, G. P. (1967). *Flume experiments on the transport of a coarse sand, U.S. Geological Survey Professional Paper 562-B*, Washington, DC.

Williams, G. P. (1978). *Hydraulic geometry of river cross-sections-Theory of minimum variance.* U.S. Geological Survey Professional Paper 1029, Washington, DC.

Wolman, M. G. (1955). *The natural channel of Brandywine Creek, Pennsylvania.* U.S. Geological Survey Professional Paper 271, Washington, DC.

Yang, C. T. (1972). Unit stream power and sediment transport. *Journal of Hydraulics Division, ASCE*, Vol. 98, No. HY10, pp. 1805–1826.

Yang, C. T. (1986). Dynamic adjustment of rivers. *Proceedings of the 3rd International Symposium on River Sedimentation*, Jackson, MS, pp. 118–132.

Yang, C. T. (1996), *Sediment Transport Theory and Practice*, McGraw-Hill Comp., Inc., New York.

21

Theory of Stream Power

Whenever there is flow of water in an alluvial channel, the water entrains and transports sediment. The entrainment and transport of sediment require work that flow must perform, and to perform the work requires energy. The water has potential energy due to elevation, which is transformed to kinetic energy, part of which is utilized to do this work and part is dissipated to overcome boundary and bed friction. The time rate of potential energy expenditure is the stream power, which plays a fundamental role in the evolution of a fluvial system. Using the theory of stream power, this chapter derives constraints for hydraulic geometry.

Notation

a = constant of proportionality or coefficient

A = cross-sectional area

a_d = coefficient

a_1 = parameter

a_2 = parameter

b = exponent

b_0 = parameter

b_1 = parameter

c = constant of proportionality or coefficient

C = resistance coefficient

C_a = coefficient

C_j = coefficient

C_k = coefficient

C_v = a constant

C_e = exponent

C_* = basic regime coefficient

---

d = flow depth
D = sediment size
f = exponent
f_d = Darcy–Weisbach friction factor
g = acceleration due to gravity
j = exponent
k = constant of proportionality or coefficient
K = large bed element size
L = distance from the drainage divide
m = exponent
p = exponent
P = wetted perimeter
Q = discharge
R = hydraulic radius
S = energy gradient, commonly approximated as bed slope or slope of the water surface
S_0 = initial channel slope upstream
SP_{cr} = critical shear power of unit channel length
SF_{cs} = cross-sectional mean shear force
SP = Stream power per unit length
SP_m = Stream power per unit wetted area of the bed or mean stream power
v = mean flow velocity
w = channel width
x = distance along the flow direction
γ = specific weight of the fluid
γ_s = specific weight of sediment
ρ = density of the fluid (water-sediment mixture)
τ = mean boundary shear stress ($= \gamma RS$)
τ_{cr} = mean critical stress
Θ_{cr} = shields critical stress

21.1 Introduction

Stream power refers to the time rate of potential energy expenditure. The water possesses potential energy due to elevation. When flowing downstream, it performs work for entraining and transporting sediment by converting the potential energy to kinetic energy, most of which is dissipated to overcome boundary and bed friction, and the remaining kinetic energy is utilized to erode and transport sediment or doing the geomorphic work. Thus, there is the force–resistance interplay, and the stream morphology responds to the balance between the force of flowing water and the resistance of the channel bed and bank sediment to the movement of water. A number of elements of the fluvial system are influenced by stream power, such as sediment transport (Bagnold, 1966), channel pattern (Schumm and Khan, 1972), aggradation/degradation thresholds (Bull, 1979), channel shape (Mosley, 1981), channel migration (Hickin and Nanson, 1984), riffle and pool characteristics (Wohl et al., 1993), and sediment delivery ratio (Phillips, 1989). This chapter discusses the theory of stream power for deriving constraints for hydraulic geometry.

21.2 Definition

Stream power (*SP*) per unit length can be expressed as

$$SP = \gamma QS, \tag{21.1}$$

where Q is discharge, γ is the specific weight of the fluid $= \rho g$, ρ is the density of the fluid (water-sediment mixture), g is the acceleration due to gravity, and S is the energy gradient commonly approximated as bed slope or slope of the water surface. *SP* is also referred to as cross-sectional stream power and can be regarded as a physically based index of sediment transport capacity of flow. *SP* has units of N/s (Newton per second) with dimensions of kg-m/s^3. *SP* per unit area is the energy per unit time expended in overcoming friction per unit length. This describes the force applied by a mass of moving water to the channel per unit time.

Stream power is also defined as the stream power per unit wetted area of the bed and is referred to as mean stream power. Thus, from Equation (21.1), one can write

$$SP_m = \frac{SP}{w} = \gamma RSv = \tau v = \gamma dSv, \tag{21.2}$$

where R is the hydraulic radius (A/P, A = cross-sectional areas, and P = wetted perimeter), approximated for rivers by flow depth d; w is the channel width, v is the mean flow velocity, and τ is the mean boundary shear stress ($= \gamma RS$). SP_m has the units of N/m/s (Newton per meter per second) with dimensions of kg/s^3. SP_m is the energy expended per unit time per unit stream bed area. This describes the force exerted by a mass of water over unit length of river reach per unit of time. It can be applied to determine if the stream is aggrading (sediment accumulation) or degrading (erosion).

In physics, power is defined as the product of specific weight, discharge, and head and a constant for unit conversion. This formula is often used for computing mechanical power measured in terms of horsepower or for calculating hydroelectric power measured in terms of kilowatts. As force times distance per unit time, it is essentially an expression of energy expenditure per unit time. If applied to a stream, then head will be the elevation difference over a given length.

21.3 Two Postulates

For deriving hydraulic geometry relations, two postulates are made. The first postulate states that along a channel within a river system the power expended per unit stream bed area tends to be constant along the river length. This implies that the stream power is equally distributed per unit area. Conceptually, the width of the stream is being used as an index of the work being done in eroding the stream banks and widening the channel.

The second postulate states that the stream power expended per unit length of the channel tends to be the same along the river length. This implies that the stream power is equally distributed per unit length.

For equal stream power distribution per unit length or per unit area, the implication is that the river system will transport the maximum possible water and sediment at each section with the least expenditure of power, which also implies the least amount of work to be performed.

21.4 Constraints for Hydraulic Geometry

From Chapter 4, the hydraulic geometry relation are expressed as

$$w = aQ^b \tag{21.3}$$

$$d = cQ^f \tag{21.4}$$

$$v = kQ^m, \tag{21.5}$$

where Q is the discharge; w is the water-surface width; d is the mean depth; v is the mean velocity, and b, f, and m are the exponents; and a, c, and k are the constants of proportionality or coefficients. To these relations one can also add

$$A = C_a Q^{C_e} \tag{21.6}$$

and

$$S = C_j Q^j, \tag{21.7}$$

where A is the cross-sectional area; C_a and C_j are the coefficients; and C_e and j are the exponents. For stream power to be equally distributed per unit length or to be constant per unit area along the stream, $SP = $ constant. Since Q varies along the stream length, one can write

$$\frac{dSP}{dx} = \frac{\partial SP}{\partial Q}\frac{\partial Q}{\partial x} = 0. \tag{21.8}$$

In Equation (21.8), $\partial Q/\partial x \neq 0$ and $\partial SP/\partial Q = 0$. This implies that the minimization of stream power is synonymous with the equalization of stream power throughout the length of the stream. Thus, with the use of Equation (21.7), we can express the change in stream power with the change in discharge as

$$\frac{\partial SP}{\partial Q} = \frac{\partial(\gamma QC_jQ^j)}{\partial Q} = 0. \tag{21.9}$$

Equation (21.9) yields the constraint:

$$j + 1 = 0 \quad \text{or} \quad j = -1. \tag{21.10}$$

Equation (21.10) states the condition that must be satisfied in order for the equal distribution of stream power to be satisfied or the stream power to be constant along the stream length.

Now consider the stream power per unit stream bed area. For constant stream power per unit stream bed area along the river length, one can write

$$\frac{dSP_m}{dx} = \frac{\partial SP_m}{\partial Q}\frac{\partial Q}{\partial x} = 0. \tag{21.11}$$

In Equation (21.11), $\partial Q/\partial x \neq 0$ and $\partial SP_m/\partial Q = 0$. This implies that the minimization of stream power per unit width is synonymous with the equalization of stream power per unit width throughout the length of the stream. Thus,

$$\frac{\partial SP_m}{\partial Q} = \frac{\partial(\gamma QS)}{w\partial Q} = 0. \tag{21.12}$$

With the use of Equations (21.3) and (21.7), we can express

$$\frac{\partial(\gamma QS/w)}{\partial Q} = \frac{\partial(\gamma QC_jQ^j/aQ^b)}{\partial Q} = 0. \tag{21.13}$$

Equation (21.13) yields the constraint:

$$1 + j - b = 0 \tag{21.14}$$

or

$$j = b - 1. \tag{21.15}$$

Equation (21.15) states the condition that must be satisfied in order for the equal distribution of stream power per unit stream bed area to be satisfied.

Looking at Equations (21.10) and (21.15), it is interesting to observe that both equations cannot be satisfied at the same time. If one condition is satisfied the other cannot. With the channel slope decreasing in the downstream direction, the stream bed profile will be concave upward and the value of j will be negative. Leopold et al. (1964) observed the bed profile to be slightly concave for constant SP_m and highly concave for constant SP. They also noted that both conditions could be satisfied simultaneously. To resolve this conflict, it may be reasonable to assume that the bed profile occupies a position between the two extremes.

To further examine, sediment size plays a fundamental role in channel evolution and shaping the channel. Large bed elements are critical in the channel evolution. Let K denote the large bed element size and let it be related to discharge through a power function:

$$K = C_k Q^p, \tag{21.16}$$

where C_k is the coefficient and p is the exponent. It can now be surmised that the bed material size K remaining in the stream can be regarded as an index of the material size being eroded.

Two possibilities can now be explored. First, the stream power per unit length per unit of large bed sediment size (or erosion) should be constant along the stream, that is,

$$\frac{SP}{K} = \frac{\gamma Q S}{K} = \text{constant}. \tag{21.17}$$

Following the same line of reasoning for the minimization of stream power, one can express

$$\frac{d\left(\frac{SP}{K}\right)}{dx} = \frac{\partial\left(\frac{SP}{K}\right)}{\partial Q}\frac{\partial Q}{\partial x} = 0. \tag{21.18}$$

In Equation (21.18), $\partial Q/\partial x \neq 0$ and $\partial(SP/K)/\partial Q = 0$. This implies that the minimization of stream power is synonymous with the equalization of stream power per unit of large bed sediment size (or erosion) throughout the length of the stream. Thus, with the use of Equation (21.16), we can express

$$\frac{\partial\left(\frac{SP}{K}\right)}{\partial Q} = \frac{\partial\left[\gamma Q C_j Q^j/(C_k Q^p)\right]}{\partial Q} = 0. \tag{21.19}$$

Equation (21.19) yields the constraint:

$$1 + j - p = 0. \tag{21.20}$$

Second, the stream power per unit stream bed area per unit of erosion (large bed element size) is constant along the stream, that is,

$$\frac{SP_m}{K} = \frac{\gamma QS/w}{K} = \text{constant.} \tag{21.21}$$

For the minimization of mean stream power, one can express

$$\frac{d\left(\frac{SP_m}{K}\right)}{dx} = \frac{\partial\left(\frac{SP_m}{K}\right)}{\partial Q}\frac{\partial Q}{\partial x} = 0. \tag{21.22}$$

In Equation (21.22), $\partial Q/\partial x \neq 0$ and $\partial(SP_m/K)/\partial Q = 0$. This implies that the minimization of mean stream power is synonymous with the equalization of mean stream power per unit of erosion throughout the length of the stream. Thus, with the use of Equation (21.16), we can express

$$\frac{\partial\left(\frac{SP_m}{K}\right)}{\partial Q} = \frac{\partial\{\gamma QC_jQ^j/[(C_kQ^p)(aQ^b)]\}}{\partial Q} = 0. \tag{21.23}$$

Equation (21.23) yields the constraints:

$$1 + j - b - p = 0. \tag{21.24}$$

Looking at Equations (21.20) and (21.24) it appears that both these equations cannot be satisfied at the same time.

In order to establish the relationship between the exponents of hydraulic geometry equations, a hydraulic constraint can be invoked through a hydraulic resistance equation. It is noted that the velocity, depth, bed size, and slope are interrelated. The Chezy equation can be used for the resistance relationship:

$$v = \frac{C}{\sqrt{g}}\sqrt{RS}, \tag{21.25}$$

in which C is the resistance coefficient having the dimensions of $\sqrt{length}/time$, and g is the acceleration due to gravity. Hence, $\frac{C}{\sqrt{g}}$ is dimensionless, which Judd (1963) expressed in the large bed element range as a power function of relative roughness d/K:

$$\frac{C}{\sqrt{g}} \propto \left(\frac{d}{K}\right)^{a_d}. \tag{21.26}$$

Using Equation (21.26), Equation (21.25) can be written for wide streams $(R \sim d)$ as

$$v \sim \left(\frac{d}{K}\right)^{a_d}\sqrt{dS} \tag{21.27}$$

$$\text{or} \quad v = C_v \left(\frac{d}{K}\right)^{a_d} \sqrt{dS}, \tag{21.28}$$

in which C_v is a constant. Inserting Equations (21.4) for d, Equation (21.5) for v, Equation (21.7) for S, and Equation (21.16) for K in Equation (21.28), the result is

$$kQ^m = C_v \left(\frac{cQ^f}{C_k Q^p}\right)^{a_d} \sqrt{cQ^f C_j Q^j}. \tag{21.29}$$

Equating the exponents on both sides of Equation (21.29), we obtain the constraint:

$$m = (a_d + 0.5)f + 0.5j - a_d p. \tag{21.30}$$

Equation (21.30) defines the velocity constraint to be satisfied.

Now recall the continuity equation

$$Q = Av = wdv. \tag{21.31}$$

Substituting Equations (21.3) for w, Equation (21.4) for d, and Equation (21.5) for v, the result is

$$Q = aQ^b cQ^f kQ^m. \tag{21.32}$$

Equating the powers of Q on both sides of Equation (21.32), we get the constraint:

$$b + f + m = 1. \tag{21.33}$$

Equation (21.33) defines the constraint for the continuity equation to be satisfied. Replacing m in Equation (21.33) with Equation (21.30), the result is

$$(a_d + 1.5)f + 0.5j - a_d p + b = 1. \tag{21.34}$$

Equation (21.34) defines another constraint to be satisfied. Thus, there are two sets of constraints for stream power minimization criteria: one set without hydraulic constraint and the other with hydraulic constraint, as shown in Table 21.1.

For wide gravel bed rivers with no bed movement and stable banks, Rogers and Thomas (1969) expressed the flow equation as

$$v = 8.2 \left(\frac{d}{K}\right)^{\frac{1}{6}} \sqrt{gdS}. \tag{21.35}$$

Equation (21.35) indicates the value of a_d as 1/6. Judd (1963) reported a value of $a_d = 1/3$. For channels of very large relative roughness ($K/R) > 0.1$, such as cobble and boulder rivers, Ackers (1958) expressed the flow equation as

Table 21.1. *Criteria for stream power minimization*

Condition to be satisfied	Criteria with no hydraulic constraint	Criteria with hydraulic constraint
Continuity equation	$b + f + m = 1$	$(a_d + 1.5)f + 0.5j - a_d p + b = 1$
Stream power per unit length (SP)	$1 + j = 0$	$(a_d + 1.5)f + 1.5j - a_d p + b = 0$
Stream power per unit stream bed area (SP_m) per unit length	$1 + j - b = 0$	$(a_d + 1.5)f + 1.5j - a_d p = 0$
Stream power per unit stream length per large bed element size (SP/K)	$1 + j - p = 0$	$(a_d + 1.5)f + 1.5j - a_d p - p + b = 0$
Stream power per unit stream length per unit stream be area per large bed element size (SP_m/K)	$1 + j - b - p = 0$	$(a_d + 1.5)f + 1.5j - a_d p - p = 0$

$$v = \frac{39.8}{K^{1.4}} R^{\frac{3}{4}} \sqrt{S}. \tag{21.36}$$

Equation (21.36) can be cast as

$$v = \frac{39.8}{K^{1.4}} R^{\frac{1}{4}} \sqrt{dS}. \tag{21.37}$$

For wide streams, R can be replaced by d, and Equation (21.37) can be written as

$$v = 7 \left(\frac{d}{K} \right)^{\frac{1}{4}} \sqrt{gdS}. \tag{21.38}$$

Equation (21.38) gives a value of a_d as ¼. Thus, the range of the a_d values is 1/6 to 1/3.

21.5 Regime River Geometry

If river meandering is not considered then by virtue of continuity a river has three degrees of freedom in width, depth, and slope to adjust its geometry in response to discharge and sediment load it receives from its watershed. This means three independent equations are needed to determine the three unknowns. To that end, sediment discharge and resistance equations have been used. For the third equation, Chang (1979a) proposed the hypothesis of minimum stream power, which states that the necessary and sufficient condition for an alluvial river to reach equilibrium defined by its width, depth, and slope is when the stream power is a minimum subject to constraints defined by discharge and sediment load imposed on it. From the definition of stream power, Q is given, so the minimum stream power will be attained when S becomes minimum. This means that minimum stream power is tantamount to minimum slope. Using this hypothesis together with sediment transport and resistance formulas, Chang (1979a) presented graphical relations for predicting width, depth, and either slope or sediment discharge. Stream power or slope increase as the bed roughness decreases, suggesting a greater value is required to transport the same rates of water and sediment. The channel geometry corresponding to minimum stream power possesses the best hydraulic efficiency for the specified water and sediment rates.

Employing the minimum stream power hypothesis, Chang (1979b) discussed river patterns. Depending on the geometry of a river and its channel slope relative to the valley slope, channel patterns can be straight, meandering, and braided. A straight channel does not follow a sinuous course, and its slope is the same as the valley slope. In such a river, the thalweg wanders back and forth from bank to bank (Leopold et al., 1964). Rivers of this kind are found on very flat slopes or on steep slopes, and are wide and shallow if the bed material is of fairly uniform size and can be straight over long distances.

Meandering channels have slope less than the valley slope and the slope range is between the two slopes of straight channels: greater than flat slopes but less than steep slopes. Meandering channels occurring on relatively flat slopes tend to be more sinuous and have smaller width–depth ratios than those on steep slopes. A meandering channel with multiple configurations has less stream power and sediment load than the straight channel with the same valley slope. A meandering channel is more stable since it corresponds to the minimization of stream power as well as sediment load. A braided channel has a large width–depth ratio, so the flow gets divided through the development of islands or bars. Such channels are usually not sinuous and occur on steep slopes. Braiding is caused by greater supply of sediment than the river can transport, leading to stream-bed aggradation or steep slopes. Leopold and Wolman (1957) reported a line separating braid channels from meandering ones as $S = 0.06Q^{-0.44}$. For a given discharge, meandering will occur on smaller slopes, but at the same slope a braided river will have greater discharge than the meandering one. However, steepness of slope is a sufficient condition for braiding.

21.6 Stream Power and Probability

Regime equations are derived using stream power and Laplace's principle of insufficient reason, which is consistent with the principle of maximum entropy. We follow the work of Cao (1996), which is a special case of the work of Singh et al. (2003a, b) discussed in the previous chapter. Cao and associates (Cao and

Knight, 1995, 1996, 1998a, 1998b) used the entropy approach for design of alluvial channels. Based on the gradient of stream power, it is hypothesized that the probabilities of width adjustment and depth adjustment are equal to each other in a regime channel having uniform material on both the bed and banks. The theory then involves this hypothesis, a suitable resistance relation, continuity equation, and a sediment transport formula.

21.6.1 Governing Equations

The gradient of stream power per unit length in the streamwise direction (x) can be expressed as

$$\frac{dSP}{dx} = \frac{d(\gamma QS)}{dx}. \tag{21.39}$$

The energy slope S (taken as an absolute value) can be expressed using the Darcy–Weisbach relation:

$$S = \frac{f_d v^2}{8gd}, \tag{21.40}$$

where f_d is the Darcy–Weisbach friction factor. The mean velocity can be expressed from the continuity equation:

$$v = \frac{Q}{dw}. \tag{21.41}$$

Inserting Equations (21.40) and (21.41) in Equation (21.39), the result is

$$\frac{dSP}{dx} = \frac{\rho f_d Q^3}{8} \left(\frac{-3}{d^4 w^2} \frac{dd}{dx} + \frac{-2}{d^3 w^3} \frac{dw}{dx} \right). \tag{21.42}$$

Dividing the right side of Equation (21.42) by the left side, we obtain

$$1 = \frac{\frac{\rho f_d Q^3}{8} \frac{-3}{d^4 w^2} \frac{dd}{dx}}{\frac{dSP}{dx}} + \frac{\frac{\rho f_d Q^3}{8} \frac{-2}{d^3 w^3} \frac{dw}{dx}}{\frac{dSP}{dx}} = P_d + P_w, \tag{21.43}$$

where

$$P_d = \frac{\frac{\rho f_d Q^3}{8} \frac{-3}{d^4 w^2} \frac{dd}{dx}}{\frac{dSP}{dx}} \tag{21.44}$$

$$P_w = \frac{\frac{\rho f_d Q^3}{8} \frac{-2}{d^3 w^3} \frac{dw}{dx}}{\frac{dSP}{dx}}. \tag{21.45}$$

Equation (21.42) states that the change in stream power per unit length is accommodated by the depth adjustment and the width adjustment. Equation (21.44) can be interpreted as the probability of depth adjustment and Equation (21.45) as the probability of width adjustment. Invoking the Laplacian principle of insufficient reason with constraints imposed on the channel, the two probabilities are equal. For a regime channel with uniform material on the bed and banks and in a state of dynamic equilibrium (in which scour and deposition are in balance), one can write

$$P_d = P_w. \tag{21.46}$$

Inserting Equations (21.44) and (21.45), the result is

$$\frac{dw}{dx} = \frac{3w}{2d}\frac{dd}{dx}. \tag{21.47}$$

Integration of Equation (21.47) yields

$$\ln w = \ln\left(Cd^{\frac{3}{2}}\right), \tag{21.48}$$

where C is the constant of integration. Equation (21.48) can be written as

$$w = Cd^{\frac{3}{2}} \quad \text{or} \quad \frac{w^{\frac{2}{3}}}{d} = C_*. \tag{21.49}$$

Equation (21.49) can be regarded as the basic regime equation with C_* as the basic regime coefficient. This equation can also be considered as the modified aspect ratio and can be combined with the continuity equation, flow resistance equation, and sediment transport equation to derive regime equations of width, depth, slope, and mean cross-sectional velocity for a stable alluvial channel. It is interesting to note that Ackers (1964) proposed an empirical aspect ratio equation as the third independent equation for deriving stream geometry as

$$\frac{d}{w} = C_1 d^{-\frac{1}{2}}, \tag{21.50}$$

where C_1 is the coefficient relating width to depth. Equation (21.50) can be expressed as

$$\frac{w^{\frac{2}{3}}}{d} = \frac{1}{C_1^{\frac{2}{3}}}. \tag{21.51}$$

Comparison of Equations (21.49) and (21.51) yields

$$C_* = \frac{1}{C_1^{\frac{2}{3}}}. \tag{21.52}$$

21.6.2 Derivation of Regime Equations

Combining the continuity equation and the Darcy–Weisbach equation, we obtain

$$Q = \sqrt{\frac{8gdS}{f_d}}dw. \tag{21.53}$$

Coupling Equations (21.53) and (21.49), the result is

$$w = C_w Q^{\frac{1}{2}}S^{-\frac{1}{4}} \tag{21.54}$$

$$d = C_d Q^{\frac{1}{3}}S^{-\frac{1}{6}}, \tag{21.55}$$

where

$$C_w = \left(\frac{C_*^3 f_d}{8g}\right)^{\frac{1}{4}} = C^{-\frac{1}{2}}C_*^{\frac{3}{4}}, \tag{21.56}$$

in which $C = \left(\frac{8g}{f_d}\right)^{\frac{1}{2}} =$ Chezy resistance coefficient. Substituting Equations (21.53) and (21.55) into the continuity Equation (21.41) one gets

$$v = C_v S^{\frac{5}{12}}Q^{\frac{1}{6}}, \tag{21.57}$$

in which

$$C_v = \left(\frac{8g}{f_d C_*^{\frac{3}{5}}}\right)^{\frac{5}{12}} = C^{\frac{5}{6}}C_*^{\frac{-1}{4}}. \tag{21.58}$$

For slope regime equation, the Engelund and Hansen (1972) sediment transport equation can be employed:

$$f_E \Phi = 0.1(\Theta)^{\frac{5}{2}}, \tag{21.59}$$

where

$$f_E = \left(\frac{2gdS}{v^2}\right) \tag{21.60}$$

$$\Phi = \frac{q_s}{\gamma_s\left[(r-1)gD^3\right]^{\frac{1}{2}}} \tag{21.61}$$

$$\Theta = \left[\frac{\tau_b}{(\gamma_s - \gamma)D}\right]^{\frac{5}{2}}. \tag{21.62}$$

$$\tau_b = \rho gdS. \tag{21.63}$$

Comparing Equation (21.60) with Darcy–Weisbach equation,

$$f_E = \frac{f_d}{4}$$ (21.64)

Then, Equation (21.59) can be written as

$$f_d \Phi = 0.4(\Theta)^{\frac{5}{2}}.$$ (21.65)

Inserting Equation (21.60) in Equation (21.63),

$$S = C_s Q^{-\frac{2}{5}},$$ (21.66)

where

$$C_s = (2.5 f_d \Phi)^{0.48} \left[\frac{(r-1)D}{C_d} \right]^{\frac{1}{2}}.$$ (21.67)

21.6.3 Basic Regime Coefficient

The regime coefficient C_* can be derived in two ways (Cao, 1996): (1) regime empirical equations and (2) stream power methodology. Lacey's regime equations for channels with width much greater than flow depth can be expressed as

$$w = 4.84 Q^{0.5}$$ (21.68)

$$d = \left(\frac{Q}{9.15 f_l} \right)^{\frac{1}{3}}.$$ (21.69)

Equations (21.68) and (21.69) can be couple by eliminating Q as

$$\frac{w^{\frac{2}{3}}}{d} = C_{Lacey},$$ (21.70)

in which

$$C_{Lacey} = 5.98 f_l^{\frac{1}{3}}$$ (21.71)

$$f_l = (2,500D)^{0.5}.$$ (21.72)

Combining Equations (21.71) and (21.72),

$$C_{Lacey} = 22.28 D^{\frac{1}{6}}.$$ (21.73)

Equation (21.71) depends only on sediment size and should hold for the range of data used in the regime theory.

Now we derive C_* using the stream power methodology. Cao (1996) proposed what he termed as the Fluvial Processes Index (FPI), which accounts for both flow and boundary conditions. Let the cross-sectional mean shear force be denoted by SF_{cs} and the critical shear power of unit channel length by SP_{cr}, which is assumed to be the constraint to resist boundary changes. Then, we can write

$$SP_{cr} = SF_{cs}v. \tag{21.74}$$

For a threshold condition, the cross-sectional mean shear force should equal the product of mean critical stress (τ_{cr}) and wetted perimeter (P):

$$SF_{cs} = \tau_{cr}P, \tag{21.75}$$

in which the critical shear stress can be expressed in terms of the Shields critical stress (Θ_{cr}) to be obtained from the Shields diagram:

$$\tau_{cr} = \Theta_{cr}D(\gamma_s - \gamma). \tag{21.76}$$

Then, FPI can expressed as

$$FPI = \frac{SP}{SP_{cr}} = \frac{\gamma QS}{\Theta_{cr}D(\gamma_s - \gamma)Pv}. \tag{21.77}$$

The regime coefficient C_* can now be written as a function of *FPI*:

$$C_* = \varphi\left(\frac{SP}{SP_{cr}}\right) = \varphi\left(\frac{Q\gamma S}{\Theta_{cr}D(\gamma_s - \gamma)d^{1.5}v}\right). \tag{21.78}$$

For wide channels P can be approximated by w, so Equation (21.78), with the aid of Equation (2.49), becomes

$$C_* = 3\left[\frac{Q\gamma S}{\Theta_{cr}D(\gamma_s - \gamma)d^{1.5}v}\right]^{0.4}. \tag{21.79}$$

Cao (1996) reported the coefficient of determination for an equation as 0.867 using a wide range of data covering discharge varying from 0.005 to 7,140 m^3/s, slope from 0.00069 to 0.0059, sediment size from 1.3 to 117 mm, mean depth from 0.031 to 6.83 m, and width from 0.52 to 542 m.

21.7 Variation of Stream Power

Stream power is spatially distributed throughout the basin (Lecce, 1997). One way to visualize the spatial distribution is by inserting hydraulic geometry equations in the definition of stream power. To that end, let the hydraulic geometry relations be expressed as

$$w \propto Q^b \tag{21.80}$$

$$d \propto Q^f \tag{21.81}$$

$$v \propto Q^m \tag{21.82}$$

$$S \propto Q^j. \tag{21.83}$$

Insertion of Equations (21.80)–(21.83) into Equations (21.1) and (21.2) permits the prediction of downstream changes in stream power with changes in discharge. In perennial streams, Bull (1979, 1991) stated that cross-sectional stream power should increase $\left(SP \propto Q^{0.2}\right)$ and mean stream power should decrease $\left(SP_m \propto Q^{0.3}\right)$ downstream. However, in ephemeral streams, he expressed that both cross-sectional stream power $\left(SP \propto Q^{-1.8}\right)$ and mean stream power $\left(SP_m \propto Q^{0.3}\right)$ should decrease, because discharge would decrease due to transmission losses.

A similar argument can be made for longitudinal profiles, which, following Leopold and Langbein (1962) and Langbein and Leopold (1964), correspond to a compromise between two opposing tendencies, including uniform expenditure of energy and minimization of total energy expenditure. Theoretically, the profile for uniformly distributed work tends to be less concave $\left(S \propto Q^{-0.5}\right)$, suggesting constant mean stream power. The profile corresponding to least work would tend to be more concave $\left(S \propto Q^{-1}\right)$, suggesting constant cross-sectional power. For an intermediate condition $\left(S \propto Q^{0.75}\right)$ balancing the uniform energy expenditure with minimum total energy expenditure, the result would be an increase in cross-sectional stream power and a decrease in mean stream power.

To describe downstream changes along individual streams, Lawler (1992) replaced discharge with distance as an index of downstream change $(Q = a_1 L^{a_2})$ and expressed slope by an exponential function $(S = S_0 \exp(-b_0 L))$. Here L is the distance from the drainage divide; S_0 is the initial channel slope upstream; and a_1, a_2, and b_0 are the parameters. Inserting these in Equation (21.1), the result is

$$SP = \gamma(a_1 L^{a_2})(S_0 \exp(-b_0 L)). \tag{21.84}$$

Equation (21.84) shows that the cross-sectional stream power would increase downstream to a peak value at $L = a_2/b_0$ prior to decreasing from then on downstream. In a similar vein, inserting a power function for channel width $(w = b_1 L^{b_2})$ in Equation (21.2), the result is

$$SP_m = \frac{\gamma(a_1 Q^{a_2})(S_0 \exp(-b_0 L))}{b_1 L^{b_2}}. \tag{21.85}$$

Equation (21.85) shows that the mean stream power would increase downstream to a peak value but closer to the divide. These inferences seem to be supported by field observations (Lecce, 1997). Knighton (1987) suggested that cross-sectional stream power should increase downstream but mean stream power should remain relatively constant. On the other hand, sediment delivery ratios decrease downstream, commonly dropping below 10% in drainage areas over 1,000 km^2 due to decreasing hillslope and channel slopes and increasing deposition in larger watersheds. In watersheds, stream power changes nonlinearly in the downstream direction (Walling, 1983). Describing downstream changes in stream power in the Henry mountains, Utah, Graf (1983) noted that the total stream power did not necessarily increase systematically in the downstream direction because of conflicting influences of channel slope, width, and depth.

The total stream power given by Equation (21.1) can be expressed as (Bull, 1979)

$$SP \propto wdvS. \tag{21.86}$$

Substituting Equations (21.80)–(21.83) in Equation (21.86), the result is

$$SP \propto Q^b Q^f Q^m Q^j \propto Q^{b+f+m+j}. \tag{21.87}$$

The values of exponents in Equations (21.80)–(21.83) depend on whether the stream is ephemeral or perennial.

The hydraulic geometry under consideration is the downstream hydraulic geometry, and changes in stream power can be compared for both types of streams by using their average exponent values. For ephemeral streams, the average values of exponents given by Leopold et al. (1964) are: $b = -0.5, f = -0.3, m = -0.2$, and $j = -0.8$. This leads stream power to

$$SP \propto Q^{-1.8}. \tag{21.88}$$

Likewise, the stream power per unit width will be

$$SP_m \propto Q^{-1.3}. \tag{21.89}$$

On the other hand, for perennial streams, $b = 0.5, f = 0.4, m = 0.1$, and $j = -0.8$. The stream power then becomes

$$SP \propto Q^{0.2}. \tag{21.90}$$

Likewise, the stream power per unit width will be

$$SP_m \propto Q^{-0.3}. \tag{21.91}$$

Equations (21.88)–(21.91) provide an indication of how stream power and stream power per unit width increase or decrease along the stream reach.

21.8 Computation of Variation in Stream Power

In watershed engineering, the impact of anthropogenic changes, such as urbanization, deforestation or afforestation, channelization, and erosion due to landscape changes on stream channels is of much interest. Computation of stream power is critical to understand changes in streams for which measurements of channel geometry change, bed load movement, deposition of sediment at the channel reach scale, shear stress, and tractive force are often made. Stream power changes can be computed using flow duration curves (Fitzerland and Bowden, 2006).

A flow duration curve (FDC) can be constructed using gaged streamflow data or simulated streamflow data. An FDC is a graphical portrayal of streamflow values arranged from the highest value to the lowest value on the ordinate (*y*-axis) and percent exceedance at each interval on the *x*-axis. It can be used to compute return periods of different stream flow events, including large magnitude events causing flooding. With FDC data, time component allows to calculate stream power for specific durations, such as one-day return flow. Hence, it provides streamflow values needed to compute stream power. The FDC data under virgin and altered conditions allows for the computation of stream power under the two conditions.

References

Ackers, P. (1958). *Resistance of fluids flowing in channels and pipes*. Research Paper No. 1, U.K. Hydraulic Research Station, Wallingford.

Ackers, P. (1964). Experiments on small streams in alluvium. *Journal of the Hydraulics Division, ASCE*, Vol. 90, No. HY4, pp. 1–37.

Bagnold, R. A. (1966). *An approach to sediment transport from general physics*. U.S. Geological Professional Paper, 422(I), pp. 37, Washington, DC.

Bull, W. B. (1979). Threshold of critical power in streams. *Geological Society of America Bulletin*, Vol. 90, pp. 453–464.

Bull, W. B. (1991). *Geomorphic Responses to Climate Change*. Oxford University Press, New York.

Cao, S. (1996). *Fluvial Hydraulic Geometry*. Chengdu University of Science and Technology Press, Chengdu, Sichuan, China.

Cao, S. and Knight, D. W. (1995). Design of threshold channels. *HYDRA*, Vol. 1, pp. 516–521.

Cao, S. and Knight, D. W. (1996). Regime theory of alluvial channels based upon the concepts of stream power and probability. *Proceedings, Institution of Civil Engineers, Waterways, Maritime & Energy*, Vol. 116, pp. 160–167.

Cao, S. and Knight, D. W. (1998a). Entropy-based design approach of threshold alluvial channels. *Journal of Hydraulic Research*, Vol. 35, No. 4, pp. 505–524.

Cao, S. and Knight, D. W. (1998b). Design for hydraulic geometry of alluvial channels. *Journal of Hydraulic Engineering*, Vol. 124, No. 5, pp. 484–492.

Chang, H. H. (1979a). Geometry of rivers in regime. *Journal of Hydraulics Division, ASCE*, Vol. 105, No. HY6, pp. 691–706.

Chang, H. H. (1979b). Minimum stream power and river channel patterns. *Journal of Hydrology*, Vol. 41, pp. 303–327.

Engelund, F. and Hansen, E. (1972). *A monograph on sediment transport in alluvial streams*. Teknisk Foriag-Copenhagen, 62 pp., Denmark.

Fitzerland, E. and Bowden, B. (2006). Quantifying increases in stream power and energy using flow duration curves to depict streamflow values. Stormwater, March/April, pp. 88–94.

Graf, Q. L. (1983). Downstream changes in stream power in the Henry Mountains, Utah. *Annals of the Association of American Geographers*, Vol. 13, No. 3, pp. 373–387.

Hickin, E. J. and Nanson, G. C. (1984). Lateral migration rates of river bends. *Journal of Hydraulic Engineering*, Vol. 110, pp. 1557–1567.

Judd, H. E. (1963). *A study of bed characteristics in relation to flow in rough, high gradient, natural channels*. Ph.D. dissertation, Utah State University, Logan.

Knighton, A. D. (1987). River channel adjustment: The downstream dimension. In: *River Channels: Environment and Process*, edited by K. Richards, pp. 95–128, Blackwell, Oxford.

Langbein, W. B. and Leopold, L. B. (1964). Quasi-equilibrium states in channel morphology. *American Journal of Science*, Vol. 262, pp. 782–794.

Lawler, D. M. (1992). Process dominance in bank erosion systems. In: *Lowland Floodplain Rivers: Geomorphological Perspectives*, edited by P. A. Carling and G. E. Petts, pp. 117–141, John Wiley & Sons, Chichester.

Lecce, S. A. (1997). Nonlinear downstream changes in stream power on Wisconsin's Blue River. *Annals of the Association of American Geographers*, Vol. 87, No. 3, pp. 471–486.

Leopold, L. B. and Langbein, W. B. (1962). *The concept of entropy in landscape evolution.* U.S. Geological Survey Professional Paper 500-A, pp. 20, Washington, DC.

Leopold, L. B. and Wolman, M. G. (1957). *River channel patterns: Braided, meandering and straight*. U.S. Geological Survey Professional Paper 282-B, Washington, DC.

Leopold, L. B., Wolman, M. G., and Miller, J. P. (1964). *Fluvial Processes in Geomorphology*. W.H. Freeman, San Francisco.

Mosley, M. P. (1981). Semi-determinate hydraulic geometry of river channels, South Island, New Zealand. *Earth Surface Processes and Landform*, Vol. 6, pp. 127–137

Phillips, J. D. (1989). Fluvial sediment storage in wetlands. *Water Resources Bulletin*, Vol. 25, pp. 867–873.

Rogers, F. C. and Thomas, A. R. (1969). Regime canals. Section 6, in: *Handbook of Applied Hydraulics*, edited by C. V. Davis and K. E. Sorensen, McGraw-Hill Book Company, New York.

Schumm, S. A. and Khan, H. R. (1972). Experimental study of channel patterns. *Geological Society of America Bulletin*, Vol. 83, pp. 1755–1770.

Singh, V. P., Yang, C. T., and Deng, Z. Q. (2003a). Downstream hydraulic geometry relations: 1. Theoretical development. *Water Resources Research*, Vol. 39, No. 12, 1337, doi:10.1029/2003WR002484.

Singh, V. P., Yang, C. T., and Deng, Z. Q. (2003b). Downstream hydraulic geometry relations: II. Calibration and Testing. *Water Resources Research*, Vol. 39, No. 12, 1337, doi:10.1029/2003WR002484.

Walling, D. E. (1983). The sediment delivery problem. *Journal of Hydrology*, Vol. 65, pp. 209–237.

Wohl, E. E., Vincent, K. R., and Merrittis, D. J. (1993). Pool and riffle characteristics in relation to channel gradient. *Geomorphology*, Vol. 6, pp. 99–110.

22

Regional Hydraulic Geometry

Relationships relating channel width, depth, and cross-section as well as discharge to drainage basin area are regional relationships, because they are developed at the basin scale. These relationships constitute regional hydraulic geometry, which is utilized in stream assessments, evaluating channel characteristics, identifying field indicators of bankfull discharge, and delineation of regional boundaries, hydrologic regions, and ecoregions. This chapter presents these regional relationships.

Notation

a = parameter

A = cross-sectional area

A_d = drainage area

a_0 = coefficient

$a_i, i = 0, 1, 2$ = empirical constants

a'_0 = constant

a'_1 = constant

a'_2 = exponent

A_{bf} = bankfull channel cross-section

a_{w1} = coefficient associated with bankfull width

a_{d1} = coefficient associated with bankfull depth

a_{a1} = coefficient associated with cross-sectional area

a_d = coefficients associated with drainage area

a_e = exponents associated with bankfull discharge

b = exponent

b'_0 = exponent

Continued

b_{w1} = exponent associated with bankfull width
b_{d1} = exponent associated with depth
b_{a1} = exponent associated with cross-sectional area
b_i, $i = 0, 1, 2$ = empirical constants
c = constant or parameter
C_o^* = Chezy's roughness coefficient
c_i, $i = 0, 1, 2$ = empirical constants
d = channel depth
D = particle size or diameter
D_T = topological fractal dimension
d_{bf} = bankfull flow depth
d_i, $i = 0, 1, 2$ = empirical constants
e_i, $i = 0, 1, 2$ = empirical constants
f = exponent
F = frequency of occurrence
g = acceleration due to gravity
j = exponent
k = exponent
L = stream length
m = exponent
n = Manning's friction coefficient
N = number of streams
o = stream order
Q = discharge
Q_{bf} = bankfull discharge
Q_2 = 2-year discharge
$Q_{2.33}$ = mean annual flow
r = constant
R = mean runoff rate per unit area from the hillslopes along a stream network
R^2 = log correlation coefficient squared
R_A = area ratio defining Horton's law of stream areas
R_B = bifurcation ratio defined by Horton's law of stream numbers
R_Q = discharge ratio defining Horton law's for discharge
R_i, $i = 1, 2, \ldots, 6$ = six dimensionless variables (or ratios or numbers)
R_L = length ratio
R_n = roughness ratio
R_d = depth ratio
R_S = slope ratio
R_v = velocity ratio
S = channel slope
S_l = suspended load
t = constant

$T_k, k = 2, 3, \ldots = $ tributaries
$v = $ cross-sectionally averaged flow velocity
$w = $ channel width
$w_{bf} = $ bankfull channel flow width
$y = $ exponent
$z = $ exponent
$\alpha = $ anomalous scaling exponent
$\beta = $ anomalous scaling exponent
$\rho = $ mass density of water
$\upsilon = $ kinematic viscosity of water
$\tau_o = $ shear stress per unit area

22.1 Introduction

Streams and rivers are dynamic systems and represent the integrated result of the interaction between climate, geology, and anthropogenic changes. Their shape, size, and spatial characteristics are determined by their watersheds and have been observed to follow laws of geomorphology. Likewise, the formation and development of floodplains also follow certain laws of nature. Because of the interaction between rivers and their floodplains, rivers and drainage basin characteristics are interrelated. Hydraulic geometric characteristics of rivers can be extended from point scale to reach scale to regional scale, at which scale the characteristics will be average. This chapter discusses regional hydraulic geometric characteristics, which are important for watershed restoration and protection, regional planning, and ecosystem management.

22.2 Relation between Discharge, Channel Shape, and Drainage Area

From Horton's law of stream areas, it is seen that within a given watershed the drainage area increases as the stream order increases. Stall and Fok (1968) related drainage area to stream order as a power relation. Using the flow duration curve, which relates discharge as a function of frequency of occurrence, Virmani (1973) expressed discharge as a function of frequency of occurrence and stream order. Combining these two relationships, discharge can be expressed as a function of frequency and drainage area as

$$\ln Q = a_0 - a_1 F + a_2 A_d, \tag{22.1}$$

in which a_i, $i = 0, 1, 2$, are empirical constants; Q is the discharge; F is the frequency of occurrence, and A_d is drainage area.

Noting that $Q = Av$ and $A = wd$, one can write

$$\ln Q = \ln A + \ln v \tag{22.2}$$

and

$$\ln A = \ln w + \ln d, \tag{22.3}$$

where A is the cross-sectional area, v is the cross-sectionally averaged flow velocity, w is the channel width, and d is the channel depth.

Similar to Equation (22.1), channel width, depth, velocity, and cross-sectional area can be expressed with the use of Equations (22.2) and (22.3) as

$$\ln w = b_0 - b_1 F + b_2 \ln A_d \tag{22.4}$$

$$\ln d = c_0 - c_1 F + c_2 \ln A_d \tag{22.5}$$

$$\ln A = d_0 - d_1 F + d_2 \ln A_d \tag{22.6}$$

$$\ln v = e_0 - e_1 F + e_2 \ln A_d, \tag{22.7}$$

where b_i, c_i, d_i, and e_i, $i = 0, 1, 2$, are empirical constants. Stall and Yang (1970), Singh and Broeren (1989), and McConkey and Singh (1992) also developed similar equations.

For Bear River basin, which occupies parts of Wyoming, Idaho, and Utah, Virmani (1973) expressed Equations (22.1) and (22.4)–(22.7) as

$$\ln Q = 1.396 - 3.808F + 0.866 \ln A_d, \quad R^2 = 0.90 \tag{22.8}$$

$$\ln w = 1.002 - 0.493F + 0.531 \ln A_d, \quad R^2 = 0.853 \tag{22.9}$$

$$\ln d = -0.907 - 1.655F + 0.341 \ln A_d, \quad R^2 = 0.892 \tag{22.10}$$

$$\ln A = 0.067 - 2.416F + 0.878 \ln A_d, \quad R^2 = 0.942 \tag{22.11}$$

$$\ln v = 1.282 - 1.663F - 0.0046 \ln A_d, \quad R^2 = 0.733, \tag{22.12}$$

in which Q is in cfs (cubic feet per second), A_d is square miles, w is in feet, d is in feet, v is in feet per second, F is in percent of days, and R^2 is the log correlation coefficient squared. To evaluate the validity of Equations (22.8)–(22.12), consider an example where the flow occurs 50% of the days per year ($F = 0.5$) and the drainage area is 100 mi^2 and one wants to know the channel cross-sectional area, width, depth, velocity, and discharge. Then, cross-sectional area is given by Equation (22.11) as

$$\ln A = 0.067 - 2.416 \times 0.50 + 0.878 \ln (100)$$
$$= 0.067 - 1.073 + 0.878 \times 2 \times 2.303$$
$$= 0.067 - 1.073 + 4.044 = 3.038.$$

This yields $A = 20.88$ ft^2, which is close to the observed value. Likewise, the channel width can be expressed as

$$\ln w = 1.002 - 0.493F + 0.531 \ln A_d$$
$$= 1.002 - 0.493 \times 0.50 + 0.531 \times \ln (100)$$
$$\ln d = -0.907 - 1.655F + 0.341 \ln A_d$$
$$= -0.907 - 1.655 \times 0.50 + 0.341 \times \ln (100)$$
$$\ln v = 1.282 - 1.663F - 0.0046 \ln A_d$$
$$= 1.282 - 1.663 \times 0.50 - 0.0046 \times \ln (100)$$
$$\ln Q = 1.396 - 3.808F + 0.866 \ln A_d$$
$$= 1.396 - 3.808 \times 0.50 + 0.866 \times \ln (100).$$

Equation (22.8) can be expressed for Q as a power function of drainage area:

$$Q = \exp (1.396 - 3.808F)A_d^{0.866}. \tag{22.13}$$

Equation (22.13) can be expressed for A_d as a function of discharge as

$$A_d = \exp (-1.612 + 4.397F)Q^{1.155}. \tag{22.14}$$

Inserting Equation (22.14) in Equations (22.9)–(22.12), the result is

$$w = \exp (0.147 + 1.84F)Q^{0.612} \tag{22.15}$$

$$d = \exp (-1.457 - 0.15F)Q^{0.394} \tag{22.16}$$

$$A = \exp (-1.348 + 1.70F)Q^{1.0142} \tag{22.17}$$

$$v = \exp (1.289 - 1.60F)Q^{-0.0053}. \tag{22.18}$$

Equations (22.15)–(22.18) can be regarded as regional hydraulic geometry equations. Similar equations were derived by Stall and Fok (1968) for Illinois streams.

Change in discharge moving downstream is caused by the increase in drainage area. The drainage area (A_d) dominantly influences the mean annual flow ($Q_{2.33}$). Flood Studies Report (1975) reported for the British Isles that

$$Q_{2.33} = a_0 A_d^{0.73}, \tag{22.19}$$

where coefficient a_0 has a range between 0.22 and 1.62.

22.3 Regional Hydraulic Geometry

Wilkerson (2008) developed regional hydraulic geometry relationships by relating bankfull discharge and channel width (w_{bf}), depth (d_{bf}), and cross-section (A_{bf}) to drainage basin area (A_d). Usually, the bankfull hydraulic geometry relations are expressed as

$$Q_{bf} = a_d A_d^{a_e} \tag{22.20}$$

$$w_{bf} = a_{w1} A_d^{b_{w1}} \tag{22.21}$$

$$d_{bf} = a_{d1} A_d^{b_{d1}} \tag{22.22}$$

$$A_{bf} = a_{a1} A_d^{b_{a1}}, \tag{22.23}$$

where a_{w1}, a_{d1}, a_{a1}, and a_d are the coefficients associated with drainage area corresponding to bankfull width, depth, cross-sectional area, and bankfull discharge, respectively; and b_{w1}, b_{d1}, b_{a1}, and a_e are the exponents associated with drainage area for bankfull width, depth, cross-sectional area, and bankfull discharge, respectively. The advantage of regional relations lies in the ease with which drainage area can be computed (Dunne and Leopold, 1978). These relations can be utilized in stream assessments, planning studies, evaluating channel characteristics, and identifying filed indicators of bankfull discharge. Regional boundaries can be delineated using physiographic provinces, hydrologic regions, and ecoregions. Regionalization schemes can also be used to demarcate regional boundaries.

Wilkerson (2008) improved the estimation of bankfull discharge by also using two-year discharge (Q_2), arguing that Q_2 contains substantial information about basin characteristics and that it is closer to Q_{bf}. He thus proposed two relations for Q_{bf}:

$$Q_{bf} = c_1 Q_2^{c_2} \tag{22.24}$$

and

$$Q_{bf} = c_3 Q_2^{c_4} A_d^{c_5}. \tag{22.25}$$

Wilkerson (2008) noted that the ratio Q_{bf}/Q_2 ranged from 01.0 to 3.0. Using data from 41 sites in western Montana, Lawlor (2004) proposed a regional relationship at ungaged sites as

$$Q_{bf} = 0.84 Q_2. \tag{22.26}$$

22.4 Relation between Mean Annual Discharge and Drainage Area

Using data from 16 selected streams in Central Pennsylvania, Brush (1961) found a strong relation between mean annual discharge ($Q_{2.33}$) in cfs and drainage area (A_d) in square miles as

$$Q_{2.33} = aA_d^b, \tag{22.27}$$

where discharge was estimated from bankfull stage in regression analysis and discharge was calculated by frequency analysis with the use of partial duration series, $a = 80$, and $b = 0.8$. For different drainage areas, the mean annual flood ($Q_{2.33}$) has been expressed as shown in Table 22.1 (Knighton, 1987). It is interesting to note that exponent b varies between 0.55 and 0.85 with a mean value of 0.70, but the relation can vary not only from one basin to another but also within the same basin. Hence, it should be taken as a first-order approximation only. The value of exponent being less than one reflects the influence of valley storage on the downstream transmission of discharge.

Table 22.1. *Relation between discharge and drainage area [Q_b = bankfull discharge; $Q_{2.33}$ = mean annual flood; Q_2 = flood discharge having a return period of two years]*

Source	Location	Drainage area (km²)	Discharge (m³/s)	Constant a	Exponent b
Nixon (1959)	29 gaging stations, England	111–7,278	Q_b	0.24	0.85
Brush (1961)	Pennsylvania, USA	20–162,300	$Q_{2.33}$	0.50	0.80
Benson (1962)	164 stations, New England, USA	4–25,070	$Q_{2.33}$	0.56	0.85
Nash and Shaw (1966)	57 stations, Great Britain	8–9,990	$Q_{2.33}$	0.76	0.74
Thomas and Benson (1970)	Potomac River basin	15–30,000	Q_2	0.86	0.86
Thomas and Benson (1970)	Sacramento and San Joaquin River basins	19–4,380	Q_2	0.64	0.83
Thomas and Benson (1975)	Louisiana	14–4,410	Q_2	4.49	0.56
Emmett (1975)	Upper Salmon River, Idaho, USA	0.4–4,670	Q_b	0.42	0.69
Flood Studies Report (1975)	533 stations, British Isles	0.05–9,868	$Q_{2.33}$	0.68	0.77

The drainage area may be computed from stream length (L) (in miles) using Hack's (1957) equation:

$$L = 1.43 \, A_d^{0.6}. \tag{22.28}$$

22.5 Leopold–Miller Extension of Horton Laws

Following Horton (1945), Leopold and Miller (1956) expressed a relation between stream order (o) and channel geometric characteristics, such as number of streams (N), stream length (L), drainage area (A_d), and channel slope (S):

$$o \propto \log N \tag{22.29}$$

$$o \propto \log L \tag{22.30}$$

$$o \propto \log S \tag{22.31}$$

$$o \propto \log A_d. \tag{22.32}$$

Equations (22.30)–(22.32) show that any two variables, other than the channel order, can be related to each other as power functions, such as

$$L \propto A_d^k; S \propto A_d^k; S \propto L^k, \tag{22.33}$$

where k is an exponent. Because discharge Q is related to drainage area as

$$Q \propto A^k. \tag{22.34}$$

Thus, one can also write

$$o \propto \log Q. \tag{22.35}$$

Recalling hydraulic geometry relations:

$$w \propto Q^b \tag{22.36}$$

$$d \propto Q^f \tag{22.37}$$

$$v \propto Q^m \tag{22.38}$$

$$S \propto Q^z \tag{22.39}$$

$$n \propto Q^y \tag{22.40}$$

$$S_l \propto Q^j, \tag{22.41}$$

where S_l is suspended load. Keeping Equation (22.35) in view, it follows from Equations (22.36)–(22.41) that one can write

$$o \propto (w, d, v, S, n, S_l). \tag{22.42}$$

Also, several of the variables in Equations (22.29)–(22.32) as well as (22.35)–(22.41) have been intercorrelated. For example,

$$S_l \propto Q^j; \text{ and } Q \propto A_d^k; \text{ so } S_l \propto A_d^{jk}. \tag{22.43}$$

Thus, a series of hydraulic variables and drainage parameters can be inter-related as power functions.

Another example is $o = k \log w$; $o = k \log S$, so S can be related as power function of width. Leopold and Miller (1956) found this function as

$$S = 0.12w^{-0.5}. \tag{22.44}$$

They noted that measurements of width, slope, and order are relatively easy to obtain, and the relationships among these can be combined with those factors that are easy to obtain and then approximate relations for those that cannot be easily measured can be obtained. An example is one of relating discharge to order. Since discharge increases with order, it provides a link between Horton's law and hydraulic geometry. If discharge has a relation to the position of the cross-section along the channel length, then an approximation to constant frequency might emerge. Since width and order are related and discharge increases downstream with increasing width, it may be possible to combine the two and express discharge as a function of order.

In another way, the discharge–order relation makes use of flood-frequency analysis, where the discharge of a flood of a given frequency increases somewhat less rapidly than drainage basin size. The discharge–drainage area relation is expressed as

$$Q \propto A_d^k, \tag{22.45}$$

where the value of k is about 0.7 to 0.8. By combining the discharge–area relation with the order–area relation, discharge can be related to order.

Now consider Manning's equation for cross-sectionally averaged velocity,

$$v = \frac{1}{n} d^{\frac{2}{3}} S^{\frac{1}{2}}. \tag{22.46}$$

Wolman (1954) expressed slope and roughness as functions of discharge as

$$S = tQ^z \tag{22.47}$$

and

$$n = rQ^y. \tag{22.48}$$

Then, one can write

$$kQ^m = \frac{(cQ)^{\frac{2f}{3}}(tQ)^{\frac{z}{2}}}{rQ^y}. \tag{22.49}$$

From Equation (22.49), it follows that

$$m = \frac{2}{3}f + \frac{1}{2}z - y. \tag{22.50}$$

If the width exponent (b) is 0.5 then because $b + f + m = 1, f + m = 0.5$. This means that as m increases f decreases. Then, from equation one can also write:

$$f = 0.3 - 0.3z + 0.6y \tag{22.51}$$

$$m = 0.2 + 0.3z - 0.6y. \tag{22.52}$$

It was noted that in many rivers, the roughness downstream remains more or less constant, that is, $y = 0$. Then, for $b = 0.5, f = 0.4$, and $m = 0.1$, yielding

$$S \propto Q^{-0.33}. \tag{22.53}$$

Leopold and Miller (1956) noted that roughness depended not only on particle size (as implied by Strickler equation, $n = cD^{\frac{1}{6}}, D$ = particle size or diameter) but also on channel configuration, dunes and antidunes, pools and riffles, composition of sediment load, vegetation, and other factors. They reported

$$D_{50} \propto Q^{-0.08}. \tag{22.54}$$

Another way of deriving the slope exponent is using the Hack (1957) equations relating channel slope and channel length and length and drainage area as

$$S \propto L^{-1} \tag{22.55}$$

$$L \propto A_d^{0.6}. \tag{22.56}$$

The bankfull discharge was related to drainage area as

$$Q_{2.33} \propto A_d^{0.70}. \tag{22.57}$$

Combining Equations (22.55), (22.56), and (22.57),

$$S \propto Q_{2.33}^{-1.16}. \tag{22.58}$$

22.6 Gupta–Mesa Theory of Hydraulic Geometry

Gupta and Mesa (2014) developed an analytical theory to develop Horton's laws for six hydraulic geometry variables, including stream discharge Q, width w, depth d, velocity v, slope S, and Manning's friction coefficient n, for self-similar networks.

22.6.1 Physical Variables and Parameters

The theory starts with defining six dimensionless ratios or numbers. To that end, physical variables and parameters governing the hydraulic geometry are defined at the bottom of a stream of Horton–Strahler order $o \geq 1$. These variables and parameters include: Q_o is the discharge (L^3/T), A_o is the cumulative drainage area (L^2), d_o is the channel depth (L), w_o is the channel width (L), v_o is the flow velocity (L/T), E_o is the elevation difference between the beginning and end of a complete Strahler stream (L), L_o is the stream length (L), S_o is the slope defined as $S_o = E_o/L_o$, R (L/T) is the mean runoff rate per unit area from the hillslopes along a stream network, g (L/T^2) is the acceleration due to gravity, ρ (kg/L^3) the mass density of water, and v is the kinematic viscosity of water (L^2/T). It is assumed that the basin is homogeneous because R is spatially uniform.

22.6.2 Dimensionless Ratios

These are 12 variables containing three dependent variables: $Q_o = v_o w_o d_o$; $S_o = E_o/L_o$; and $L_o = A_o^\alpha$. The remaining nine independent variables involve three basic dimensions: length (L), time (T), and mass (M). Thus, six dimensionless variables (or ratios or numbers), denoted as R_i, $i = 1, 2, \ldots, 6$, can be defined as follows.

1. Discharge ratio, R_1 can be defined as

$$R_1 = \frac{Q_o}{RA_o}.$$

(22.59)

2. Runoff ratio, R_2, can be defined as

$$R_2 = \frac{R\sqrt{A_o}}{d_o v_o}.$$

(22.60)

Equation (22.60) involves the ratio of runoff per unit width of drainage basin to the discharge per unit channel width.

3. Basin Froude number R_3 can be defined as

$$R_3 = \frac{v_o}{\sqrt{gE_o}} = \frac{v_o}{\sqrt{gS_o L_o}}.$$

(22.61)

4. Channel slope, R_4, can be defined as

$$R_4 = S_o = \frac{E_o}{L_o}.$$

(22.62)

5. Reynolds number, R_5, can be defined as

$$R_5 = \frac{v_o d_o}{v}.$$

(22.63)

6. Force ratio, R_6, can be defined as

$$R_6 = \frac{v_o^2}{g d_o S_o}.$$

(22.64)

It is interesting to note that the denominator in Equation (22.64) is the square of the shear velocity. For the numerator, the total frictional force along the channel can be expressed as $\tau_o(2d_o + w_o)L_o \approx \tau_o w_o L_o$, where τ_o is the shear stress per unit area and is proportional to the square of the velocity. The gravitational force for the mass of water along the channel length is given by $\rho g w_o d_o L_o S_o$. Taking the ratio of these forces leads to Equation (22.64). It is interesting to note that R_6 is proportional to the Darcy–Weisbach friction factor, and $1/\sqrt{R_6}$ is linearly related to the logarithm of relative roughness defined by the ratio of flow depth to the roughness height (Leopold et al., 1964).

22.6.3 Mass Conservation

The next element of the analytical theory is the mass conservation in a self-similar network, by which it was shown that in the limit

$$R_Q = \frac{Q_{o+1}}{Q_o} = R_B = \frac{N_o}{N_{o+1}} = R_A = \frac{A_{o+1}}{A_o}$$

(22.65)

and

$$R_Q = R_B = R_A = \frac{(2 + a + c) + \sqrt{(2 + a + c)^2 - 8c}}{2},$$

(22.66)

where R_Q is the discharge ratio defining Horton law's for discharge; R_B is the bifurcation ratio defined by Horton's law of stream numbers; R_A is the area ratio defining Horton's law of stream areas; and parameters a and c are from the relation for tributaries (T) as

$$\frac{T_k}{T_{k-1}} = c, \quad k = 2, 3, \ldots \tag{22.67}$$

and $T_1 = a$. Let the mean number of streams be defined (o-k) that join the stream of order o. Then, it was shown (Dodds and Rothman, 1999) that

$$T_k = ac^{k-1}. \tag{22.68}$$

Equation (22.65) shows that

$$Q_{o+1}N_{o+1} = Q_oN_o. \tag{22.69}$$

22.6.4 Derivation of Horton Laws

From the definition of R_1, it may be possible to write in the limit

$$RR_1 = \frac{R_1(o+1)}{R_1(o)} = 1, o = 1, 2, \ldots. \tag{22.70}$$

The five hydraulic geometry variables – w, d, v, S, and n – have all been expressed as power functions of discharge Q (Leopold and Miller, 1956) as:

$$w_o \propto Q_o^b \tag{22.71}$$

$$d_o \propto Q_o^f \tag{22.72}$$

$$v_o \propto Q_o^m \tag{22.73}$$

$$S_o \propto Q_o^z \tag{22.74}$$

$$n_o \propto Q_o^y. \tag{22.75}$$

Following Gupta and Waymire (1998), for self-similarity, if $v = f_1(Q)$ then $f_1(Q_1 \cdot Q_2) = f_1(Q_1)f_2(Q_2)$ whose solution is a power law. Thus,

$$v = f_1(Q) \propto Q^m. \tag{22.76}$$

Following Equation (22.65), the Horton law can be expressed for velocity as

$$\frac{v_{o+1}}{v_o} = \frac{Q_{o+1}^m}{Q_o^m} = R_Q^m = R_v. \tag{22.77}$$

In a similar manner, one can write for width as

$$w = f_1(Q) \propto Q^b. \tag{22.78}$$

Then, in the limit the Horton law can be expressed as

$$\frac{w_{o+1}}{w_o} = \frac{Q_{o+1}^b}{Q_o^b} = R_Q^b = R_w. \tag{22.79}$$

Likewise, for depth, one can express

$$\frac{d_{o+1}}{d_o} = \frac{Q_{o+1}^f}{Q_o^f} = R_Q^f = R_d. \tag{22.80}$$

22.6.5 Determination of Width Exponent

From the ratio R_2 given by Equation (22.60), one can write

$$RR_2(o) = \frac{\sqrt{A_{o+1}}}{\sqrt{A_o}} \cdot \frac{d_o v_o}{d_{o+1} v_{o+1}}. \tag{22.81}$$

The quantity $d_o v_o$ in Equation (22.81) can be replaced by Q_o / w_o as

$$RR_2(o) = \frac{\sqrt{A_{o+1}}}{\sqrt{A_o}} \cdot \frac{Q_o w_{o+1}}{Q_{o+1} w_o}. \tag{22.82}$$

In the limit Equation (22.82) can be expressed as

$$\frac{\sqrt{A_{o+1}}}{\sqrt{A_o}} \cdot \frac{Q_o w_{o+1}}{Q_{o+1} w_o} = \frac{R_A^{\frac{1}{2}} R_w}{R_Q}. \tag{22.83}$$

This leads in the limit to

$$RR_1(o) = RR_2. \tag{22.84}$$

From Equation (22.65), $R_Q = R_A$. Combining Equations (22.79) and (22.84),

$$R_w = RR_2 R_Q^{1/2} = R_Q^b. \tag{22.85}$$

Equation (22.85) shows that $RR_2 = 1$ and exponent $b = 0.5$.

22.6.6 Determination of Velocity and Depth Exponents

Here the Horton ratio for the basin Froude number is expressed. Using Equation (22.61), one can write in the limit:

$$RR_3 = \frac{R_3(o+1)}{R_3(o)} = \frac{R_v}{\sqrt{R_L R_S}}. \tag{22.86}$$

In a similar fashion, one can write for slope.

$$RR_4 = \frac{S_0(o+1)}{S_0(o)} = R_S = \frac{Q^z(o+1)}{Q^z(o)} = R_Q^z. \tag{22.87}$$

Peckham (1995) has shown that the topological fractal dimension D_T for Tokunaga networks can be expressed in terms of bifurcation ratio R_B and link ratio R_C as

$$D_T = \frac{\log R_B}{\log R_C}. \tag{22.88}$$

Because $R_B = R_A$, and link lengths are constant, $R_C = R_L$, where R_L is the length ratio. It then follows that

$$R_L = R_A^{\frac{1}{D_T}}. \tag{22.89}$$

For typical river networks, $1.7 < D_T < 1.8$. Peckham and Gupta (1999) found that the sum of area exponent (α_T) and the inverse of the fractal dimension $(\beta_T = 1/D_T)$ was equal to 1. Now inserting the velocity ratio $R_v = R_Q^m$ into Equation (22.86), one obtains

$$R_v = R_Q^m = RR_3 R_Q^{\frac{1}{4}} R_Q^{\frac{z}{2}}. \tag{22.90}$$

Equation (22.90) shows that $RR_3 = 1$ and

$$m = \frac{1}{2}\left(z + \frac{1}{2}\right). \tag{22.91}$$

In like manner, the Horton ratio for dimensionless number proportional to the Darcy–Weisbach resistance factor given by Equation (22.64) can be considered in the limit:

$$\frac{R_6(o+1)}{R_6(o)} = RR_6 = \frac{R_v^2}{R_d R_S}. \tag{22.92}$$

Equation (22.92) can be recast for obtaining the depth exponent as

$$R_d = R_Q^f = \frac{R_v^2}{RR_6 R_S} = \frac{R_Q^{2m}}{RR_6 R_Q^z}. \tag{22.93}$$

Equation (22.93) shows that $RR_6 = 1$, and

$$f = 2m - z. \tag{22.94}$$

Solution of Equations (22.91) and (22.94) yields $f = 0.5$, $m = 0$, and $z = -0.5$.

Gupta and Mesa (2014) noted that slope tends to 0 as order tends to be very large and cannot be eliminated from Equations (22.61) and (22.64). Therefore, they generalized the dimensional analysis for the hydraulic geometry theory. They renormalized ratios 3 and 6 as

$$R_{3m}(o) = \frac{v_o}{\sqrt{gL_o S_o^\alpha}} \tag{22.95}$$

$$R_{6m}(o) = \frac{v_o^2}{gd_o S_o^\beta}, \tag{22.96}$$

where R_{3m} is the modified R_3, R_{6m} is the modified R_6, and α and β are the anomalous scaling exponents. Now the dimensionless Horton ratios corresponding to Equations (22.95) and (22.96) can be expressed in the limit as

$$RR_{3m} = \frac{R_{3m}(o+1)}{R_{3m}(o)} = \frac{R_v}{\sqrt{R_L R_S^\alpha}} \tag{22.97}$$

$$RR_{6m} = \frac{R_{6m}(o+1)}{R_{6m}(o)} = \frac{R_v^2}{R_d R_S^\beta}. \tag{22.98}$$

Recalling that $R_L = R_A^{\frac{1}{D_T}} = R_Q^{\frac{1}{D_T}}$ and $R_S = R_A^z = R_Q^a$, and $R_v = R_Q^m$, Equation (22.97) yields

$$R_v = R_Q^m = RR_{3m} R_Q^{\frac{1}{2D_T}} R_Q^{\frac{za}{2}}. \tag{22.99}$$

Equation (22.99) shows that $RR_{3m} = 1$, and

$$m = \frac{1}{2}\left(za + \frac{1}{D_T}\right). \tag{22.100}$$

Since $m + f = 0.5$, the depth exponent follows from Equation (22.100) as

$$f = \frac{1}{2}\left(1 - za - \frac{1}{D_T}\right). \tag{22.101}$$

Another expression for the depth exponent can be obtained by recasting Equation (22.98) as

$$R_d = R_Q^f = \frac{R_v^2}{RR_{6m} R_S^\beta} = \frac{R_Q^{2m}}{RR_{6m} R_Q^{z\beta}}. \tag{22.102}$$

Equation (22.102) shows that $RR_{6m} = 1$, and with the use of Equation (22.100),

$$f = 2m - z\beta = z(\alpha - \beta) + \frac{1}{D_T}. \tag{22.103}$$

The slope exponent can be obtained by equating Equations (22.101) and (22.103) as

$$z(3\alpha - 2\beta) = 1 - \frac{3}{D_T}.$$ (22.104)

Thus, the hydraulic geometry exponents can now be summarized as

$$b = \frac{1}{2}; z = \frac{1 - \frac{3}{D_T}}{3\alpha - 2\beta}; m = \frac{z\alpha + \frac{1}{D_T}}{2}; f = \frac{1}{2} - m.$$ (22.105)

22.6.7 Manning's Roughness Coefficient and Its Exponent

Equation (22.96) can be expressed as

$$R_{6m}(o) = \frac{v_o^2}{gd_o S_o^\beta} = \frac{v_o^2}{gd_o S_o S_o^{-1+\beta}}.$$ (22.106)

Equation (22.106) can be expressed as Chezy's equation

$$v_o = C_o^* \sqrt{d_o S_o} = (gd_o S_o)^{\frac{1}{2}} S_o^{\frac{-1+\beta}{2}} (R_{6m}(o))^{\frac{1}{2}}.$$ (22.107)

Equation (22.107) suggests an expression for Chezy's roughness coefficient as

$$C_o^* = [R_{6m}(o)]^{\frac{1}{2}} g^{\frac{1}{2}} S_o^{\frac{-1+\beta}{2}}.$$ (22.108)

Since the slope ratio tends to R_S, RR_{6m} tends to 1. The Manning fiction factor and Chezy's roughness coefficient are related (Leopold et al., 1964), one can write

$$v_o = C_o^* \sqrt{d_o S_o} = \frac{1}{n_o} d_o^{\frac{1}{6}} \sqrt{d_o S_o}.$$ (22.109)

Equation (22.109) yields

$$n_o = \frac{d_o^{\frac{1}{6}}}{C_o^*}.$$ (22.110)

The friction factors can be expressed as a ratio as

$$\frac{C_{o+1}^*}{C_o^*} = \left[\frac{d_{o+1}}{d_o}\right]^{\frac{1}{6}} \left[\frac{n_o}{n_{o+1}}\right].$$ (22.111)

Inserting Equation (22.108) and taking the limit, Equation (22.111) can be written as

$$R_n = R_d^{\frac{1}{6}} R_S^{\frac{1-\beta}{2}}.$$ (22.112)

Using Equation (22.75) and Horton's law for velocity given by Equation (22.77), one gets $R_n = R_Q^y$ and using $R_S = R_Q^z$, one can write

$$R_n = R_Q^y = R_Q^{\frac{f}{6}} R_Q^{\frac{z(1-\beta)}{2}}. \tag{22.113}$$

Equation (22.113) yields

$$y = \frac{f}{6} - z\left(\frac{-1+\beta}{2}\right). \tag{22.114}$$

The constraints on β are $\beta < 1$ and $\beta < 3\alpha/2$. From Equation (22.105), one obtains

$$\beta = \frac{1-3f}{z}. \tag{22.115}$$

Introducing Equation (22.115) in Equation (22.114), the result is

$$y = \frac{f}{6} - z\frac{-1+\beta}{2} = \frac{5f}{3} + \frac{z}{2} - \frac{1}{2}, \tag{22.116}$$

which is independent of D_T, but because m and z depend on D_T, y also depends on it.

22.7 Channel Geometry Method

Employing the relationship between stream channel geometry and river discharge, the channel geometry method can be used for flood discharge estimation at ungagged sites. This method, reported by Wharton (1995), involves empirically developing relationships between streamflow data measured at gaging stations and channel dimensions measured from natural river reaches in the vicinity of gages. Then, these relationships are regionalized. Thereafter, flood flow characteristics can be estimated at a given site by knowing only channel width or capacity or at ungagged sites. The relationships take on the power function form:

$$Q = a_0' w^{b_0'}; \quad Q = a_1' A^{a_2'}, \tag{22.117}$$

in which A is the cross-sectional area, and b_0' and a_2' are exponents, and a_0' and a_1' are constants. Here discharge Q is the dominant channel-forming discharge. Wharton (1995) suggested the following guidelines for developing and applying the channel-geometry method.

(1) **Selecting reaches for measuring river channel geometry:** Natural reaches of straight or stable meandering channels with length of at least four to five widths should selected. These reaches should not be subject to any kind of human

intervention, such as deepening, widening, realigning, lining with riprap or concrete; alteration by natural processes, such as natural linings, obstructions like debris dams, bank cutting or bar deposition, large pools or locally steep inclines; joining of tributaries; change in bank or bed-material characteristics; and sudden adjustment of cross-sectional form.

(2) **Selecting cross-sections:** At least three evenly spaced cross-sections should be selected from the reaches. These cross-sections should be such that they are close to being rectangular or trapezoidal, not of unusual shape, not located either upstream or downstream of tributaries, flow velocities are approximately symmetrical, and the channel geometry reference level on both banks and along the reach is about the same elevation.

(3) **Measuring river channel dimensions:** Channel dimensions should be measured in the field at consistently selected geomorphic reference levels. For data from archive sources, three geomorphic reference levels to represent the range of channel morphologies are recommended in the United States: arid, semi-arid, and montane. However, the distinction between the three levels is not always clear. Cross-sections should be surveyed along the shortest cross-channel distance. Depositional bar level or lower limit of permanent vegetation has been used for both bankfull or active channel. The bankfull level may be used to define the level of valley flat.

(4) **Computing flood discharges:** Channel cross-section more accurately describes river size than a single width value, and remains relatively constant from pool to riffle sections even if the width varies markedly. Different return year floods may be computed by combining the mean annual flood obtained from channel geometry with the appropriate regional flood frequency curve.

Channel geometry relations have been developed for estimating flood frequency characteristics of rivers in arid, semi-arid, humid temperate, and tropical environments, as shown in Table 22.2 (after Wharton, 1995).

Besides flood estimation at ungagged sites, the channel geometry method can also be employed for flood management. First, the channel geometry equations help understand channel adjustment for a variety of environments. Second, changes in discharge characteristics along a channel over time can be deciphered from these equations. Third, these equations provide regional average of channel geometry; local deviations serve as a source of information for river management. Fourth, the equations can serve as a basis for estimating channel dimensions and river design.

Wharton (1992) plotted mean annual flood (m³/s) against bankfull channel width (m) and found the value of width exponent as 1.97 and the constant

Table 22.2. *Channel geometry relations [* denotes that discharge is measured in cubic feet per second, and channel dimensions in feet.]*

Country/authors	Streams	Relation	Notation
USA			
Osterkamp and Hedman (1982)	252 streams in Missouri River basin 151 streams in western US.	$*Q_{ma} = 0.027w_a^{1.71}$	A_b = channel cross-section at bankfull (m^2)
Hedman and Osterkamp (1982)	Alpine and pine-forested	$*Q_2 = 1.3w_a^{1.62}$	A_{ot} = channel cross-section at the overtopping level (m^2)
	Northern plains and intermontane areas east of Rocky Mountains	$*Q_2 = 4.8w_a^{1.60}$	A_s = aspect ratio [maximum depth (m) to hydraulic ratio at bankfull level (m)] (m^2)
	Southern plains east of Rocky mountains	$*Q_2 = 7.8w_a^{1.70}$	d_b = mean channel width at bankfull (A_b/w_b) (m)
	Plains and intermontane areas west of Rocky Mountains	$*Q_2 = 1.8w_a^{1.70}$	d_m = mean channel depth at the depositional bar level (m)
Osterkamp and Hedman (1979)	Surface mine areas of arid and semi-arid regions of western US.	$*Q_{10} = 4.14w_a^{1.63}$	D_m = mean diameter of bed sediment (m)
Osterkamp and Hedman (1977)	32 high gradient streams in Montana, Wyoming, South Dakota, Colorado, and New Mexico	$*Q_2 = 0.0017w_a^{1.98}$	Q_{ma} = mean annual flood (m^3/s)
Hedman et al. (1974)	120 streams in Kansas	$*Q_2 = 22.3w_a^{1.40}$	Q_{med} = median annual flood (m^3/s)
Hedman et al. (1972)	53 perennial streams in the mountain region of Colorado	$*Q_2 = 0.991w_a^{1.797}$	Q_r = mean annual runoff (acre-feet per year)
Hedman (1970)	Arid and sub-humid parts of California		Q_2 = 2-year return period flood (m^3/s)
	28 perennial streams	$*Q_r = 186w_d^{1.54}d_m^{0.88}$	Q_{10} = 10-year return year flood (m^3/s)
	20 ephemeral streams	$*Q_r = 258w_{kern}^{0.80}d_m^{0.60}$	S = bankfull channel slope
	Perennial, intermittent, and ephemeral streams in southeastern Montana: 38 streams in flat plains land affected by intense summer thunderstorms		w_a = channel width at the active channel reference level (m)
Omang et al. (1983)	28 streams on flat plains but flood peaks not as variable	$*Q_2 = 10.0w_a^{1.16}$	w_b = channel width at the bankfull level (m)

w_d = channel width at the depositional bar reference level (m)

w_{ot} = channel width at the overtopping level (m)

Reference	Description	Equation
Canada Bray (1975) **New Zealand** Mosley (1979)	12 streams in forested mountain area, runoff mainly from snowmelt	$*Q_2 = 3.52w_a^{1.59}$ $*Q_2 = 10.7w_a^{1.14}$
	71 gravel bed rivers in Alberta	$*w_b = 4.75Q_2^{0.527}$
	73 alluvial river channels, South Island	$Q_{ma} = 1.6A_b^{0.9}A_r^{-0.376}S^{-0.392}D_m^{0.278}$
Great Britain Nixon (1959)	22 rivers in England and Wales	$*w_b = 1.65Q_b^{0.50}$
Wharton (1992)	Rivers in England, Wales, and Scotland	
	75 bankfull sites	$Q_{ma} = 0.20w_b^{1.97}$ $Q_{ma} = 1.16A_b^{1.31}$
	109 overtopping sites	$Q_{ma} = 0.34w_{ot}^{1.68}$ $Q_{ma} = 1.2A_{ot}^{1.14}$
Italy Caroni and Maraga (1984)	30 rivers in the Po basin, northern Italy	$Q_{med} = 0.045w_b^{2.09}$
Java Wharton and Tomlinson (1999)	24 rivers	$*Q_{ma} = 0.87w_b^{1.60}$
Burundi Wharton and Tomlinson (1999)	8 rivers	$*Q_{ma} = 0.96w_b + 1.44$
Ghana Wharton and Tomlinson (1999)	12 rivers	$*Q_{ma} = 1.8w_b^{1.50}$

of proportionality as 0.20, and mean annual discharge (m³/s) against bank with full cross-sectional area (m²), where the value of proportionality constant was 1.16 and the cross-sectional area exponent as 1.31. The relationships plotted were satisfactory and can be very useful where gaged discharge data is missing. Wharton (1995) noted that channel depth, although easy to measure, was not a good predictor because river bed profiles were highly variable at both cross-sections and along channel reaches. Further, it is difficult to measure velocity if flow characteristics are evaluated using a geomorphic feature.

The channel geometry method has been shown to be reliable in estimating flood discharges in humid environments (Wharton, et al. 1989) where vegetation and material properties exercise less influence because channel banks are relatively stable and channels adjust to the prevailing flow regime, thus permitting a stable mean channel geometry. The method can also be applied to catchment where measured data are available, for it provides a quick and inexpensive way to compute flood discharge. It is also noted that river channels adjust their cross-sections in response to discharge and sediment loads received from the catchment, implying that they integrate the effect of a range of catchment characteristics and climate. Since the method does not require slope and roughness estimates, it can be regarded as an improvement over slope-area methods. However, significant errors may occur if equations developed for specific hydrologic, geologic, and climatic regions are applied to other areas. The method is especially useful as a reconnaissance method that may be valuable before using a physically based hydrological model and in more arid environments where traditional rainfall-runoff models prove to be unreliable.

22.8 Variation of Channel Geometry and Drainage Area

Empirical evidence shows that at-a station hydraulic geometry parameters (exponents of width, depth and velocity) vary with drainage area and the downstream hydraulic geometry parameters vary with the frequency of discharge (Dodov and Foufoula-Georgiou, 2004). The flow characteristics of a drainage basin vary as the drainage area varies, and the variation in flow parameters is not necessarily linear. For example, for downstream hydraulic geometry the relation between channel width and discharge is expressed as a power function, which plots as a straight line on a log-log paper. The slope of this line or the exponent of discharge, denoted by b, is found to be about 0.5. However, basin hydrologic characteristics change at an area of about 250 square kilometers. Therefore, for drainage areas greater than 250 km², the value of b exponent may be different or there may be another straight line, suggesting that one line may not suffice. It has been shown that the b value for small and very large basins is smaller than the b

value for intermediate basins. Thornes (1970) suggested a two-line relationship, because the change in width is very small for low discharges but is much greater for large discharges. Thornes (1974) suggested three types of width variation with distance: (1) null, (2) linear, and (3) negative exponential. The third type is observed in arid areas. Carlston (1969) pointed out that very large rivers, such as Mississippi River, Nile River, or Yangtze River, accommodate downriver increase in discharge mainly through the increase in depth, but lesser rivers usually accommodate the downriver increase in discharge mainly through an increase in width. Stall and Fok (1968) reported a consistent relationship between discharge and drainage area. Since hydraulic geometry variables – width, depth and velocity – are expressed as power functions of discharge, it would seem logical that these variables should be expressible as power functions of drainage area.

Much of the morphological activity occurs within a short period of time during which the hydrograph peak occurs. Without requiring flood frequency analysis, Klein (1981) defined a basin's peakedness index as the ratio of mean flow to peak flow. By definition, as the index gets smaller the rate of change is higher. This index is, however, sensitive to peak flow and period of flow, and varies with drainage area. For drainage areas up to about 250 km^2 (in humid area) the index values are low and the rate of change of the index is low, and for larger drainage areas the index values and the rate of change are higher.

In small basins the peakedness is high but the peakedness index is low, which will lead to the deepening of the channel, resulting in greater values of the depth exponent f and velocity exponent m. On the other hand, for larger basins, the peakedness is low but the peakedness index is high; the result is the widening of the channel and the increase in the b exponent. However, for very large basins, such as the Mississippi River, the downriver increase in discharge is accommodated through an increase in depth (increase in the exponent f) and a decrease in the b exponent.

References

Bray, D. I. (1975). Representative discharges for gravel-bed rivers in Alberta, Canada. *Journal of Hydrology*, Vol. 27, pp. 143–153.

Benson, M. A. (1960). Characteristics of frequency curves based on a theoretical 1000-year record. In: *Flood Frequency Analyses: manual of Hydrology. Part 3. Flood Flow Techniques.* pp. 51–73, U.S. Geological Survey Water Supply Paper 1543-A, Washington, DC.

Benson, M. A. (1962). Factors influencing the occurrence of floods in a humid region of diverse terrain: U.S. Geological Survey Water-Supply Paper 1580-B, 64 p., Washington, DC.

Brush, L. M. (1961). *Drainage basins, channels and flow characteristics of selected streams in central Pennsylvania.* USGS Professional Paper 282F, pp. 145–181, Washington, DC.

Carlston, C. W. (1969). Downstream variations in the hydraulic geometry of streams: Special emphasis on mean velocity. *American Journal of Science*, Vol. 67, pp. 499–504.

Caroni, E. and Maraga, F. (1984). Flood prediction from channel width in the Po River basin. *Progress in Mass Movement and Sediment Transport Studies: Problems of Recognition and Prediction, Torini*, 5–7, Italy.

Dodds, P. S. and Rotham, D. H. (1999). Unified view of scaling laws for river networks. *Physical Review E*, Vol. 59, 4865, doi:10.1103/PhysRevE.59.4865.

Dodov, B. and Foufoula-Georgiou, E. (2004). Generalized hydraulic geometry: Derivation based on a multiscaling formalism. *Water Resources Research*, Vol. 40, W06302, doi: 10.1029/2003WR002082.

Dunne, T. and Leopold, L. B. (1978). *Water in Environmental Planning*. W.H. Freeman, San Francisco.

Emmett, W. W. (1975). *The channels and waters of the upper Salmon River area, Idaho: U.S. Geological Survey Professional Paper 870-A*, P. 115, Washington, DC.

Flood Studies Report (1975). *Natural Environment Research Council (NERC)*, London.

Gupta, V. K. and Mesa, O. J. (2014). Horton laws for hydraulic-geometric variables and their scaling experiments in self-similar Tokunaga river networks. *Nonlinear Processes in Geophysics*, Vol. 21, pp. 1007–1025.

Gupta, V. K. and Waymire, E. (1998). Spatial variability and scale invariance in hydrologic regionalization. In: *Scale Dependence and Scale Invariance in Hydrology*, edited by G. Sposito, Cambridge University Press, London, pp. 88–135.

Hack, J. T. (1957), Studies of longitudinal stream profiles in Virginia and Maryland: U. S. Geological Survey Prof. Paper 294-B, Washington, DC.

Hedman, E. R. (1970). *Mean annual runoff as related to channel geometry of selected streams in California*. U.S. Geological Survey Water Supply Paper199-E, Washington, DC.

Hedman, E. R. and Osterkamp, W. R. (1982). *Streamflow characteristics related to channel geometry of streams in Western United States*. U.S. Geological Survey Water Supply Paper 2193, Washington, DC.

Hedman, E. R., Moore, P. O., and Livingstone, R. K. (1972). *Selected streamflow characteristics as related to channel geometry of perennial streams in Colorado*. U.S. Geological Survey, Open-File Report (200), H358s, Washington, DC.

Hedman, E. R., Kastner, W. M., and Hejl, H. R. (1974). *Selected stream characteristics as related to active channel geometry of streams in Kansas*. State of Kansas Water Resources Board Technical Report No.10, Kansas.

Horton, R. E. (1945). Erosional development of streams and their drainage basins: Hydrophysical approach to quantitative morphology. *Bulletin of the Geological Society of America*, Vol. 56, pp. 275–370.

Klein, M. (1981). Drainage area and the variation of channel geometry downstream. *Earth Surface Processes and Landforms*, Vol. 6, pp. 589–593.

Knighton, A. D. (1987). River channel adjustment: The downstream dimension. In: *River Channels: Environment and Process*, edited by K. S. Richards, pp. 95–128, Basil Blackwell, Oxford.

Lawlor, S. M. (2004). *Determination of channel morphology characteristics, bankfull discharge, and various design-peak discharges in western Montana*. U.S. Geological Survey Scientific Investigations Report 2004-5263, pp. 26, Washington, DC.

Leopold, L. and Miller, J. (1956). *Ephemeral streams: Hydraulic factors and their relation to the drainage net*. USGS Professional Paper 282-A, pp. 37, Washington, DC.

Leopold, L., Wolman, M. G., and Miller, J. P. (1964). *Fluvial Processes in Geomorphology*. H.H. Freeman Press, San Francisco.

McConkey, S. A. and Singh, K. P. (1992). Alternative approach to the formulation of basin hydraulic geometry equations. *Water Resources Bulletin*, Vol. 28, No. 2, pp. 305–312.

Mosley, M. P. (1979). Prediction of hydrologic variables from channel morphology, South Island rivers. *Journal of Hydrology (New Zealand)*, Vol. 18, No. 2, pp. 109–120.

Nash, J. E. and Shaw, B. L. (1966). Flood frequency as a function of catchment characteristics. *Proceedings, River Flood Hydrology Symposium, Institution of Civil Engineers*, London, pp. 115–136.

Nixon, M. (1959). A study of bankfull discharges of rivers in England and Wales. *Proceedings of Institution of Engineers*, Vol. 12, No. 2, pp. 157–174.

Omang, R. J., Parret, C., and Hull, J. A. (1983). *Mean annual flood and peak flow estimates based on channel geometry of streams in southeastern Montana*. U.S. Geological Survey Water Resources Investigations, 82-4092, Washington, DC.

Osterkamp, W. R. and Hedman, E. R. (1977). Variation of width and discharge for natural high-gradient stream channels. *Water Resources Research*, Vol. 13, No. 2, pp. 256–258.

Osterkamp, W. R. and Hedman, E. R. (1979). Discharge estimates in surface-mine areas using channel geometry techniques. *Proceedings of the Symposium on Surface Mining Hydrology, Sedimentology and Reclamation*, University of Kentucky, Lexington.

Osterkamp, W. R. and Hedman, E. R. (1982). *Perennial streamflow characteristics related to channel geometry in Missouri River basin*. U.S. Geological Survey Professional Paper 1242, Washington, DC.

Peckham, S. D. (1995). New results of self-similar trees with applications to river networks. *Water Resources Research*, Vol. 31, pp. 1023–1029.

Peckham, S. D. and Gupta, V. K. (1999). A reformulation of Horton's laws for large river networks in terms of statistical self-similarity. *Water Resources Research*, Vol. 35, pp. 2763–2777.

Singh, K. P. and Broeren, S. M. (1989). Hydraulic geometry of streams and stream habitat assessment. *Journal of Water Resources Planning and Management*, Vol. 115, No. 5, pp. 583–597.

Stall, J. B. and Fok, Y. S. (1968). *Hydraulic geometry of Illinois streams*. Research Report No. 15, pp. 47, University of Illinois Water Resources Center, Urbana.

Stall, J. B. and Yang, C. T. (1970). *Hydraulic geometry of 12 selected stream systems of the United States*. Research Report No. 32, pp. 73, University of Illinois Water Resources Center, Urbana.

Thomas, D. M. and Benson, M. A. (1975). *Generalization of Streamflow Characteristics From Drainage-Basin Characteristics*. Geological Survey Water-Supply Paper 1975, 55 p., U.S. Geological Survey, Washington, DC.

Thornes, J. B. (1970). The hydraulic geometry of stream channels in the Xingu-Araguaia headwaters. *Geographical Journal*, Vol. 130, No. 3, pp. 376–382.

Thornes, J. B. (1974). *Speculation on the behavior of stream channel width*. Discussion Paper No. 49, Graduate School of Geography, London School of Economics, London, England.

Virmani, J. K. (1973). *The relationship between channel forming flows and the cross-section shape, slope, and bed materials in large bed element streams*. Unpublished Ph.D. dissertation, Utah State university, Logan.

Wharton, G. (1992). Flood estimation from channel size: guidelines for using the channel geometry method. *Applied Geography*, Vol. 12, No. 4, pp. 339–350.

Wharton, G. (1995). The channel geometry method: Guidelines and applications. *Earth Surface Processes and Landforms*, Vol. 20, pp. 649–680.

Wharton, G. and Tomlinson, J. (1999). Flood discharge estimation from river channel size in Java, Ghana, and Burundi. *Hydrological Sciences Journal*, Vol. 44, No. 1, pp. 97–111.

Wharton, G., Arnell, N. W., Gregory, K. J., and Gurnell, A. M. (1989). River discharge estimated from river channel dimensions. *Journal of Hydrology*, Vol. 106, pp. 365–376.

Wilkerson, G. V. (2008). Improved bankfull discharge prediction using 2-year recurrence-period discharge. *Journal of American Water Resources Association*, Vol. 44, No.1, pp. 243–258.

Wolman, M. G. (1954). A method of sampling coarse material. *Transactions of the American Geophysical Union*, Vol. 35, pp. 951–956.

Index